Illumination Engineering—From Edison's Lamp to the Laser

Illumination Engineering—From Edison's Lamp to the Laser

Joseph B. Murdoch
University of New Hampshire

Macmillan Publishing Company
NEW YORK

Collier Macmillan Publishers
LONDON

Macmillan Publishing Company
866 Third Avenue, New York, N. Y. 10022

Collier Macmillan Canada, Inc.

Printed in the United States of America

printing number
1 2 3 4 5 6 7 8 9 10

Library of Congress Cataloging in Publication Data

Murdoch, Joseph B., 1927-
Illumination engineering—from Edison's lamp to the laser.

Includes bibliographies and index.
1. Electric lighting. 2. Lighting. I. Title.
TK4175.M87 1985 621.32'1 84-20134
ISBN 0-02-948580-0

To Ann and
Joe, Pete, Amy, Annie and Tommy
and Grandma Jessie
Who, as in the past, continue to make it
all so very worthwhile

Contents

Preface

When I was writing a book on electrical network theory several years ago, the publisher's representative told me "Joe, one does not complete a book, one abandons it." This statement has a certain applicability here, because, as I wrote this book, it became clear to me that I could not possibly cover all of the myriad of topics in the field of lighting. Choices had to be made, and were.

The book is a scientific and technical treatment of lighting, rather than an architectural and design treatment. It presumes that the reader has a working knowledge of algebra, trigonometry, and introductory calculus. It further presumes a background in physics, such as would be obtained from a one-year college-level experience. It is intended both for students in college courses and for self-study by practitioners in the field.

The book evolved from notes for a course in illumination engineering which I have taught regularly at the University of New Hampshire since the 1950s. Typically, juniors and seniors in electrical and mechanical engineering elect the course, but I have had students from other engineering disciplines, and from physics, chemistry, mathematics, and psychology as well. It seems to me that students in architecture also could take the course, but we do not have an architecture program at UNH.

The purposes of the book (and of the course) are severalfold:

1. To provide a foundation in the science and mathematics of light, sight, and lighting.
2. To develop the analytical tools and skills required for quantitative lighting calculations.
3. To present lighting measurement techniques and instruments.

4. To give an historical perspective—the scientific and engineering achievements of the past relevant to lighting.
5. To present the current state of knowledge in the field of illumination.
6. To provide a reference and updating base for those whose knowledge of lighting is rusty or incomplete.
7. To include sufficient detail, examples, problems (with answers), and references that self-study is possible without formal instruction.
8. To point out that, although the book is largely technical and quantitative, there is another important aspect to lighting, namely considerations of quality and aesthetics. A good lighting design is a blend of both.

The book contains too much material to be covered in a one-semester course. Ideally, two semesters should be allocated, with the first devoted to the science of lighting and the second to lighting calculations and applications. If this is not possible, then a choice must be made as to which aspect to feature. If the science of lighting is to be paramount, it is suggested that Chaps. 1, 2, 3, 4, 5, 6, and 10 be stressed. If lighting calculations and applications are of prime importance, then Chaps. 1, 2, 5, 7, 8, 9, and 11 should receive emphasis. Whichever emphasis is chosen, there is an extensive set of problems for each chapter, with answers to selected problems, at the back of the book.

I think it has been my desire to write a book on lighting ever since my first exposure to the subject while I was attending Case Institute of Technology in the late 1940s. Along with many other Case graduates, I owe a special debt of gratitude to Professor Russell Putnam for getting me started in lighting and for being my friend and mentor in those earlier days.

In recent years, it has been my good fortune that UNH is only 60 miles away from Dr. Robert Levin of GTE Lighting Products. Bob has given most generously of his time in helping me with the lighting program here. Also, as the prime reviewer for the book, he has read and critiqued each chapter in detail, for which I am most grateful. The book is much better than it would have been because of Bob's painstaking review and comments.

I would also like to thank Mr. Calvin Gungle of GTE Lighting Products for his careful review of the electrical discharge material for Chaps. 4 and 6, Mr. John Kaufman, Technical Director of the Illuminating Engineering Society, for his reading of the first several chapters and his encouragement, Mr. William Pierpoint of the Naval Construction Battalion Center at Point Hueneme, California, for his comments on the Daylighting chapter, and Dr. Gary Yonemura of the National Bureau of Standards for his review of the material on vision. Finally, I am very much indebted to Mrs. Alice Greenleaf of our Word Processing Center for her patience and diligence in typing the manuscript. It will seem strange not to be taking something to type down the hall to Alice each day.

Joseph B. Murdoch

Illumination Engineering– From Edison's Lamp to the Laser

Chapter 1

Introducing Light and Seeing

1.1 PRE-ELECTRIC LIGHTING

It is likely that interest in lighting had its beginnings when one of our prehistoric ancestors gathered up (carefully) the blazing embers of a forest fire and brought them into his cave to provide warmth and light. Much later, in medieval times, metal fire baskets called *cressets* were developed. Holding glowing pine knots, these containers were suspended from ceilings or hung on wall brackets. Pine wood was also used for light in colonial times. It was called *candlewood* and bundles of slivers were placed in metal stands near the hearths in homes.

One could say that the flaming torch was the first "portable" lamp. American Indians used pine knots as torches, whereas early colonial settlers found that certain reeds or rushes, when dried and dipped in grease or fat, could be burned and carried about to provide illumination. These were called *rushlights*.

For a variety of obvious reasons, wood did not turn out to be very satisfactory as a source of light. Down through recorded history four other basic light sources have emerged. These are, in chronological order, oil lamps, candles, gas lighting, and electric lamps. We will consider the rise and fall of the first three of these in the remainder of this section and address the early development of electric lighting in the section to follow.

The ancient civilizations of Babylonia and Egypt (around 3000 B.C.) had crude oil lamps. These were in the form of small open stone or sea shell bowls containing animal fat or fish or vegetable oil. Later the bowls were made of clay and, still later, of metal, particularly bronze, pewter, and tin. Around 500 B.C., the Greeks and Romans developed the oil reservoir lamp, which had a wick projecting from a channel and a cover to keep the fuel in the reservoir clean.

Early colonists used virtually the same type of oil lamp as did the Romans. It is true that their lamps were made of metal and were constructed better, particularly with regard to the wick (the most widely used was the Betty lamp, brought over from Germany), but they were basically the same lamp that had been developed 1000 years earlier. Then, in the mid 1700s, the rise of the whaling industry provided a better fuel. Whale oil lamps produced steadier and brighter light, but the lamps were still smoky and rather dangerous. About this time Benjamin Franklin discovered that placing two wicks side by side and slightly apart produced more light than two single-wick lamps. Lamps employing the Franklin principle were used in the United States until after the Civil War.

The discovery of oil by Colonel Drake in 1858 provided a new liquid fuel for lighting. Kerosene (coal oil) lamps were safer, gave better light, and were less expensive to operate. Three other developments aided their acceptance. One was the invention of the friction match, which made it much easier to light the lamp. Second was the development of the flat wick and associated mechanical button for turning it in and out of the lamp. Last was the creation of the lamp chimney. In 1783, a Swiss chemist named Ami Argand discovered that placing a glass cylinder over a lamp flame caused the flame to stop smoking and become steadier and brighter. Thomas Jefferson, writing from Europe in 1784, states "There has been a lamp called a cylinder lamp lately invented here. It gives a light equal to six or eight candles. This improvement is produced by forcing the wick into a hollow cylinder so that there is a passage of air through it." When it is realized that lamps of that time generally provided the equivalent light of one candle per wick, the significance of Argand's discovery is appreciated.

We turn next to the development of the candle, which is closely associated with the development of the Christian church and dates from about 400 A.D. Few candles were used in nonchurch settings until the fourteenth century. Church candles were made of beeswax (the bee was the symbol of purity) and were quite expensive. Inexpensive, smelly, and smoky tallow candles were the lot of the common people, if they were able to obtain them.

Most early American colonists fashioned candles from tallow also, using crude wicks of flax. The wealthier people could afford wax from the bayberry bush, which was cleaner and more fragrant than tallow.

The rise of the whaling industry produced not only a better fuel for oil lamps but also a better wax for candles. Spermaceti, a translucent and slightly crystalline wax which vaporizes readily, is derived from oil from the head cavity of the sperm whale. It is superior to all other candle materials and thus spermaceti candles were used in the past as standards for basic lighting measurements.

In the 1850s it was discovered that much of the smoke and odor of tallow could be eliminated by removing the glycerine from animal fats. The result was the stearine candle. At nearly the same time paraffin, a petroleum product, was introduced. Most inexpensive candles today are made of molded paraffin.

The Chinese are believed to be the first people to use natural gas for lighting, piping it from salt mines through bamboo tubes. Around 1800, several people discovered how to distill coal to obtain natural gas and gas lighting was introduced. By 1823 London had 40,000 gas lamps. However it was the mid 1800s before the gaslight era began in the United States with fuel for both light and heat distributed from central station facilities to residential and commercial users.

The first gas burners were simply iron pipes with holes in the end. These clogged, and a lava tip for gas burners was developed. Later the fishtail burner, with two jets at 90° forming a flat flame, was introduced.

Throughout most of the nineteenth century, gas lighting was provided by the light from an open gas flame, rather than by the burning gas heating an object to incandescence. Then, in 1886, C. A. Von Welsbach, in an effort to improve gas lighting to compete with the infant electric lighting industry, invented the Welsbach burner. This "mantle" consisted of a cone of knit cotton fabric soaked in nitrates (later oxides). When heated for the first time, the cotton was burned leaving a thin fish net of ash that became incandescent when placed over a gas flame. The mantle gave up to three times as much light as the open gas flame. Refinements of the Welsbach mantle are still in use today, especially in camping lanterns.

The invention of the incandescent lamp by Thomas Edison in 1879 triggered a 30-year battle between the approximately 500 gas companies existing in the United States at that time, with their central stations and distribution networks, and the fledgling electric industry. Obviously the latter won the battle, for gas lighting devices were severely limited. They had to be burned upright; nothing inflammable could be near them; they required air but could not tolerate drafts; they were difficult to start; and, above all, they were dirty. But, despite electric lighting's merits, nearly one-quarter of the people in the United States in the mid 1900s still depended on flame sources for light.

1.2 EARLY DEVELOPMENTS IN ELECTRIC LIGHTING

The first recorded man-made electric light was by Otto Von Guericke, a German scientist, in 1650. Von Guericke's apparatus consisted of a chain attached to a spinning sulfur ball. When he held his hand against the ball, sparks jumped from the chain.

In 1710, Sir Francis Hauksbee, an Englishman, produced a glow inside a glass globe from which air had been evacuated and mercury added. He called the glow "electric light" and claimed that his experiment proved that electricity could produce light.

Benjamin Franklin, with his famous kite experiment in 1752, gave further proof that electricity could produce light by collecting electric charge from clouds into Leyden jars during a thunderstorm.

As the nineteenth century began, the development of electric lighting took three rather distinct paths, two of which were initiated by the work of Sir Humphrey Davy, a British scientist. In 1802, Davy showed that if thin strips of metal were raised to high temperatures by passing electricity through them, they would emit light. This observation ultimately led to the development of the incandescent lamp.

Then, in 1808, before members of the Royal Institution, Davy demonstrated the first carbon arc, a luminous discharge about 4 in long between pieces of charcoal connected to a battery of 2000 cells. Davy's source was the forerunner of the carbon arc lamp used extensively in outdoor lighting in the late 1800s.

Davy's observation of incandescence led to a crude incandescent lamp by De la Rue in England in 1820. De La Rue's lamp consisted of a coiled filament of platinum wire mounted in a glass tube. The ends of the coil were connected to brass caps at the ends of the tube.

Early experiments had shown that a metal filament has a very short life if heated in air—it literally burns up. Thus it is likely that De la Rue's lamp was evacuated, but it is also likely that the vacuum was not very good and thus that the lamp had a very short life, because the mercury vacuum pump, which was so important in later incandescent lamp development, was not invented until 1865 (by Herman Sprengel).

From De la Rue's lamp in 1820 until Edison lit his first successful lamp in 1879, many people worked on incandescent lamp development. However, the lamps they created did not prove practical, largely because of unreliability, short life, and excessive operating expense. William Sawyer, Moses Farmer, and Hiram Maxim in the United States and Frederick DeMoleyns and Sir Joseph Swan in England all developed lamps during this period. Farmer, a native of New Hampshire, used the platinum filament incandescent lamp he developed to light the parlor in his home in Salem, Massachusetts, in 1859.

All these experimenters used thick filaments, of either platinum or carbon, which had low electrical resistance and required large currents to heat them to incandescence. It remained for Edison to show that a fine hairlike filament, which had high resistance and thus required a small current to cause it to emit light, was the answer for a commercially practical incandescent lamp.

Concurrent with early incandescent lamp development was the development of the arc light, again stemming from Davy's early work. The first commercial arc lamp was installed in the Dungeness Lighthouse in England in 1862. The earliest arc lamp systems in the United States were developed by Edward Weston, Elihu Thomson, William Wallace, and Charles Brush in the 1870s and 1880s. These were open carbon arc systems, as contrasted to the enclosed gas discharge arc lamps that we have today. Weston made the first carbon arc street lighting installation in Philadelphia in 1879, whereas Brush was responsible for providing the first all-night street lighting system in Cleveland, also in 1879. Two sets of 8-h carbons were installed in each Brush arc lamp, the second set automatically going into operation after the first set was used up.

The third path of light source development in the 1800s was that of enclosed gaseous discharge lamps. The German scientist, Herman Geissler, had observed in 1856 that a tube containing a small amount of air would glow if a high voltage were placed across its ends. In 1860, John Way, an Englishman, created the first mercury arc by breaking a stream of mercury connected to a battery and observing its characteristic blue-green color. Arons developed a mercury vapor arc in 1892 which led Peter Cooper Hewitt to produce a commercial tubular mercury lamp with a mercury pool cathode and an iron anode in 1901. When the lamp was tilted, a column of mercury bridged the gap between the anode and cathode. As the lamp was righted, the mercury column broke and the electric arc discharge began. The Cooper-Hewitt lamp was widely used in the early 1900s and led to the development of high-pressure mercury lamps in the 1930s and metal halide and sodium lamps in the 1950s and 1960s.

None of the lamps discussed thus far makes use of fluorescence. In 1852, Sir George Stokes discovered the basic principle of transforming ultraviolet radiation into light. He found that a quinine sulfate solution glowed when irradiated by ultraviolet energy. Seven years later, A. E. Becquerel discovered that certain powdered materials when placed in an evacuated tube to which a high voltage was applied would fluoresce. These early discoveries led to extensive research on fluorescent materials in the early 1900s and to the development of the fluorescent lamp in the 1930s.

1.3 THE WAVE NATURE OF LIGHT

Light is defined as visually evaluated radiant energy. This definition suggests that light is first of all energy; secondly that it is transmitted by radiation, and lastly that it is a form of radiant energy to which the eye is sensitive. We need to explore all of these facets of light.

Until the middle of the seventeenth century, most scientists, particularly in the field of optics, accepted the corpuscular theory of light as enunciated by Sir Isaac Newton (1642–1727). According to Newton, all hot bodies emitted elastic corpuscles, each having the same, very high velocity and each having a size dependent upon its color. These corpuscles were considered to be minute particles that traveled in straight lines and could be reflected and refracted. Reflection was explained much as one would explain the bouncing of a ball against a wall. Refraction, the bending inward of light as it enters a denser medium, was justified on the basis of increased gravitational attraction of the more dense medium, causing a particle entering it to bend toward the perpendicular to the surface of the medium. Because the bending indicated an acceleration toward the perpendicular, it was postulated that the velocity of the particles would be greater in the denser medium. Newton explained diffraction, the bending of light around a sharp edge, in the same way, but he could not explain the phenomena of polarization and interference of light with his corpuscular theory.

In 1670, Christian Huygens (1629–1695), Dutch astronomer, physicist and mathematician, developed a wave theory of light in opposition to the corpuscular theory of Newton. Huygens described the bending of light as a fundamental principle of wave motion, Huygens' principle, which states that every point on an advancing wavefront behaves as a source of secondary wavelets which are sent out radially. Thus a wavefront encountering an obstruction will send out a secondary wave in a direction different from that of the incident wave. The similarity to the action of ripples on a lake encountering an obstruction should immediately come to mind. Huygens showed that the laws of reflection and refraction could be explained by his wave theory. He also proposed that color was a function of wavelength, again in contradiction of Newton, and conducted many experiments involving polarization of light.

The English physicist Thomas Young (1773–1829) reinforced Huygens' wave theory with his famous experiment on interference in 1827. Young showed that a narrow beam of light passing through a pinhole and then through an additional pair of pinholes to a screen created alternating light and dark bands on the screen. The light bands occurred when the secondary waves from the pair of pinholes reinforced each other by arriving in time phase at the screen; the dark bands occurred when the waves arrived out of time phase and thus interfered with each other.

It remained until 1862 for the controversy between the Newton corpuscular theory and the Huygens-Young wave theory to be resolved. In that year, Jean Bernard Leon Foucault (1819–1868), a French physicist known for the Foucault pendulum, conducted a critical experiment concerning the speeds of light in air and water. Foucault showed that the speed of light was less in the denser medium, conclusively disproving the corpuscular theory of Newton.

Thus, in the mid 1800s light was considered as wave motion by most scientists. The waves vibrated at right angles to the direction of propagation of the wavefront (transverse waves, as opposed to longitudinal waves which vibrate along the direction of propagation) and light "rays" were nothing more than lines showing the paths followed by different portions of the wavefronts. Reflection, refraction, interference, diffraction, and polarization could all be explained in terms of wave theory.

It remained for James Clerk Maxwell (1831–1879), the noted Scottish physicist, to unify these concepts in his Treatise on Electricity and Magnetism, published in 1873. Essentially, Maxwell showed that electromagnetic waves consist of electric and magnetic fields vibrating at right angles to each other in space and at right angles to the direction of propagation. These fields are in phase with each other in time and have the same frequency. His theory made it unnecessary to have a medium, such as the fictitious "ether" of the early nineteenth century, for transmission of a wave. He further theorized that the velocity of electromagnetic waves could be computed from purely electric and magnetic measurements. He calculated this velocity as very nearly the measured velocity of light and thus was able to infer that light consisted of electromagnetic waves of short wavelength.

Maxwell's work paved the way for Heinrich Hertz (1857–1894), a German physicist, to produce electromagnetic waves experimentally in the microwave region of the spectrum in 1887, to show that they had the properties of light waves, and to verify that Maxwell's theories were indeed valid and universally applicable to all electromagnetic radiation. Hertz's work proved Maxwell's inference that the common denominator in all electromagnetic radiation is the velocity of light.

It is useful to examine Maxwell's electromagnetic waves in further detail. The diagrams in Fig. 1.1 depict the electric field intensity $\mathscr{E}$ of an electromagnetic wave propagating through space. Figure 1.1*a* shows the variation in $\mathscr{E}$ with time at a fixed point in space ($s = 0$ is assumed). We can write an expression for this variation as

$$\mathscr{E} = E \sin 2\pi f t \tag{1.1}$$

What this equation says is that the electric field intensity at $s = 0$ is a sinusoidal function of time with period $T = 1/f$ seconds, where f is frequency in hertz, and amplitude E volts/meter. Once the wave is established, the waveshape at any other point in space would be the same except that it would be shifted in time and it would likely be attenuated. If an observer stands a fixed distance from a radio antenna, light source, or some other kind of electromagnetic wave generator and holds an "electric field intensity detector" attached to some kind of graphical read-out device, such as a chart recorder or an oscilloscope, he sees the waveform in Fig. 1.1*a* after the wave first reaches him.

Now suppose our observer could move speedily outward from the source with time fixed. He would see a pattern such as that in Fig. 1.1*b* after the wave is established, where s is the distance from the source. When he reaches the point at which the wave begins to repeat itself, he has traversed 1 wavelength, λ, in space, as labeled in Fig. 1.1*b*.

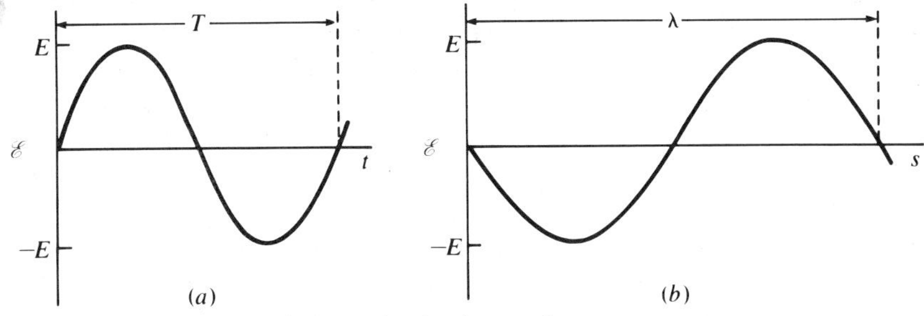

Figure 1.1 Electric field intensity in time and space.

We can write an equation for this variation of the electric field intensity with distance at a fixed point in time after the wave is established.

$$\mathscr{E} = -E \sin 2\pi \frac{s}{\lambda} \tag{1.2}$$

This is also a sinusoidal variation, but as a function of s rather than t, with wavelength λ playing the same role that period did in Eq. (1.1).

The time and space variations in Eqs. (1.1) and (1.2) can be combined to obtain an overall expression for the electric field intensity, given by

$$\mathscr{E} = E \sin 2\pi\left(ft - \frac{s}{\lambda}\right) \tag{1.3}$$

At $s = 0$, Eq. (1.3) becomes Eq. (1.1); at $t = 0$, it becomes Eq. (1.2).

What is the velocity of propagation of the wave? This will be the velocity of a fixed point on the wave and is obtained by requiring that the quantity $ft - (s/\lambda)$ in Eq. (1.3) remain constant as time passes. Thus as t changes from t_1 to t_2, s must change from s_1 to s_2 such that

$$ft_1 - \frac{s_1}{\lambda} = ft_2 - \frac{s_2}{\lambda} \tag{1.4}$$

Rearranging this expression gives

$$\frac{s_2 - s_1}{t_2 - t_1} = \lambda f \tag{1.5}$$

The left side of Eq. (1.5) is indeed the velocity of the wave, the rate of change of position of a point on the wave with time. The minus sign in Eq. (1.3) can now be explained by noting that if we had chosen a plus sign, the velocity in Eq. (1.5) would have been negative and the wave would have been traveling in the wrong direction (from observer to source).

Let us summarize what we have shown. The electric field intensity of an electromagnetic wave is sinusoidal in both time and space. At a fixed point in space, it is sinusoidal in time with a period T. At a fixed point in time, it is sinusoidal in space with a wavelength λ. The wave propagates with a velocity given as the product of wavelength and frequency.

The magnetic field intensity H of an electromagnetic wave is similar to the electric field intensity, except that the two waves lie at right angles to each other in space. Both lie at right angles to the direction of propagation. The fields are in time phase in that, at a given point in space, $\mathscr{E}$ reaches a positive maximum at the same instant of time as does H. The ratio of the two field intensities is always the same and can be shown to be given by

$$\frac{\mathscr{E}}{H} = \frac{\mu_0}{\epsilon_0} \tag{1.6}$$

where μ_0 is the permeability of free space and ϵ_0 is its permittivity. These two quantities occur throughout the study of electromagnetic theory. Maxwell's calculation in 1873 of the velocity of electromagnetic waves

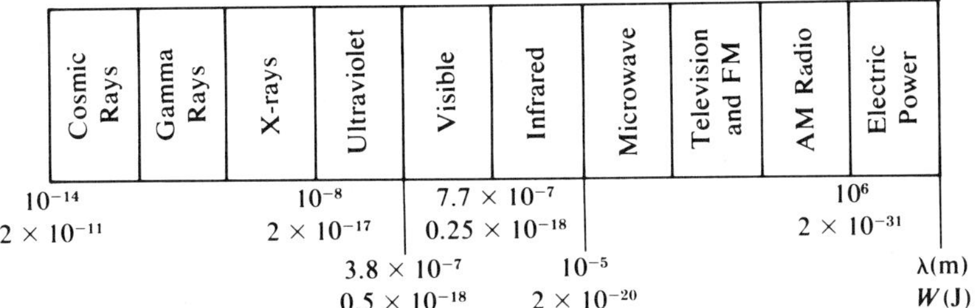

Figure 1.2 The electromagnetic spectrum: wavelengths and photon energies.

radiating from an oscillating electric circuit depended on them in the form of the following simple equation

$$v = \sqrt{\frac{1}{\mu_0 \epsilon_0}} \tag{1.7}$$

where

$$\mu_0 = 4\pi \times 10^{-7}\ \text{N} \cdot \text{s}^2/\text{C}^2 \quad \text{and} \quad \epsilon_0 = \frac{1}{36\pi \times 10^9}\ \text{C}^2/\text{N} \cdot \text{m}^2 \tag{1.8}$$

Inserting these values in Eq. (1.7) gives $v = 3 \times 10^8$ m/s, which is the speed of light. This led Maxwell to conclude that light consisted of electromagnetic waves and lets us revise Eq. (1.5) to show that, for all electromagnetic radiation, frequency and wavelength are related by

$$c = \lambda f \tag{1.9}$$

where c is the velocity of light. If c is in meters per second, then λ is in meters and f in cycles per second or hertz. The full gamut of the electromagnetic spectrum is shown in Fig. 1.2. Note the immense range of wavelengths present, from the very short cosmic rays (10^{-14} m) to the very long 60-Hz electric power waves (3100 mi).*

The portions of the electromagnetic spectrum with which we will be concerned are the ultraviolet, visible, and infrared, as shown in Fig. 1.3, with emphasis on the visible. It is seen that light waves occupy a very small portion of the spectrum, extending roughly from 380 to 770 nm. A nanometer (nm) is 10^{-9} m, about 50 billionths of an inch.† We generally associate six basic colors with the visible spectrum, as indicated in Fig. 1.3. These colors are not distinct bands, as implied by Fig. 1.3, but rather blend together, as anyone who has looked at a rainbow or examined the light emerging from a prism realizes.

Why is is that this narrow band of wavelengths is of such importance?

*The energies (W) associated with Fig. 1.2 will be discussed in Sect. 1.5.

†Light wavelengths in the past were also measured in Angstroms and microns. An Angstrom (Å) is 10^{-10} m and a micron (μ) is 10^{-6} m. The Angstrom unit is named for A. J. Angstrom (1814–1874), a Swedish physicist noted for his atlas of the solar spectrum.

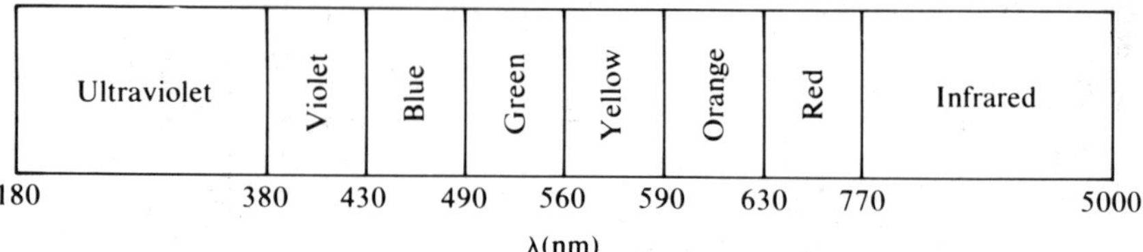

Figure 1.3 The visible and near-visible spectra.

The answer, of course, is that our eyes respond to these wavelengths, and to these wavelengths only. This concept of sensitivity of a receptor to certain wavelengths and not to others is an important one and is not unique to the eye. For example, skin sensitivity for tanning centers at about 300 nm in the ultraviolet. Bacteria are critically sensitive to radiation centered at 265 nm, and germicidal lamps provide radiation near this wavelength (at 253.7 nm) for product sanitation and to kill air-borne bacteria in sanitariums and nurseries. In a different field, resonant circuits in radios and TV sets are "tuned" to frequencies in the 300-m and 3-m regions, respectively, of the spectrum.

Although we are interested in the various spectral colors, our primary interest is in the sensation of whiteness. Equal-energy white light can be defined as the simultaneous presence, in equal energy amounts, of all the wavelengths within the visible spectrum. Incandescent, fluorescent, and high-intensity discharge lamps do not provide equal-energy white light. Incandescent lamps are much stronger in the reds and yellows than they are in the blues and greens. Standard fluorescent lamps are somewhat weak in blues and reds. Mercury lamps have only blues, greens and yellow-greens and sodium lamps are strongest in the yellows. Sunlight provides a nearly uniform spectral distribution but is somewhat deficient in blues and violets. However each of these sources, by itself, appears to the eye to produce white light, even though one white light may seem "warmer" or "cooler" than another. Thus it is not necessary to have equal-energy white to create the sensation of whiteness. It is only when these lamps are used to illuminate colored objects that we can detect the differences in spectral content between them.

1.4 WORK, ENERGY, AND POWER

Because so much of the subject of illumination engineering has to do with power and energy, it is important to understand at the outset what these two quantities are, how they relate to each other, and how they are connected to the concept of work. Let us consider work first. Let a constant force F be applied to a body such that the body moves a distance s in a direction making an angle θ with the line of application of the force. The work done by the force F is given by $F \cos \theta \times s$, the product of the component of the force in the direction of motion and the displacement. For work to be done, there must be motion and there must be a force component in the direction of the motion (work is not done if you hold a baby still in your arms but work is done if you rock the baby to and fro).

In Systeme Internationale (SI) units, the unit of force is the newton and the unit of displacement is the meter. Thus the unit of work is the newton-meter and is called a joule. One joule of work is done when a force of one newton in the direction of motion moves a body one meter.

Now let us determine how work done on a body relates to energy gained by that body. In introductory physics courses it is customary to consider the situation in which an applied force moves a block up an inclined plane in opposition to both a friction force and a component of force due to the mass of the block, as illustrated in Fig. 1.4*a*. The work done by the external force is shown to be equal to the sum of the work done against friction and the gains in potential energy and kinetic energy of the block.

In our case, we will choose a different illustration which yields the same conclusions but which is more relevant to the subject matter of this book, in particular to the energy exchanges which occur within gaseous discharge light sources. Consider the situation shown in Fig. 1.4*b* where we have two parallel plates separated by a distance d, with an applied voltage V, positive at the left plate. The region between the plates is assumed to be filled with gas atoms and a single electron sits just outside the positive plate. We apply a force so as to accelerate the electron toward the negative plate and we wish to relate the work done by this force to the energy gains of the electron.

There are three forces in Fig. 1.4*b*. One is the applied force F which we must somehow exert on the electron to move it toward the negative plate. The second is the force of the applied electric field, which resists our efforts to move the electron. The electric field is given by V/d and its force by $q_e V/d$, where q_e is the charge on the electron (-1.60×10^{-19} C). The third force is a "friction" force f imposed by the medium between the plates. In a gaseous discharge device this is the effect of the collisions of the electron with the many gas atoms in the space. Most of these are collisions in which the kinetic energy of the two colliding bodies is conserved, called *elastic collisions*,

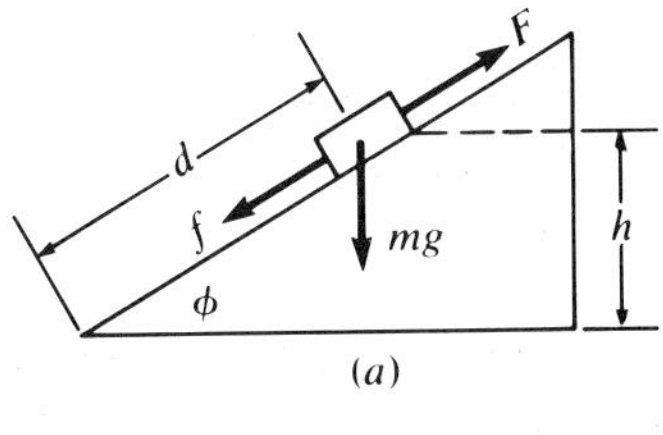

(*a*)

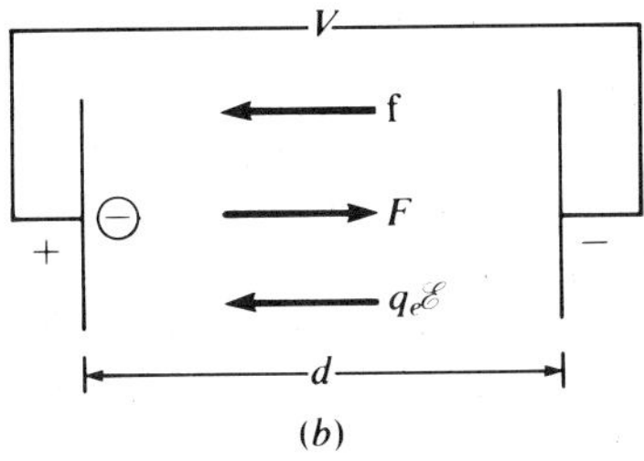

(*b*)

Figure 1.4 Relating work and energy.

and result in the exchange of a small amount of kinetic energy between the electron and the gas atom.

With the forces defined, we can write Newton's second law of motion for the situation in Fig. 1.4*b* as

$$F - f - \frac{q_e V}{d} = m_e a \tag{1.10}$$

where m_e is the mass of the electron (9.11×10^{-31} kg) and a is the acceleration.

Equation (1.10) is a force equation. To convert it to a work equation, we multiply each side by d.

$$Fd - fd - q_e V = m_e a d \tag{1.11}$$

Since the forces in Fig. 1.4*b* are assumed constant, the acceleration is constant and is given by

$$a = \frac{v^2}{2d} \tag{1.12}$$

where v is the velocity of the electron as it reaches the negative plate.

Inserting Eq. (1.12) into Eq. (1.11) and rearranging yields

$$Fd = fd + q_e V + \tfrac{1}{2} m_e v^2 \tag{1.13}$$

Let us carefully examine each term in Eq. (1.13). The term on the left side is the work done by the external force applied to the electron. This is the "input" to the system. The "output" consists of the three terms on the right side of the equation. The leftmost of these is the work done against the "friction" between the electron and the medium. This will cause the gas atoms to increase in temperature, so this work ultimately manifests itself as heat energy imparted to the system (mechanical energy converted to thermal energy). There are, of course, other ways of imparting heat to a body. For example, passing an electric current through a wire, as in an incandescent lamp, causes the wire to become hot. This is the conversion of electric energy to thermal energy.

The middle term on the right side of Eq. (1.13) is the increase in potential energy of the electron. This energy is called potential energy because it is determined by the position of the electron, rather than by its motion. Note that the increase in potential energy is independent of the path taken in going from one plate to the other. It depends only on the difference between the final and initial positions. Note also that the electron at the negative plate has a greater potential to do work than it did at the positive plate. Thus the work going into potential energy is recoverable, unlike the friction work which is lost as heat to the gas atoms.

The rightmost term in Eq. (1.13) is the increase in the kinetic energy of the electron and is governed by the increase in its velocity in going from one position to another.

To sum up, the work done by the external force in Fig. 1.4*b* manifests itself as an increase in temperature due to friction, an increase in the potential energy of the electron due to a change in its position, and an increase in the kinetic energy of the electron due to an increase in its velocity. For Eq. (1.13) to balance dimensionally, the unit of energy must be the same as that of work, namely the joule in the SI system. When work is done on a body it gains energy; when work is done by a body, it loses energy. We will refer back to the result in Eq. (1.13) in subsequent chapters when we discuss the mechanisms by which light sources process energy and produce visible radiation.

Thus far we have excluded the time element from our discussions. If we exert a force of one newton on a body and move it a distance of one meter, we do one joule of work, whether the task takes a second, an hour, or some other period of time to perform. We need an additional concept to describe the rate at which work is done or energy is produced or consumed. This concept is power, which is defined as the time rate of doing work.

$$W = \int_{t_1}^{t_2} P\,dt \tag{1.14}$$

$$P_{\text{av}} = \frac{W}{t_2 - t_1} \qquad \left(\text{Average power} = \frac{\text{work done or energy produced}}{\text{time interval}}\right)$$

In the SI system, the unit of power is the watt, which is a joule per second. Thus the joule is a watt-second.

Electric energy is generally measured in kilowatthours, where 1 kilowatthour $= 3.6 \times 10^6$ J. Another common unit of energy measurement is the British thermal unit (Btu), which is the amount of heat to raise a pound of water one degree Fahrenheit. The unit of power in this system is the Btu per hour which is equivalent to 0.293 W. Thus each watt represents 3.413 Btu/h.

In these days of worldwide concern about energy, there is much laypublic confusion about what the terms power and energy mean. We all wish to conserve energy, but there is no particular virtue in saving power.* For example, if reducing the lighting level in a large office by 30% reduces the amount of work done to the extent that an extra shift is required to maintain work output level, more lighting energy will be used than before. It is not only that the lights are on that is important, it is how long they are on.

1.5 THE PARTICLE NATURE OF LIGHT

It might appear from the foregoing that Maxwell's treatise provided a final and unified theory of light. As a matter of fact, most scientists at the close of the nineteenth century felt that little new knowledge concerning the nature of light would be forthcoming in the future. However they were quite wrong.

*Except for the possibility of a demand charge to users who have a large kilowatt connected load.

Despite its great accomplishments, electromagnetic wave theory could not explain the radiation spectrum from a blackbody or photoelectric emission from the surfaces of metals. We will discuss each of these phenomena in detail in Chaps. 3 and 4, covering the work of Planck, Einstein, and the other physicists who made such notable contributions during the first part of the twentieth century. At this point we should note that in 1905, Einstein, drawing on earlier work by Planck, postulated that the energy in a light beam was concentrated in small packets (quanta) called photons, instead of being distributed throughout space in its electric and magnetic fields. Each photon has an energy in joules given by

$$W = hf = \frac{hc}{\lambda} \tag{1.15}$$

where h is Planck's universal constant (6.63×10^{-34} J · s).

We should also note a parallel development with which the reader who has taken a chemistry course is presumably familiar. In 1911, the New Zealand physicist Ernest Rutherford (1871–1937), while working in England, verified the existence of the atomic nucleus. He proposed a model for the atom consisting of a nucleus, positively charged, surrounded by orbiting electrons, negatively charged, very similar in concept to our sun and planetary system.

Specifically the Rutherford model had these characteristics:

1. Positively charged nucleus of relatively large mass surrounded by negatively charged electrons of much lesser mass.
2. Electrically neutral atom: the number of protons in the nucleus equals the number of orbiting electrons.
3. Diameter of atom is 10^5 times diameter of nucleus.
4. Electrons attracted toward the nucleus by Coulomb's law:

$$F = \frac{-q_e^2}{4\pi\epsilon_0 r^2} \tag{1.16}$$

   where r is the distance of the electron from the nucleus.
5. Electrons revolving in fixed orbits about the nucleus, requiring that each electron have a centripetal acceleration toward the nucleus and thus a centripetal force given by Newton's second law as

$$F = \frac{m_e v^2}{r} \tag{1.17}$$

   where v is the electron's velocity. This force is provided by the Coulomb's law attraction.

This model came into immediate conflict with Maxwell's electromagnetic theory, which asserts that an accelerating electron radiates energy, causing its total energy to decrease. Since an electron orbiting a nucleus is accelerated continually toward that nucleus, it should continually radiate electromagnetic energy until all of its energy is gone and it collapses into the nucleus. Furthermore, as the electron spirals in to rest, its angular velocity ($\omega = v/r$) should continually increase. This predicts a continuous spectrum in contradiction to the observed line spectra of hydrogen, mercury, etc.

Niels Bohr (1885–1962), the Danish physicist who had studied atomic physics under Rutherford in England, attempted to resolve this difficulty. In a series of papers from 1913 to 1915, he developed a theory of atomic structure which came to be known as the Bohr atom and earned for him the Nobel prize in physics in 1922. His theory consists of three fundamental postulates:

1. An electron in an atom can exist only in certain specified orbits. When in one of these orbits, the electron is in a stable state and does not emit radiation.
2. An electron may jump suddenly from one stable state to another of lower energy. When it does so, a single photon is emitted, whose energy is given by

$$\frac{hc}{\lambda} = W_i - W_f \tag{1.18}$$

   where W_i is the energy of the initial orbit and W_f the energy of the final orbit.
3. Only those orbits are permitted in which the angular momentum of the election is an integral multiple of $h/2\pi$. Stated mathematically

$$m_e vr = n \frac{h}{2\pi} \quad \text{J} \cdot \text{s} \tag{1.19}$$

   where n is an integer.

The spectral lines of hydrogen, the simplest element, were very well known at the time Bohr set forth his theories. Thus his postulates were tested thoroughly on hydrogen and were found to predict the frequencies of its spectral lines within one part per million.

Let us obtain the radii of the orbits, the permitted levels, and the resulting spectral lines of hydrogen. From Rutherford's third and fourth postulates, the centripetal force causing the electron to travel in a circle is provided by the Coulomb's law attraction between the electron and proton. Thus

$$\frac{q_e^2}{4\pi\epsilon_0 r^2} = \frac{m_e v^2}{r} \tag{1.20}$$

Equations (1.19) and (1.20) may be solved simultaneously for r and v, since all other quantities in these equations are presumed known. The results are:

$$r = \frac{n^2 h^2 \epsilon_0}{\pi m_e q_e^2} \qquad v = \frac{q_e^2}{2nh\epsilon_0} \tag{1.21}$$

The only variable in these two equations is n, the integer which quantizes the orbits.

Unlike the situation in Fig. 1.4b and Eq. (1.13), there is no friction force associated with an electron rotating about a nucleus. Thus the energy of an orbiting electron consists of just two parts. First is its potential energy, which is the work done in bringing it from infinity to the orbit in question. This is given by

$$\mathrm{PE} = \int_{\infty}^{r} \mathbf{F} \cdot dr = \int_{\infty}^{r} \frac{q_e^2}{4\pi\epsilon_0 r^2}\, dr = -\frac{q_e^2}{4\pi\epsilon_0 r} \tag{1.22}$$

The second energy of the electron is kinetic and is obtained from Eq. (1.20) as

$$\mathrm{KE} = \tfrac{1}{2} m_e v^2 = \frac{q_e^2}{8\pi\epsilon_0 r} \tag{1.23}$$

Thus the total energy of the electron in an orbit is

$$\mathrm{TE} = \mathrm{KE} + \mathrm{PE} = -\frac{q_e^2}{8\pi\epsilon_0 r} \tag{1.24}$$

Equations (1.21) through (1.24) show that as n decreases, r decreases, PE decreases, v increases, KE increases, and TE decreases. Thus an electron closest to the nucleus ($n = 1$) has the least potential energy, the greatest kinetic energy, and the least total energy. Also, all energies are zero at $r = \infty$.

A tabulation of radii and energies for the single hydrogen electron appears in Table 1.1. The radii vary directly as n^2; the total energies inversely as n^2.

Equation (1.18) gives the energy of a photon resulting from a transition from one energy level in an atom to a lower one. Assume an electron in a hydrogen atom has been lifted from its normal state ($n = 1$) to a higher energy state ($n = 3$) due to a collision of the atom with a fast-moving external electron. Assume further that, following the collision, the atom's electron re-

TABLE 1.1 HYDROGEN: RADII AND ENERGIES

n	r (nm)	TE (J $\times 10^{-19}$)
1	0.053	−21.65
2	0.21	−5.41
3	0.48	−2.40
4	0.85	−1.36
5	1.33	−0.86
∞	∞	0

turns to the $n = 2$ state. A photon will be emitted whose wavelength is given by

$$\lambda = \frac{hc}{W_i - W_f} = \frac{6.63 \times 10^{-34} \times 3 \times 10^8}{(-2.40 + 5.41) \times 10^{-19}}$$
$$= 6.61 \times 10^{-7} = 661 \text{ nm}$$

This is indeed one of the observed spectral lines of hydrogen.

The work of Planck, Einstein, Rutherford, and Bohr led to many years of controversy in physics, culminating in the reluctant acceptance of the fact that light appears to be dualistic in nature. At a source or at a receiver, the concepts of quantum theory appear to explain better what is going on, whereas in the medium between source and receiver electromagnetic wave theory seems to provide the better explanation. This is illustrated in Fig. 1.2, where we have included the photon energies throughout the spectrum, obtained from Eq. (1.15). Thus Fig. 1.2 may be considered as either an energy spectrum or a wave spectrum, in keeping with the dualistic nature of electromagnetic radiation.

1.6 THE PROCESS OF SEEING

Figure 1.5 depicts the process of seeing. The first prerequisite is a source of visible radiant energy—the sun, a flame, or one of the many sources of electric light available. The second requirement is some kind of a light modifier—a natural or manufactured object that reflects or transmits light to the eye (it may also absorb light). This may be a window, the moon, the page of a book, a gravel pit, etc. We call such an object a modifier because in most instances it will alter the spectral character of the radiation from the source.* This is illustrated in the extreme in Fig. 1.6. In Fig. 1.6*a*, white light, containing all spectral colors, impinges on red, green, and blue surfaces. Three light waves are reflected to the eye, a red wave from the red surface, and green and blue waves from the green and blue surfaces, respectively. The red surface appears red to the viewer, the green surface green and the blue surface blue, assuming adaptation to a neutral (gray) background.

In Fig. 1.6*b*, the source has only green wavelengths. Green light is reflected to the eye from the green surface, which then appears green to the viewer. The other two surfaces appear dark, because no light is reflected to the eye from them. Thus it is the spectral character of both the source and the modifier that determines what the viewer perceives.

Returning to Fig. 1.5, we note that the third ingredient of seeing is the eye, which serves as both the receiver of the light "messages" and also the

*To be rigorous, we should also include the media between the source and modifier and between the modifier and receiver. These can also alter the spectral content of the source radiation.

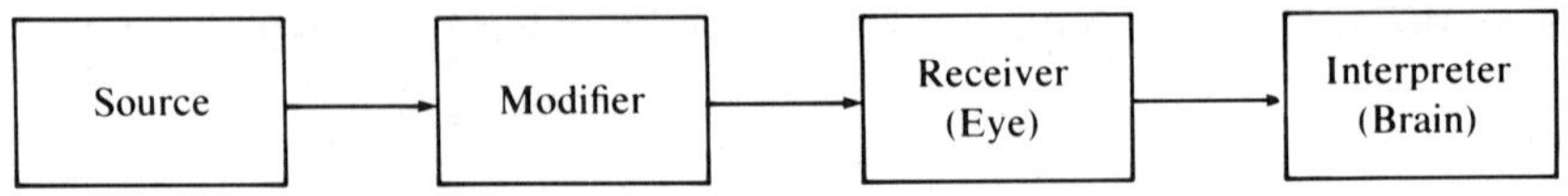

Figure 1.5 The process of seeing.

processor of them for transmission to the brain. The brain is the fourth element, the deprocessor and interpreter link in the seeing process. It receives the signals from the optic nerve, decodes them, and thus provides the viewer with perception and understanding of the object being viewed.

The eye is shown diagrammatically in Fig. 1.7. Let us consider each of its parts in some detail.

1. *Sclera and choroid.* These are the outer jacket of the eye and its lining. The sclera maintains the eye in nearly spherical shape and about 1 in in diameter. In front, it comprises the white of the eye. The choroid is the black coating which lines the sclera. It absorbs all stray light inside the eye and thus makes vision more distinct, serving the same function as the black interior of a camera.
2. *Cornea.* This is the first of the three major parts of the optical system of the eye. It is a bulging transparent membrane at the eye's front, consisting of five layers. The cornea has a radius of curvature of about 7.5 mm, a thickness of 0.4 mm, and a refractive index of 1.37. It has a fixed focus and supplies 70% of the converging power of the eye. It absorbs all wavelengths less than 295 nm, the wavelength in the ultraviolet at which air becomes opaque.

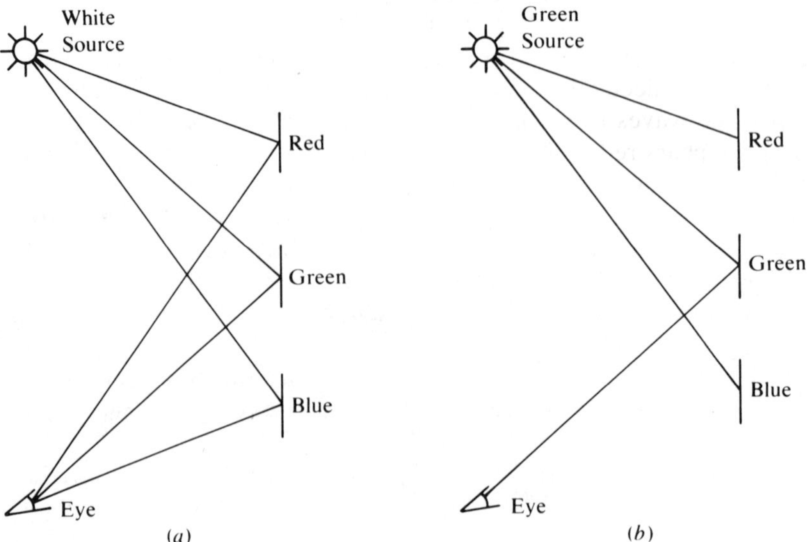

Figure 1.6 Spectral modification of light.

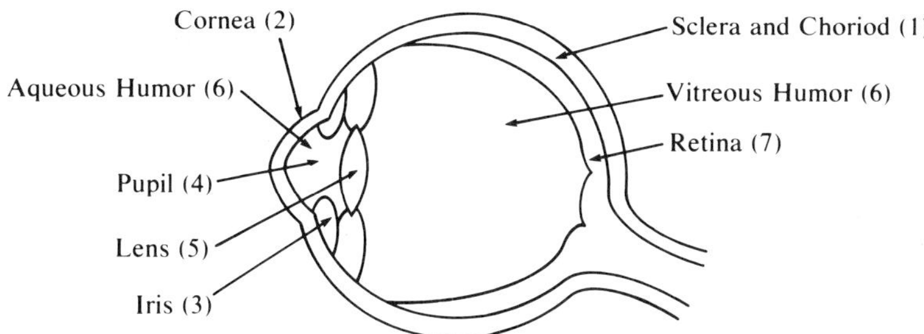

Figure 1.7 The eye. (Numbers in parentheses refer to text.)

3. *Iris*. This is the continuation of the choroid in the front of the eye. It is the colored portion of the eye (blue, brown, etc.) and it serves the same role as a diaphragm in a camera by controlling the size of the aperture (pupil) through which light is admitted, thus protecting the eye against excessive light (the eye is well-shielded from excess light from above by the eyebrow, eyelash and eyelid, but it not at all well-shielded from below).
4. *Pupil*. This is the small circular opening in the iris. It is the second part of the eye's optical system, controlling the amount of light entering the eye and confining the light to the central area of the lens to minimize distortion of whatever is being viewed. Under normal illumination levels, pupil diameter is 3 to 4 mm and ranges from 8 mm under dim light to 2 mm under bright light. Thus the optical adaptation range of the eye is about 16 to 1. But everyday illumination levels vary over a 10^7 to 1 range. Thus the eye must also adapt photochemically to changes in illumination level, since its optical system is not designed to do anywhere near the entire job of adaptation. It has been shown that 10^{-15} W of green light, equivalent to a candle at a distance of 2 mi, can produce the sensation of sight, so the eye is indeed a very sensitive instrument.
5. *Lens*. This is the third part of the eye's optical system. It is crystalline, located just behind the pupil, and suspended in place by the ciliary muscle in back of the iris. In a camera, accommodation (focusing) is accomplished by moving the lens; in the eye the ciliary muscle alters the shape of the lens to change focus.

   The outer surface of the lens has a radius of curvature of 10 mm; the inner suface 6 mm. The lens has an average index of refraction of 1.44, having a higher index than this at its center and lower at its periphery. It is the outer surface of the lens which changes most in shape during accommodation. When the eye is viewing a distant object, the ciliary muscle relaxes, allowing the ligaments attached to the lens to pull it into a more flattened shape. The reverse occurs for viewing near objects.

   The lens absorbs all wavelengths less than 350 nm. It tends to blur an object which is multicolored because the lens cannot focus accurately more than one wavelength at a time.

6. *Aqueous and vitreous humors.* These are the transparent liquids filling the two inner chambers of the eye. The aqueous humor lies between the cornea and the lens and is a waterlike substance having an index of refraction of 1.34. The vitreous humor, located behind the lens, is more dense and jelly-like and also has an index of refraction of 1.34. Both these media serve to bend and focus, as well as transmit, the incident light.
7. *Retina.* This is the inner lining at the back of the eye and is the "film of the camera." Contained in it are the two types of photoreceptors, the rods and the cones. Also present are two spots: the fovea, a small pit 0.5 mm in diameter subtending an angle of 2° at the pupil, where most of the distinct three-dimensional seeing takes place and where color is distinguished, and the blind spot on the nasal side of the fovea, which is the connection of the eye to the optic nerve.

    The retina is less than 0.25 mm thick and is composed of 10 layers, with the rods and cones residing in the second layer in from the vitreous humor. The radius of curvature of the retina is 12.0 mm and the distance from the lens to the retina is 14.6 mm.
8. *Extraocular muscles.* These enable the eye to focus on an object and also to look in any forward direction. Each eye is provided with a set of six extraocular muscles, which are coordinated by the brain to work in harmony with each other so that in the majority of persons neither set dominates.
9. *Rods and cones.* These are the two types of receptors in the retina that transform the radiant energy entering the eye into chemical energy, which, in turn, causes electric impulses to be produced and sent out through the optic nerve to the brain. Their properties are summarized in Table 1.2.

It was mentioned that the adaptation range of the eye is 10^7 to 1 but that the optical system of the eye accounts for only a 16 to 1 range. Thus another mechanism must account for the major part of the eye's ability to adapt to changing light levels. This mechanism is photochemical in nature.

Research indicates that the photochemical cycle depicted in Fig. 1.8 occurs in the retina. In the rods there is a watery, light-sensitive fluid called visual purple or rhodopsin. When light impinges on this fluid, it spreads out over the fovea, is bleached, and decomposes chemically, first into a yellow substance called retinene and a protein called opsin and, after that, into vitamin A. During the absence of light the rhodopsin builds up again and is a maximum when the rods are fully dark adapted.

It is not clear how this chemical process leads to electric impulses to the optic nerve. One theory holds that certain molecular changes in the retinene and opsin produce changes in the electric energy in the rods which

TABLE 1.2

	Rods	Cones
Number	130×10^6	7×10^6
Shape	Cylindrical, 0.07 mm long and 0.002 mm in diameter.	Conical, 0.005 mm base diameter and 0.07 mm high.
Distribution	Located outside fovea. Proportion of rods increases as radial distance from fovea increases.	150,000 per square millimeter in fovea, none outside (if the concentration were uniform it would be impossible to focus attention on a part of the field of view).
Adaptation time	30–40 min	Less than 2 min (seeing would cease if the field of view were completely uniform and the eye did not wander continuously).
Sensitivity*	Scotopic, <0.034 cd/m², high luminance sensitivity, nighttime vision. Poor sensitivity to contrast.	Photopic, >3.4 cd/m², low luminance sensitivity, daytime vision. Excellent sensitivity to contrast.
Connection to optic nerve	Connected in groups, providing poor visual acuity (fuzzy vision) but high sensitivity.	Individually connected, providing sharp imaging but low sensitivity.
Characteristics of vision	Achromatic, all objects appear gray. More sensitive to blue than red. Vision is peripheral over a 190° field with poor depth perception. Poor for detecting fine detail.	Chromatic, with the peak sensitivity in the yellow-green. Vision is central and binocular over a 120° field giving good depth perception. Excellent for resolving fine detail.

*There is a third category, called mesopic vision, which overlaps the photopic and scotopic categories and covers the luminance range from 0.034 to 3.4 cd/m².

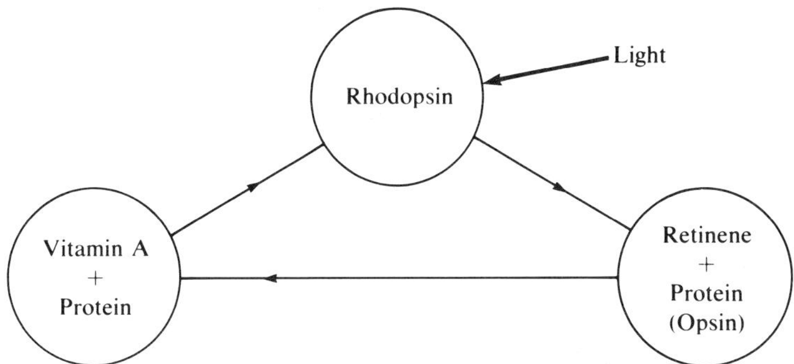

Figure 1.8 Photochemical process in retina.

eventually result in the development of nerve impulses in the ganglion nerve cells adjacent to the optic nerve. These impulses then travel to the visual cortex of the brain at a speed of about 4 mi/min, taking less than 2 ms to get there (there is also a brief period between stimulation of the retina and generation of the nerve impulse). It is held generally that similar reactions take place in the cones.

We have yet to discuss the eye's spectral sensitivity. Recall that we defined light as radiant energy that has been visually evaluated. The eye-brain combination performs the visual evaluation and "weights" each incoming visual wavelength in the manner shown in Fig. 1.9. The eye is most sensitive to light in the yellow-green portion of the spectrum at 555 nm. The spectral luminous efficiency $V(\lambda)$ of the eye at that wavelength is assigned the value 1. At all other wavelengths the eye is less sensitive and it requires a greater amount of energy to produce the same visual sensation. We will review the manner in which this curve was obtained in Chap. 3.

1.7 LIGHTING TERMS, SYMBOLS, AND UNITS

Our purpose in this section is to introduce, in a brief idealized way, the six major lighting terms and their units of measure and to show how these terms are interrelated. Then, in Chap. 2, we will derive more rigorous definitions and descriptions of these quantities and develop procedures for calculating them. The reader may wonder why it is necessary to visit the lighting terms twice. It has been the author's experience, from many years of teaching, that these terms and units are difficult to master and that they are best understood when presented in an idealized way first.

The illumination quantities, along with their symbols and units, are listed in Table 1.3. The basic unit, from which the others are derived, is the candela (cd) which is the unit of luminous intensity (I) of a source of light in a specified direction. The "standard" candela, as will be discussed in Chap. 3, is derived from a blackbody radiator operating at a prescribed temperature and pressure.

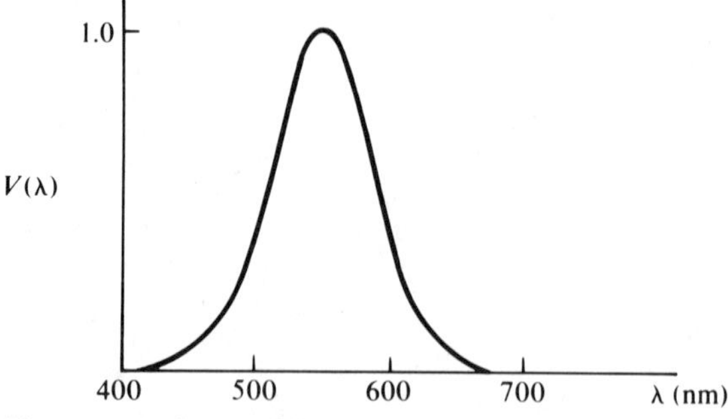

Figure 1.9 Spectral luminous efficiency.

TABLE 1.3 ENTITIES IN THE ILLUMINATION SYSTEM

Symbol	Concept	English unit	Metric unit
I	Luminous intensity or candlepower	Candela (cd)	Candela (cd)
Φ	Luminous flux*	Lumen (lm)	Lumen (lm)
E	Illuminance	Lumen per square foot [footcandle (fc)]	Lumen per square meter [lux (lx)]
M	Luminous exitance	Lumen per square foot (lm/ft^2)	Lumen per square meter (lm/m^2)
L	Luminance†‡	Candela per square foot (cd/ft^2)	Candela per square meter (cd/m^2)
Q	Quantity of light	Lumen-second (lm · s)	Lumen-second (lm · s)

*Luminous flux corresponds to power in the radiation system and quantity of light corresponds to energy. Thus the lumen and the watt are the same dimensionally, as are the lumen-second and the joule (watt-second).

†Luminance is also often expressed in candelas per square inch and candelas per square centimeter.

‡The footlambert is a unit of luminance equal to $1/\pi$ candelas per square foot. Its use is deprecated, but it does still appear in much of the lighting and vision literature.

It is not surprising that a name sounding like candle emerged as the unit of luminous intensity. In the early days of lighting in the United States, the whale oil lamp and the spermaceti candle were the major sources of light. In 1860, a unit of luminous intensity known as the "candle" was established using a spermaceti candle of specific weight burning at a specified rate as the basic standard. Later gas flames were calibrated in terms of this definition. The usual gas flame had a luminous intensity of 16 candles. The first incandescent lamps were burned at such voltage that they also had a luminous intensity of 16 candles. This voltage was approximately 110 V, but varied over a rather wide range depending on the length and diameter of the lamp filament, which was not as closely controlled during manufacture as it is today.

With the candela established, we can proceed to define the other lighting terms and units. In Fig. 1.10*a* we have placed a uniform point source of 1 cd at the center of a 1-m-radius sphere. By uniform, we mean that the source has the same intensity in every direction. By point we mean that the source is much smaller than any other dimension present. We think of this source as emitting luminous flux (Φ) uniformally in all directions. The analogy shown in Fig. 1.10*b* is a small sphere perforated with holes and connected to a garden hose. The holes are uniformly spaced and of the same size. When the faucet valve is turned on, fine streams of water emerge from the holes and travel radially outward (at least they do so close to the sphere until gravitational effects take over). This is *fluid flux*.

The unit of luminous flux is the lumen (lm) and is the rate at which luminous energy is incident on a 1-m^2 surface 1 m away from a uniform point source of 1 cd intensity. Since the surface area of a sphere is $4\pi r^2$, the total luminous flux emitted by the source in Fig. 1.10*a* is 4π lm. Thus a uniform 1-cd point source emits 4π lm.

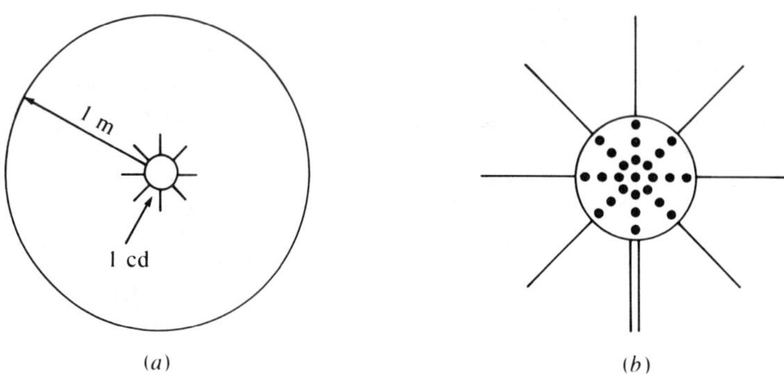

Figure 1.10 Luminous flux and a water analogy.

From its definition, we see that luminous flux is the time rate of flow of luminous energy. This makes the lumen dimensionally equivalent to the watt, which is the time rate of flow of energy. The two can be related through the spectral luminous efficiency curve of the eye, as will be shown in Chap. 3.

As luminous flux travels outward from a source, it ultimately impinges on objects, where it is reflected, transmitted, and absorbed. The illuminance (E) on a surface is the density of luminous flux incident on that surface. Thus,

$$E = \frac{\Phi}{A} \tag{1.25}$$

where A is surface area.

The unit of illuminance is the lux (lx) in the SI system and the footcandle (fc) in the English system, a lux being a lumen per square meter and a footcandle a lumen per square foot.

The inside of the sphere in Fig. 1.10*a* receives 4π lm from the 1-cd source. Since the sphere's surface area is 4π m^2, the illuminance on the inside surface of the sphere is 1 lx. There are 10.76 ft^2 in a square meter, so the illuminance is also 0.093 fc.

Knowing the density of luminous flux arriving at a surface, we now need a measure of the density of luminous flux leaving the surface. This requires two concepts, one to describe the total luminous flux density leaving the surface and the second to describe the luminous flux density leaving the surface in a particular direction. The former concept is provided by luminous exitance (M), the latter by luminance (L).

Assume that the sphere in Fig. 1.10*a* is made of translucent glass or plastic and that it transmits 80% of the luminous flux it receives, reflecting none back to the inside surface and absorbing the remaining 20%. Thus $3.2\pi (= 0.8 \times 4\pi)$ lm leave the sphere, and, recalling that the sphere surface area is 4π m^2, the density of luminous flux leaving the sphere is 0.8 lm/m^2. We say that the sphere has a luminous exitance of 0.8 lm/m^2 or 0.074 lm/ft^2.

Now let us stand far away from the sphere and look toward it. What we see is no longer a luminous sphere but rather a luminous circular plane, much as we see the full moon as a flat luminous disc when we gaze at it from Earth. The luminance of the sphere is defined as the density of luminous intensity in the direction of viewing. It is obtained by dividing the intensity of the sphere in the direction of viewing by the projected area of the sphere in that direction. The sphere appears from outside as though it has an intensity of 0.8 cd. Its projected area is π m^2. Thus its luminance is $0.8/\pi$ cd/m^2 or $0.074/\pi$ cd/ft^2.

It may appear from the foregoing that luminous exitance and luminance are related (through a π factor). This is not generally the case. It is true here only because we have assumed that the translucent sphere emits lumens uniformly in all directions. Such a surface is called "perfectly diffusing" and we will derive the conditions for this somewhat restricted behavior in Chap. 2.

The final term we will consider is quantity of light (Q). Quantity of light is luminous energy and is related to luminous flux, which is luminous power, throught the parameter of time. We use the lumen as the measure of luminous flux and the lumen-second or lumen-hour as the measure of light. The lumen-second is useful when we are concerned with bursts of lumens lasting for short periods of time, as in flashtube photography. We use the lumen-hour, much as we use the kilowatthour, in evaluating the consumption of energy, in this case luminous energy, over extended periods of time.

EXAMPLE 1.1

A glass globe is shown in Fig. 1.11. The upper hemisphere is silvered on the inside so that 90% of the lumens hitting it are reflected to the lower hemisphere. The latter is translucent glass and has a transmittance of 75% and an absorptance of 25%. The sphere has a radius of 0.5 m with a 100-cd lamp at its center. The lamp emits lumens uniformly in all directions, as does the lower hemisphere. Find the illuminance on the inside of the lower hemisphere and the luminous exitance and luminance of the outside of the lower hemisphere.

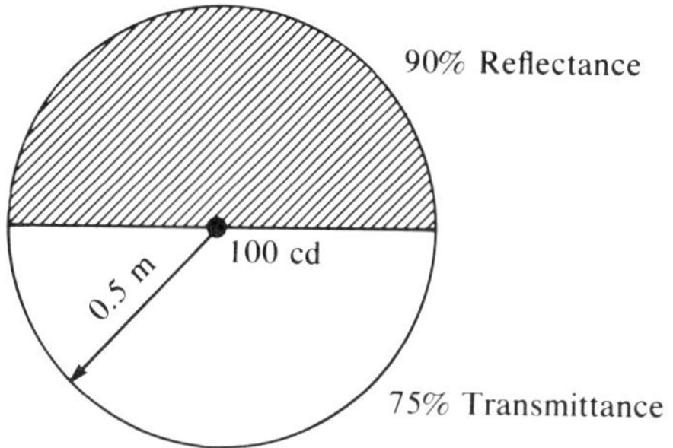

Figure 1.11

Solution. A 1-cd lamp emits 4π lm, so a 100-cd lamp emits 400π lm. Half of these, 200π lm go upward. Of these upward lumens, 90% or 180π lm are reflected downward. In addition, 200π lm go directly downward from the lamp. This gives a total of 380π lm incident on the inside of the lower hemisphere. The surface area of the sphere is π m^2; the area of the lower hemisphere is 0.5π m^2. Thus the illuminance on the inside of the lower hemisphere is $380\pi/0.5\pi = 760$ lx.

The lower hemisphere has a transmittance of 75%. Therefore 285π lm leave the lower hemisphere and its luminous exitance is $285\pi/0.5\pi = 570$ lm/m^2. Finally, a 1-cd source provides 2π lm in a hemisphere. Thus the outside of the lower hemisphere appears to have a luminous intensity of $285\pi/2\pi = 142.5$ cd. The projected area of this hemisphere from afar is 0.25π m^2. Dividing intensity by projected area gives the luminance of the bottom hemisphere as $142.5/0.25\pi = 570/\pi$ cd/m^2. Note that, in this case, because the lower hemisphere is assumed to be perfectly diffusing, $\pi L = M$.

REFERENCES

American National Standards Institute: "Nomenclature and Definitions for Illuminating Engineering," ANSI/IES RP-16, New York, 1980.

Cox, J.A.: *A Century of Light,* The Benjamin Co., New York, 1979.

General Electric Co.: "Light and Color," TP119, Nela Park, Cleveland, Ohio, 1968.

———: "Light Measurement and Control," TP118, Nela Park, Cleveland, Ohio, 1971.

Kaufman, J.E.: *IES Lighting Handbook—Reference Volume,* IESNA, New York, 1981, chaps. 1 and 3.

Luckiesh, M.: *Light, Vision and Seeing,* D. Van Nostrand, New York, 1944.

Millman, J., and C.C. Halkias: *Electronic Devices and Circuits*, McGraw-Hill, New York, 1967.

Oetting, R.L.: "Electric Lighting in the First Century of Engineering," *Transactions of AIEE,* November 1952.

Sears, F.W., and M.W. Zemansky: *University Physics*, Addison-Wesley, Cambridge, Mass. 1955.

Thwing, L.: *Flickering Flames,* Charles E. Tuttle, Rutland, Vt., 1958.

Chapter 2

Lighting Calculations and Measurements

2.1 SI UNITS

In most areas of human endeavor, it is necessary to have standardized units of measure. For each area of the physical sciences and engineering, four basic units are required. Three of these are common to all areas; the fourth is a chosen unit unique to the particular area being examined.

In Systeme International (SI) units, the three common base units are mass in kilograms, length in meters, and time in seconds. In the caloric system, the fourth basic unit is temperature in kelvins; in electromagnetics, it is electric current in amperes. For the science of light, it is luminous intensity in candelas.

Once the fourth basic unit is identified for a particular area, all other units in that area may be defined. Thus, with the candela defined, we were able in Sec. 1.7 to define the other basic lighting units, such as the lumen and the lux, in a rather idealized way. In this chapter we will define the lighting terms more rigorously and utilize them in calculations and measurements relating to point and area sources of light.

For those not familiar with the SI system, a few comments about it may be useful before proceeding. Until relatively recently there have been two systems of units throughout the world, namely the English system, used by the United States and the British Commonwealth, and the old metric system, used by the rest of the world. Over the past several years, the world has been moving slowly toward a new third system of units, the international system called SI.

The SI system is based on seven fundamental entities, whose units, definitions, and symbols have been agreed to internationally. These are length in meters, mass in kilograms, time in seconds, electric current in amperes, temperature in kelvins, the amount of a substance in moles, and luminous intensity in candelas. In addition, there are two supplementary units, namely the radian as the measure of a plane angle and the steradian as the measure of a solid angle. There are 17 derived units, obtainable from the seven fundamental and two supplementary units. Among these are force in newtons, energy in joules, and power in watts.

There are three major advantages of the SI system over its predecessors. First, SI is a coherent system in that the factors relating the various units are all unity. For example, as noted in Sec. 1.4, a force of 1 N moving a mass of 1 kg a distance of 1 m does 1 J of work. Second, the SI system is absolute. It does not have the same units for different quantities as we do, for example, in the English system, where force and mass can each be expressed in pounds. Third, the SI system is unique in that each quantity has only one unit. Thus energy, for example, is always in joules whether we are talking about an electrical system, a mechanical system, or a thermal system.

There are a few exceptions to SI which, through custom and conditioning, are not likely to change. The major of these are the use of liter, rather than cubic meter, to represent liquid volume, the more-prevalent use of degree, rather than radian, the use of minutes, hours, days and years to represent time increments, and the use of degrees Celsius to represent temperature.

2.2 THE LIGHTING TERMS REVISITED

The introduction of the six lighting terms in Sec. 1.7 was based on the notion of a 1-cd point source of light radiating uniformly at the center of a 1-m translucent sphere. That description was, by intention, oversimplified. In practice, most light sources are not point sources nor do they have the same intensity in all directions. Also, receiver surfaces are not usually oriented perpendicularly to the light rays impinging on them and sources do not emit light rays only in directions perpendicular to their surfaces. Thus we need more general definitions and expressions for the lighting terms which take these real-world constraints into account.

Nonuniform, nonpoint light sources can be treated in one of two ways. Either we can consider them as composed of small differential elements, each of which may be considered as a point source, or we can observe the entire source at a distance which is large compared with the maximum source dimension, so that its luminous flux may be considered as radiation from a point. With either approximation, the notion of conical tubes of flux radiating outward from a point is useful. This requires that we develop the concept of a solid angle.

Refer to Fig. 2.1*a*, in which the radian measure of a two-dimensional angle is presented. When s, the arc length on the circumference of the circle, is equal to r, the radius of the circle, the angle θ is 1 rad. Since the circumference of the circle is $2\pi r$, there are 2π rad in a plane around a point.

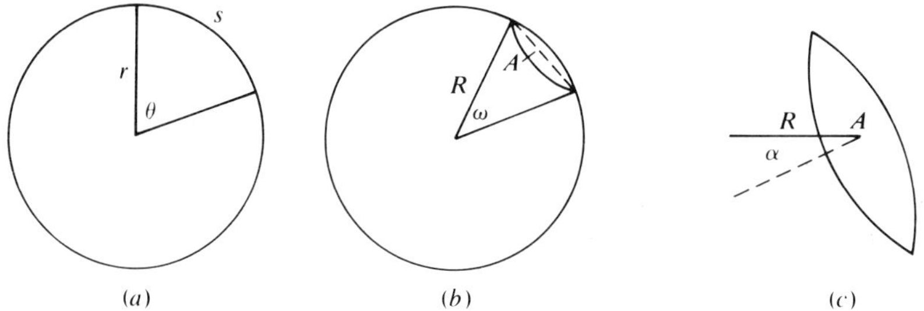

Figure 2.1 Illustrating the radian and steradian.

The extension of the two-dimensional case to three dimensions is shown in Fig. 2.1*b*. When A, the intercepted area on the sphere, equals R^2, where R is the radius of the sphere, the angle ω, which is referred to as a solid angle, is 1 steradian (sr). We say that the area A subtends a solid angle of 1 sr at the center of the sphere. Since the surface area of the sphere is $4\pi R^2$, there are 4π sr surrounding a point in space. If the area A is tilted at an angle α with respect to a radius vector through its center, as shown in Fig. 2.1*c*, it is the projected area of A, given by $A \cos \alpha$, that determines the solid angle.

The simplest geometry for a solid angle is a cone whose apex is at the center of a sphere. This is illustrated in Fig. 2.2. In this case, it is easy to relate the solid angle ω to the cone angle ψ_c.

We note that the differential area intercepted on the sphere surface is

$$dA = 2\pi R \sin \psi \; R \, d\psi \tag{2.1}$$

which, when integrated, yields

$$\begin{aligned} A &= \int_0^{\psi_c} 2\pi R^2 \sin \psi \, d\psi \\ &= 2\pi R^2 \, (1 - \cos \psi_c) \end{aligned} \tag{2.2}$$

Then ω and ψ_c can be related by

$$\omega = \frac{A}{R^2} = 2\pi(1 - \cos \psi_c) \tag{2.3}$$

When $\omega = 1$ sr, solution of Eq. (2.3) yields $\psi_c = 32.8° = 0.57$ rad.

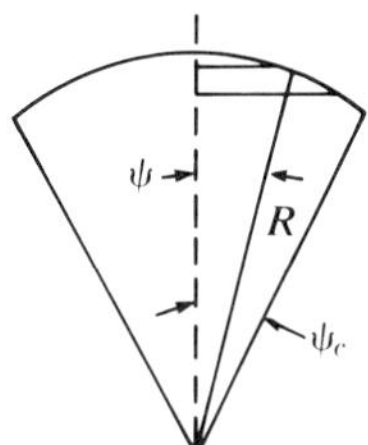

Figure 2.2 Comparison of solid angle and cone angle.

In Sec. 1.7, the lighting terms were developed with luminous intensity as the fundamental entity, in accord with the SI system in which the candela is one of the seven base units. For several reasons, it is preferable to use luminance, rather than intensity, as the fundamental quantity from which the other lighting terms are derived. Luminance relates better to fundamental concepts in optics; it is the parameter of a blackbody radiator that is specified in developing a lighting standard; it is a quantity that relates to a ray of light and thus it can be used to describe both source and receiver surfaces as well as the transmitting medium in between; it is a density quantity and thus is much more suitable for describing differential source areas.

Consider the situation in Fig. 2.3*a*. Luminous flux is traveling from left to right and encounters two apertures, one of which is tilted at an angle θ with respect to the direction of propagation. An amount of flux $\Delta\Phi$ diverges from the right aperture.

Shrink the left aperture so that it nearly closes, as shown in Fig. 2.3*b*. Now a differential amount of flux $d\Phi$ emerges from the left aperture within a differential solid angle $d\omega$. This small bundle of rays diverges toward the right aperture. Next shrink the right aperture so that it nearly closes, as shown in Fig. 2.3*c*. This permits one ray of the bundle to pass. The projected area of the right aperture in the direction of the ray is

$$dA_p = dA \cos\theta \tag{2.4}$$

and we define the luminance of the ray by

$$L = \frac{d^2\Phi}{d\omega\, dA \cos\theta} \tag{2.5}$$

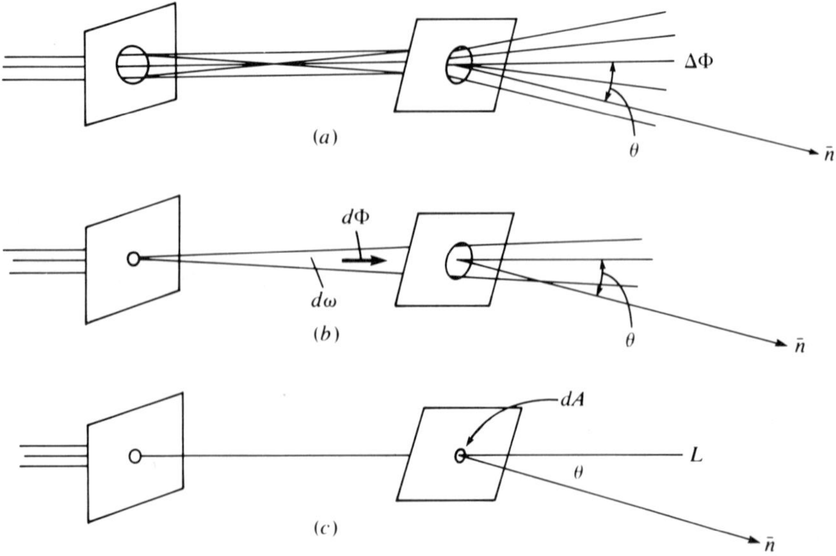

Figure 2.3 The definition of luminance.

Thus luminance (at a point and in a direction) is the quotient of the differential luminous flux at a differential element of the surface surrounding the given point and propagated in a differential cone containing the given direction by the product of the differential solid angle of the cone and the projection of the differential area on a plane perpendicular to the given direction. The differential luminous flux may be arriving, leaving or passing through the differential surface element. Luminance is thus an inherent characteristic of a ray, meaning that the luminous flux associated with a ray possessing a certain luminance is invariant.

With the fundamental definition of luminance in hand, we can use Eq. (2.5) and Fig. 2.3 to derive the remaining lighting terms. Consider intensity first. It may be defined, using Fig. 2.3*b*, as the ratio of the differential luminous flux to the differential solid angle. Thus

$$I = \frac{d\Phi}{d\omega} \tag{2.6}$$

Since the steradian is dimensionless, Eq. (2.6) shows that the candela and the lumen are dimensionally the same.

Intensity is the property of a point source in a given direction. This shows up in Eq. (2.6) because a solid angle must have a point as its apex. In practice, as was noted previously, finite source size is accommodated by either observing a source at a distance which is large compared with the largest source dimension so that the luminous flux may be treated as though it were diverging from a point or by breaking the source into small elements, each of which may be considered as a point source with its own intensity.

In either case, the applicability of Eq. (2.6) is determined by

1. Whether or not the average luminance of the projected source area remains constant as distance is varied

and **2.** Whether or not the solid angle subtended by the *source* at the *receiving point* varies inversely with the square of the distance.

This latter constraint can be justified through Fig. 2.4. The solid angle subtended by the differential source element dA at the receiver point is

$$d\omega = \frac{dA \cos \theta}{d^2} \tag{2.7}$$

If source to receiver distance is large compared with source area, $\cos \theta$ and

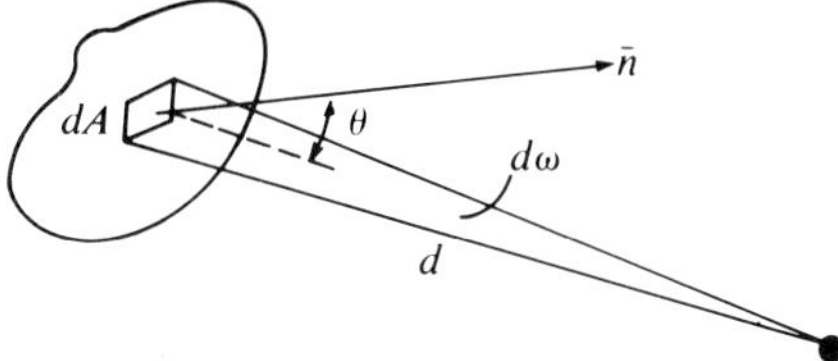

Figure 2.4 Test of Eq. (2.6).

d^2 will remain essentially constant as source area is traversed. This gives

$$\omega = \int_s \frac{\cos\theta}{d^2}\,dA = \frac{\cos\theta}{d^2}\int_s dA = \frac{A\cos\theta}{d^2} \tag{2.8}$$

Thus ω is seen to vary inversely as distance squared, the source may be considered as a point source, and the concept of intensity in Eq. (2.6) is valid.

Turning next to illuminance, we manipulate Eq. (2.5) to obtain

$$L = \frac{d\left(\dfrac{d\Phi}{dA}\right)}{d\omega\cos\theta} = \frac{dE}{d\omega\cos\theta} \tag{2.9}$$

Here illuminance appears as a differential area density of luminous flux, as depicted in Fig. 2.5, where the luminous flux is shown approaching dA and the luminance is from an external source.

The flux in Fig. 2.5 impinges on the receiver surface at an angle. This is usually the case. If dA were rotated so as to be perpendicular to the direction of L, more flux would be intercepted and E would increase. In general, we can write

$$E = E_n\cos\theta \tag{2.10}$$

where E_n is the illuminance when the receiver surface is normal to the incident flux. This result is known as Lambert's cosine law.

Equation (2.9) is useful in calculating the illuminance at a point from a large area source, such as the sky, whose distance from the receiver surface is undefined. To this end we can write

$$\begin{aligned} dE &= L\cos\theta\,d\omega \\ E &= \int_s L\cos\theta\,d\omega \end{aligned} \tag{2.11}$$

where the integral is taken over the entire source area.

Equation (2.5) can be manipulated in a second way to display intensity as the differential solid-angle density of luminous flux:

$$L = \frac{d\left(\dfrac{d\Phi}{d\omega}\right)}{dA\cos\theta} = \frac{dI}{dA\cos\theta} \tag{2.12}$$

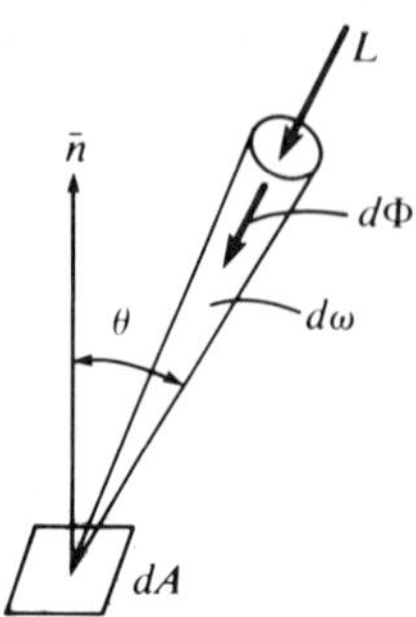

Figure 2.5 Interpretation of Eq. (2.9).

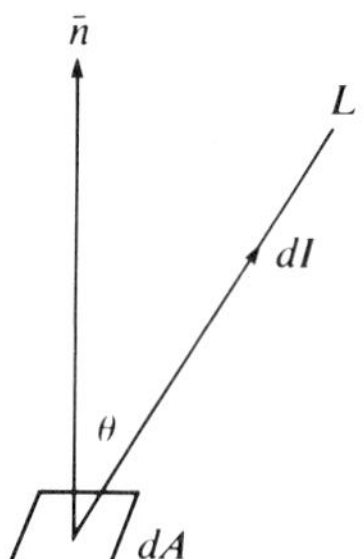

Figure 2.6 Interpretation of Eq. (2.12).

The interpretation of this form is shown in Fig. 2.6. Equation (2.12) ties together intensity and luminance by showing that the former is the latter multiplied by the projected differential source area.

We have yet to revisit luminous exitance. Illuminance was redefined in differential form through Eq. (2.9) when the flux $d\Phi$ approached the area dA in Fig. 2.5. Similarly, we redefine luminous exitance as

$$M = \frac{d\Phi}{dA} \tag{2.13}$$

where it is understood now that the differential flux is leaving the differential area in Fig. 2.5.

Finally, quantity of light can be defined more rigorously by

$$Q = \int_0^T \Phi(t)\, dt \tag{2.14}$$

which gives the luminous energy emitted or received during a period of time T from a source whose luminous flux output varies with time.

2.3 THE INVERSE-SQUARE LAW (ISL)

The general situation of a finite, nonuniform source radiating luminous flux to a point on a receiver surface is shown in Fig. 2.7. The source element dA is assumed to have a luminance L in the direction of point P on the receiver element dA'. The source emits $d\Phi$ lumens in the solid angle $d\omega$ which dA' subtends at the source. The line from source to receiver makes angles α with the normal to dA and β with the normal to dA'.

From Eq. (2.4), we obtain

$$d\Phi = I\, d\omega = \frac{I\, dA' \cos\beta}{R^2} \tag{2.15}$$

Thus the illuminance at P is

$$E = \frac{d\Phi}{dA'} = \frac{I \cos\beta}{R^2} \tag{2.16}$$

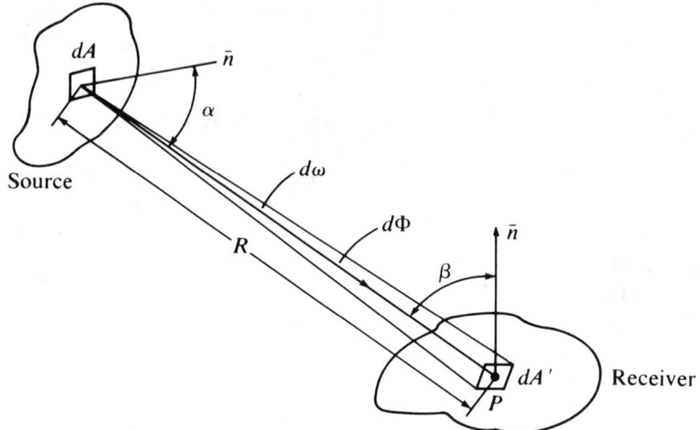

Figure 2.7 General inverse-square law situation.

This is the form of the inverse-square law for a point source. If we now consider dA to be a part of a finite-area source, Eq. (2.16) is modified to read

$$dE = \frac{dI \cos \beta}{R^2} \tag{2.17}$$

From Eq. (2.12),

$$dI = L \cos \alpha \, dA \tag{2.18}$$

which, when inserted into Eq. (2.17), yields

$$dE = \frac{L \cos \alpha \cos \beta \, dA}{R^2} \tag{2.19}$$

Equation (2.19) is the inverse-square-law formulation for an element of an area source.

To find the intensity of a finite-area source in the direction of P and the illuminance it provides at P, we integrate Eqs. (2.18) and (2.19) over the source area and obtain

$$I = \int_s L \cos \alpha \, dA \tag{2.20}$$

$$E = \int_s \frac{L \cos \alpha \cos \beta \, dA}{R^2} \tag{2.21}$$

Equation (2.21) is perfectly general and may be used to calculate the illuminance provided by sources of any shape and luminance distribution, subject only to one's ability to perform the integration, either in closed form or by computer approximation.

Equation (2.20) is not perfectly general. Intensity is a property of a point source in a given direction. If the distance R in Fig. 2.7 is large compared with the maximum dimension of the source (far-field case), Eq. (2.20) will yield the intensity of the source. In this case the receiver is far enough away so that the lines drawn from each source element to the point may be assumed parallel. If the distance R is not large compared with the maximum source dimension (the near-field case), Eq. (2.20) fails to yeild the source intensity. Thus Eq. (2.20) can be used with nonpoint sources, but only if the receiver point is relatively distant from the source. We will illustrate this quantitatively with an example in Sect. 2.8 when we consider inverse square law approximations for area sources.

EXAMPLE 2.1

A globe street light has a uniform intensity of 2000 cd. Find the illuminance on the vertical stop sign shown in Fig. 2.8.

Solution. From the geometry, the angle of incidence at the sign is $\beta = 16.7°$. Then Eq. (2.16) gives

$$E = \frac{2000 \cos 16.7}{15^2 + 50^2} = 0.7 \text{ fc}$$

EXAMPLE 2.2

If the pavement in Fig. 2.8 has a reflectance of 35%, what is the luminous exitance at a point on the pavement halfway between the pole and the sign?

Solution. The illuminance at the designated point is

$$E = \frac{2000 \cos 51.3}{20^2 + 65^2} = 1.2 \text{ fc}$$

By reflectance, we mean the fraction of the luminous flux arriving at a point which departs from the point. Thus the luminous exitance is

$$M = 0.35 \times 1.2 = 0.4 \text{ lm/ft}^2$$

EXAMPLE 2.3

A flat circular lighting fixture 15 cm in diameter is mounted on a ceiling in a recreation room as shown in Fig. 2.9. Its luminance in the direction of point P on a table is 40 cd/cm². Find its intensity in the direction of P and the horizontal illuminance on the table at P.

Solution. The luminous surface area of the fixture, ignoring its edges, is 56.25π cm². Angle α is 56.3°. Thus the intensity of the source toward

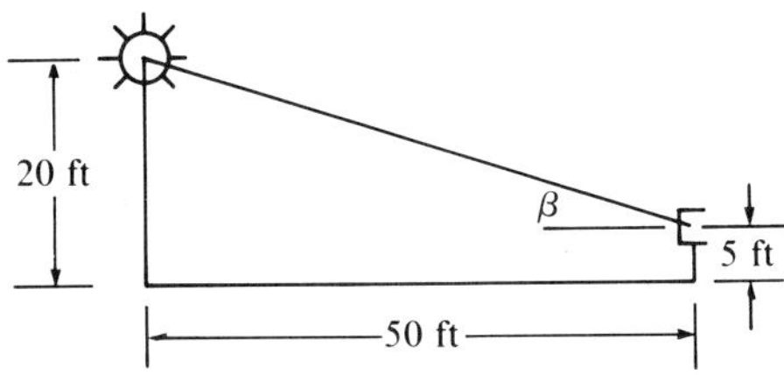

Figure 2.8

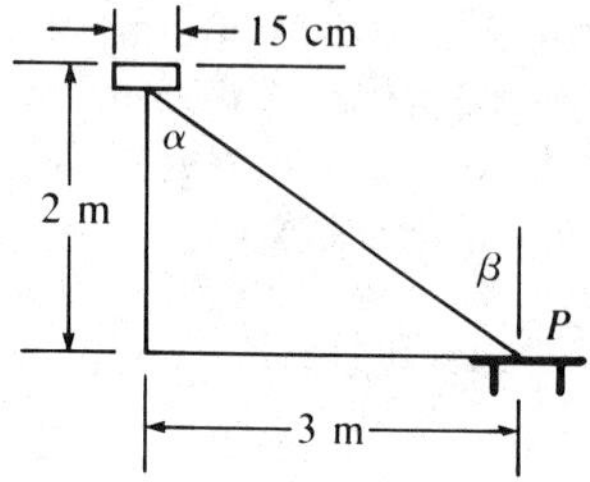

Figure 2.9

P is, from Eq. (2.20) (source size is small compared with source to receiver distance).

$$I = 40 \times 56.25\pi \times 0.555 = 3920 \text{ cd}$$

Then, noting that $\beta = \alpha$, the illuminance at P is given by Eq. (2.16) as

$$E = \frac{3920}{13} \times 0.555 = 167 \text{ lx}$$

EXAMPLE 2.4

A spot lamp is aimed at point Q (see Fig. 2.10). It has an average luminance of 300 cd/in^2 in the direction of point P and a face diameter of 4.75 in. Find the horizontal illuminance at point P.

Solution. The luminous surface area of the lamp is

$$A = \pi \times (\tfrac{19}{8})^2 = 17.72 \text{ in}^2$$

From P, however, we see a projected area of $A \cos \alpha$. For the given geometry, $\alpha = 19.4°$ and

$$A_P = 17.72 \cos 19.4 = 16.7 \text{ in}^2$$

From Eq. (2.20) (again the source size is small compared with the

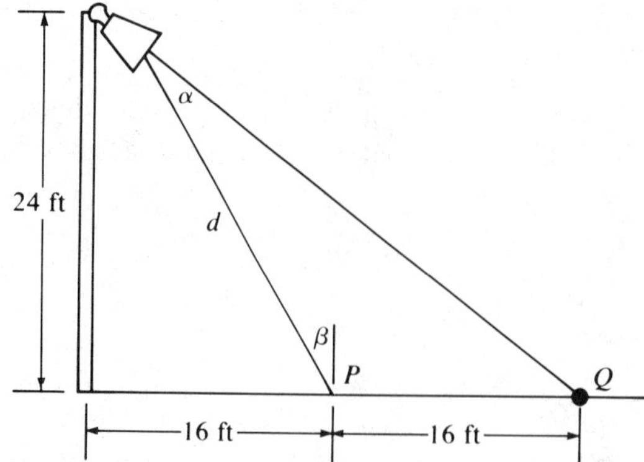

Figure 2.10

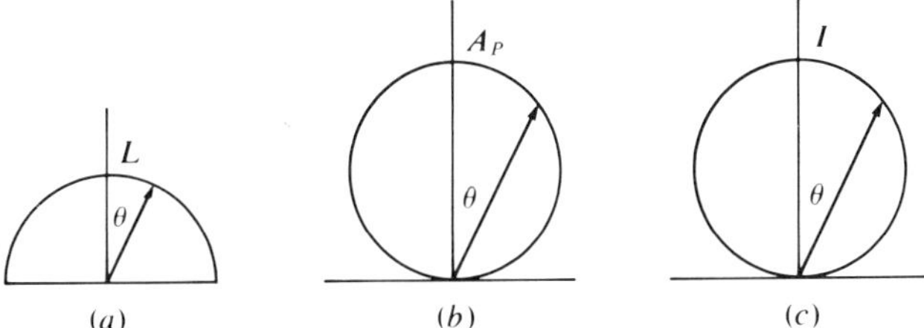

Figure 2.11 Interpretation of perfectly diffusing surfaces.

source to receiver distance), the intensity of the lamp in the direction of P is

$$I = 300 \times 16.7 = 5010 \text{ cd}$$

The angle of incidence at point P is calculated from the given geometry to be 33.7°. Then Eq. (2.16) yields

$$E = \frac{5010 \cos 33.7}{24^2 + 16^2} = 5.0 \text{ fc}$$

2.4 PERFECT DIFFUSION

A perfectly diffusing surface is one which emits or reflects luminous flux in such a way that it has the same luminance regardless of viewing angle. Such a surface is called *Lambertian*.*

No surface is perfectly diffusing, but many surfaces, especially nonglossy ones, are nearly so. An example of a surface which is not Lambertian is a mirror. Such a surface lies at the other extreme from a perfectly diffusing surface and is called specular.

The constraint of luminance being independent of angle is illustrated in Fig. 2.11*a*. What does this require of luminous intensity? In Eq. (2.12) we found that luminous intensity and luminance are related through projected area. The projected area of a flat surface varies as the cosine of the angle with the normal, as shown in Fig. 2.11*b*. Thus intensity must also follow a cosine law, as shown in Fig. 2.11*c*, and is given by

$$I(\theta) = I \cos \theta \tag{2.22}$$

where I is the intensity of the surface when viewed normally.

It was mentioned in Sec. 1.7 that, in the case of a perfectly diffusing surface, luminous exitance and luminance are related through a π factor. To explore that relationship further, consider Fig. 2.12 which shows a source element dA radiating from the center of a hemisphere. The differential

*Named for J. H. Lambert who made many early contributions to photometry.

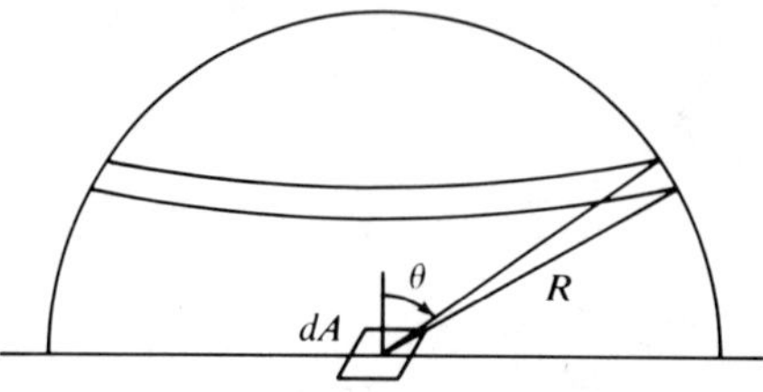

Figure 2.12 Perfectly diffusing source element.

illuminance on the hemispherical ring shown is, from Eq. (2.19),

$$dE_r = \frac{L \cos \theta \, dA}{R^2} \tag{2.23}$$

where r denotes *ring*.

Multiplying dE_r by the area of the ring gives the luminous flux which the ring receives.

$$\begin{aligned} d\Phi_r &= \frac{L \cos \theta \, dA}{R^2} (2\pi R^2) \sin \theta \, d\theta \\ &= 2\pi L \, dA \sin \theta \cos \theta \, d\theta \end{aligned} \tag{2.24}$$

For a perfectly diffusing surface, L is independent of θ and the total luminous flux emitted by dA is

$$\begin{aligned} d\Phi &= 2\pi L \, dA \int_0^{\pi/2} \sin \theta \cos \theta \, d\theta \\ &= \pi L \, dA \end{aligned} \tag{2.25}$$

Thus the luminous exitance of dA is, from Eq. (2.13),

$$M = \frac{d\Phi}{dA} = \pi L \tag{2.26}$$

If L is in candelas per square meter, M is in lumens per square meter. If L is in candelas per square foot, M is in lumens per square foot.

The key to the development leading to Eq. (2.26) is being able to move L outside the integral in Eq. (2.25). For nonperfectly diffusing surfaces, the procedure just described could still be used to find M, but the integration would be more difficult and the result in Eq. (2.26) would not be valid.

EXAMPLE 2.5

A spherical glass globe 1 ft in diameter is perfectly diffusing. At its center is a uniform-intensity point source of 50 cd. The luminance of the globe is 0.3 cd/in^2. Find the transmittance of the globe (the percent of the lumens arriving on the inside of the hemisphere which depart from the outside).

Solution. The problem will be worked in two ways:

1. *Lumen procedure.* The lamp emits $50 \times 4\pi = 200\pi$ lm. The surface area of the sphere is π ft^2. Thus the illuminance on the inside of the

sphere is $200\pi/\pi = 200$ fc. The luminance of the sphere is $0.3 \times 144 = 43.2$ cd/ft^2 and thus its luminous exitance is $43.2\pi = 135.7$ lm/ft^2. The transmittance is the ratio of M to E, giving

$$\tau = \frac{135.7}{200} = 0.68$$

2. *Candela procedure.* The projected area of the globe is $\pi r^2 = 36\pi$ in^2. Thus the apparent intensity of the globe is $0.3 \times 36\pi = 33.9$ cd. The ratio of the apparent intensity of the globe to the intensity of the lamp inside (for the perfect spherical diffuser) is also the transmittance. The result is

$$\tau = \frac{33.9}{50} = 0.68$$

EXAMPLE 2.6

The luminance of a small flat differential source element dA is $L = 10 \cos\theta$ cd/m^2, where θ is the angle with the normal to dA. Find the luminous exitance of dA.

Solution. For this situation, Eq. (2.25) becomes

$$d\Phi = 20\pi\, dA \int_0^{\pi/2} \sin\theta \cos^2\theta \, d\theta$$

which yields $d\Phi = (20\pi/3)\, dA$. Thus the luminance exitance is $M = 20\pi/3$ lm/m^2 and is less than it would be if dA were perfectly diffusing with a luminance of 10 cd/m^2.

2.5 THE SKY AS A SOURCE

Suppose it is desired to determine the illuminance at a point on the ground from the sky. Attempts to use the ISL [Eq. (2.16) or (2.19)] appear to fail because we have no values for the intensity of the sky and do not know how far away it is.

Consider Fig. 2.12 again, but this time let the ring on the hemisphere be a portion of the sky (the source) and dA be the receiver. Assume the luminance of the hemisphere in the direction of dA can vary with the polar angle θ.

We desire to find the horizontal illuminance on dA due to the entire hemisphere. The differential illuminance produced by the ring is

$$dE = \frac{L(\theta)\cos\theta}{R^2}(2\pi R^2)\sin\theta\, d\theta \tag{2.27}$$

Assume $L(\theta)$ to be constant between the limits θ_1 and θ_2. Then integration of Eq. (2.27) yields

$$E_{\theta_1\to\theta_2} = 2\pi\ L_{\theta_1\to\theta_2}\int_{\theta_1}^{\theta_2} \sin\theta \cos\theta\ d\theta$$
$$= \pi\ L_{\theta_1\to\theta_2}(\sin^2\theta_2 - \sin^2\theta_1) \qquad (2.28)$$

If $L(\theta)$ is constant over all the hemisphere, Eq. (2.28) becomes

$$E_{0\to 90} = \pi L \qquad (2.29)$$

Thus the horizontal illuminance produced by a sky of constant luminance is π times that luminance. Note that the distance of the sky from the observer does not appear in either Eq. (2.28) or (2.29).

There is an alternative way of arriving at Eq. (2.27) through the use of Eq. (2.9). The solid angle formed by the differential ring is $d\omega = 2\pi \sin\theta\ d\theta$ and inserting this into Eq. (2.9) gives Eq. (2.27).

EXAMPLE 2.7

Assume an overcast sky has the following luminances:

θ, degrees	L, cd/in^2
0–40	4.0
40–60	3.0
60–90	2.0

Find the horizontal illuminance produced at a point on the ground.

Solution. From Eq. (2.28),

$$E = 144\pi[4(\sin^2 40 - \sin^2 0) + 3(\sin^2 60 - \sin^2 40) + 2(\sin^2 90 - \sin^2 60)]$$
$$= 144\pi(1.65 + 1.01 + 0.50) = 1430 \text{ fc}$$

The 144 is necessary to convert cd/in^2 to cd/ft^2 so that the illuminance will be in lm/ft^2, or fc.

2.6 MEASURING LUMINOUS FLUX

An incandescent lamp is shown mounted horizontally on a stand in Fig. 2.13. We will walk slowly around the lamp, keeping the distance d constant and following the path shown. Every 10°* we will stop, point an illuminance meter at the lamp and take a reading. After all readings from 0 to 180° are taken, we will use Eq. (2.16) (with $\beta = 0$) to find the intensity of the lamp at each viewing position. We will then plot these intensity values versus viewing angle θ on a polar diagram. The result is shown in Fig. 2.14 and is the intensity distribution curve for the given source. The semicircles labeled 500, 1000, 1500 are intensity loci in candelas. To find the intensity at any angle,

*In modern photometers, readings are taken at least every 5°, and preferably every $2\frac{1}{2}$°, especially for lamps with rapidly changing intensity with angle.

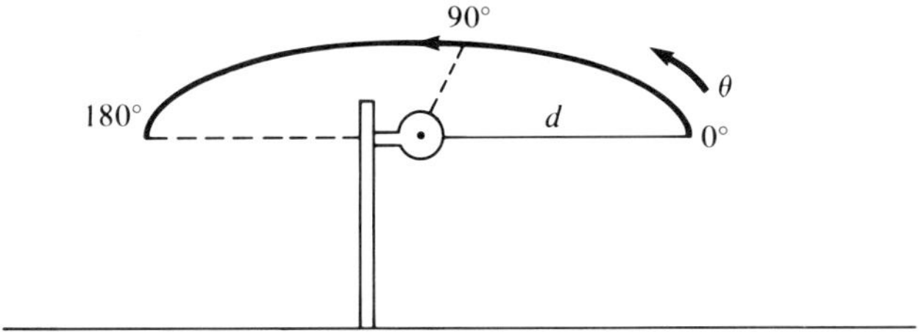

Figure 2.13 Laboratory setup for measuring lumens.

one interpolates between these loci. For example, the intensity at 30° is approximately 1250 candelas.

It should be noted that the graph contains values from 0 to 180° only. This is because the lamp is presumed symmetrical and thus its intensity at 330°, for example, is the same as that at 30°. If symmetry is not assumed, the incandescent lamp may be rotated about the horizontal in Fig. 2.13 to obtain an average intensity in each angular direction.

Lamp and luminaire (lighting fixture) manufacturers obtain intensity distribution curves using devices called goniophotometers, where the prefix "gonio" means directional. These devices are very similar in principle to that shown in Fig. 2.13. Each consists of a calibrated light-sensitive cell for measuring illuminance mounted at the end of an arm which can be rotated about a source at a fixed distance from the source. A considerable amount of space is required for such an instrument because, for the ISL to hold, the distance between source and meter must be large compared with the maximum dimension of the source. Generally this distance is kept greater than five times this dimension, which gives an accuracy of the order ±1% for

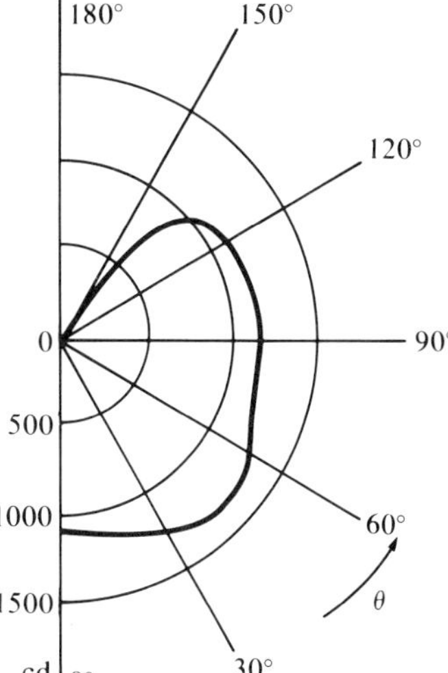

Figure 2.14 Intensity distribution curve of an incandescent lamp.

many lighting fixtures. Thus, for an 8-ft-long fluorescent lamp, the longest commercially available source, the minimum distance is 40 ft, requiring an 80-ft-diameter circle through which the meter rotates and thus an 80-ft-high room. This distance can be reduced by two means:

1. Mounting the meter near the source and using a mirror to reflect the source rays to the meter. Such an arrangement can provide a 40-ft light path in a 20-ft space.
2. Mounting the lamp on the ceiling to obtain the 0 to 90° portion of the candlepower distribution curve and then lowering the lamp to the floor for the 90 to 180° portion of the curve. Instead of an 80-ft-high room, one only 40-ft high is needed.

A much greater than 5:1 ratio is required for certain luminaires, such as spotlights. In such cases the minimum distance for accuracy is a function of the reflector and lens characteristics of the luminaire.

Goniophotometers are calibrated with incandescent lamps which have, in turn, been calibrated against standards from the National Bureau of Standards in Washington, D.C. Modern goniophotometers are completely automated and computerized so that the movement of the meter arm, the recording of data, and the plotting of the candlepower distribution curve are all done automatically. Some recent models place a light cell at each required angle, rather than using a single movable cell.

The intensity distributions of many fluorescent lamps and luminaires are nonsymmetrical and thus a family of intensity distribution curves is required. Generally candlepower data is taken in at least three vertical planes, oriented at 0°, 45°, and 90° with respect to the lamp or luminaire axis, as shown in Fig. 2.15. In more complete photometry, 22.5° and 67.5° planes are also used.

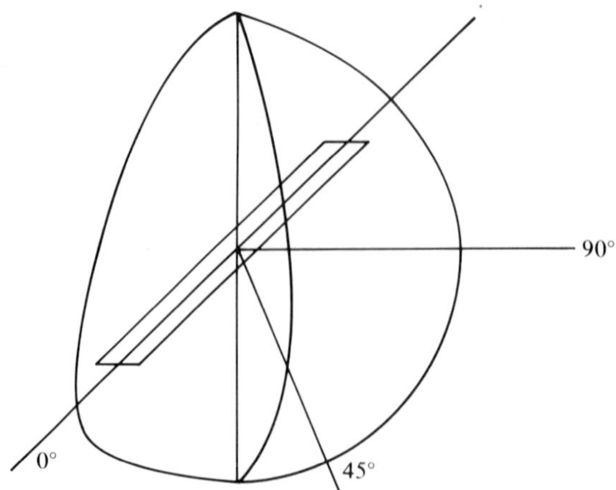

Figure 2.15 Fluorescent photometry.

The lamp manufacturer must be able to tell the customer how many lumens a given lamp emits. For an ideal source, which has uniform intensity in all directions, the answer is easily found by multiplying the intensity by 4π. But when the intensity is nonuniform, which is usually the case, the problem is more difficult and calculations from the candlepower distribution curve must be made.

Let a source have an intensity $I(\theta)$ in a small angular band $d\theta$, as shown in Fig. 2.16. Place a sphere of radius r around the source. Then, from the ISL, the illuminance on the ring intercepted by $d\theta$ on the sphere is everywhere

$$E = \frac{I(\theta)}{r^2} \tag{2.30}$$

To obtain the lumens received by the ring, it is necessary to multiply E by the ring area. This yields

$$d\Phi = \frac{I(\theta)}{r^2}\,(2\pi r^2)\sin\theta\,d\theta = 2\pi I(\theta)\sin\theta\,d\theta \tag{2.31}$$

If $I(\theta)$ is approximately constant in an angular band from θ_1 to θ_2, integration of Eq. (2.31) yields

$$\Phi_{\theta_1 \rightarrow \theta_2} = 2\pi I(\theta)\,(\cos\theta_1 - \cos\theta_2) \tag{2.32}$$

where $I(\theta)$ is generally taken as the intensity at the midpoint of the zone between θ_1 and θ_2. If $I(\theta)$ has a constant value I for all θ, then Eq. (2.32) gives $\Phi = 4\pi I$, as would be expected.

The quantity $2\pi(\cos\theta_1 - \cos\theta_2)$ is called a zonal constant. The zonal constants for the 10° zones appear in Table 2.1.

Note that the zonal constants for angles greater than 90° are the same as for angles less than 90°; for example, the 30°–40° zone has the same zonal constant (same area) as the 140°–150° zone.

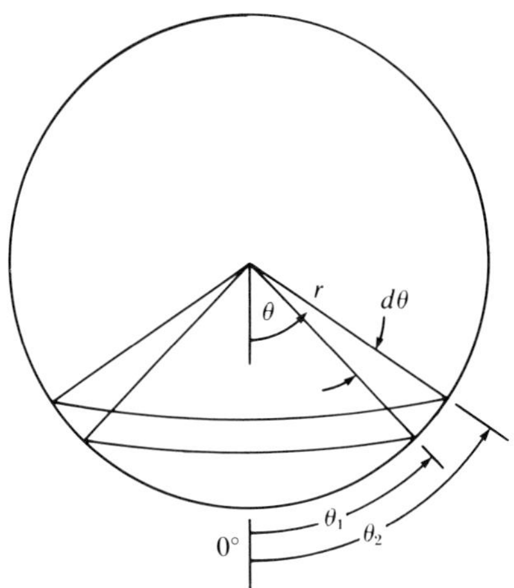

Figure 2.16 Zonal lumens.

TABLE 2.1

Zone (degrees)	Zonal constant	Zone (degrees)
0–10	0.095	170–180
10–20	0.283	160–170
20–30	0.463	150–160
30–40	0.628	140–150
40–50	0.774	130–140
50–60	0.897	120–130
60–70	0.993	110–120
70–80	1.058	100–110
80–90	1.091	90–100

With the concept of zonal constants, we can rewrite Eq. (2.32) to read

$$\Phi_{\theta_1 \rightarrow \theta_2} = I(\theta)\ ZC_{\theta_1 \rightarrow \theta_2} \tag{2.33}$$

where ZC means zonal constant. Equation (2.33) says that the lumens in the zone bounded by θ_1 and θ_2 are obtained by multiplying the intensity in that zone (presumed constant throughout the zone) by the zonal constant of the zone.

To get the total lumens emitted by the lamp, we must sum the contributions from all zones. Using Eq. (2.32), this sum can be written as

$$\Phi = \sum_{0^\circ}^{180^\circ} 2\pi I(\theta)\ (\cos\theta_1 - \cos\theta_2) \tag{2.34}$$

or, using Eq. (2.33), as

$$\Phi = \sum_{0^\circ}^{180^\circ} I(\theta) ZC_{\theta_1 \rightarrow \theta_2} \tag{2.35}$$

EXAMPLE 2.8

Find the total lumens given off by the lamp whose candlepower distribution curve is given in Fig. 2.14.

Solution. We will work this problem using the zonal constants and assuming that the intensity in each 30° zone may be considered constant in that zone. This assumption is fairly good, except for the 120 to 150° zone, where it is an obvious approximation.

Zone	Average I (candelas)	ZC	Zonal lumens
0–30	1200	0.841	1009
30–60	1250	2.299	2874
60–90	1100	3.142	3456
90–120	1050	3.142	3299
120–150	850	2.299	1954
150–180	0	0.841	0
		Total lumens = 12,592 ≃	12,600

EXAMPLE 2.9

A lamp having a uniform candlepower of 1000 cd in all directions is placed in a reflector, as shown in Fig. 2.17. The reflector intercepts all of the lumens in the 60 to 180° zone and reflects 80% of these. All of these reflected lumens enter the 0 to 60° zone. Find the total lumens in the 0 to 60° zone and the average intensity of the beam.

Solution. Since I is constant, we consider the region from 60 to 180° as a single zone and use Eq. (2.32) to obtain

$$\begin{aligned}\Phi_{60\text{-}180} &= 2\pi 1000(\cos 60 - \cos 180) \\ &= 2000\pi(0.5 + 1) = 3000\pi \text{ lm}\end{aligned}$$

Of these lumens, 80% are reflected into the 0 to 60° zone. Thus

$$\Phi_{\text{refl.}} = 0.8(3000\pi) = 2400\pi \text{ lm}$$

The lumens going directly from the lamp to the 0 to 60° zone are given by

$$\begin{aligned}\Phi_{0\text{-}60} &= 2\pi 1000(\cos 0 - \cos 60) \\ &= 2000\pi(1 - 0.5) = 1000\pi \text{ lm}\end{aligned}$$

Thus the total lumens in the 0 to 60° beam are

$$\Phi_{\text{beam}} = 2400\pi + 1000\pi = 3400\pi \text{ lm}$$

Now, to find the beam intensity, we insert Φ_{beam} into Eq. (2.32) and solve for I_{beam}.

$$\begin{aligned}3400\pi &= 2\pi I_{\text{beam}} (\cos 0 - \cos 60) \\ I_{\text{beam}} &= 3400 \text{ cd}\end{aligned}$$

We note that the effect of the reflector is to convert a 1000-cd source, uniform in all directions, into a directed beam of 3400-cd intensity and 120° spread (60° each side of the center line), with no intensity in any other direction. This, of course, is the principle behind all spot and flood lamps, auto headlights, searchlights, and projection lamps.

Intensity data are not always displayed on a polar diagram, as in Fig. 2.14. For lamps with a directed beam, a rectangular plot is often more useful.

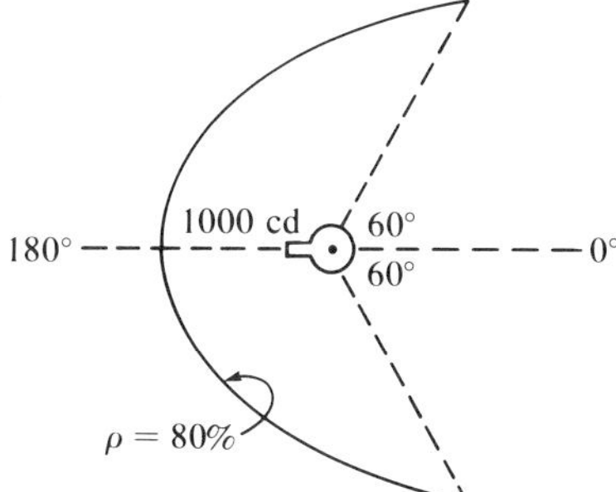

Figure 2.17

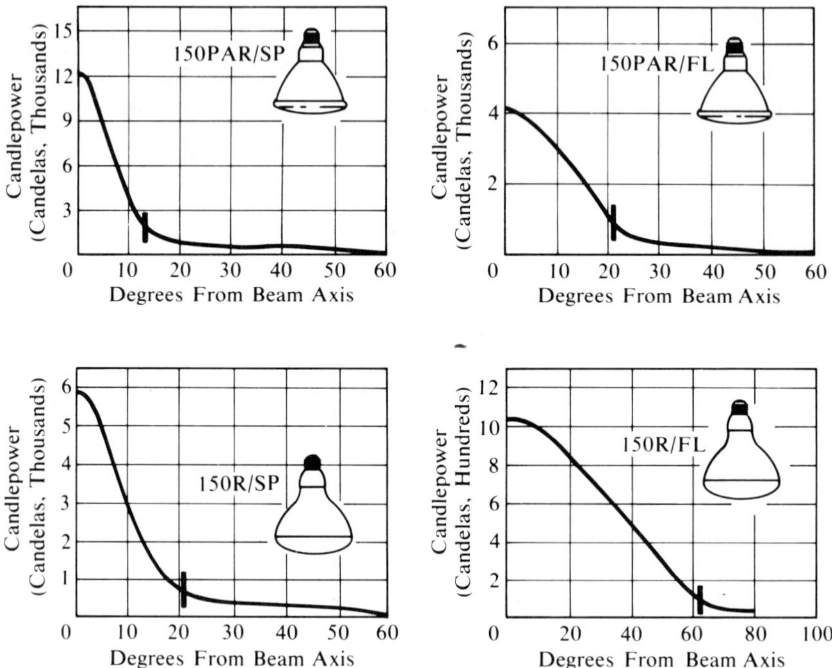

Figure 2.18 Intensity data on a rectangular plot. (*Source:* General Electric Co., "Incandescent Lamps," Bulletin TP110R1, 1980. Reprinted with permission.)

Such a presentation is shown in Fig. 2.18 for four 150-W PAR and R lamps. The notation PAR means *parabolic aluminized reflector*. These lamps are of two-piece heat-resistant glass construction, one piece for the parabolic reflector and the other for the lens. R (reflector) lamps consist of a one-piece blown glass bulb, again with the reflector in the shape of a parabola. Both lamps are widely used and are available in both spot and flood distributions, with the latter providing about twice the beam spread of the former, where beam spread is defined as twice the angle from the beam axis at which the intensity is 10% of its maximum value. Each of these lamps will be discussed in detail in Chap. 6.

There is another method of measuring the total lumens emitted by a light source, which involves the use of an integrating or Ulbricht sphere. Consider Fig. 2.19, which shows an incandescent lamp at the center of a hollow sphere of radius r. The inside of the sphere is coated with a matte-finish white paint to make it perfectly diffusing and very highly reflecting. What we desire to show is that the total luminous flux emitted by the lamp is directly related to the illuminance at a small opening in the sphere wall.

The illuminance at any point on the inside of the sphere is composed of two components: a direct component due to luminous flux traveling directly from the source to the point and an indirect component due to the luminous flux reaching the point by reflection from all other points within the sphere.

Let the sphere in Fig. 2.19 have a reflectance ρ, where ρ is the ratio of the reflected to incident lumens at any point on the inside of the sphere. Further, let the luminance at point X be L. Then, from Eq. (2.19), the illuminance at a point Y on the inside of the sphere due to a small element of sphere area dA at point X is

$$dE_Y = \frac{L\ dA \cos^2\theta}{d^2} \tag{2.36}$$

Noting that $d = 2r\cos\theta$, Eq. (2.36) becomes

$$dE_Y = \frac{L\ dA}{4r^2} \tag{2.37}$$

We observe that the illuminance at point Y due to a small element at point X is independent of the position of point Y (independent of θ). Thus Eq. (2.37) gives the illuminance at any point on the inside of the sphere due to the element at point X. Also the illuminance at point Y due to any other element on the inside of the sphere is given by Eq. (2.37), again because that equation does not depend on θ. We conclude that every element on the inside of the sphere illuminates every other element equally and that the indirect illuminance is uniform throughout the sphere.

Now assume that the source at the center of the sphere emits Φ lumens. Of these, $\rho\Phi$ are reflected. When these reflected lumens hit the inside of the sphere, $\rho^2\Phi$ are reflected, and so on. The total reflected lumens are

$$\Phi_{\text{refl.}} = \rho\Phi\,(1 + \rho + \rho^2 + \cdots + \rho^n) \tag{2.38}$$

Since ρ is a number less than 1, the sum in the parentheses has a finite value of $1/(1 - \rho)$ and thus we can rewrite Eq. (2.38) as

$$\Phi_{\text{refl.}} = \frac{\rho}{1 - \rho}\,\Phi \tag{2.39}$$

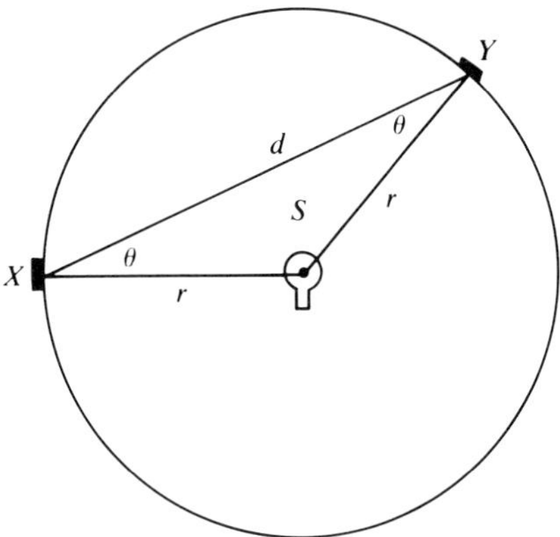

Figure 2.19

Noting that the sphere surface area is $4\pi r^2$, the illuminance due to reflected lumens (the indirect illuminance) is

$$E = \frac{\rho}{4\pi r^2(1-\rho)} \Phi \tag{2.40}$$

This is the relation we seek. It shows that, if we place an illuminance meter in the sphere wall (at point Y in Fig. 2.19) and put a small shield between the source and the meter to block out direct illuminance (shown at point S in Fig. 2.19) the reading of the meter will be proportional to the total lumens emitted by the lamp.

In practice, the sphere is calibrated by measuring a standard source whose total lumen output is known. This enables us to establish the calibration constant of the sphere, $\rho/[4\pi r^2(1-\rho)]$. Calling this constant k, the luminous flux of any source placed within the sphere is obtained from

$$\Phi = \frac{1}{k} E \tag{2.41}$$

where E is the reading of the illuminance meter placed in the sphere wall.

EXAMPLE 2.10

An incandescent lamp is placed inside a 2-ft-radius integrating sphere whose reflectance is 0.95. The meter in the sphere wall reads 300 fc. How many lumens does the lamp emit and what is its average intensity?

Solution. From Eq. (2.40),

$$\Phi = \left[\frac{4\pi(4)(0.05)}{0.95}\right] 300 = 794 \text{ lm}$$

Then the average intensity of the lamp is

$$I = \frac{794}{4\pi} = 63 \text{ cd}$$

2.7 LIGHTING A SIGN

A 10-ft-long, 2-ft-high sign at the entrance to a housing development is to be lighted by 150-W PAR lamps (see Fig. 2.18) staked into the ground as shown in Fig. 2.20. The problem is to decide what type of lamp to use (spot or flood), how many to use, and how far apart to space them in order to obtain adequate illuminance on the sign.

As a first step, let us find the illuminance provided by a single lamp at various points. The situation is shown in Fig. 2.21. The lamp is aimed at point C, a distance x_2 laterally from point A, where a horizontal line drawn through the middle of the sign and a perpendicular to this line from the lamp intersect. We desire to find the illuminance on the sign at point B, a distance

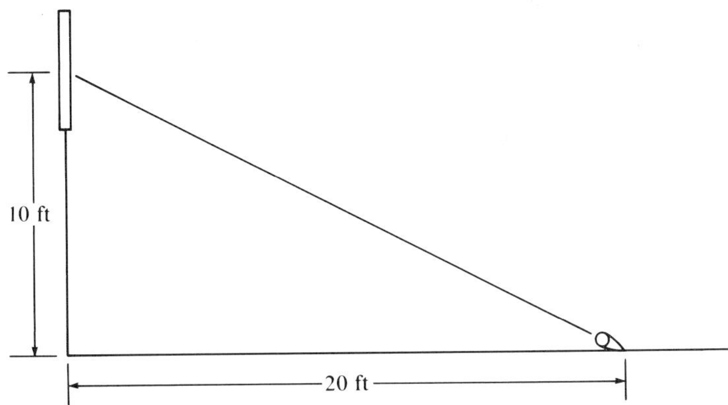

Figure 2.20 Lighting a sign.

x_1 laterally from point A. To calculate E, we use Eq. (2.16) and read values of intensity from the curves in Fig. 2.18. Thus the illuminance at point B is

$$E_B = \frac{I(\alpha)\cos\beta}{x_1^2 + h^2 + q^2} = \frac{I(\alpha)\, q}{(x_1^2 + h^2 + q^2)^{3/2}} \tag{2.42}$$

For the diagram in Fig. 2.21, α is the sum of the angles γ and δ. For the other case, where point C is to the left of point A, α is the difference of these two angles. We can write

$$\gamma = \tan^{-1}\frac{x_1}{\sqrt{h^2 + q^2}} \qquad \delta = \tan^{-1}\frac{x_2}{\sqrt{h^2 + q^2}} \tag{2.43}$$

Then, using the law of tangents,

$$\alpha = \tan^{-1}\frac{(x_1 \pm x_2)\sqrt{h^2 + q^2}}{h^2 + q^2 \mp x_1 x_2} \tag{2.44}$$

where the upper signs apply to the situation shown in Fig. 2.21. For the

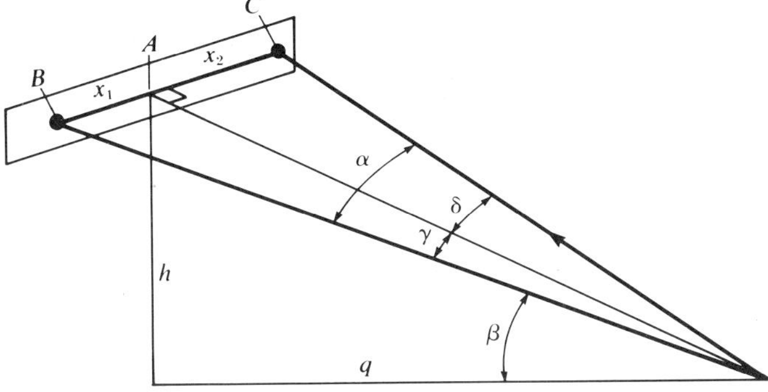

Figure 2.21 Geometry for one lamp.

TABLE 2.2

x (ft)	$\gamma = \alpha$ (degrees)	β (degrees)	I_{SP} (candelas)	I_{FL} (candelas)	E_{SP} (footcandles)	E_{FL} (footcandles)
0	0	26.6	12,000	4,100	21.5	7.3
1	2.6	26.7	11,800	4,000	21.0	7.1
2	5.1	27.0	8,500	3,800	15.0	6.7
3	7.6	27.6	6,000	3,600	10.4	6.3
4	10.1	28.3	3,200	3,000	5.5	5.1
5	12.6	29.2	2,000	2,500	3.3	4.2
6	15.0	30.2	1,400	2,000	2.3	3.2
7	17.4	31.4	1,200	1,500	1.9	2.3
8	19.7	32.6	1,000	1,000	1.5	1.5

given problem, $h = 10$ ft and $q = 20$ ft. For simplicity, let us assume the lamp is aimed at point A, making $\delta = 0$. Inserting the given dimensions into Eqs. (2.42) and (2.44) for various values of x_1 yields the results in Table 2.2 and the graphs in Fig. 2.22.

We are now ready to address the problems of which lamp to choose, how many are needed, and what spacing is best to achieve a reasonably uniform illuminance level. This is done largely by trial and error. Two solutions are shown in Fig. 2.23. Each uses two lamps spaced 6 ft apart. The floods provide a bit better uniformity but only about half the illuminance of the spots. Also the floods will give more "spill light," light that does not fall on the sign and thus can produce unwanted luminance on adjacent surfaces and possibly glare to those residents of the housing development living behind the sign. In this case the better choice of the two would probably be to use spots.

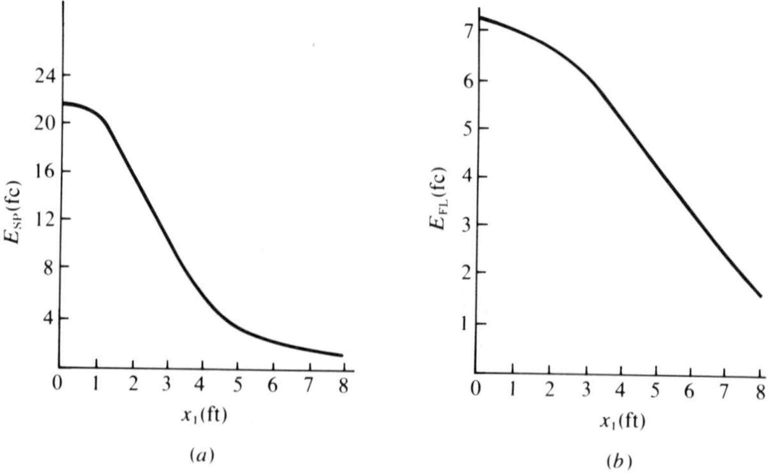

Figure 2.22 E_{sp} and E_{fl} versus x_1 for one lamp.

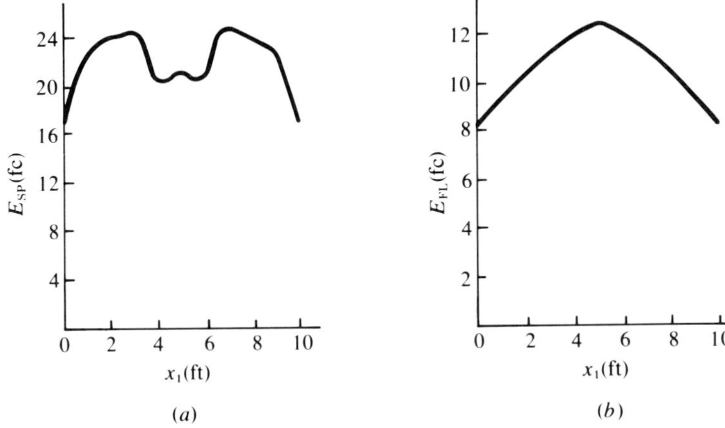

Figure 2.23 Two sign-lighting solutions.

We have by no means exhausted all of the possibilities for properly lighting the sign. We should try different aiming angles, different spacings, different lamp-to-sign distances, and different lamps. Before doing this, however, we should either computer-program the entire procedure and let the computer do the iterations or develop graphical techniques, such as plotting the illuminances on tracing paper so that we may superimpose curves to obtain the total illuminance from more than one source. We will discuss other calculation procedures for flood and spotlighting in Chap. 11.

2.8 INVERSE-SQUARE-LAW APPROXIMATIONS

In Sec. 2.6, it was noted that we have no way of directly measuring luminous intensity or luminous flux. Rather we measure illuminance and then use the ISL to obtain these quantities. This leads us to inquire further about the accuracy of ISL calculations.

Recall that the ISL is based on a point source. For a nonpoint source, we reasoned that the ISL could still be used if the source size were small compared with the distance from the source to the receiver. We noted that the ratio of this distance to the maximum source dimension should be at least 5:1 for the ISL to hold within the order of $\pm 1\%$. Let us investigate this constraint more fully by considering the disc source in Fig. 2.24. The source has a radius $R = 1$ ft and a constant luminance of L cd/ft^2 in the direction of P. Assume that D, the distance from source to receiver, is not large compared with $2R$, the maximum source dimension. Thus the ISL does not hold for the entire source.

However it does hold for the small ring in Fig. 2.24, for at all points within that ring, α and β and the distance from the ring to P are constant and thus the differential form of the ISL in Eq. (2.19) may be applied, yielding

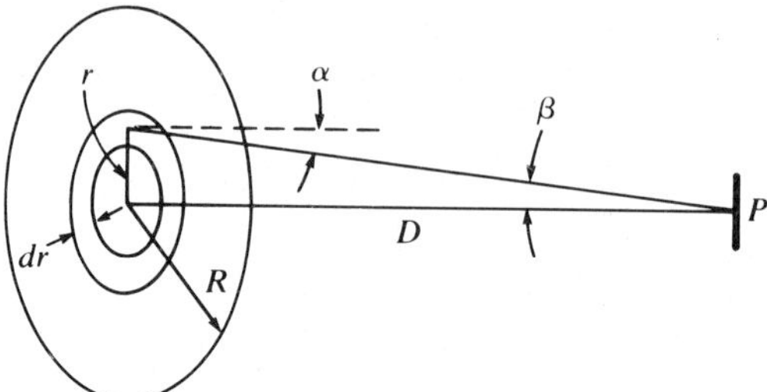

Figure 2.24 A nonpoint source.

$$dE = \pi D^2 L \frac{2r\,dr}{(D^2 + r^2)^2} \tag{2.45}$$

The right-hand portion of Eq. (2.45) is directly integrable, giving

$$E = \pi D^2 L \int_0^R \frac{2r\,dr}{(D^2 + r^2)^2} = \frac{\pi L R^2}{D^2 + R^2} \tag{2.46}$$

The differential intensity of the source in the direction of P is, from Eq. (2.18),

$$dI = \frac{L(2\pi\,r dr) D}{(D^2 + r^2)^{1/2}} \tag{2.47}$$

To obtain the intensity of the source from afar, we write

$$I = 2\pi L \int_0^R \frac{r\,dr\,D}{(D^2 + r^2)^{1/2}} \tag{2.48}$$

The quantity $D/(D^2 + r^2)^{1/2}$ is the cosine of the source angle. Since I, by definition, is in a given direction, α must be constant, in this case zero, during the integration. Stated another way, $D >> r$ for all r. Thus

$$I = 2\pi L \int_0^R r\,dr \tag{2.49}$$

yielding

$$I = \pi R^2 L = LA \tag{2.50}$$

Thus, from afar, the source intensity is simply the product of luminance times projected source area, as we would expect.

Suppose now we hold an illuminance meter perpendicularly to the source at P, as we did in the simple photometer in Fig. 2.13, and measure E at P for different values of D. The values we get will be those given by Eq. (2.46). Let us also calculate E from Eq. (2.16) (with $\beta = 0°$), using the erroneus assumption that the ISL holds for the entire source. The results are presented in Table 2.3 for $R = 1$ ft, $L = 500$ cd/ft^2, and thus $I = 1570$ cd [from Eq. (2.50)]. We see that for the actual illuminance and the approximate (ISL) illuminance to be within 1% of each other, the ratio of D to $2R$ (the maximum source dimension) must be 5:1 or greater. However, if a 3%

TABLE 2.3

D (ft)	Actual E [Eq. (2.46)]	ISL E [Eq. (2.16)]	% Error ($100R^2/D^2$)	Apparent I [(Actual E)D^2]
0.5	1260	6280	400	314
1	785	1570	100	785
2	314	393	25	1257
4	92.4	98.2	6.25	1478
6	42.5	43.6	2.78	1528
8	24.2	24.5	1.56	1547
10	15.6	15.7	1.00	1555
15	6.95	6.98	0.44	1564
20	3.92	3.93	0.25	1567
30	1.74	1.75	0.11	1569
40	0.981	0.982	0.06	1570

error can be tolerated, a ratio of only 3:1 is required. Although we cannot generalize directly to other source shapes from this example, the pattern is much the same if the source luminance distributions remain nearly Lambertian.

If we were given the source in Fig. 2.24 in the laboratory and asked to measure its intensity as viewed from far away, we would use an illuminance meter and obtain the values of "Actual E" in Table 2.3. Then we would compose the "Apparent I" column in Table 2.3 by multiplying each actual E by D^2, as though the ISL in Eq. (2.16) held for the entire source. As D increases, it is seen that the apparent intensity approaches the actual intensity of the whole source from afar. In limit notation

$$I = \lim_{D\to\infty} ED^2 \tag{2.51}$$

We can also arrive at these results from Eq. (2.46). For $D >> R$,

$$E = \frac{\pi L R^2}{D^2} = \frac{LA}{D^2} = \frac{I}{D^2} \tag{2.51}$$

2.9 STRIP, TUBE, AND RECTANGULAR SOURCES

In computing the illuminance at a point produced by a source of irregular shape or by many sources of regular shape but separated from one another, it is useful to use the "building-block" approach. The way this method works is as follows: We define a few standard shapes of sources and work out formulas for the illuminances they produce. The formulas are expressed with distances, angles, and luminances as variables, so that the user can insert the values applicable to a particular lighting situation. We then use the principle of superposition and sum the illuminances produced by whatever combination and number of standard source shapes are required to match the installation under consideration.

These ideas of modeling a physical system, of using several simple entities to replicate a complex system, and of employing the principle of superposition to obtain the aggregate answer are powerful ideas which are widely used in physical science and engineering. The principle of superposition states that the response of a system to several stimuli is the sum of the responses of the system to the individual stimuli applied separately. The only constraint on the system is that it must be linear, that is, if a stimulus S produces a response R, then a stimulus kS must produce a response kR, where k is any number. This constraint assures that a system will not saturate under the increased stimulus and thus give no greater response than it did before (we know that this happens in sound systems that are not very high fidelity—turning up the volume control too far produces little change in volume and introduces distortion).

Strip Source

Consider Fig. 2.25 which shows a strip source of luminance L and width W in the yz plane. We desire the illuminance at point P opposite the middle of the strip, on a plane parallel to the yz plane. The technique we will use is to find the illuminance at P produced by the small segment of width W and length dy and then integrate to find the contributions of all such segments.

We need Eq. (2.19). The distance from the source segment to P is $\sqrt{y^2 + q^2 + r^2}$. The angles α and β are equal and their cosines are $q/\sqrt{y^2 + q^2 + r^2}$. Inserting these quantities into Eq. (2.19) yields

$$dE_H = \frac{LWq^2\,dy}{(y^2 + q^2 + r^2)^2} \tag{2.53}$$

which when integrated between the limits of $y = \pm h$, where $2h$ is the length of the strip, gives

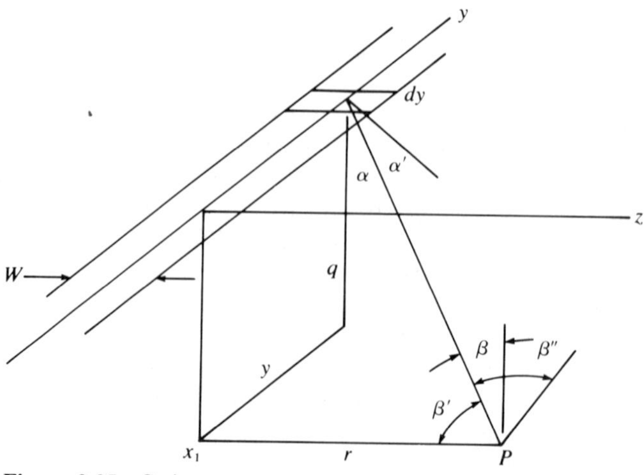

Figure 2.25 Strip source.

$$E_H = \frac{LWq^2}{q^2 + r^2}\left(\frac{h}{q^2 + r^2 + h^2} + \frac{1}{\sqrt{q^2 + r^2}} \tan^{-1} \frac{h}{\sqrt{q^2 + r^2}}\right) \tag{2.54}$$

where the subscript H denotes that the illuminance is on a horizontal plane in Fig. 2.25. If the strip extended from 0 to h and point P were opposite the end of the strip, we would divide Eq. (2.54) by 2.

If h becomes infinite, indicating that the strip source is very long compared with the distance from it to the point P, the first term inside the parentheses in Eq. (2.54) vanishes and the second term becomes $\pi/(2\sqrt{q^2 + r^2}$. The result is

$$E_H = \frac{\pi LWq^2}{2(q^2 + r^2)^{3/2}} \tag{2.55}$$

If h becomes very small compared with the distance to P, Eq. (2.54) becomes

$$E_H = \frac{LWq^2}{q^2 + r^2}\left(\frac{h}{q^2 + r^2} + \frac{h}{q^2 + r^2}\right) = \frac{2LWq^2h}{(q^2 + r^2)^2} \tag{2.56}$$

where we have used the approximation that for small values of an angle, the tangent of the angle equals the angle in radians. Noting that $2Wh$ is the area of the small strip, Eq. (2.56) is simply the ISL value of E_H for the entire source.

The illuminances produced on planes other than those parallel to the plane of the strip are often of interest. An example is the illuminance produced on a wall by a lighting fixture on a ceiling. In Fig. 2.25, to find the vertical illuminance at P on a plane parallel to the xy plane requires only that β be replaced by β'. Thus in Eq. (2.53), the cosine term is changed from $q/\sqrt{y^2 + q^2 + r^2}$ to $r/\sqrt{y^2 + q^2 + r^2}$. Since q and r are constant, there is no change in the integration and the result is

$$E_{V_\parallel} = E_H \frac{r}{q} \tag{2.57}$$

where $V_\parallel$ means the illuminance at P on a vertical plane parallel to the strip.

To find the vertical illuminance at P on a plane parallel to the xz plane, β becomes β'' in Fig. 2.25 and the cosine term in Eq. (2.53) must be changed to $y/\sqrt{y^2 + q^2 + r^2}$. Since y is a variable, this does change the integration. Equation (2.53) becomes

$$dE = \frac{LWqy\,dy}{(y^2 + q^2 + r^2)^2} \tag{2.58}$$

which gives

$$\begin{aligned} E_{V_\perp} &= \frac{LWq}{2} \int_0^h \frac{2y\,dy}{(y^2 + q^2 + r^2)^2} \\ &= \frac{LWq}{2}\left(\frac{-1}{y^2 + q^2 + r^2}\right)\Bigg|_0^h = \frac{LWqh^2}{2(q^2 + r^2)(q^2 + r^2 + h^2)} \end{aligned} \tag{2.59}$$

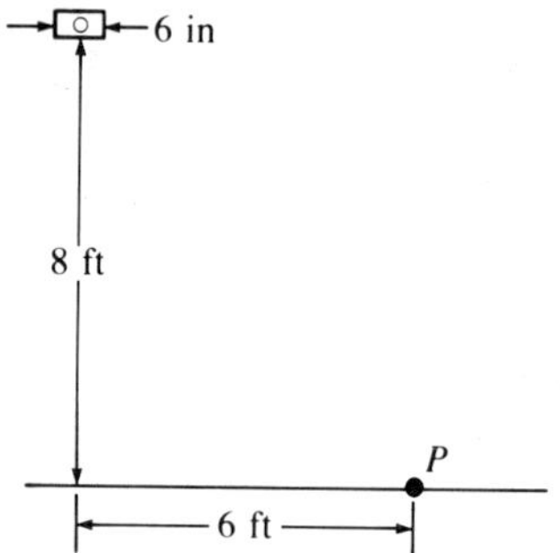

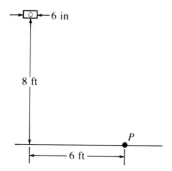

Figure 2.26

Note that V ⊥ is used to indicate that the illuminance at P is on a vertical plane perpendicular to the strip. Note also that the limits of the integration are from 0 to h, rather than from $-h$ to h because the surface at P would only see one-half of the strip in this case.

EXAMPLE 2.11

Ten 8-ft-long single-lamp fluorescent fixtures are mounted end-to-end on the ceiling of a long corridor in Fig. 2.26. We desire the illuminance at point P at the midpoint of the corridor (between fixtures 5 and 6) and 6 ft out from directly beneath the fixtures. The fixtures have a width of 6 in, negligible thickness, and a constant luminance of 1 cd/in^2.

Solution. Even though the array of fixtures is not infinite in length, it is long enough (40 ft each side of the midpoint) compared with the minimum distance from the fixtures to point P (10 ft) that Eq. (2.55) can be used with reasonable accuracy. Inserting values, we have

$$E_H = \frac{\pi(1)(144)(\frac{6}{12})(64)}{2(1000)} = 7.24 \text{ fc}$$

If we use the exact relationship in Eq. (2.54), the answer is

$$E_H = \frac{1(144)(\frac{6}{12})(64)}{100}\left(\frac{40}{1700} + \frac{1}{10}\tan^{-1}\frac{40}{10}\right) = 7.19 \text{ fc}$$

Thus the error using Eq. (2.55) is less than 1%.

In Table 2.4 we have tabulated the illuminances obtained from strip sources (Fig. 2.26) of different lengths. We see that the illuminance does not differ appreciably from the value for an infinite strip until the number of units is reduced to four, where an error of about 8% results. At this point the length of the strip each side of the midpoint is 16 ft, giving a ratio of 1.6:1 of this length to the minimum distance to point P (10 ft). We can reasonably conclude that if the ratio of strip half-length to the minimum distance from fixture to point of measurement is 2:1 or greater, there will be no serious error introduced if Eq. (2.55), rather than Eq. (2.54), is used to compute E_H.

TABLE 2.4

Number of 8-ft units	Value of *h* (feet)	Footcandles from Eq. (2.54)	% Error
∞	∞	7.24	0
10	40	7.19	0.7
8	32	7.17	1.0
6	24	7.06	2.5
4	16	6.73	7.6
2	8	5.36	35.1

Tube Source

The only difference between a tube source and a strip source is that the angle α changes. The tube source always presents a projected width of W towards point P in Fig. 2.25, whereas a strip source does not. Thus, for a tube source, α changes to α' in Fig. 2.25 and cos α changes from $q/\sqrt{y^2+q^2+r^2}$ to $\sqrt{q^2+r^2}/\sqrt{y^2+q^2+r^2}$ in Eqs. (2.53) and (2.58). Since q and r are not variables, none of the integrations change and we obtain the general result

$$E_{\text{tube}} = \sqrt{1+\frac{r^2}{q^2}}\, E_{\text{strip}} \tag{2.60}$$

When $r = 0$, that is, when the point P is directly under the source, $E_{\text{tube}} = E_{\text{strip}}$. Otherwise $E_{\text{tube}} > E_{\text{strip}}$.

EXAMPLE 2.12

Assume the fluorescent fixtures in Fig. 2.26 are tubular with a diameter of 6 in. What will be the horizontal illuminance at point P?

Solution. Using Eq. (2.60) and the value of E_H from Example 2.11, we have

$$E_H = 7.19\sqrt{1+\tfrac{36}{64}} = 8.99 \text{ fc}$$

Rectangular Plane

The final standard source shape that will be considered is the rectangular plane, shown in Fig. 2.27. The plane has a luminance L, dimensions h and w, and lies in the xy plane. The point P lies a distance q outward from a corner of the plane. We have selected a small differential element of dimensions dx and dy for consideration. The illuminance at P produced by this element is, from Eq. (2.19),

$$dE = \frac{Lq^2 dx\, dy}{(q^2+x^2+y^2)^2} \tag{2.61}$$

There are two variables in Eq. (2.61), unlike the previous cases where only

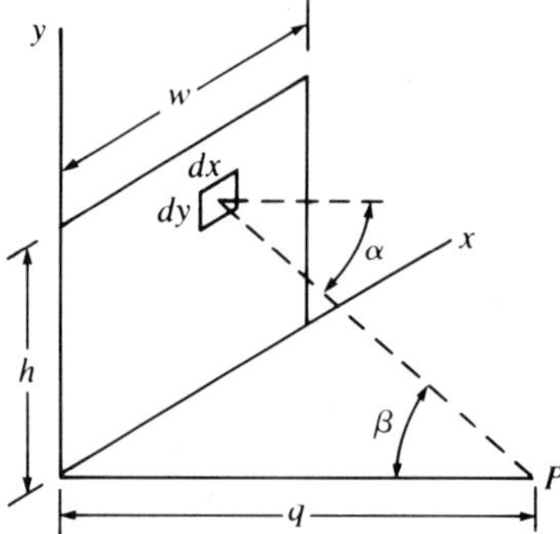

Figure 2.27 Rectangular plane.

one variable was present in the dE expression. Thus we must integrate twice to obtain E. The result is

$$E_{\parallel} = \frac{L}{2}\left(\frac{h}{\sqrt{h^2+q^2}}\tan^{-1}\frac{w}{\sqrt{h^2+q^2}} + \frac{w}{\sqrt{w^2+q^2}}\tan^{-1}\frac{h}{\sqrt{w^2+q^2}}\right) \tag{2.62}$$

where the symbol $\parallel$ indicates that the illuminance is on a plane parallel to the source plane.

When computing the illuminance caused by daylight coming through windows, we alter Fig. 2.27 so that the plane of interest at point P is horizontal instead of vertical, because the windows are vertical and we are interested in the horizontal illuminance on desk tops, tables, etc. With this change, Eq. (2.61) is modified to read

$$dE = \frac{Lqy\ dx\ dy}{(q^2+x^2+y^2)^2} \tag{2.63}$$

which, when integrated, yields

$$E_{\perp} = \frac{L}{2}\left(\tan^{-1}\frac{w}{q} - \frac{q}{\sqrt{h^2+q^2}}\tan^{-1}\frac{w}{\sqrt{h^2+q^2}}\right) \tag{2.64}$$

where the symbol $\perp$ indicates that the plane of P is perpendicular to the plane of the source.

EXAMPLE 2.13

Find the illuminance at point P produced by the rectangular source shown in Fig. 2.28. The source has a luminance of 2 cd/in^2 and is located in the xy plane.

Solution. In Fig. 2.28, we have labeled four areas. What we want is the illuminance produced by area a. To get it, we need to find the illuminance produced by area $abcd$ and subtract out the illuminance produced by areas b, c, and d. Specifically,

$$E_a = E_{abcd} - E_{bc} - E_{cd} + E_c \tag{2.65}$$

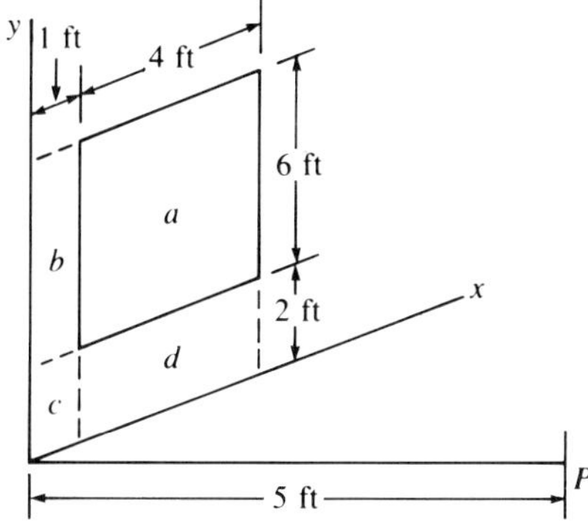

Figure 2.28

The reason for the $+E_c$ is that we have removed area c twice by subtracting E_{bc} and E_{cd}. Let us tabulate the data we need.

Area	h	w	E [Eq. (2.62)]
$abcd$	8	5	146
bc	8	1	41
cd	2	5	68
c	2	1	20

Then, from Eq. (2.65), $E = 146 - 41 - 68 + 20 = 57$ fc.

2.10 MEASURING ILLUMINANCE

Most modern instruments for measuring illuminance and related quantities make use of the photoelectric effect, first discovered by Hertz in 1887. There are four basic forms of photoelectric phenomena, namely photoemissive, photoconductive junction, photoconductive bulk, and photovoltaic.

The photoemissive effect is the liberation of electrons from a photosensitive surface when light impinges on that surface. As will be discussed in Chap. 4, it was this effect that could not be explained by electromagnetic wave theory and which led Planck and others to postulate the photon. Vacuum and gas phototubes and multiplier phototubes operate in the photoemissive mode.

The photoconductive effect is manifested by certain semiconductor devices. The basic phenomenon is that the conductivity of a piece of semiconductor material is altered drastically by incident radiation. The radiant energy causes covalent bonds to be broken within the semiconductor, thus creating many hole-electron pairs in excess of those created normally by thermal generation. The resulting increase in current carriers decreases the resistance of the material.

There are two types of photoconductive devices. Photoconductive

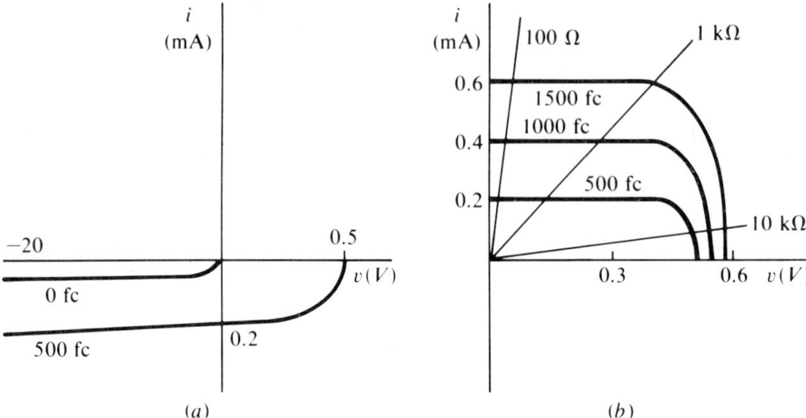

Figure 2.29 Silicon cell light meter.

junction devices, such as the photodiode and phototransistor, are made of germanium or silicon. These devices exhibit very fast response but relatively low sensitivity (mA/fc). Photoconductive bulk devices utilize crystals, such as cadmium sulfide, which are doped with impurities of silver, antimony, or indium. These devices have very high sensitivity; their resistances may vary from 2 MΩ in absolute darkness to 10 Ω under strong light.

The fourth type of photoelectric phenomena, which is utilized in illuminance measuring devices, is the photovoltaic effect. A typical photovoltaic cell is shown in Fig. 2.29. The P-type silicon is doped with boron; the N-type silicon with arsenic. When light strikes the P-type material, excess hole-electron pairs are created. The holes migrate across the junction into the P-type material; the electrons move to the N-type material. The resulting unbalance of charge carriers creates a voltage at the diode terminals. This is the photovoltaic voltage and is approximately 0.5 V for silicon and 0.1 V for germanium.

The volt-ampere characteristic of a typical silicon photocell is shown in Fig. 2.30*a*. In the third quadrant, where both current and voltage are negative, the cell is acting in the photoconductive mode. In this quadrant, an external voltage source (battery) is required to operate the device. In the fourth quadrant, with current negative and voltage positive, it is operating in the photovoltaic mode and no external battery is needed. It is this latter situation that applies when we are interested in using the cell to measure illuminance.

The fourth quadrant in Fig. 2.30*a* is inverted and expanded in Fig. 2.30*b*. We see that the open-circuit output voltage varies nonlinearly (logarithmically) with illuminance level. This is also nearly the case with high-resistance (10-kΩ) loads. On the other hand, the short-circuit current varies linearly with illuminance level, and the current for low-resistance (100-Ω) loads nearly so. This latter condition is utilized in an illuminance meter, that is, the output current in the low-resistance microammeter is di-

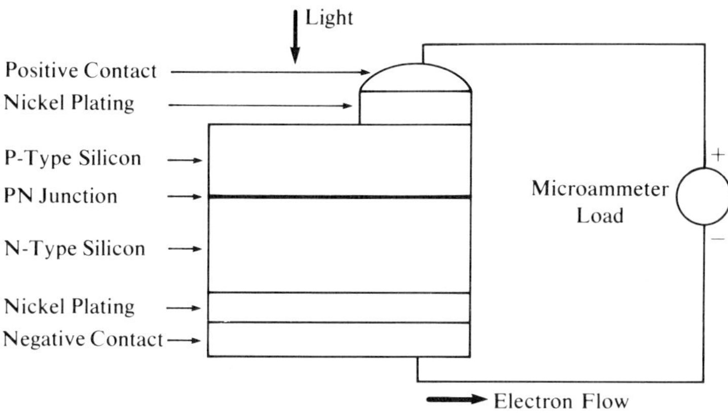

Figure 2.30 Silicon cell volt-ampere characteristics.

rectly proportional to the illuminance level, and thus, since the deflection of the microammeter needle is directly proportional to current, it is possible to calibrate the microammeter scale directly in footcandles or lux.

The first photovoltaic cell light meter was developed by the Weston Instrument Company in 1931 and led to the lighting industry's first portable illuminance meters and photographic exposure meters. Modern illuminance meters are more sophisticated than their predecessors but embody the same basic principles. Selenium, which was used as the cell material until quite recently, has given way to silicon. The latter is much more linear and less temperature dependent than the former.

Most contemporary meters are color and cosine corrected. In Fig. 2.31 we show spectral sensitivity curves of an uncorrected selenium cell, a corrected cell, and the human eye. The problem, of course, is that we desire an illuminance meter to evaluate luminous flux in the same way that the eye does. This means that the light cell's spectral sensitivity curve should match that of the eye. As can be seen in Fig. 2.31, an uncorrected cell has a much greater spectral sensitivity to wavelengths in the violet-blue and red-orange

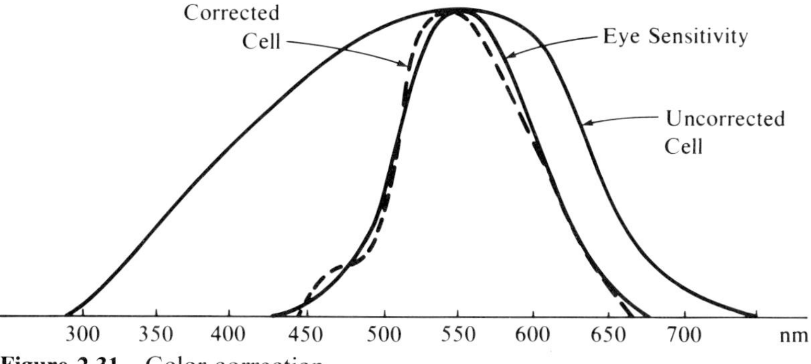

Figure 2.31 Color correction.

regions of the spectrum than does the eye. Also the cell responds to ultraviolet radiation whereas the eye does not. What the meter manufacturers have done is to integrate a carefully designed color filter with the cell so that the filter-cell spectral sensitivity curve nearly matches that of the human eye, as shown in Fig. 2.31.

Cosine correction is also necessary for accurate illuminance measurement. When luminous flux strikes the meter at an angle, the meter should evaluate this flux so that its reading is the product of the reading that would be obtained with the cell held perpendicular to the incoming light times the cosine of the angle of incidence.

A meter which is not cosine corrected generally has a glass plate covering the cell. Light rays striking the glass plate at a grazing angle are for the most part reflected and thus do not enter the cell and contribute to the meter reading. Cosine correction is achieved by replacing the glass plate with a diffusing cover of white plastic, designed to intercept wide-angle rays and direct them to the cell.

The effect of cosine correction is shown in Fig. 2.32. The plot of cos θ versus θ is shown dotted. This is how the cell should respond. The response of the uncorrected cell with a glass cover is the dashed curve, and this results in illuminance readings that are lower than they should be. The solid curve is for the cell with cosine correction and matches closely the cosine curve.

Another issue that should be considered in choosing a light meter is its accuracy. The calibration of light meters is often "traceable to the National Bureau of Standards," but the number of links in the chain of calibration between the meter and NBS is not always clear. Laboratory meters should be accurate within ±5% and portable meters within ±15%. An accuracy of about ±2% is the current state-of-the art limit.

It is possible to measure crudely luminous exitance, reflectance, and transmittance with an illuminance meter. To measure luminous exitance, the luminous flux density leaving a surface, the cell is held close to the surface

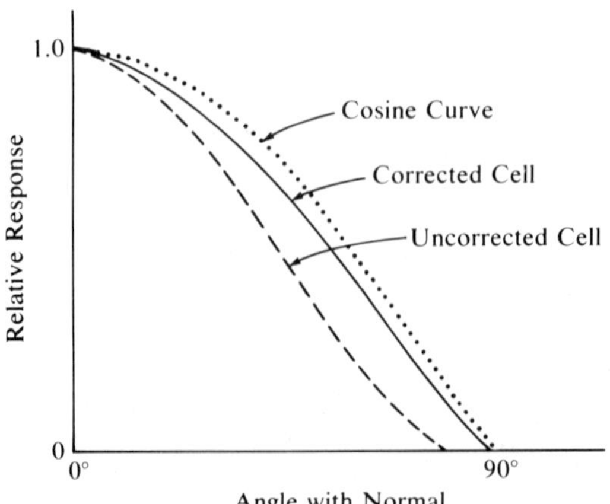

Figure 2.32 Cosine correction.

and facing it and then withdrawn a short distance until the meter reading becomes constant. The reading of the meter is now in lumens per square meter or lumens per square foot of luminous exitance. Care must be taken that neither the meter nor the measurer creates a shadow on the surface being measured.

To measure the reflectance of a surface, the cell is held against the surface facing outward and the illuminance reading is noted. Then the luminous exitance of the surface is measured and the ratio of the two is the reflectance. To measure the transmittance of a piece of glass or plastic, the sample is placed over the cell for one reading and then a second reading is taken with the sample removed. The ratio of the two readings is the transmittance.

Frequently, the light cell of an illuminance meter is in the form of a separate light paddle, connected to the display by an extension cord. This enables the user to read the meter without imposing a shadow on the cell face. Such meters are available in a variety of lux or footcandle ranges, one example being ranges of 0 to 6, 0 to 12, 0 to 60, 0 to 120, and 0 to 600 fc. Meters are also available with scales of 0 to 0.5 and 0 to 2 fc for low-level illuminance measurements, such as in roadway lighting, and with 0 to 6000, or 0 to 12,000 fc scales for high-level illuminance measurements of sunlight and extremely bright discharge light sources. Meters with a wide range of luminous exitance scales are also available. In many contemporary illuminance meters the microammeter has been replaced by solid-state amplifiers and light-emitting diode digital readout.

Early luminance meters, such as the Luckiesh-Taylor brightness meter, required the user to make a visual photometric match between the luminance of the object being measured and that produced by a small calibrated incandescent lamp within the instrument. Such matches were difficult to make, largely because of color temperature differences between the object and calibrated lamp fields, and several measurements of a given luminance by the same or different observers were likely to show quite a wide spread.

Contemporary luminance meters do not require visual matching. Their optics permit luminance measurements of less than a 1° field (21 in at 100 ft). One simply aims the device at the area of interest by viewing through a sight tube and then pushes a button to display the luminance digitally. The meters are color corrected and have ranges from 0.1 to 20,000 cd/m^2.

A luminance meter is susceptible to all of the errors of an illuminance meter and, in addition, its optics are very dust sensitive. Luminances outside the intended field of view will be included in the luminance reading if the lenses are dirty.

REFERENCES

American National Standards Institute: "Nomenclature and Definitions for Illuminating Engineering," ANSI/IES RP-16, New York, 1980.

Boast, W.B.: *Illumination Engineering,* McGraw-Hill, New York, 1953.

General Electric Co.: "Light Measurement and Control," TP118, Nela Park, Cleveland, Ohio, 1971.

Hewitt, H., and A.S. Vause: *Lamps and Lighting,* American Elsevier, New York, 1966.

IES Lighting Design Practice Committee: "The Determination of Illumination at a Point in Interior Spaces," *Journal of IES,* January 1974.

Johnston, N.B.: "Introducing the Modern Metric System into Engineering Education," *Engineering Education,* April 1977.

Kaufman, J.E.: *IES Lighting Handbook*, Reference Volume, Illuminating Engineering Society, New York, 1981.

Levin, R.E.: "The Photometric Connection—Parts 1-4," *Lighting Design and Application,* September–December 1982.

Millman, J., and C.C. Halkias: *Electronic Devices and Circuits,* McGraw-Hill, New York, 1967.

Murdoch, J.B.: "Inverse Square Law Approximations of Illuminance," *IES Journal,* 1981.

Chapter 3

Radiant Energy as Waves

3.1 INTRODUCTION

In Secs. 1.3 and 1.5, we reviewed the theories which have been developed since Newton's time to explain the nature of light, reaching the somewhat discomforting conclusion that light is dualistic in nature, with the properties of both waves and particles.

This chapter and the next will be devoted to exploring this wave-particle duality of light, with concentration on the wave nature of light in this chapter and its particle nature in Chap. 4. We begin with a discussion of power density and spectral power density, quantities used to describe the total energy content of a wave and the energy content in selected spectral bands. This leads us to blackbody radiation and the laws of Planck, Stefan, and Wien governing that radiation.

We then discuss a luminous intensity standard based on the blackbody radiator and show how the lumen and the watt can be related through this standard and the spectral luminous efficiency curve of the eye.

The chapter concludes with a discussion of selective radiators (of which the tungsten filament is an example), the concept of the color temperature of a light source, methods of making spectral power measurements, and the radiation properties of the sun.

3.2 POWER DENSITY AND SPECTRAL POWER DENSITY

In much of the discussion on radiation to follow we will be as much concerned with power density as with power, that is, we will wish to know the watts per unit surface area emitted by a radiator or received by a surface, as well as the total watts emitted or received. We define radiant exitance (M) as emitted, transmitted, or reflected radiant power density and we define irradiance (E) as received radiant power density. Thus M is radiation leaving a surface whereas E is radiation arriving at a surface.*

We will also be interested in power density within a narrow wavelength band, called spectral power density. Spectral power density leaving a surface is called spectral radiant exitance, M_λ;† that arriving at a surface is called spectral irradiance, E_λ. To illustrate these new quantities, consider Fig. 3.1 in which the spectral radiant exitance of a fictitious source is shown. The graph is a plot of M_λ in microwatts per square centimeter of source surface area per 10 nm of wavelength. What this plot gives us is the radiant exitance in any 10-nm band centered about the wavelength that we are considering. For example, at 400 nm, the radiant exitance in a 10-nm band from 395 to 405 nanometers is 40 μW/cm^2. In a 2-nm band centered at 400 nm (from 399 to 401 nm) the radiant exitance is $40(0.2) = 8$ μW/cm^2. The factor 0.2 is necessary because the wavelength band is now 2 nm wide whereas M_λ is based on a 10-nm width.

It is possible to find the total radiant exitance M emitted by the source in Fig. 3.1. To do this, we blanket the region beneath the curve with narrow rectangles, each of the same width. In the case illustrated we have chosen 20-nm widths and three of these rectangles are shown. The height of each is the value of M_λ at the midpoint wavelength of that rectangle. Thus for the leftmost of the three rectangles in Fig. 3.1, the height is the value of M_λ at 380 nm, which is 33 μW/(cm$^2 \cdot$ 10 nm). This rectangle contributes $33(2) = 66$ μW/cm^2 to the total radiant exitance of the source, the factor 2 being necessary because of the 20-nm width. The other two rectangles contribute 80 and 96 μW/cm^2 to the total. Summing the values for all rectangles would give the total radiant exitance.

What we have done above is find the approximate area beneath the curve in Fig. 3.1. If the equation for M_λ as a function of λ were known, we could find the total emitted power density exactly through integration as

$$M = \int_0^\infty M_\lambda \, d\lambda \tag{3.1}$$

To sum up, the vertical distance to a spectral power density curve at a specified wavelength gives the spectral power density at that wavelength and the area bounded by the curve and the horizontal axis gives the total power density.

*It is sometimes confusing that the symbols for radiant and luminous quantities are the same in the standards literature. The context in which a given symbol is used should clarify which entity we are talking about. Otherwise we will use the subscript e for radiant quantities and v for luminous ones.

†The subscript λ indicates the derivative of M with respect to λ, whereas parentheses indicate that the entity is a function of λ, for example $V(\lambda)$.

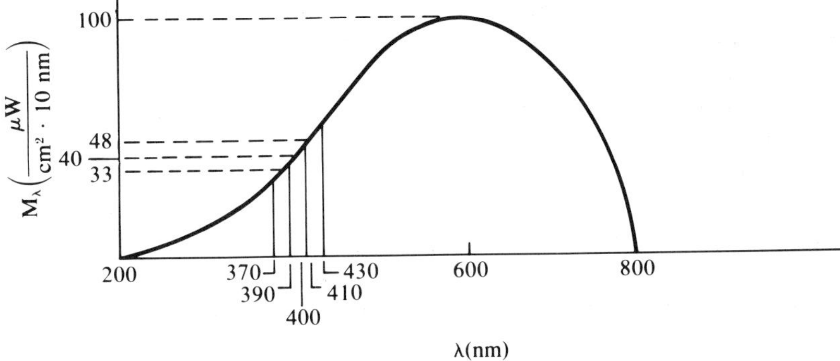

Figure 3.1 Emission from a radiant source.

EXAMPLE 3.1

To illustrate the concepts of a spectral distribution curve, consider Fig. 3.2, which is a graph of spectral population density (people per square mile per 2-year age interval) plotted against age. We desire to find the total number of people per square mile.

Solution. Let us break the horizontal axis up into 4-year intervals and read off the spectral population density at the midpoint of each interval. For example from 16 to 20 years, the midpoint is 18 years and the value is 300 people per square mile per 2-year interval or 600 people per square mile per 4-year interval. The total population density is the sum of all such values. Starting with the 0- to 4-year age interval, we have (240 + 380 + 390 + 330 + 300 + 340 + 390 + 410 + 410 + 400 + 370 + 300 + 240 + 170 + 120 + 80 + 50 + 30 + 10)2 = 9920 people per square mile.

3.3 BLACKBODY RADIATION AND PLANCK'S EQUATION

There are three major ways in which heat energy is transferred from one body to another, namely conduction, convection, and radiation.

Assume you are standing near a campfire. If you (foolishly) reach in

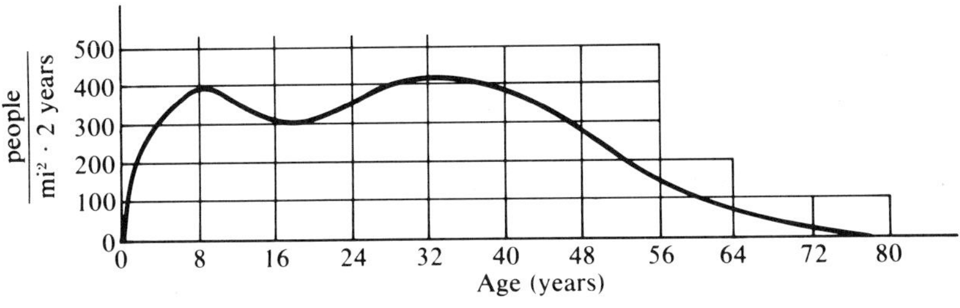

Figure 3.2

and touch one of the glowing embers, heat reaches your hand by conduction. If you hold your hands above the fire, heat reaches your hands by convection, the upward flow of air that has been heated. Now if you stand back from the fire, you still feel the sensation of heat. Energy is now reaching you by radiation.

Radiation is the emission of energy in the form of electromagnetic waves from all bodies. We have learned that such energy travels with the speed of light and does not require a medium for its passage; in fact it travels best in a vacuum. When such radiation encounters a body, it is absorbed, transmitted, or reflected. That which is absorbed may be reradiated, increase the temperature of the body, or be converted to other energy forms.

Light, as we have seen, is radiant energy in the narrow band of wavelengths from 380 to 770 nm. It is customary to divide the generation of light into two categories—incandescence and luminescence. Incandescence is visual radiation from the surface of a radiator that is directly related to the temperature of that radiator. The entire family of incandescent lamps falls into this category; each lamp containing a filament that is heated to incandescence by the passage of electric current through it. Luminescence is visual radiation not directly related to temperature. Glow lamps, "neon" signs, fluorescent lamps, high-intensity discharge lamps, electroluminescent panels, light-emitting diodes, and lasers are included in this grouping. In the remainder of this chapter we will consider incandescence. The various forms of luminescence will be discussed in later chapters.

In dealing with incandescent, or temperature, radiation, it is convenient to consider an ideal radiator known as a blackbody. A blackbody radiator absorbs all of the radiant energy reaching it, transmitting and reflecting none. Because it is a perfect absorber, it is also a perfect radiator. A blackbody radiates more power at each wavelength and more total power than any other incandescent radiator of the same surface area and operating at the same temperature.

From 1897 to 1900, Lummer and Pringsheim obtained blackbody radiation curves experimentally for a variety of temperatures, using the open end of a specially constructed and uniformly heated tube as their blackbody source. Then in 1900, Lord Rayleigh and J. H. Jeans attempted to derive an expression for the radiation from a blackbody radiator using classical means. These men proceeded directly from the assumptions of classical statistical and Newtonian mechanics and Maxwell's electromagnetic equations. They assumed that a blackbody cavity contains electromagnetic waves bouncing back and forth between the cavity walls and that the superposition of these waves creates stationary waves in space within the cavity. With respect to energy, each of these waves behaves like an ordinary mechanical harmonic oscillator to which the laws of statistical mechanics are applicable. These are:

1. The average kinetic energy of the collection of harmonic oscillators is $\frac{1}{2}kT$ times the number of degrees of freedom that they have inside

the box, where k is the Boltzmann constant* and T is temperature in kelvins.

2. The average potential energy equals the average kinetic energy. Thus the average total energy of the oscillators is kT times the number of degrees of freedom.
3. The number of stationary waves (which is the number of degrees of freedom) per unit volume and per unit wavelength is $8\pi/\lambda^4$, which gives an average total energy per unit volume and per unit wavelength of $(8\pi/\lambda^4)\ kT$.

From these basic relations, Rayleigh and Jeans derived the following expression for the spectral power density distribution of a blackbody radiator:

$$M_\lambda = \frac{2\pi kTc}{\lambda^4} \frac{\text{W}}{\text{cm}^2 \cdot \text{cm}} \tag{3.2}$$

where c (the velocity of light) is in centimeters per second and λ is in centimeters. This expression is known as the Rayleigh-Jeans law.

It was found that Eq. (2.3) did not agree at all well with the experimental results of Lummer and Pringsheim. At short wavelengths it is extremely in error and it leads to the obviously incorrect result that the total emitted power density of a blackbody radiator is infinite. It does agree with experimental values for λT greater than $1.5 \times 10^5\ K \cdot \mu\text{m}$ (long-wavelength infrared radiation).

In the year 1901, Max Planck (1858–1947), the famous German physicist and originator of quantum theory, developed a purely empirical formula to represent the Lummer-Pringsheim results. The equation bears Planck's name and is

$$M_\lambda = c_1 \lambda^{-5} (e^{c_2/\lambda T} - 1)^{-1} \tag{3.3}$$

where M_λ is the spectral power density in watts per square centimeter radiated by the blackbody per micrometer of wavelength band, λ is in micrometers, T is in kelvins, and c_1 and c_2 are constants given by

$$c_1 = 2\pi hc^2 = 37{,}418\ (\text{W} \cdot \mu\text{m}^4)/\text{cm}^2 \qquad c_2 = \frac{hc}{k} = 14{,}388\ \mu\text{m} \cdot \text{K} \tag{3.4}$$

In these equations c is again in centimeters per second, k is the Boltzmann constant, and h is Planck's constant (6.63×10^{-34} J · s).

Planck's constant, like the Boltzmann constant, is a universal constant of nature. In his work with blackbody radiators, Planck discovered that the emitted spectral radiation could not be described in terms of harmonic oscillators possessing all possible energies, as Rayleigh and Jeans had tried to do. Rather he found that it was necessary to consider the emitted energy as

*The Boltzmann constant is a universal constant relating the average kinetic energy of an aggregate of molecules to their temperature. Its value is 1.38×10^{-23} joules per degree Kelvin (J/K). It is named after Ludwig Boltzmann (1844–1906), professor of physics at Vienna, who made many contributions to kinetic theory and thermodynamics.

occurring in discrete packets, which he called quanta. As was noted in Sec. 1.5, each quantum has an energy in joules given by*

$$W = hf = h\,\frac{c}{\lambda} \tag{3.5}$$

Thus, Planck's constant is the proportionality constant relating the frequency and energy of a quantum of radiation.

Returning now to Eq. (3.3), the only change Planck made in what Rayleigh and Jeans had done was to assume that the stationary waves in the blackbody cavity could have only integrally related discrete energies given by 0, hf, $2hf$, $3hf$. . . . He further assumed that the probability that a given stationary wave would have a particular one of these energies could be calculated from classical statistical mechanics, as before. Thus he showed the average total energy per oscillator to be

$$W = kT\,\frac{hc/\lambda kT}{e^{hc/\lambda kT} - 1} \quad \text{joules} \tag{3.6}$$

instead of kT joules, as used by Rayleigh and Jeans. Applying this extra factor to the Rayleigh-Jeans law gives Planck's equation,

$$M_\lambda = \frac{2\pi kTc}{\lambda^4}\,\frac{hc/\lambda kT}{e^{hc/\lambda kT} - 1} = \frac{2\pi hc^2\lambda^{-5}}{e^{hc/\lambda kT} - 1} \tag{3.7}$$

Planck's assumption that harmonic oscillators can have only discrete energies opened the door to quantum theory. His assumption implies that matter does not radiate energy continuously, as had been previously assumed, but rather in very small discrete amounts, For example, the energy of a visible quantum at 600 nm is

$$hf = \frac{(6.63 \times 10^{-34})\,(3 \times 10^{8})}{6 \times 10^{-7}} = 33 \times 10^{-20}\ \text{J}$$

which is indeed very small. In large-scale, macroscopic mechanics the energies involved are so large that we cannot observe the tiny discrete changes that are occurring with time. Thus the energy changes appear smooth and continuous. This is not the case in atomic, microscopic mechanics, which will be demonstrated when we consider particle radiation and luminescent sources in later chapters.

3.4 STEFAN, BOLTZMANN, AND WIEN

In Fig. 3.3, blackbody radiation curves for several values of temperature are shown. The visible spectrum is labeled to show that portion of each curve to which our eyes are sensitive.

*The preferred symbols for energy and power are Q and Φ, respectively. However, W for work and energy and P for power are well-established in the engineering and physics literature and we will continue to use them. We will use M and E to represent both radiant and luminous power density.

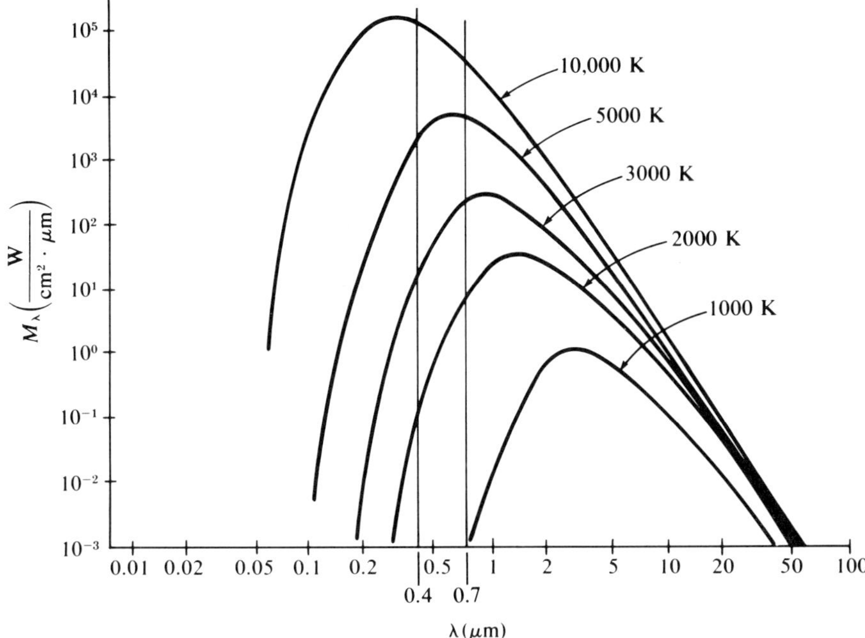

Figure 3.3 Blackbody radiation curves.

Two important characteristics of blackbody radiation can be seen in Fig. 3.3. First the total radiated power density from a blackbody increases dramatically as temperature increases. Second, as temperature goes up, the peak of the radiation curve shifts toward the shorter wavelengths. Let us examine each of these phenomena in detail.

To obtain the total radiated power density from a blackbody radiator as a function of temperature, we must obtain the areas under the curves in Fig. 3.3. This is done by integrating Planck's equation over all λ for each temperature.

$$M = \int_0^\infty c_1 \lambda^{-5} (e^{c_2/\lambda T} - 1)^{-1} d\lambda \tag{3.8}$$

To evaluate Eq. (3.8), we must first expand the term in parenthesis into a Taylor series:

$$\frac{1}{e^{c_2/\lambda T} - 1} = e^{-c_2/\lambda T} + e^{-2c_2/\lambda T} + e^{-3c_2/\lambda T} + \cdots \tag{3.9}$$

Then integration term by term and insertion of the limits yields

$$\mathbf{M} = \frac{6c_1 T^4}{c_2^4} \left(\frac{1}{1^4} + \frac{1}{2^4} + \frac{1}{3^4} + \cdots \right) \tag{3.10}$$

The quantity in parentheses is an infinite series whose sum can be shown to be $\pi^4/90$. Thus

$$M = \frac{4c_1 T^4 \pi^4}{90 c_2^4} \equiv \sigma T^4 \tag{3.11}$$

which is the Stefan*-Boltzmann law.

The value of σ, the Stefan-Boltzmann constant, is 5.6032 $\times 10^{-12}$ W/(cm$^2 \cdot$ K^4), which gives M in watts per square centimeter.

The result in Eq. (3.11) that the radiated power density of a blackbody radiator depends only on the temperature raised to the fourth power is remarkably simple. Thus if we know M at one temperature, call it T_1, we can find M at a second temperature T_2 by writing

$$M_{\text{at } T_2} = \frac{T_2^4}{T_1^4} \, (M_{\text{at } T_1}) \tag{3.12}$$

EXAMPLE 3.2

The filament of an incandescent lamp is 0.006 cm in diameter and 60 cm long. It consumes 100 W. Assuming that the filament can be considered a blackbody radiator,† at what temperature is it operating? How many watts would it consume at a temperature of 3000 K?

Solution. The surface area of the filament is $2\pi rl = 2\pi(0.003)(60) = 1.13$ cm^2. Thus the radiated power density is $M = 100/1.13 = 88.5$ W/cm^2.

From Eq. (3.11),

$$88.5 = 5.67 \times 10^{-12} \, T^4$$

giving

$$T = 1990 \text{ K}$$

For a temperature of 3000 K, Eq. (3.12) yields

$$P = \frac{3000^4}{1990^4}(100) = 517 \text{ W}$$

Note that input power, must be increased more than five times to raise the temperature by 50%.

We stated that two things happen as the temperature of a blackbody radiator increases. The second is that the peaks of the blackbody radiation curves occur at shorter wavelengths. We might expect this result from personal experience. Suppose we set the surface unit of an electric stove on low heat. In addition to getting warm, it glows a deep red. On medium heat, the unit gets hotter and brighter, but also the color has more orange in it. On high

*J. Stefan (1835–1893), Austrian physicist.

†An incandescent lamp filament is actually a selective radiator, to be discussed in Sec. 3.7. Also a filmament 60 cm long would require supports, which would produce temperature gradients along its length.

heat the color is definitely orange. Thus the shift toward the visible spectrum is indeed noticeable.

We can derive a quantitative measure of this shift of the peak to shorter wavelengths by reconsidering Eq. (3.3). For temperatures (2000 to 4000 K) and wavelengths (0.30 to 0.80 μm) of interest to us in dealing with incandescent light sources and the visible and near-visible portions of the spectrum, Planck's equation can be simplified to

$$M_\lambda = c_1 \lambda^{-5} e^{-c_2/\lambda T} \tag{3.13}$$

which is known as Wien's* Radiation law. What we have done to arrive at Eq. (3.13) is to note that, for the temperatures and wavelengths indicated, $e^{c_2/\lambda T} >> 1$ and thus the 1 may be neglected in Planck's equation. For example at $T = 4000$ K and $\lambda = 0.8$ μm, $e^{c_2/\lambda T} \sim 90$. For lower temperatures and shorter wavelengths, this quantity will be still larger.

To find the value of λT in Eq. (3.13) at which M_λ is a maximum, we set the derivative of M_λ with respect to λ equal to zero. The result is

$$-5c_1 \lambda^{-6} e^{-c_2/\lambda T} - \frac{c_2}{T}\left(\frac{-1}{\lambda^2}\right) c_1 \lambda^{-5} e^{-c_2/\lambda T} = 0 \tag{3.14}$$

yielding

$$\lambda_{max} T = \frac{c_2}{5} = 2878 \ \mu\text{m} \cdot \text{K} \tag{3.15}$$

Inserting this result into Eq. (3.13) gives

$$M_{\lambda_{max}} = \frac{c_1 T^5 e^{-5}}{(2878)^5} = 1.28 \times 10^{-15} T^5 \quad \text{W/(cm}^2 \cdot \mu\text{m)} \tag{3.16}$$

Equations (3.15) and (3.16) show that as T increases, the point at which M_λ is maximum occurs at shorter wavelengths and the $M_{\lambda_{max}}$ increases markedly. For example, when T is 2900 K, which is approximately the temperature of an incandescent lamp filament, M_λ is a maximum at $\lambda = 1$ μm, well into the infrared region of the spectrum. This is why an incandescent lamp is rich in reds and yellows, but deficient in blues and greens. It is also why it is an inefficient light source. To cause the M_λ peak to occur in the red region of the spectrum, say at 0.7 μm, requires that $T = 2878/0.7 = 4110$ K. But tungsten, the filament material in an incandescent lamp, melts at 3655 K, so operating at this temperature would cause the lamp to burn up immediately. This is the light versus life tradeoff that we are always faced with in incandescent lamp design. Normally we want at least a 1000-h life for general lighting lamps, so we accept less light. Where maximum light output and more nearly balanced white light are necessary, as in the case of photo flood lamps for studio photography, we sacrifice life and operate the filament at 3100 to 3400 K, significantly higher than the normal 2900 K.

*Wilhelm Wien (1864–1928), German physicist noted for his work in optics and radiation.

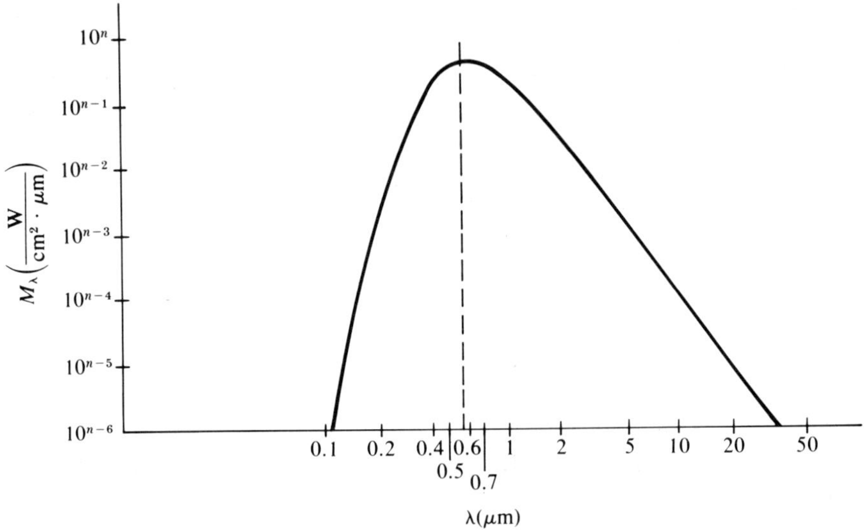

Figure 3.4

EXAMPLE 3.3

The spectral power density curve of a blackbody radiator is shown in Fig. 3.4.

(a) At what temperature is it operating?
(b) How many watts per square centimeter is it emitting?
(c) What is the maximum value of M_λ?
(d) What is the value of n?

Solution

(a) From Fig. 3.4, $\lambda_{max} = 0.57\ \mu m$. Then Eq. (3.15) yields

$$T = \frac{2878}{0.57} = 5050\ K$$

(b) From Eq. (3.11)

$$M = (5.67 \times 10^{-12})(5.05^4 \times 10^{12}) = 3688\ W/cm^2$$

(c) From Eq. (3.16)

$$M_{\lambda max} = (1.28 \times 10^{-15})(5.05^5 \times 10^{15}) = 4200\ W/cm^2 \cdot \mu m)$$

(d) From this result and Fig. 3.4, the value of n must be 4.

3.5 LUMINOUS INTENSITY STANDARD

In Sec. 1.7, it was pointed out that, in the days of pre-electric lighting, a spermaceti candle and later a gas flame were used to define the candle as the unit of luminous intensity.

In 1909, the U.S. National Bureau of Standards (NBS) redefined the candle in terms of a group of carbon filament incandescent lamps having precise filament dimensions and operating at a closely controlled voltage. This standard was accepted in both the United States and Europe, but, while an improvement over the candle and the gas flame, was still not wholly satisfactory because of the difficulty in controlling the various variables involved.

In 1937, the International Committee on Weights and Measures (CIPM) adopted the following resolution: "From January 1, 1940, the unit of luminous intensity shall be such that the brightness of a blackbody radiator at the temperature of solidification of platinum is 60 candles per square centimeter." Their choice of 60 was to make the "new candle" and the "old candle" nearly the same in brightness.

World War II intervened at this time and little was done to implement the new definition until 1948, when the International Commission on Illumination (CIE) recommended that the new unit of luminous intensity be named the candela. In that same year, the General Conference on Weights and Measures (CGPM) refined the prewar definition to read, "The candela is the unit of luminous intensity. The magnitude of the candela is such that the luminance of a blackbody radiator at the temperature of soldification of platinum is 60 candelas per square centimeter."

The resulting Waidner and Burgess blackbody standard is illustrated in Fig. 3.5. The thorium oxide test tube is surrounded by platinum in a thermally insulated housing. The platinum is maintained at a pressure of 101,325 N/m^2 and is heated by an applied electromagnetic field until it melts (the melting temperature of thorium oxide is much higher than that of platinum). Then the platinum is allowed to cool and the luminance of the thorium oxide is measured down the sight tube at the temperature (2042 K) at which the platinum returns to solid form. This luminance is assigned the value of 60 cd/cm^2 and thus each square centimeter has a luminous intensity of 60 cd.

Although the blackbody radiator was much superior to the carbon filament lamp as a standard, because its radiant power density could be

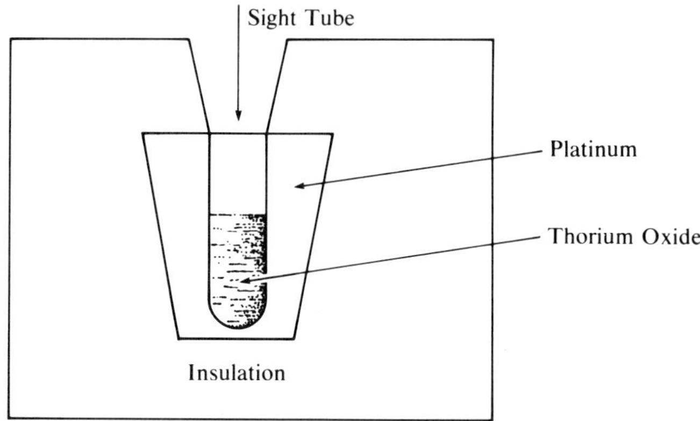

Figure 3.5 Waidner and Burgess standard.

calculated accurately and because the variables associated with its measurement could be closely controlled, it still suffered from being based, in part, on a photometric measurement (luminance). There was considerable interest worldwide in having an illumination standard which depended only on radiometric quantities (watts). In 1979, the CGPM adopted the following definition: "The candela is the luminous intensity in a given direction of a source which is emitting monochromatic radiation of frequency 540×10^{12} Hz (555 nm) and whose radiant intensity in that direction is $\frac{1}{683}$ W/sr."

What this definition does is relate the candela to a single spectral line of radiation at the peak of the eye's spectral luminous efficiency curve (555 nm). Thus it eliminates the need for a photometric measurement and defines the candela strictly in radiometric terms. It remains for us for us to show where the value $\frac{1}{683}$ comes from and how the earlier (1948) definition of the candela is consistent with the new (1979) definition.

3.6 RELATING THE LUMEN AND THE WATT

The spectral luminous efficiency curve of the human eye was introduced in Sec. 1.6. This curve was developed in the early 1920s by K. S. Gibson and E. P. T. Tyndall of the National Bureau of Standards. These men reasoned that the sensation of luminance could be separated from that of color in a comparison test of two radiations of different wavelenth if the wavelengths differed only slightly. During the period 1921 to 1923 they had 52 observers make luminance comparisons. Each observer was required to adjust the power density input to a source at one wavelength until its luminance balanced that of a source at a second wavelength 10 nm away. The luminances were compared in a circular field photometer, with one luminance in the left half-circle and the other in the right. The process was repeated for one of the original wavelengths and a third wavelength 10 nm from it, and so on until the entire visible spectrum was covered.

Gibson and Tyndall's work was summarized in a 1923 paper entitled "Visibility of Radiant Energy." They found that the minimum power density input to give a prescribed constant luminance occurred at 555 nm. Thus it becomes possible to define the spectral luminous efficiency at any other wavelength in terms of that at 555 nm by the relation.

$$V(\lambda) = \frac{M_{555}}{M_\lambda} \tag{3.17}$$

where M_{555} is the spectral power density at wavelength λ required for a luminance balance. Values for $V(\lambda)$ at 10-nm intervals were adopted by the CIE in 1924 and by the CIPM in 1933 (Table 3.1).

Gibson and Tyndall's work was done at photopic levels of luminance. Photopic vision is dominated by cones and requires an adaptation luminance of at least 3.4 cd/m^2. By contrast, scotopic vision, which is dominated

TABLE 3.1 SPECTRAL LUMINOUS EFFICIENCY VALUES FOR PHOTOPIC VISION

Wavelength (nm)	Spectral luminous efficiency	Wavelength (nm)	Spectral luminous efficiency
380	0.00004	580	0.870
390	0.00012	590	0.757
400	0.0004	600	0.631
410	0.0012	610	0.503
420	0.0040	620	0.381
430	0.0116	630	0.265
440	0.023	640	0.175
450	0.038	650	0.107
460	0.060	660	0.061
470	0.091	670	0.032
480	0.139	680	0.017
490	0.208	690	0.0082
500	0.323	700	0.0041
510	0.503	710	0.0021
520	0.710	720	0.00105
530	0.862	730	0.00052
540	0.954	740	0.00025
550	0.995	750	0.00012
555	1.000	760	0.00006
560	0.995	770	0.00003
570	0.952		

by rods, occurs at luminance levels below 0.034 cd/m^2. The region in between, where both rods and cones are operative, is called mesopic vision.

Spectral luminous efficiency curves for both photopic, $V(\lambda)$, and scotopic, $V'(\lambda)$, vision are shown in Fig. 3.6*a*. For scotopic vision, the curve is shifted 50 nm to the left from the photopic curve. This phenomenon is termed the Purkinje effect, after Johannes von Purkinje, and means qualitatively that the eye is relatively less sensitive to red and more sensitive to blue at night than during the day. Von Purkinje first observed the effect while walking in a field. He noted that blue flowers viewed at dawn were brighter than red ones, but that the reverse was true in midday.

The results in Fig. 3.6*a* should not be interpreted to mean that every human being has the same spectral luminous efficiency curves. In fact, a given observer's curves are likely to vary markedly from the standards. For photopic vision, work by several researchers has shown that at 490 nm where $V(\lambda)$ is listed as 0.208, actual values will vary from 0.10 to 0.38. At 590 nm, where $V(\lambda)$ is 0.757, the observed range is from 0.62 to 0.95, and at 640 nm, with $V(\lambda) = 0.175$, values range from 0.09 to 0.32. Thus we should interpret the curves in Fig. 3.6*a* as average spectral luminous efficiency curves for a "standard" observer and recognize that a given observer's curves may differ rather widely from them.

The watt has been described as the time rate of flow of radiant energy;

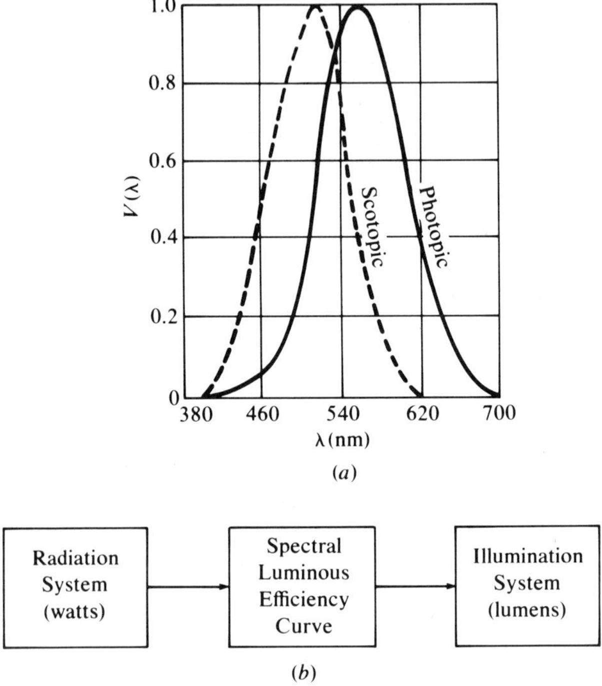

Figure 3.6 (*a*) Spectral luminous efficiency curves. (*b*) Relating watts to lumens.

the lumen as the time rate of flow of luminous energy. To link these two units, we need the $V(\lambda)$ function and the blackbody standard radiator. This linkage through the eye is illustrated in Fig. 3.6*b*.

The spectral luminous efficiency curve is the screen that enables us to evaluate the visual effect of radiant energy. For the purpose of illustrating how such a calculation is made, consider the linearized spectral power density and spectral luminous efficiency curves in Fig. 3.7. To obtain the visual effect of a given source, we multiply the spectral power density curve of the source (M_λ versus λ), point by point, by the spectral luminous efficiency curve of the eye [$V(\lambda)$ versus λ]. In Fig. 3.7 this multiplication is accomplished easily because we have used straight-line representations for both the source and the eye curves. The resulting product curve, $M_\lambda V(\lambda)$ versus λ, is shown in Fig. 3.7*c*. We call the quantity $M_\lambda V(\lambda)$ spectral light power density, with units of light watts (lW) per square centimeter per micrometer. The area under this curve is the total light watts emitted by the source, and for the example, is $\frac{1}{2}(1)(0.1) + \frac{1}{2}(3)(0.1) = 0.20$ lW/cm^2.

We must now relate light power density [$M_\lambda V(\lambda)$] to lumen density. To do this we return to the blackbody standard and recall that is has a luminance of 60 cd/cm^2 at 2042 K. Since it is perfectly diffusing, its luminous exitance is 60π lm/cm^2 at that temperature. If we plot the blackbody spectral power curve for 2042 K and multiply it point-by-point by the spectral lumi-

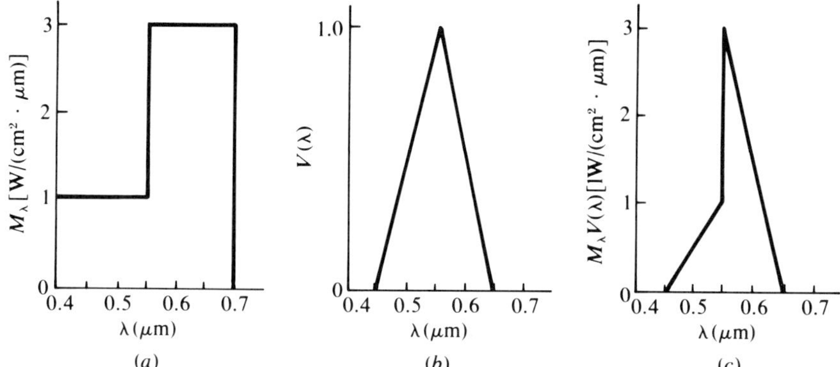

Figure 3.7 Illustrating the calculation of light watts.

nous efficiency curve, as we did in Fig. 3.7, we obtain the spectral light power density curve for the blackbody at that temperature. We then find the area under this curve to get the total blackbody light power density. This figure has been calculated accurately to be 0.27598 lW/cm².

We now require that 60π lm/cm² be equivalent to 0.27598 lW/cm². This means that each light watt must generate $60\pi/0.27598 = 683$ lm. Thus, for any source, it is necessary to multipy the number of light watts present by a constant $k = 683$ lm/lW* to get the number of lumens produced. The fictitious source in Fig. 3.7 provides $683(0.20) = 137$ lm/cm².

What k tells us is that if radiant power could be concentrated at the single wavelength of 555 nm, each watt of such power would yield one light watt and thus 683 lm of yellow-green light. Because radiant power is customarily distributed over a wide band of wavelengths, much of it in the infrared, and because the eye has less response at other wavelengths than it has at 555 nm and has essentially zero response outside the visible spectrum, the luminous efficacy in lumens per watt (lm/W) is generally far below the theoretical maximum. The source in Fig. 3.7 emits 0.6 W/cm² (the area under the M_λ graph). Since it also emits 137 lm/cm², its luminous efficacy is 228 lm/W.

For a radiant power density such that M_λ is constant throughout the visible spectrum and zero at all wavelengths outside the visible spectrum, the luminous efficacy is approximately 187 lm/W. To obtain this value, we divide the visible spectrum from 380 to 770 nm into thirty-nine 10-nm bands and assign $\frac{1}{39}$ W to each band. Then we can write

$$\Phi = 683 \sum_{380}^{770} \frac{1}{39} V(\lambda)\, \Delta\lambda$$
$$\equiv 683\, V(\lambda)_{\text{av}} \tag{3.18}$$

The summation of $V(\lambda)$'s is obtained from Table 3.1 as 10.6858. Then

*A similar constant has been developed for the scotopic case with the value 1754 scotopic lumens per watt and was adopted by the CIE in 1951.

$V(\lambda)_{av} = 10.6858/39 = 0.2740$ and $\Phi = 683(0.2740) = 187$ lm. Thus a single watt of power, uniformly distributed between 380 and 770 nm, yields 187 lm. In practice, even though M_λ may be reasonably constant throughout the visible spectrum, thus giving white light, M_λ is not zero in the infrared and the luminous efficacy is below the 187-lm/W value.

For general service incandescent lamps, so much of the radiation is in the infrared that only about 20 lm/W can be achieved in practice. For tungsten-halogen lamps, up to 35 lm/W can be obtained. Greater luminous efficacies can be realized with white luminescent sources. Mercury lamps yield up to 65 lm/W, fluorescent lamps up to 100 lm/W, metal halide lamps up to 125 lm/W, and high-pressure sodium lamps up to 140 lm/W.*

The development leading to the constant k allows us to derive expressions for the illumination quantities in terms of their radiant power equivalents. Let M_λ be the average watts per unit of surface area per nanometer emitted by a radiator in the neighborhood of wavelength λ. Then $M_\lambda \, d\lambda$ is the watts per unit area emitted in a range of wavelengths $d\lambda$ wide centered at λ and $M_\lambda V(\lambda) \, d\lambda$ gives the light watts per unit area in this range of wavelengths. The total light watts per unit area are obtained by integration over the visible spectrum. Then multiplying by k converts the result to lumens per unit area, which is luminous exitance M. We have

$$\mathrm{M} = k \int_{380}^{770} M_\lambda V(\lambda) \, d\lambda \tag{3.19}$$

Similar expressions follow for luminous flux, illuminance, and quantity of light.

$$\Phi = k \int_{380}^{770} P_\lambda V(\lambda) \, d\lambda \tag{3.20}$$

$$E = k \int_{380}^{770} E_\lambda \, V(\lambda) \, d\lambda \tag{3.21}$$

$$Q = k \int_{380}^{770} W_\lambda V(\lambda) \, d\lambda \tag{3.22}$$

3.7 SELECTIVE RADIATORS AND COLOR TEMPERATURE

No radiator of a practical sort is truly a blackbody. We define a graybody radiator as one whose radiation curve is the same in shape as that of a blackbody for a given temperature but has a lower radiated spectral power density at each wavelength. For a graybody radiator, Eq. (3.11) becomes

$$M = \varepsilon \, M_{BB} = \varepsilon \sigma T^4 \tag{3.23}$$

where M_{BB} is the spectral power density of a blackbody and ε is called the

*Luminescent sources require auxiliary equipment which consumes power and thus reduces overall efficiency.

emissivity of the graybody and is a number between 0 and 1. If the radiator is almost a blackbody, ε is almost 1. If the source is a poor radiator, ε is near 0.

A selective radiator is one whose radiation curves differ in shape from those of a blackbody. For example, tungsten is a selective radiator, as shown in Fig. 3.8. It matches a graybody radiator in the visible and near infrared regions quite well but falls considerably below the graybody in the far infrared. Equation (3.23) does not apply to a selective radiator. Rather we must return to Eq. (3.3) and rewrite Planck's equation to read

$$M_\lambda = \varepsilon(\lambda) M_{\lambda\text{BB}} = c_1 \varepsilon(\lambda) \lambda^{-5} (e^{c_2/\lambda T} - 1)^{-1} \tag{3.24}$$

where $\varepsilon(\lambda)$ is the spectral emissivity of the selective radiator. Thus a selective radiator has a different emissivity at each wavelength whereas the graybody radiator has a single emissivity for all wavelengths.

It is possible to define a weighted-average emissivity for a selective radiator as the ratio of the total radiated power density of the selective radiator to the total radiated power density of a blackbody operating at the same temperature.

$$\bar{\varepsilon} = \frac{\int_0^\infty \varepsilon(\lambda)\, M_{\lambda\text{BB}}\, d\lambda}{\int_0^\infty M_{\lambda\text{BB}}\, d\lambda} \tag{3.25}$$

Then the total radiated power density of the selective radiator can be expressed in the form of Eq. (3.23) as

$$M = \bar{\varepsilon}\sigma T^4 \tag{3.26}$$

For tungsten operating at 2900 K, $\bar{\varepsilon} = 0.346$, as indcated in Fig. 3.8. Inserting this value into Eq. (3.26) gives $M = 139$ W/cm² for tungsten at this temperature.

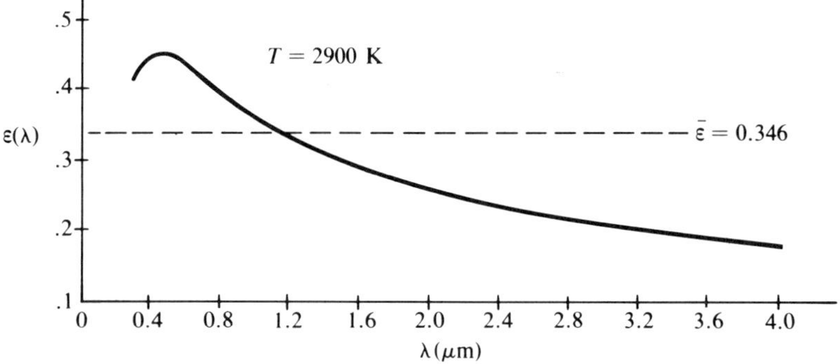

Figure 3.8 Spectral emissivity of tungsten.

EXAMPLE 3.4

An idealized radiation curve of a selective radiator operating at 3000 K is shown in Fig. 3.9. Find the watts per square centimeter emitted by the radiator and its emissivity.

Solution. To find M, the radiant exitance from the radiator, we need the area under the radiation curve. This can be found, in this straight-line case, as the area of two triangles and one rectangle.

$$M = 30[(0.3)(0.5) + (0.8)(1) + (8.6)(0.5)] = 30(5.25) = 157.5 \text{ W/cm}^2$$

From Eq. (3.11), a blackbody at the given temperature would emit

$$M_{\text{BB}} = (5.67 \times 10^{-12})(81 \times 10^{12}) = 459 \text{ W/cm}^2$$

Thus the weighted-average emissivity, from Eq. (3.25), is

$$\bar{\varepsilon} = \frac{157.5}{459} = 0.34$$

In describing the color of radiated power, it is convenient to use the concept of color temperature (CT). We define this quantity as the absolute temperature of a blackbody radiator which provides a color match between the blackbody and the given radiator. A blackbody is black at room temperature (300 K), red at 800 K, yellowish-white at 3000 K, white at 5000 K, bluish-white at 8000 K, and blue at 60,000 K.

Generally, an exact color match between a blackbody and an arbitrary radiator is impossible. This is because, as will be shown in Chap. 5, two degrees of freedom are necessary to color match two radiators. In this case, we have only one variable, temperature. However, if the color to be matched does not differ appreciably from that of a blackbody, such as in the case of many selective radiators, the concept of correlated color temperature (CCT) can be used. CCT is defined as the absolute temperature of a blackbody at which the closest possible color match with the given radiator is obtained. For example, a 100-W incandescent lamp operating at a filament temperature of 2850 K has a CCT of 2870 K. This means that a blackbody must be operated at 2870 K to provide the closest possible color match to

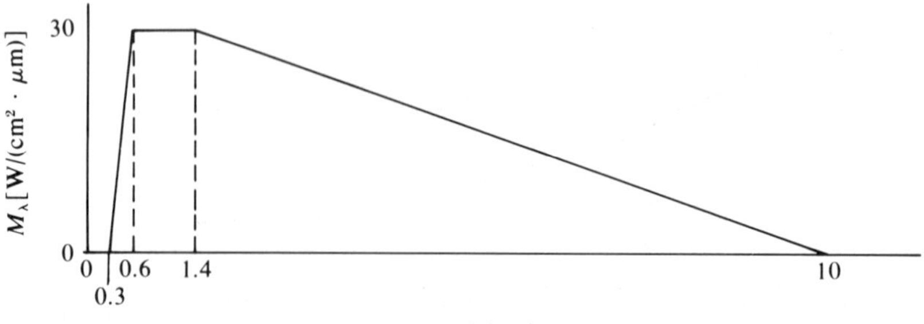

Figure 3.9 Idealized radiator spectral distribution.

the 100-W lamp at 2850 K. The reason for this is that the spectral emissivity of tungsten increases slightly as wavelength decreases in the visible portion of the spectrum.

When we consider luminescent sources and daylight, we are not dealing with heated objects giving off light. Still the concept of CCT may be used. For example, northwest blue sky has a CCT of 25,000 K, which means that a blackbody would have to operate at this temperature to best match the color of this sky. In the area of luminescent sources, a warm white fluorescent lamp has a CCT of 3000 K, whereas the CCT of a cool white fluorescent lamp is 4200 K. The former gives yellowish-white light; the latter more bluish-white.

3.8 SPECTRORADIOMETRIC MEASUREMENTS

In previous sections of this chapter, we have shown how to calculate the power density and spectral power density of a light source. Let us now consider how to measure these quantities.

An elementary radiometer for measuring power density is shown in Fig. 3.10. This device is a radiation thermocouple and consists of two junctions (J_1 and J_2) connected by fine wire and including an indicating instrument such as a galvanometer G. The junction J_2 is maintained at a constant temperature. The junction J_1 is heated by the incoming radiation and its temperature is directly proportional to the power density of that radiation. The temperature difference between the two junctions causes a small voltage to be induced in the circuit which then produces a current in the galvanometer causing its needle to deflect. The deflection is a measure of the power density and the galvanometer scale can be labeled to give this density directly in microwatts per square centimeter.

The radiation thermocouple is an example of a radiometer whose detector is relatively nonselective in wavelength response. Many modern radiometer/photometer combinations use solid-state sensors, such as photodiodes and phototransistors, which are wavelength selective. Also, their overall response is often modified by optical filters, either to produce a

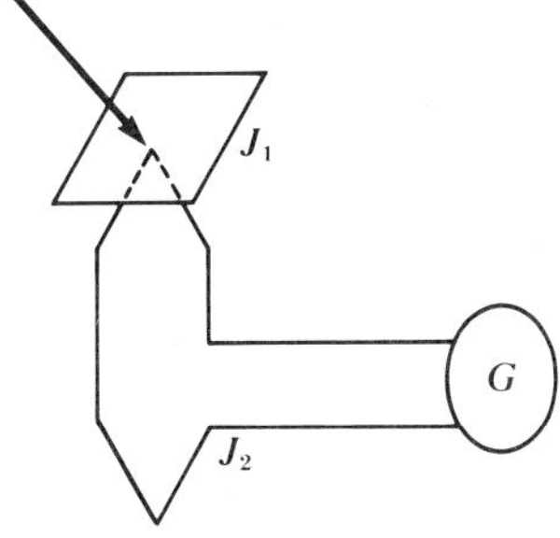

Figure 3.10 A radiation thermocouple.

TABLE 3.2

Detector	Wavelength range
Photomultiplier tube	125–1100 nm
Silicon photodiode	200–1200 nm
Intrinsic germanium	0.9–1.5 μm
Lead sulfide	1.0–6.5 μm
Indium arsenide	1.0–3.6 μm
Indium antinomide	2.0–5.4 μm
Doped germanium	2.0–38 μm
Mercury cadmium telluride	8–13 μm

specified response over a narrow range of wavelengths, such as duplicating the eye's spectral luminous efficiency curve, or to level the detector's response over a wide range of wavelengths. The overall range of spectral response of common detector types is shown in Table 3.2.

Contemporary radiometer/photometers employ solid-state digital logic and operational amplifier technology throughout. This provides very high sensitivity, excellent linearity over a wide range, stability and reliability over extended periods of time, and good signal-to-noise ratio.

If it is desired to measure spectral power density, a spectroradiometer or spectrophotometer is required. The former measures radiant flux as a function of wavelength whereas the latter measures the reflectance and transmittance of surfaces and media as a function of wavelength. The instruments are very similar, with the former designed to measure the spectral content of a source and the latter the spectral reflectance or transmittance of a sample in the presence of a known source.

All spectrophotometers are composed of four basic components:

1. A light source
2. A light-dispersing system, called a monochromator
3. A detector system
4. A sample-processing system

Item 1 is absent if the device is used as a spectroradiometer.

An elementary spectroradiometer is shown in Fig. 3.11. Radiation from a source enters the instrument at the left. It is imaged at the slit S_1 by a condensing lens system L_1. Then it enters a glass prism P, where two things take place. First the entire beam is refracted from the path it previously followed. Second, the phenomenon of dispersion occurs, in which the radiation is separated into its spectral components. The reason for the dispersion lies of course in the fact that the prism glass has a different refractive index for each wavelength, bending red wavelengths the least and violet wavelengths the most. The rays emerging from the prism are somewhat diverging, so a focusing lens system L_2 converges them so that they are parallel. They then enter the slit S_2 and those that pass through impinge on a detecting

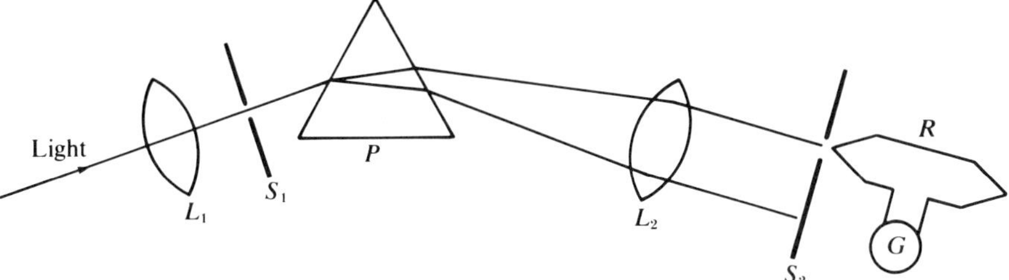

Figure 3.11 A spectroradiometer.

device, such as a radiation thermocouple R. The slit is adjustable in width but is generally set at 2 to 4 nm so that the incoming spectrum can be analyzed in reasonably fine detail. The width used depends on such factors as detector sensitivity, signal strength, scan speed, and required spectral resolution. The slit and sensor are movable through a fine gear control.

The galvanometer is calibrated to read in watts per square centimeter per nanometer or per 10 nm, regardless of the actual slit width. This is to provide easy comparison of data from one spectral curve to another.

Let us discuss the parts of a spectrophotometer in greater detail. First consider choosing a source. The most commonly used sources and their wavelength ranges are shown in Table 3.3. Xenon sources have a high source luminance and small size and thus, when used with relatively narrow slits, produce a high monochromator output. Mercury sources are similar in intensity to xenon sources but, because they produce line spectra, are not suitable for wavelength scanning applications. Tungsten-halogen lamps have a wider wavelength range than xenon lamps, but have much larger source size and less luminance. Deuterium sources are ideal for ultraviolet spectroradiometry since they have small source size and very little long wavelength radiation. Silicon carbide resistors provide approximate blackbody radiation at 1000 to 1200 K and are useful for infrared spectroradiometry.

The function of the monochromator is to separate the source radiation into intense narrow wavelength bands of 10 nm or less. A typical monochromator is shown in Fig. 3.12. The input lens-aperture system is designed to perform two main functions:

1. Send as much radiation as possible through the slit S_1.

TABLE 3.3

Source	λ Range
Xenon	200–700 nm
Mercury	Line spectrum
Tungsten-halogen	300–2500 nm
Deuterium	180–400 nm
Silicon carbide	2–30 μm

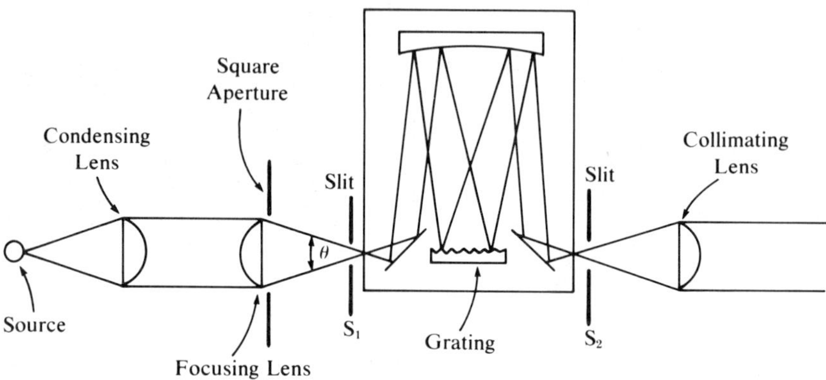

Figure 3.12

2. Just fill the acceptance cone of the monochromator. This is the cone defined by the angle θ in Fig. 3.12.

Radiation emerging from S_1 is reflected to a concave mirror and then to a diffraction grating, which is used instead of a prism in most contemporary monochromators because it produces a linear dispersion. Following reflection from the concave mirror, the beam exits through S_2 and is collimated by a lens placed one focal length from S_2. A subsequent optical system directs it to the detector.

Detector types have been presented previously in Table 3.2. In contemporary spectroradiometers, the detector system is entirely automated, with the spectral scanning and the detector output signal amplification and processing all under microcomputer control.

3.9 LINE SPECTRA

In dealing with temperature radiators such as blackbodies, continuous spectra result. There is a second type of spectrum, called a line spectrum, where the radiant power density is concentrated at certain wavelengths, with no radiant power density in between. Such spectra, which will be discussed in Chap. 4, are typical of luminescent light sources. An example is the line spectrum of a mercury vapor lamp shown in Fig. 3.13*a*. This lamp has distinct spectral lines of varying power densities at 365, 405, 436, 546, and 578 nm.

It is incompatible to plot spectral power density from a continuous spectrum and power density from a line spectrum on the same graph, since the units differ. To overcome this difficulty, lines are plotted as finite-width rectangles on the spectral power density plot, with the height of each rectangle such that its area gives the power in the line. Thus the spectral lines in Fig. 3.13*a* are shown as bands of 10 nm width in Fig. 3.13*b*, where the height of each band is set so that the area under each rectangle gives the watts per square centimeter in that spectral line. Also in that diagram, we show that a continuous spectrum and a line spectrum can exist simultaneously in a given lamp. This is true of fluorescent lamps, which have the mercury lines and, in addition, a continuous spectrum due to the phosphor on the inside of the bulb. It is also true of color-corrected high-intensity discharge lamps, where a phosphor is added to supplement the mercury or metal halide lines.

EXAMPLE 3.5

A mercury lamp has four lines in the visible spectrum at 404.7, 435.8, 546.1, and 578.0 nm with radiant power densities of 1.0, 0.8, 0.1, and 0.2 mW/cm^2, respectively. Assuming that 30% of the input power density to the lamp is in these visible lines, compute the lumens per square centimeter emitted by the lamp and the lamp's luminous efficacy in lumens per watt.

Solution. It is necessary first to obtain the light watts per square cen-

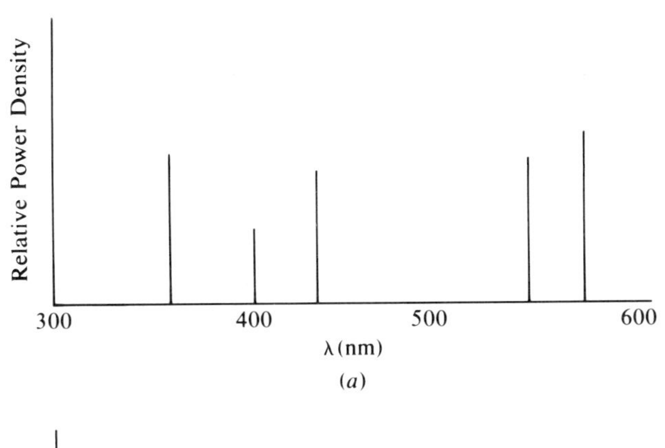

(*a*)

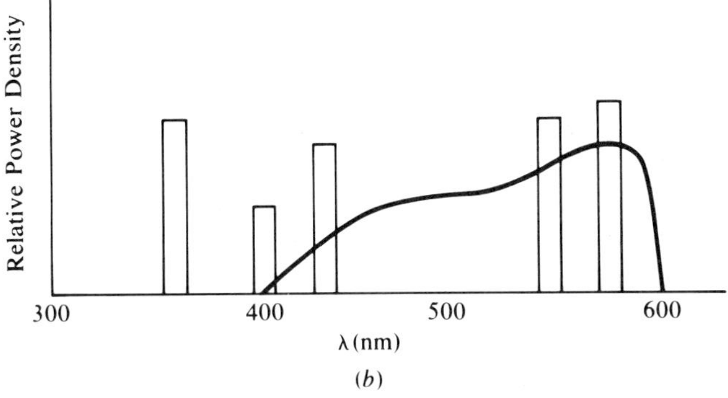

(*b*)

Figure 3.13 Line spectra.

timeter emitted by the lamp. This is done in tabular form as follows:

λ_i(nm)	M_i(mW/cm^2)	V_i	V_iM_i[(light mW)/cm^2]
404.7	1.0	0.0008	0.0008
435.8	0.8	0.018	0.0144
546.1	0.1	0.979	0.0979
578.0	0.2	0.886	0.1772
	2.1		0.2903

The values for V_i are read from Table 3.1. We note that

$$\Sigma M_i = 2.1 \text{ mW/cm}^2 \qquad \text{and} \qquad \Sigma V_iM_i = 0.2903 \text{ (light mW/cm}^2$$

Thus

$$\Phi = 683\ (0.2903 \times 10^{-3}) = 0.198 \text{ lm/cm}^2$$

With 30% of the input power density in the visible lines, the total input power density is 7.0 mW/cm^2. This yields a luminous efficacy of $\frac{198}{7} = 28$ lm/W.

EXAMPLE 3.6

An idealized spectral power distribution curve (watts per square centimeter per 10 nanometers) of a combined continuous spectrum and line spectrum source is shown in Fig. 3.14. Find the watts per square centimeter emitted by the source.

Solution. We must find the areas under the continuous and line spectrum graphs and add them.

Continuous: $M = \frac{1}{2}(10)(20) + \frac{1}{2}(10)(10) = 150$ W/cm^2
Line: $M = 8(1) + 6(1) = 14$ W/cm^2
Total: $M = 164$ W/cm^2

3.10 THE SUN

Before the Arab oil embargo in 1973, there was little interest in North America in daylighting. Energy was cheap and plentiful and thus there was next to no motivation to utilize sunlight and skylight in interior spaces to provide working levels of illumination. All of that has changed in the intervening years.

The sun is of course our source of daylight. Although a relatively small star, it nevertheless has a huge reservoir of energy. Its visible surface is 864,000 miles in diameter. It has a volume 1.3×10^6 as great as the earth and its mass is 0.3×10^6 that of the earth. The sun's average surface temperature is 5750 K. It emits radiant energy at the rate of 6.25 kW/cm^2. It's relatively high temperature gives it a luminous efficacy of 90 lm/W.

The earth receives only one part in 10^9 of the total energy radiated by

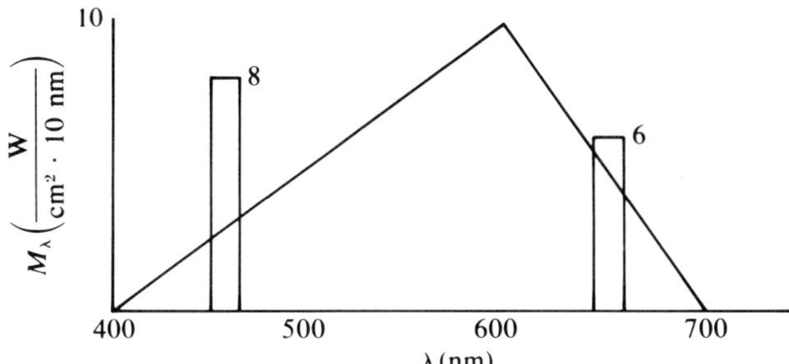

Figure 3.14

the sun. This amounts to 0.135 W/cm^2 outside our atmosphere and 0.120 W/cm^2 at sea level on a clear day, giving a yearly energy intake of over 10^{18} kWh.

The solar radiation on a horizontal surface at sea level on a clear day with the sun at zenith is shown in Fig. 3.15. About 40% of this received power is in the visible region of the spectrum and 55% is in the infrared. The peak of the sunlight radiation curve outside the atmosphere is at 475 nm. As Fig. 3.15 shows, this peak is shifted slightly, to about 500 nm, for zenith solar radiation at sea level. Curves for late afternoon and early norming would show considerably greater shift. This, of course, is due to selective absorbtion of solar radiation by the earth's atmosphere and accounts for the orange appearance of the sun early and late in the day. An additional effect is that the angle of incidence of the sun's rays is much greater early and late in the day than it is at noon. This reduces the incident power density at all wavelengths.

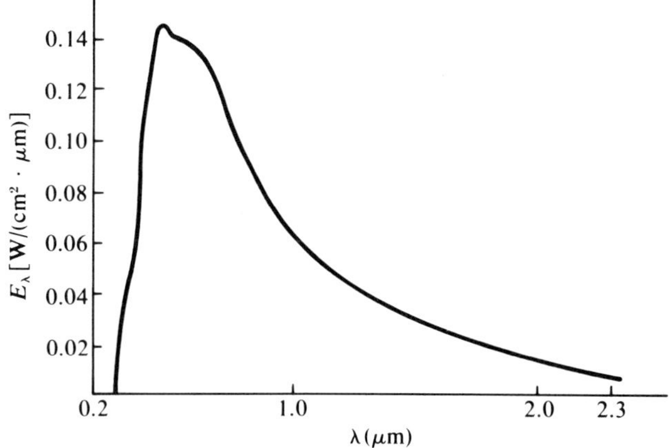

Figure 3.15 Horizontal solar radiation at sea level on a clear day with the sun at zenith. (*Source:* This figure is adapted from P. Moon, *The Scientific Basis of Illuminating Engineering,* McGraw-Hill, New York, 1936, with permission from Dover Publications, Inc.)

We can see that sunlight is an extremely variable light source, with respect to both overall power density and spectral power density distribution. When we combine this variability with that introduced by clouds, we see that the proper daylighting of an interior space is indeed a challenge. We will discuss daylighting in Chap. 9.

REFERENCES

Bitter, F.: *Currents, Fields and Particles,* Wiley, New York, 1957.

Boast, W.B.: *Illumination Engineering,* McGraw-Hill, New York, 1953.

Gilbert, R.G.: *Optimizing Monochromator Performances,* Oriel Corporation, Stanford, Conn.

Kaufman, J.E.: *IES Lighting Handbook*, Reference volume, Illuminating Engineering Society, New York, 1981.

Moon, P.: *The Scientific Basis for Illuminating Engineering,* McGraw-Hill, New York, 1936.

Sears, F.W., and M.W. Zemansky: *University Physics*, Addison-Wesley, Cambridge, Mass., 1955.

Waymouth, J.F., and R.E. Levin: *Designers Handbook—Light Source Applications,* GTE Products Corp., Danvers, Mass., 1980.

Chapter 4

Radiant Energy as Particles

4.1 DISCREPANCIES IN WAVE THEORY

In Chap. 1 we pointed out that as the year 1900 approached, the electromagnetic theory of Maxwell was widely accepted and it was felt that very little additional information would be added in the future to our knowledge of light.

However, there were two major radiation phenomena (and a few phenomena in other areas of physics) that were not explained by Maxwell's wave theory. One of these was the radiation from a blackbody. As discussed in Chap. 3, Planck's work in deriving the equation for a blackbody radiator led him to the conclusion that light energy is contained in discrete packets called quanta, with each quantum having an energy in joules given by

$$W = hf = \frac{hc}{\lambda} \tag{4.1}$$

A second phenomenon involving electricity and light that was not explained by electromagnetic wave theory was the photoelectric effect, the emission of electrons from the surfaces of metals when illuminated by visible energy. The experimental evidence concerning the photoelectric effect was as follows:

1. When a beàm of light falls on a photosensitive surface, electrons are emitted from the surface with velocities ranging from 0 to some value $v_{\max}$.
2. $v_{\max}$ is independent of the intensity of the incident light but is inversely proportional to its wavelength. Thus a weak beam of

short wavelength causes a greater emission velocity than a strong beam of long wavelength.

3. The photoelectric current increases as the intensity of the incident light increases.
4. For a given photoelectric surface, there is no emission of electrons if the wavelength of the incident light is greater than λ_c, where λ_c is defined as the critical wavelength for that surface.

This experimental evidence was contrary to electromagnetic theory. According to electromagnetic theory, the electric field of the light wave should impart energy to the electrons in the metal surface. An increase in the intensity of the light (an increase in the electric field) should increase the energy imparted to the electrons and thus increase their emission velocities from the surface. This is not at all what occurred.

In 1905, Albert Einstein (1879–1955), drawing on Planck's earlier work in 1900, extended that work to cover not only emission and absorption of radiation but also all other properties of radiant energy when such energy occurs in small packets. Einstein published three papers in 1905 which revolutionized physics and provided the foundation for the atomic age. One of these, which formed the cornerstone of what was soon to be called "quantum theory," was an explanation of the photoelectric effect. Using Eq. (4.1), Einstein stated that a single quantum of radiant energy may collide with an electron in the surface of a metal and transfer its energy to the electron in an all or none exchange. If the energy exchange takes place, the photon ceases to exist and the electron is ejected from the metal with a velocity given by

$$\tfrac{1}{2}\, m_e v^2 \leq hf - W_m \tag{4.2}$$

In this inequality, the term on the left is the kinetic energy of the departing electron and v is its ejection velocity. On the right side of the inequality, hf is the energy of the incident photon and W_m is the "work function" of the metal, the least amount of energy required to eject an electron from the surface of the metal. The inclusion of the inequality sign in Eq. (4.2) can be explained by noting that, in a metal at a given temperature, electrons exist at all possible energies from 0 to some maximum value. If a photon collides with an electron having the maximum possible energy, the equal sign in Eq. (4.2) holds. Otherwise the inequality sign holds; that is, an energy greater than W_m is required to eject the electron.

We can now explain the four pieces of experimental evidence concerning the photoelectric effect.

1. If a beam of monochromatic light with hf greater than W_m impinges on a metal surface, electrons will be emitted with velocities up to a maximum of

$$v_{\max} = \sqrt{\frac{2(hc/\lambda - W_m)}{m_e}} \tag{4.3}$$

2. $v_{\max}$ is independent of the intensity of the light, because intensity

does not appear in Eq. (4.2), but is inversely proportional to wavelength.

3. The photoelectric current is the number of electrons per second ejected from the surface. As the intensity of the beam increases, the number of photons per second increases, which increases the number of electrons per second ejected.
4. There is no ejection of electrons if the wavelength of the incident radiation is above a certain value. To find that value, require that $v = 0$ in Eq. (4.2) and obtain

$$\lambda = \frac{hc}{W_m} \equiv \lambda_c \tag{4.4}$$

It is tempting to suggest that we have come full circle—from the corpuscular (particle-oriented) theory of Newton to the electromagnetic (wave-oriented) theory of Huygens and Maxwell to the quantum (particle-oriented) theory of Planck and Einstein. But this would be an oversimplification. The quanta of Planck and Einstein are quite different from the corpuscles of Newton and, in fact, do not even obey Newton's laws of motion.

As suggested in Sec. 1.5, it turns out that we need both electromagnetic wave theory and quantum theory to adequately describe light and other radiant energy. In other words, we need to think of light as being dualistic in nature. Light propagation from source to receiver is explained best by electromagnetic wave theory, while the interaction of light and matter is best understood by quantum theory.

4.2 ELECTRONVOLTS

In dealing with energies of electrons, a more convenient measure of energy is the electronvolt (eV), rather than the joule. In electrical terms, a joule is the energy gained by a *coulomb* of electric charge as it falls through a voltage of one volt. Similarly, an electronvolt is the energy gained by an *electron* in falling through a voltage of one volt. Since $q_e = -1.60 \times 10^{-19}$ C is the charge on the electron, the joule and the electronvolt are related by

$$1 \text{ eV} = 1.60 \times 10^{-19} \text{ J} \tag{4.5}$$

With the electronvolt defined, Eq. (4.1) can be rewritten as

$$W = \frac{hc}{\lambda} \text{ joules} = \frac{hc}{q_e \lambda} \text{ eV} = \frac{1240}{\lambda} \text{ eV} \tag{4.6}$$

where λ is in nanometers.

EXAMPLE 4.1

A tungsten surface having a work function of 4.52 eV is irradiated by a quantum of radiation from mercury at 253.7 nm. Find the maximum velocity of the emitted electron.

Solution. Using Eq. (4.2), with the equal sign, and Eq. (4.6), we have

$$\tfrac{1}{2}\, m_e v^2 = \frac{1240}{\lambda} - W_m$$

where the energies are expressed in electronvolts. Inserting values gives

$$\tfrac{1}{2}\, m_e v^2 = \frac{1240}{253.7} - 4.52 = 0.36 \text{ eV} = 0.576 \times 10^{-19} \text{ J}$$

Then

$$v_{\max} = \sqrt{\frac{2(0.576 \times 10^{-19})}{9.11 \times 10^{-31}}} = 3.56 \times 10^5 \text{ m/s}$$

EXAMPLE 4.2

One-tenth of one percent of the electric energy supplied to a 50-W mercury lamp appears in the spectral line at 253.7 nm. Find the number of photons per second emitted by the lamp at this wavelength.

Solution. 0.001(50) = 0.05 W emitted at 253.7 nm. From Eq. (4.1)

$$W = \frac{hc}{\lambda} = \frac{(6.63 \times 10^{-34})\,(3 \times 10^8)}{253.7 \times 10^{-9}} = 0.784 \times 10^{-18} \text{ J/photon}$$

We now know the number of joules per second (watts) emitted and the number of joules per photon. Their ratio gives the number of photons per second.

$$\frac{0.05}{0.784 \times 10^{-18}} = 6.38 \times 10^{16} \text{ photons/s}$$

Note the immensity of this number.

In Sec. 1.5 we reviewed the Rutherford-Bohr model of the atom and developed Table 1.1 to show the energy levels of hydrogen in joules. These levels are converted to electronvolts and the 0 energy reference is shifted from $n = \infty$ to $n = 1$ in Table 4.1. This causes all energy levels to be positive,

TABLE 4.1 ENERGY LEVELS OF HYDROGEN

n	TE (eV)	TE + 13.53 (eV)
1	−13.53	0
2	− 3.38	10.15
3	− 1.50	12.03
4	− 0.85	12.68
5	− 0.54	12.99
∞	0	13.53

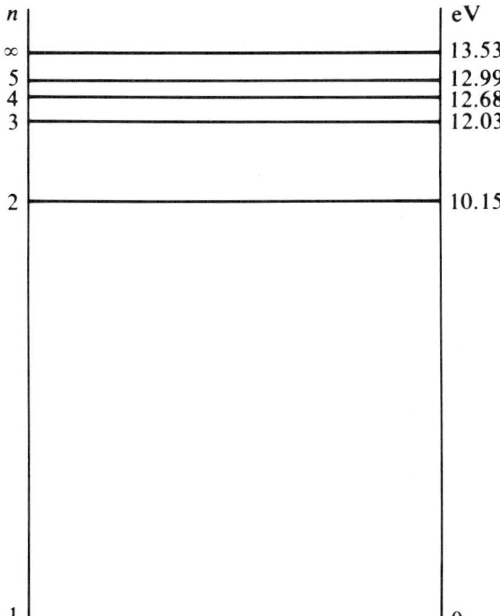

Figure 4.1 Energy levels of hydrogen.

making it easier to calculate transitions from one orbit to another. Table 4.1 can be used to construct an energy-level diagram for hydrogen as shown in Fig. 4.1.

Equation (1.18) gives the energy of a photon resulting from a transition from one energy level in an atom to a lower one. Equation (4.6) is the relationship between the energy of a photon in electronvolts and its wavelength in nanometers. Combining these two equations gives

$$\lambda = \frac{1240}{W_i - W_f} \tag{4.7}$$

as the wavelength in nanometers of a photon emitted when an electron in an orbit with energy W_i electronvolts falls toward the nucleus to an orbit of lesser energy W_f electronvolts.

EXAMPLE 4.3

How fast must an electron be traveling at the instant it collides with a mercury atom to cause a 253.7-nm photon to be emitted.

Solution. From Eq. (4.7),

$$W_i - W_f = \frac{1240}{253.7} = 4.88 \text{ eV} = 7.82 \times 10^{-19} \text{ J}$$

Equating this to the kinetic energy of the arriving election yields

$$7.82 \times 10^{-19} = 0.5(9.11 \times 10^{-31})v^2$$
$$v = 1.32 \times 10^6 \text{ m/s}$$

EXAMPLE 4.4

An electron traveling at 2×10^6 m/s collides with a hydrogen atom. Find:

(*a*) The energy level to which the valence electron is likely to be raised.
(*b*) The λ of the photon emitted upon impact.
(*c*) The velocity of the departing electron.

Solution.

(*a*) The KE of the arriving electron is

$$\text{KE} = 0.5(9.11 \times 10^{-31})\,(4 \times 10^{12}) = 1.822 \times 10^{-18}\text{ J} = 11.39\text{ eV}$$

From Fig. 4.1, this can raise the hydrogen electron to the $n = 2$ level (10.15 eV).

(*b*) Assuming the hydrogen valence electron returns to its rest state ($n = 1$),

$$\lambda = \frac{1240}{10.15} = 122\text{ nm}$$

(*c*) The KE remaining after the collision is $11.39 - 10.15 = 1.24\text{ eV} = 1.984 \times 10^{-19}$ J. Thus

$$1.984 \times 10^{-19} = 0.5(9.11 \times 10^{-31})v^2$$
$$v = 0.66 \times 10^6\text{ m/s}$$

4.3 EXTENSIONS OF BOHR'S MODEL

The Bohr model was very successful in explaining the observed spectral lines of hydrogen and a few other elements, but was unsuccessful in explaining the spectra of multi-electron atoms and of molecules. Moreover, even for hydrogen, it gave incorrect values for angular momentum.

At this point, research proceeded along two paths:

1. Extensions of the Bohr model by Sommerfeld, Zeeman, Goudsmit, and Uhlenbeck
2. Development of wave and quantum mechanics by DeBroglie, Schrodinger, Heisenberg, and Dirac

We will review the first of these developments in this section and the second in the following section.

It soon became obvious that a single quantum number n is not sufficient to describe all that goes on in each electron orbit. Thus it was necessary to modify Bohr's work in the following ways:

1. Elliptical, rather than circular, orbits were assumed, with the nucleus at one of the foci of the ellipse.

2. The spin of an electron about its axis was taken into account.
3. A three-dimensional, rather than two-dimensional, model of the atom was allowed, in which the planes of the electron orbits were inclined at certain angles with respect to the earth's magnetic field.

Each of these modifications resulted in a separate quantum number, so that each electron ended up with four quantum numbers, one to describe its main energy level, one to describe its angular momentum and the shape of its orbit, one to describe the inclination of its orbit to the earth's magnetic field, and the last to describe its spin and spin momentum. The measurable effect of these refinements in the single-quantum-number Bohr model is to split each main energy level into several levels of differing energy, thus splitting the spectral lines.

The first change in the Bohr model is to allow elliptical orbits, with the nucleus at one of the foci of the ellipse. This idea was introduced by Arnold Sommerfeld, a German physicist.

In a circular orbit, the distance of the electron from the nucleus never changes. Thus the electron has no velocity or momentum in a radial direction. With an elliptic orbit, the electron has two kinds of momentum, one angular and one linear. Sommerfeld postulated that each of these momenta must be quantized, and thus the elliptic orbit needs two quantum numbers to define it. These are specified in the following way:

1. n = principal quantum number = 1, 2, 3, ... This quantum number quantizes the main energy levels, as in the case of the Bohr circular orbit model.
2. $$l = \text{orbital quantum number} = 0, 1, 2, \ldots, n-1 \tag{4.8}$$

   This quantum number subdivides a main energy level into two or more different energy levels. Further, it defines the shape of the orbit and quantizes the electron's angular momentum. Its maximum value is $n - 1$ because it does not include the linear radial momentum and thus does not include the energy associated with this momentum.

It had been observed that an atom in a magnetic field had a more complicated emission spectra than the same atom in the absence of a magnetic field. This splitting of energy levels in the presence of a magnetic field is called the Zeeman effect after its discoverer, Pieter Zeeman, a Dutch physicist. It could not be explained by the two-dimensional Bohr model.

Actually it is quite reasonable that an atom should be considered as three-dimensional with the electron orbits at various inclination angles with respect to some reference axis. But what reference axis should be used? All atoms on earth are subject to the earth's magnetic field. An electron moving in this field will experience a force given by

$$F = q_e v B \tag{4.9}$$

where F = the force on the electron, in newtons
q_e = the charge on the electron in coulombs
v = the electron's velocity in meters per second
B = the density of the magnetic field in webers per square meter (teslas)

This force is always in a direction perpendicular to the direction of the earth's magnetic field. It constrains the orbit to assume only certain inclination angles with respect to this field. These angles are determined by the number m, which quantizes the components of orbital angular momentum in the direction of the field. It can be shown that the only permissible values of this quantum number are

$$m = 0,\ 1,\ 2,\ \ldots,\ \pm l \tag{4.10}$$

Thus for each value of l, there are $2l + 1$ values for m, each representing a slightly different energy level. For example, if $n = 2$ and $l = 1$, then $m = 0$, ± 1 and there could be three electrons in the $n = 2$, $l = 1$ energy level with slightly different total energies.

We can show diagrammatically what this quantizing of angular momentum in the direction of the earth's magnetic field means. Assume $l = 3$ and refer to Fig. 4.2 where each of the seven radial l arrows is 3 units long. The inclination of each of the arrows is such that its projection on the vertical axis, which is the direction of the earth's magnetic field is 0, ±1, ±2, or ±3. This agrees with Eq. (4.10) which says that when $l = 3$, m can have seven integer values only. To conclude, l and m should be carefully distinguished. l quantizes angular momentum in general while m quantizes that portion of angular momentum in the direction of the earth's magnetic field.

In 1925, George Uhlenbeck and Samuel Goudsmit, Dutch physicists, found that some of the earlier discrepancies in the elliptic-orbit model of Sommerfeld could be resolved if the electrons in the orbits were allowed to

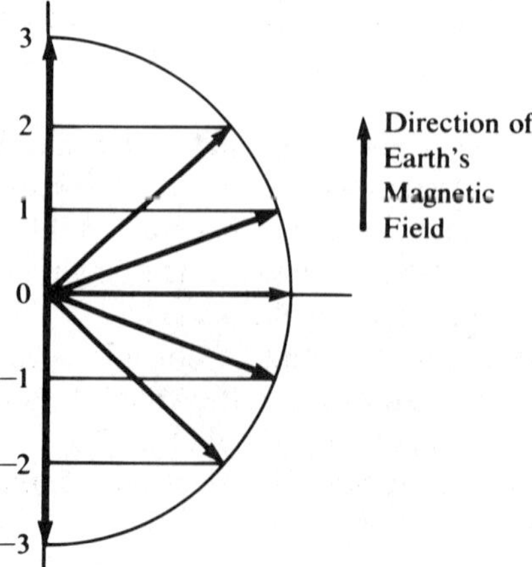

Figure 4.2

spin about their axes. This spinning produces a small magnetic field. It was assumed that the spin axis of all electrons was along the earth's magnetic field. Thus the small spin magnetic fields produced were either in the same direction as the earth's magnetic field or exactly opposite, depending on the direction of the electron's spin.

Like orbital angular momentum, spin angular momentum is quantized. The spin quantum number is s and has the two values $\pm\frac{1}{2}$, corresponding to permitted angular momenta, always along the direction of the earth's magnetic field, of $\pm\frac{1}{2}$ $(h/2\pi)$ joule-seconds.

From an energy standpoint, the inclusion of spin splits each energy level and sublevel in an atom into two energy levels very close together. The energy level for $s = +\frac{1}{2}$ is slightly higher than that for $s = -\frac{1}{2}$.

We now have four quantum numbers, n, l, m, and s, with which to describe the state of a given electron within an atom. Before we can describe the quantum states of all electrons in an atom, we need a very important principle postulated by Wolfgang Pauli, Austrian physicist (1900–1958), in 1925. This is the Pauli exclusion principle which states simply that no two electrons in an atom can have the same four quantum numbers.

Using the four quantum numbers, the restrictions of Eqs. (4.8) and (4.10), and the Pauli exclusion principle, we can construct the various shells and subshells permitted in an atom.

1. *First shell* ($n = 1$). When $n = 1$, the only possible value for l is 0. Thus m is 0. The quantum number s can be $\pm\frac{1}{2}$ so, from the Pauli principle, there can be a maximum of two electrons in the first shell.
2. *Second shell* ($n = 2$). When $n = 2$, l can be 0 or 1.
   - *(a)* For $l = 0$, $m = 0$, $s = \pm\frac{1}{2}$. This gives a maximum of two electrons in this subshell, as in the first shell.
   - *(b)* For $l = 1$, m can be $-1, 0, +1$ and s can be $\pm\frac{1}{2}$. This gives a maximum of six electrons in this subshell.

   Thus the $n = 2$ shell can have a maximum of eight electrons.
3. *Third shell* ($n = 3$). When n is 3, l can be 0, 1, or 2.
   - (a) For $l = 0$, two energy states are permissible as before.
   - *(b)* For $l = 1$, six energy states are permissible as before.
   - *(c)* For $l = 2$, m $= 2, -1, 0, 1, 2$ and $s = \pm\frac{1}{2}$. Thus there are ten permitted energy states.

   For the entire $n = 3$ shell, there can be a maximum of $2 + 6 + 10 = 18$ electrons.

From a lighting standpoint, we are interested in the atomic structure of the elements helium, neon, argon, krypton, xenon, sodium, mercury, indium, scandium, and thallium. In Table 4.2 we have summarized the distributions of the electrons among the energy shells for these elements. An atom always seeks its lowest energy state, so we should expect to find the inner shells filled before electrons are permitted in the outer shells. Thus most atoms have an inner core of tightly bound electrons surrounded by an outer shell

TABLE 4.2 DISTRIBUTION OF ELECTRONS IN THE LIGHTING ELEMENTS

Element	Atomic number	Main energy level n				
		1	2	3	4	5
Helium	2	2				
Neon	10	2	8			
Sodium	11	2	8	1		
Argon	18	2	8	8		
Scandium	21	2	8	9	2	
Krypton	36	2	8	18	8	
Indium	49	2	8	18	18+3	
Xenon	54	2	8	18	18+8	
Mercury	80	2	8	18	32	18+2
Thallium	81	2	8	18	32	18+3

containing the remaining electrons. It is generally these outer-shell electrons that are affected by collisions of the atom with fast moving electrons or protons.

To describe Table 4.2, let us consider one or two of its members. Sodium has an atomic number of 11. Thus it normally has 11 electrons and these are distributed two in the first shell, eight in the second shell, and one in the third shell. Thus sodium has a 10-electron core surrounded by one outer-shell (valence) electron. Mercury has an atomic number of 80. It has an inner core of $2 + 8 + 18 + 32 = 60$ electrons. Beyond this core is the $n = 5$, $l = 0$ level containing two electrons; the $n = 5, l = 1$ level with six electrons; and the $n = 5$, $l = 2$ level with 10 electrons. This makes a total of $60 + 18 = 78$ electrons and leaves two remaining valence electrons in the $n = 5, l = 3, m = 0, s = \pm\frac{1}{2}$ subshell of the fifth main shell.

4.4 QUANTUM MECHANICS AND QUANTUM THEORY

Quantum theory occupied the first quarter of the twentieth century, quantum mechanics the second. The work of Planck, Einstein, Rutherford, Bohr, Sommerfeld, and others, which we have previously discussed, forms the basis for quantum theory. It is not our purpose in this section to begin a lengthy discussion of quantum mechanics. Rather we will explain briefly what it entails and how it differs from its simpler, more intuitive, and more familiar predecessor.

Quantum mechanics brings together the wave mechanics of the German physicist, Erwin Schrodinger, the matrix mechanics and the uncertainty principle of another German physicist, Werner Heisenberg, and Einstein's relativity theory, as related to quantum mechanics by P. A. M. Dirac, the British physicist. Quantum mechanics differs from quantum theory in the following ways:

1. It is essentially mathematical in nature, describing atoms in terms of partial differential equations rather than solar systemlike models.
2. It makes fewer arbitary assumptions. Quantum theory contained a number of these which matched experimental evidence.
3. It deals with readily observable and measurable spectral lines rather than with less-accessible electron orbits.
4. It does not rely on Newtonian mechanics but instead is self-contained and consistent.
5. It explains the intensities of spectral lines (which quantum theory does not).

In 1924, the French physicist Louis DeBroglie (1892–) postulated that all matter possesses wavelike properties. He stated that what he called "wave-particle duality" was not confined to photons but rather is a characteristic of all fundamental particles, such as electrons, atoms, protons, and molecules. For example, an electron under the influence of an electric field experiences an acceleration due to the force of the field, which is given by Newton's second law ($f = ma$). Thus an electron obeys the laws of particle dynamics. On the other hand, when a stream of electrons is focused on a grating of atomic dimensions (a group of very closely spaced parallel lines), the electrons exhibit diffraction patterns similar to those of light waves. Thus the electron appears dualistic in nature.

Reconsider the photon. Its energy is given by Eq. (4.1) and it travels at the speed of light. Following DeBroglie's postulate, a photon should behave like a particle and thus have mass. From Einstein's relativity theory (another of his famous papers of 1905), the equivalence of mass and energy can be simply stated as

$$W = mc^2 \tag{4.11}$$

The two energies for the photon should be equal:

$$mc^2 = \frac{hc}{\lambda}$$

Thus

$$\lambda = \frac{h}{mc} = \frac{h}{p} \tag{4.12}$$

Equation (4.12) states simply that the wavelength of a photon is Planck's constant divided by its linear momentum p.

Equation (4.12) was expressed for a photon. To apply it more generally, we should not limit ourselves to the velocity of light. Thus

$$\lambda = \frac{h}{mv} = \frac{h}{p} \tag{4.13}$$

is a more general expression for any particle. To sum up, the wavelength of any moving particle is the ratio of Planck's constant to the particle's linear momentum.

EXAMPLE 4.5

What is the wavelength of an electron that is traveling with a velocity of 10^4 m/s? What is the mass of a photon having that wavelength? What is the wavelength of a 200-g baseball traveling at 30 m/s?

Solution. From Eq. (4.13),

$$\lambda = \frac{h}{m_e v} = \frac{6.63 \times 10^{-34}}{(9.11 \times 10^{-31})\ 10^4} = 0.728 \times 10^{-7}\ \text{m} = 72.8\ \text{nm}$$

This radiation is in the far ultraviolet region of the spectrum.

From Eq. (4.12),

$$m = \frac{h}{\lambda c} = \frac{6.63 \times 10^{-34}}{(72.8 \times 10^{-9})\ (3 \times 10^8)} = 3.04 \times 10^{-35}\ \text{kg}$$

Note that the photon has a very tiny mass, much less than that of the electron.

For the baseball,

$$\lambda = \frac{6.63 \times 10^{-34}}{0.2(30)} \approx 10^{-34}\ \text{m}$$

This wavelength is too small for ordinary observation.

Recall that Bohr's third postulate stated that the angular momentum of an electron must be an integral multiple of $h/2\pi$. With DeBroglie's wave-particle duality, it is possible to derive this relationship, previously given in Eq. (1.19).

According to Eq. (4.13), a particle of mass m moving with a velocity v has a distinct wavelength associated with it. If the particle is an electron in a circular orbit around the nucleus of an atom, it is reasonable to assume that the circumference of the orbit must be an integral number of these electron wavelengths, as shown in Fig. 4.3, where four wavelengths per circumference are assumed. If this were not the case, an electron's wave motion

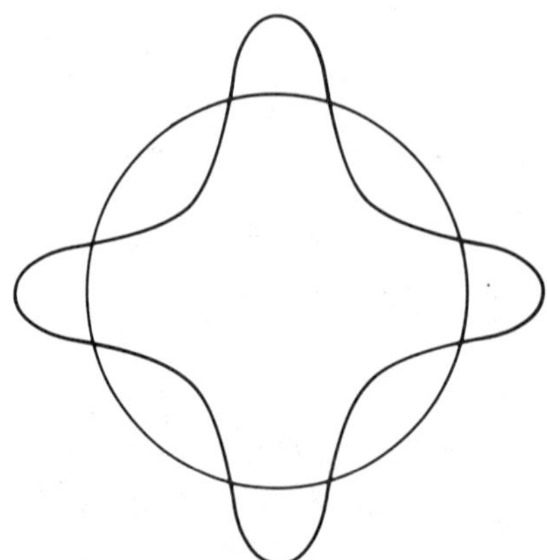

Figure 4.3

would tend to interfere with itself as the electron completed one circumference and started another. This quantization of the orbit may be stated mathematically as

$$2\pi r = n\lambda \tag{4.14}$$

Now combining Eqs. (4.13) and (4.14) yields the result in Eq. (1.19), which was anticipated by Bohr 12 years earlier.

DeBroglie's work was the first radical departure from classical mechanics. It was followed in 1926 by Schrodinger's development of a mathematical analysis of the atom which he called wave mechanics. Schrodinger essentially replaced all of quantum theory with a single, rather complex, partial differential wave equation. Schrodinger's equation had certain discrete ("eigenvalue") solutions which he showed were identical with the quantum numbers of quantum theory.

At about the same time, Heisenberg developed what he called matrix mechanics. Heisenberg's model of the atom was also mathematical, but was a matrix of rows and columns of quantities representing the locations and velocities of electrons within an atom. The rows were the quantum numbers of quantum theory. These matrices could be solved for the spectral lines of an atom.

Schrodinder subsequently showed that wave mechanics and matrix mechanics were different mathematical models of the same thing, and he blended the two theories into what has since become known as quantum mechanics.

It remained for Dirac to handle the case when electrons are moving at such speeds that relativistic effects become pronounced. In 1928 he developed a relativistic wave equation which properly treats the high-speed electron and also accounts for the spin of the electron about its axis.

Some physical interpretations of quantum mechanics can be given. In quantum mechanics, we still have the atomic nucleus. Surrounding it are groups of stationary waves, which have their maximum amplitudes at locations matching the orbits of quantum theory. An electron's location is determined probabilistically: there is a certain probability that an electron will occupy a given space at a given time. Each electron can be thought of as being somewhere between two spheres, with the mean radius between them the most probable distance of the electron from the nucleus.

4.5 EXCITATION, IONIZATION, AND RADIATION

Under normal conditions, an atom is electrically neutral. The number of orbiting electrons equals the number of protons in the nucleus and all orbiting electrons are in their lowest possible states (in those orbits nearest the nucleus).

If sufficient energy is added to an atom, through collision with an electron, another atom, or by some other means, one or more electrons in the atom may be hoisted to higher energy levels in orbits farther from the nucleus. We then say that the atom is in an *excited state* and we denote the energy so applied as the *excitation energy.*

The outermost electrons in an atom are the most likely to be excited. This is true because:

1. The next higher permitted levels are empty and thus there are no electrons tending to repel the valence electron back to its original level.
2. The difference in energy between the valence electron level and the first vacant level is small compared with the difference in energy between lower levels and the first vacant level or between lower levels and the valence electron level.

Once excited, the electron normally remains in its elevated state for a brief instant (10^{-8} s) before returning to a lower energy state and releasing a photon whose wavelength is given by Eq. (4.7). There are energy levels in atoms from which transitions cannot occur downward. These are called *metastable states.* An electron excited to or falling to a metastable state must sit there until it receives additional energy to lift it to a higher state, from which it can fall to a nonmetastable state. Residence in a metastable state generally lasts about 10^{-2} s.

If the energy supplied to the atom is sufficient to hoist an electron to the $n = \infty$ level, the electron will be separated from the atom and a positive ion will remain. We say that the atom has become *ionized* and we denote the least amount of energy required to do this as the *ionization energy.*

The ionization and first excitation energies for elements included in Table 4.2 appear in Table 4.3. Note that the ionization energies for the inert gases are higher than for the other elements. This is because the outermost level for each of the inert gases is a complete subshell and thus more energy is required to remove an electron than if this were not the case.

TABLE 4.3 EXCITATION AND IONIZATION ENERGIES OF THE LIGHTING ELEMENTS

Element	First excitation energy (eV)	Ionization energy (eV)
Helium	20.91	24.46
Neon	16.58	21.47
Sodium	2.10	5.12
Argon	11.56	15.68
Scandium	2.70	6.70
Krypton	9.98	13.93
Indium	2.75	5.76
Xenon	8.39	12.08
Mercury	4.66	10.39
Thallium	3.27	6.07

4.6 ENERGY LEVELS AND SPECTRA OF THE LIGHTING ELEMENTS

In this section we examine the energy level diagrams and emitted spectral lines of some of the elements commonly used in gaseous discharge light sources. We begin with sodium. Sodium's atomic number is 11, indicating an inner core of two electrons in the $n = 1$ shell and eight electrons in the $n = 2$ shell, with one valence electron in the $n = 3$ shell. As previously discussed, the valence electron is the one most likely to be involved in excitation or ionization of the sodium atom. For example, it requires 5.12 eV to remove this electron from sodium, whereas to remove a second electron requires 47 eV and to remove a third requires 71 eV.

In the normal state of sodium, the valence electron has the four quantum numbers, $n = 1, l = 0, m = 0, s = -\frac{1}{2}$. From this as the zero energy base level, the energy level diagram in Fig. 4.4 for sodium is built. There are several things to note about this diagram:

1. The $n = 3, l = 0$ energy level is labeled 0 eV. It is common practice to assign 0 eV to the level in which the valence electron resides when the atom is in its normal state (if there is more than one valence electron we assign 0 eV to the one with highest energy).

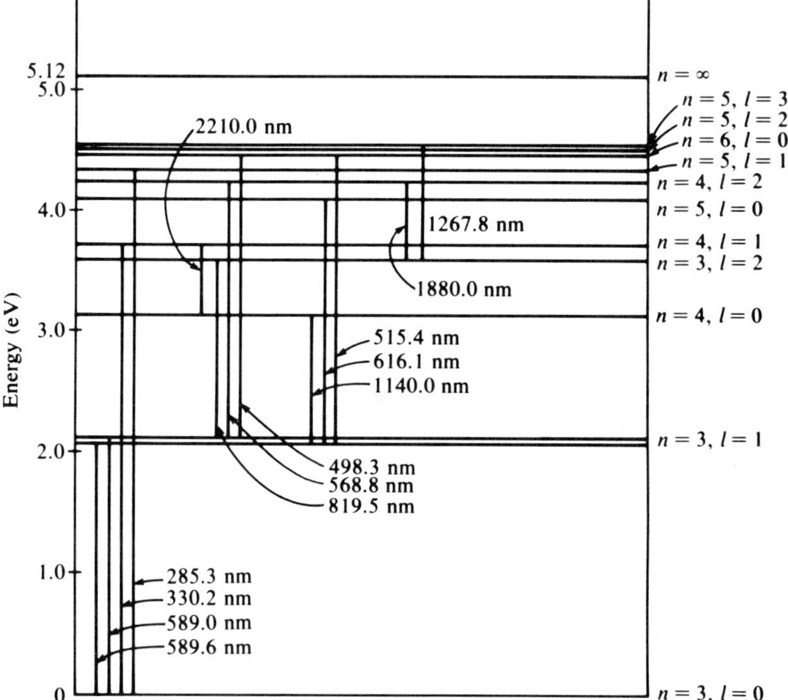

Figure 4.4 Energy level diagram of sodium.

Then all excited levels are a plus number of electronvolts above the normal state.

2. Only four $n = 3$ subshells are shown, whereas we know there are 18. Also only two $n = 3$, $l = 1$ subshells are shown, whereas we know there are 6. Actually all of the subshells do exist, but the earth's magnetic field is so weak that the energy differences between the $m = -1$, 0, and $+1$ levels are very small and they appear to merge into one level.
3. The $n = 4$, $l = 0$ level is below the $n = 3$, $l = 2$ level. This is a characteristic of the alkali metals (sodium, potassium, rubidium, caesium), which always have a single valence electron located in an $l = 0$ subshell immediately following a completed $l = 1$ subshell. What happens in many-electron atoms is that the main shells overlap. Actually the electrons are still in their lowest possible energy levels when the atom is in its normal state. Thus, the $n = 4$, $l = 0$ level is in reality a lower energy leven than the $n = 3$, $l = 2$ level for sodium.
4. In determining which transitions between energy levels can take place in an atom like sodium, certain "selection rules" apply regarding permitted changes in angular momentum, the most important of which is that in any transition, the value of l changes by either $+1$ or -1. The permitted transitions for sodium are shown in Fig. 4.4. The major transitions are the ones from $n = 3$, $l = 1$ to $n = 3$, $l = 0$ which produce the two predominant yellow lines in the sodium spectrum at 589.0 and 589.6 nm. No spectral lines occur between $n = 4$, $l = 0$ and $n = 3$, $l = 0$; between $n = 3$, $l = 2$ and $n = 3$, $l = 0$, etc., in agreement with the selection rule previously mentioned.

We turn next to mercury, whose core is a $2 + 8 + 18 + 32 + 18$ electron configuration. Outside of this are two valence electrons with opposite spins occupying the $n = 6$, $l = 0$ level. The energy diagram for mercury is shown in Fig. 4.5. We note the following:

1. The $n = 6$, $l = 0$ normal state is assigned the value 0 eV.
2. No transitions downward occur from the 5.43- and 4.66-eV levels. These are two metastable states and, as we have previously noted, if an electron arrives in one or the other of these it will sit there until boosted to a higher level by an incoming electron or other means.
3. The lowest excitation level which is not a metastable state is 4.88 eV. Transitions from this level to the normal state yield 253.7-nm radiation. This spectral line, well into the ultraviolet, is the predominant line in a low-pressure mercury spectrum. Germicidal lamps use this line directly for air sanitation in hospitals, nurseries, and the like, and to retard spoilage of food products. Fluorescent lamps depend on 253.7 nm in that these photons excite the phosphor coating on the inside of the bulb wall. Energy transitions downward

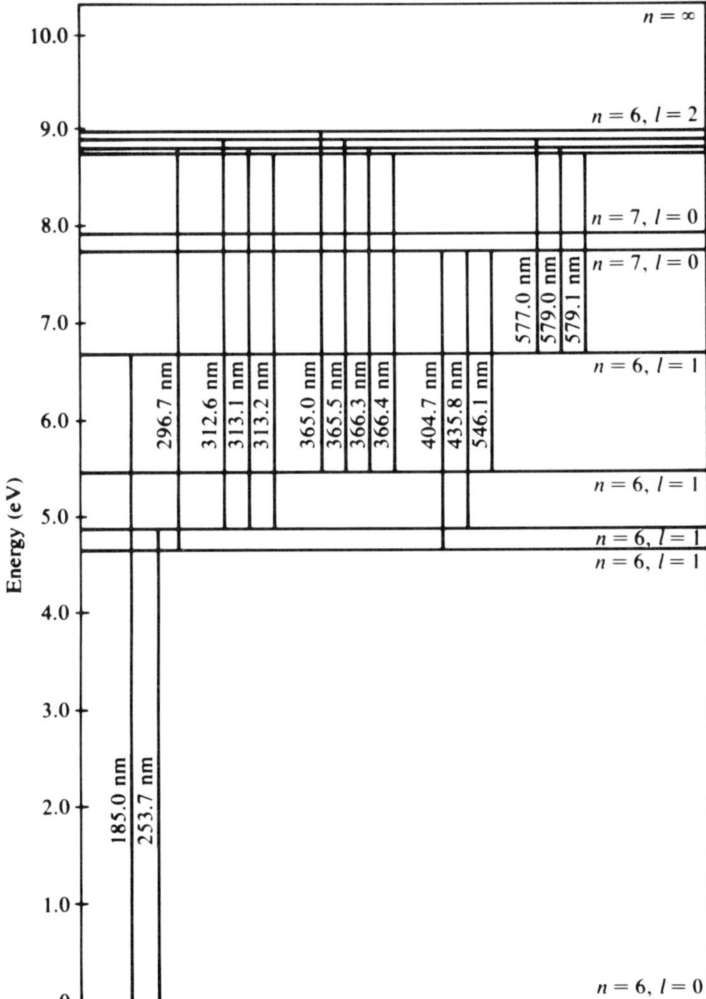

Figure 4.5 Energy level diagram of mercury.

within the phosphor then produce the visible energy that the lamp emits.

4. The three levels at 5.43, 4.88, and 4.66 eV form a so-called "triplet." With one valence electron at the ground level and the other at one of these levels, the total l of the two electrons is 1, as is the total s. The total (angular plus spin) momentum is quantized by requiring that the vector sum of $l + s$ be an integer. The vector sum of $l + s$ is 2 for the 5.43-eV level, 1 for the 4.88-eV level and 0 for the 4.66-eV level. A second selection rule states that the *total* angular momentum can change by only +1 or −1 in a transition. When both valence electrons are in their normal states, $l + s = 0$. Thus, the only permit-

ted downward transition from these three levels is from the 4.88-eV level. This is a major reason that a fluorescent lamp is so efficient in generating 253.7-nm radiation.

5. Transitions from the 6.70-eV level to the ground state produce a line at 185 nm. Radiation at this wavelength produces ozone, which is a form of oxygen having three atoms in each molecule. So-called "ozone lamps" produce this spectral line and ozone is created from the oxygen in the air around the lamp.
6. Transitions from the $n = 6$, $l = 2$ levels to the two lowest levels at 4.66 and 4.88 eV produce a series of wavelengths in the 300-nm range to which human skin is sensitive. Sun lamps are designed particularly to produce these lines.
7. Transitions from $n = 6$, $l = 2$ to the 5.43-eV metastable state produce a series of lines in the 365-nm region. This near UV radiation is called "black light." When it impinges on fluorescent paints and dyes, they appear to "glow in the dark." Applications include fluorescent markers, fluorescent signs, observation of defects in machined parts, and detection of contaminants in foodstuffs.
8. The six remaining lines shown in Fig. 4.5 are the visible lines in the mercury spectrum. These are present in a fluorescent lamp but do not contribute nearly as much to the light output as does the phosphor. However, in high-pressure mercury lamps, used in industry, in outdoor floodlighting, and in sports and street lighting, these lines provide nearly all of the visible output. High pressure favors the production of these lines. This is so because high pressure means many mercury atoms are present, and this, in turn, maximizes the opportunity for multiple excitations in which an electron in, say, the 4.66-eV state receives an additional amount of energy and is hoisted to, for example, the 7.72-eV level, from which it can fall to one of three $n = 6$, $l = 1$ levels, giving off a photon at 404.7, 435.8, or 546.1 nm. Or it could be elevated to one of the $n = 6$, $l = 2$ levels and fall to the $n = 6$, $l = 1$ level at 6.70 eV. Again three visible lines result, at 577 to 579 nm.

The inert gases, helium, neon, argon, krypton and xenon, are used extensively in lighting. Argon and krypton are used as a starting aid in fluorescent lamps, as is neon in low-pressure and some high-pressure sodium lamps. Xenon is used in flash tubes (for high-speed photography) and in high-pressure sodium lamps. Krypton and argon are used as filling gases in incandescent lamps. Neon is used in small glow lamps. All of these gases are used in gaseous discharge sign tubing (so-called "neon signs"). As an example of the inert gases, the energy level diagram of neon is shown in Fig. 4.6.

Most of the orange-red light that is characteristic of neon comes from transitions from the $n = 3$, $l = 1$ level (18.3 to 18.9 eV) to the $n = 3$, $l = 0$ level (16.6 to 16.8 eV). The first excitation potential, as in the case of mercury, is a metastable state. The energy of this state (16.5 eV) is greater than the

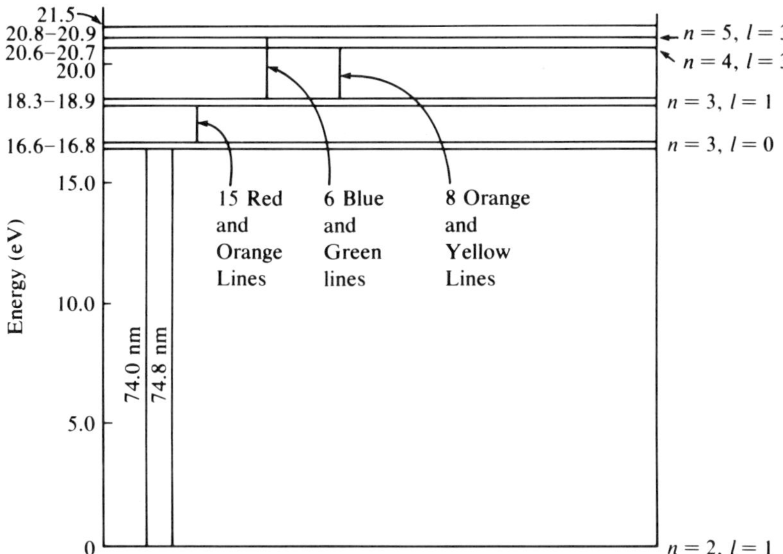

Figure 4.6 Neon.

ionizing energy of either sodium or mercury. Thus when neon is mixed with sodium or mercury in a gaseous discharge tube, the metastable neon atoms can ionize sodium or mercury atoms by collision. Such gas mixtures are called Penning mixtures and are discussed further in Sec. 6.11.

Scandium, sodium, thallium, and indium are used, along with mercury, in metal halide lamps. These elements are introduced into the lamp in the form of their halide salts and their spectra are such as to provide a better color balance than is possible in a mercury lamp. The predominant metal halide lamp in the United States is the sodium-scandium lamp, with thallium and indium used in special metal halide lamps.

As an example of one of these elements, thallium's energy level diagram is shown in Fig. 4.7. Of particular importance is the strong green

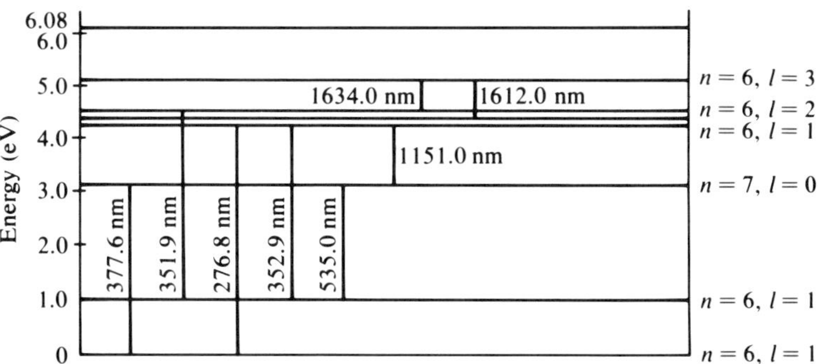

Figure 4.7 Thallium.

line at 535 nm, near the point of maximum visibility on the eye sensitivity curve. This line results from a transition to a metastable level at 0.95 eV.

4.7 TYPES OF COLLISIONS IN GASES

Collisions in gases, where one particle, such as an electron, collides with another particle, such as a gas atom, are of two general types, namely, elastic and inelastic. In an elastic collision, both momentum and kinetic energy are conserved. Thus

$$m_1 v_{1i} + m_2 v_{2i} = m_1 v_{1f} + m_2 v_{2f} \tag{4.15}$$

$$m_1 v_{1i}^2 + m_2 v_{2i}^2 = m_1 v_{1f}^2 + m_2 v_{2f}^2 \tag{4.16}$$

where the subscript i denotes prior to the collision and the subscript f after the collision. An elastic collision does not result in either excitation or ionization of the gas atom.

In an inelastic collision only momentum is conserved and Eq. (4.16) does not hold. Rather some of the kinetic energy of the incoming particle (electron) is transferred to the gas atom, causing the latter to become excited or ionized.

Let us examine several of the types of collisions that can occur in a container filled with gas atoms, gas ions, free electrons, and photons.

Electron Impact

Gaseous discharge light sources (fluorescent and high-intensity discharge lamps) are generally tubular in shape with an electrode at each end. Through heating of the electrodes and application of a voltage between electrodes, electrons are emitted and accelerated from one end of the tube to the other. When these electrons collide with gas atoms in the tube, several things may occur.

1. A relatively slow moving electron will likely have an elastic collision with a gas atom. In modeling this event, we assume that the gas atom is standing still and that its mass is very much greater than that of the electron. An analogy would be a rapidly moving marble colliding with a stationary bowling ball.

   To reduce the problem to a single dimension, let the collision be head-on, rather than grazing. Then Eqs. (4.15) and (4.16) yield

$$v_{1f} = -v_{1i} \qquad v_{2f} = 2\frac{m_1}{m_2} v_{1i} \tag{4.17}$$

   where v_{1i} is the velocity of the electron prior to the collision and v_{1f} and v_{2f} are the velocities of the electron and gas atom, respectively, after the collision. Equations (4.17) indicate that, in an elastic collision of these two particles, the electron rebounds with its kinetic energy essentially unchanged and the gas atom acquires only a very small amount of kinetic energy from the electron. For example, in an elastic collision of an electron with a mercury atom, only about 0.0005% of the electron's energy is acquired by the mercury atom. As the electron continues on its way, it will tend to acquire as much kinetic energy between collisions from the applied electric field as it loses during a collision. Since it loses very little energy during an elastic collision, its average kinetic energy will be relatively high.

2. An electron moving faster than in (1) may engage in an inelastic collision with a stationary gas atom in its ground state causing the latter to become excited. The incoming electron must possess a kinetic energy greater than the atom's first excitation potential for this to happen, and the energy balance relation is

$$\mathrm{KE}_i = \mathrm{KE}_f + W_{\mathrm{exc}} \tag{4.18}$$

where KE_i = kinetic energy of the electron just prior to collision
KE_f = kinetic energy of the electron just after collision
W_{exc} = energy of one of the atom's excitation levels

Not all collisions where the incoming electron has a kinetic energy greater than the first excitation potential of the atom result in the atom becoming excited. The probability of excitation is highest when the electron energy is just slightly greater than one of the excitation energies of the atom. Even then only about 3 out of 1000 such collisions are inelastic, resulting in excitation.

3. An electron moving more rapidly than in (2) may possess sufficient energy to ionize a gas atom. In this case the energy balance relation is

$$\mathrm{KE}_i = \mathrm{KE}_f + W_{\mathrm{ion}} + \mathrm{KE}_0 \tag{4.19}$$

where KE_i and KE_f are as before, W_{ion} is the ionization potential of the atom, and KE_0 is the kinetic energy of the electron freed from the atom. Once again, not all such collisions result in ionization. The probability of ionization is very small if the electron possesses just the ionization energy. It is a maximum if the electron's energy is $2W_{\mathrm{ion}}$.

4. Very often, in an electron-atom collision, the atom is already in an excited state. This is particularly likely if the atom has metastable states because the average lifetime in a metastable state is about 10^6 times that in any other excited state.

If an electron collides with an excited gas atom, two things may occur. The gas atom may become excited to a higher level from which it can radiate a photon and return to its rest state. This process is important in high-pressure mercury lamps. The visible lines in a mercury spectrum result from transitions from the higher energy levels. These higher energy levels are more easily reached by the two-collision process just described than by a single collision.

A second possibility is that the gas atom will become ionized. Such a result is important in low-pressure mercury arcs to create sufficient positve ions to sustain the discharge.

Positive Ion or Excited Atom Impact

Whenever an ionizing collision occurs, a positive ion is created which is accelerated by the electric field in a direction opposite to that of the flow of electrons. As it moves, the positive ion gains energy and ultimately collides with a gas atom. Such a collision is likely to be elastic if the gas atom and positive ion are the same element. The reason for this is that the positive ion and the gas atom have essentially the same mass. Thus when a moving positive ion collides with a stationary gas atom, the former loses approximately half its energy to the latter. Just as with electrons, the positive ion tends to gain as much energy between collisions as it loses during a collision. Thus its average kinetic energy is much lower than that of a free electron and sufficient for elastic collisions only. The result is that there is very little excitation or ionization of gas atoms by positive ions of the same element.

Within a mixture of gases, the situation is somewhat different. Excited atoms of one element can collide with unexcited atoms of another element, causing the latter to become ionized. This is possible only when the excitation energy of the first atom is greater than the ionization potential of the second atom. The possibility is enhanced if the excited atom is in a metastable state. For example, an argon atom at its first excitation level of 11.56 eV, which is metastable, can ionize a mercury atom, whose ionization potential is 10.39 eV.

Photon Impact

Unlike electrons and positive ions, photons are not influenced by electric fields. However, like these particles, photons can have excitation and ionization collisions with gas atoms.

1. Excitation by photon impact is similar to excitation by electron impact except for one important difference. The photon must possess exactly the difference in energy between two excitation levels in the gas atom for excitation to occur. Thus a neon photon, for example, cannot excite a mercury atom, but a mercury photon can excite a mercury atom. Suppose a mercury atom near the center of a gaseous discharge tube is excited by electron impact and emits a 253.7-nm photon. This photon impacts a second mercury atom and immediately ceases to exist, but the second mercury atom emits a 253.7-nm photon. This absorption and emission process can repeat itself

many times before a photon is emitted near the tube wall and escapes. The phenomenon is known as "imprisonment of radiation."

2. In ionization by photon impact, the photon need not possess exactly the ionization energy but it must possess at least this amount. It if possesses more, the excess energy will appear as kinetic energy of the freed electron.

Thermal Agitation

The average kinetic energy of an assemblage of gas atoms is given by

$$W = \tfrac{3}{2}\,kT \quad \text{joules} = \frac{T}{7730}\ \text{electronvolts} \tag{4.20}$$

where k is the Boltzmann constant and T is the temperature in kelvins. The most probable energy of an individual gas atom is one-third this amount. Thus at room temperature, the gas atoms have an average kinetic energy of 0.04 eV and a most probable energy of 0.013 eV, far less than the energies required for excitation and ionization. Stated another way, an electron with sufficient thermal energy to generate a 253.7-nm photon in mercury (4.88 eV) would have to be at $T = 38{,}000$ K. Thus, except in very high temperature, high-pressure gaseous discharges, thermal agitation does not play a significant part in the excitation and ionization process and the assumption we have made that the gas atoms are stationary is a reasonable one.

4.8 NUMBER OF COLLISIONS IN GASES

In the previous section we considered the types of collisions that can occur between particles in a gaseous medium. In this section we will examine how many collisions can occur, what the distance between collisions is, and how pressure, temperature and tube dimensions affect these quantities.

The kinetic theory of gases describes a gas molecule as a small sphere in constant motion due to thermal agitation and continually making elastic collisions with its neighbors and the walls of its confining chamber. An electron fired into a gas at high speed is pictured as a much smaller sphere, colliding with and rebounding from the gas atoms and following a rather random path through the gas medium.

For an ideal gas,

$$p = nkT \tag{4.21}$$

where p = pressure in newtons per square meter [pascals (Pa)]
n = concentration of gas molecules per cubic meter
k = Boltzmann constant in joules per degree kelvin
T = temperature in kelvins

This equation states that at a given temperature and pressure all gases must contain the same concentration of gas molecules. Thus a gram-molecular weight of one gas contains the same number of molecules as a gram-molecular weight of any other gas. This is Avogadro's number, 6.02×10^{23} molecules per gram-molecular weight. Also the gram-molecular weights of all gases must occupy the same volume at a given temperature and pressure. At 273 K and one atmosphere (760 nm of mercury or 1.013×10^5 Pa) this volume is 0.0224 m³. Thus there are 2.69×10^{25} gas molecules per cubic meter under these conditions. An average value for the radius of a gas molecule is approximately 10^{-10} m, giving the volmue occupied by the gas molecules in a cubic meter of gas as

$$V = (\tfrac{4}{3}\pi \times 10^{-30})\,(2.69 \times 10^{25}) = 1.1 \times 10^{-4}\ \text{m}^3 \tag{4.22}$$

This means that the gas molecules occupy about 1/10,000 of the space in a cubic meter of gas.

With this picture of an ideal gas established, we would like to calculate the average number of collisions an electron makes while traversing a certain path length within the gas and the average distance an electron travels between such collisions. This latter quantity is known as the mean free path (mfp) of the electron. To obtain these quantities, we employ the model of the electron-gas medium shown in Fig. 4.8.

As we have noted, the gas molecules have an effective radius of $r_a = 10^{-10}$ m. The diameter of an electron is $d_e = 10^{-15}$ m. We place a cylinder of diameter $2r_a$ and arbitrary length l within the gas and aim an electron into the cylinder along its center line. We assume that the gas atoms are stationary with respect to the moving electron, that the electron can be considered as a point with respect to the gas atoms, and that the actual path the electron travels may be replaced by a straight path of equal length. With these assumptions, the electron in Fig. 4.8 will collide with every gas molecule whose center lies within the cylinder. Thus we can write

$$n_c = \pi r_a^2\, ln \tag{4.23}$$

where n_c is the number of collisions the electron makes in the distance l meters and n is the number of gas molecules per cubic meter. Inserting n from Eq. (4.21) into Eq. (4.23) yields

$$n_c = \frac{\pi r_a^2\, pl}{kT} \tag{4.24}$$

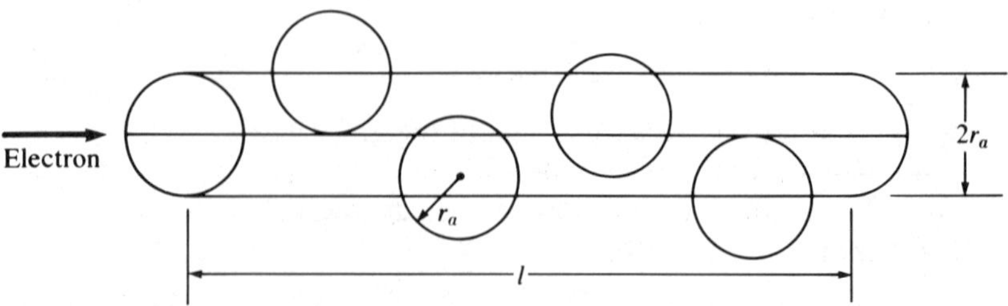

Figure 4.8 Model of electron-gas interaction.

TABLE 4.4

Gas	mfp (cm)	n_c (collisions/cm)
Mercury	0.027	37
Argon	0.043	23
Neon	0.065	15
Xenon	0.031	32

Equation (4.24) shows that the number of collisions in a distance l meters varies directly with pressure and inversely with temperature. It also shows that, with temperature held constant, the number of collisions is a function of the pl product only.

To find the mean free path, we divide l by n_c in Eq. (4.24) and obtain

$$\text{mfp} = \frac{kT}{\pi r_a^2 \, p} \tag{4.25}$$

The mfp varies directly with temperature and inversely with pressure. Values of n_c and mfp for electrons in some of the lighting gases at a pressure of 1 mm of mercury and a temperature of 273 K are shown in Table 4.4.

EXAMPLE 4.6

What must the radius of a mercury atom be for it to have the mean free path given in Table 4.4?

Solution. The temperature is 273 K and the pressure is $(1.013 \times 10^5)/760 = 133$ Pa. Inserting values in Eq. (4.25) yields

$$r_a = \sqrt{\frac{(1.38 \times 10^{-23})273}{\pi(2.7 \times 10^{-4})(133)}} = 1.8 \times 10^{-10} \text{ m}$$

4.9 TOWNSEND DISCHARGE

Three major types of gaseous discharges can be identified. These are, in order of increasing current, Townsend, glow, and arc. A most convenient way to examine these three types is through a composite volt-ampere characteristic such as shown in Fig. 4.9. The gas tube contains two unheated parallel-plane electrodes. It is connected in series with a voltage source and limiting resistor, both variable. The circuit equation is

$$I = -\frac{1}{R}V_L + \frac{V}{R} \tag{4.26}$$

and the resulting load line is drawn on the volt-ampere characteristic. What we wish to do is slowly increase the supply voltage V and observe and explain the changes that take place in I and V_L.

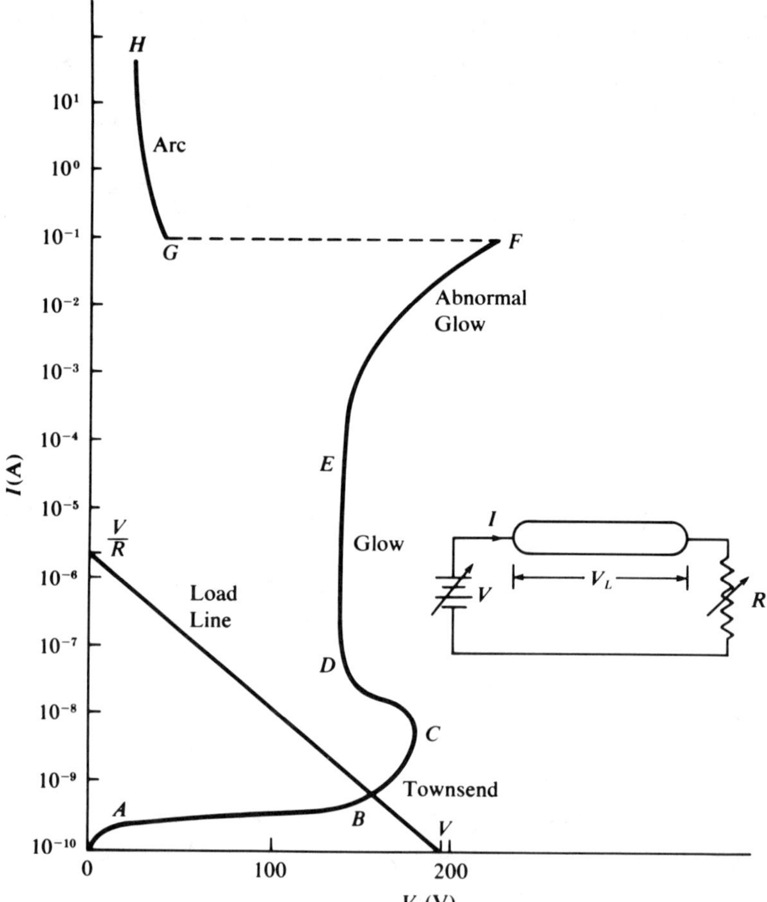

Figure 4.9

All gases are always very slightly ionized because of external radiation present in the environment. As the supply voltage is increased from zero in Fig. 4.9, the current I increases as the few electrons and positive ions in the gas tube are drawn to the electrodes. At A, these charges are being drawn away as fast as they are being created and there is negligible increase in current as the tube voltage is further increased from A to B. Note that the currents involved here are very small, of the order of nanoamperes or less.

After B, a new phenomenon begins. The voltage V_L is now large enough so that the free electrons in the gas acquire sufficient kinetic energy to ionize additional gas atoms. The process is a geometric progression from B to C and results in a large increase in current. Let us examine this avalanche build-up or current quantitatively. Assume that a single electron advancing 1 cm in the gas creates α new electrons (and positive ions). Then n

electrons per cubic centimeter advancing dx centimeters create $n\alpha\ dx$ new electrons per cubic centimeter. Thus

$$dn = n\alpha\ dx \tag{4.27}$$

and

$$\int_{n_0}^{n} \frac{dn}{n} = \alpha \int_{x_0}^{x} dx \tag{4.28}$$

giving

$$n = n_0 e^{\alpha(x-x_0)} \tag{4.29}$$

To convert Eq. (4.29) to a current equation, we multiply both sides by $q_e v/t$, where q_e is the charge on the electron, v is the volume of the gas between x_0 and x, and t is the time for an electron to move from x_0 to x. The result is

$$I = I_0 e^{\alpha(x-x_0)} \tag{4.30}$$

If we now let $x_0 \neq 0$ at one electrode (the cathode) and $x = l$ at the other electrode (the anode), we obtain

$$I_A = I_0 e^{\alpha l} \tag{4.31}$$

which expresses the electron current at the anode in terms of the electron current at the cathode.

It is impossible to have more current at one electrode than at the other in a series circuit. The current everywhere in the circuit must be I_A, the current at the anode. Whenever a free electron is created, so is a positive ion. The positive ion current at the cathode is given by

$$I_+ = E_A - I_0 = I_0(e^{\alpha l} - 1) \tag{4.32}$$

At point C, conditions are right for the discharge to become self-maintaining. For this to occur, a new phenomenon, secondary emission, must begin. As we have noted, the positive ions created by the ionization process are accelerated toward the cathode by the electric field. As the field strength increases (as V increases), some of these ions will have sufficient kinetic energy to dislodge electrons from the cathode upon impact. The condition for the discharge to become self-maintaining is that the electric field must be large enough so that, for each primary electron, enough positivie ions are created so that one of these has sufficient energy to dislodge one secondary electron from the cathode. Then the primary electron is no long needed and the discharge maintains itself.

To formulate the requirement for self-maintenance analytically, we observe from Eq. (4.32) that each primary election produces $e^{\alpha l} - 1$ positive ions. Assume that each positive ion generates β secondary electrons. Then $\beta(e^{\alpha l} - 1)$ secondary electrons will be produced by each primary electron. For the discharge to be self-maintaining,

$$\beta(e^{\alpha l} - 1) = 1 \tag{4.33}$$

To develop a new expression for the anode current when secondary electrons are present, we observe that I_0, the electron current at the cathode, is now composed of two parts. One is the primary electron current, which we call I_1, and the second is the secondary electron current, denoted by I_2. Thus

$$I_0 = I_1 + I_2 = I_1 + I_0\beta(e^{\alpha l} - 1) \quad (4.34)$$

giving

$$I_0 = \frac{I_1}{1 - \beta(e^{\alpha l} - 1)} \quad (4.35)$$

Then Eq. (4.31) yields

$$I_A = \frac{I_1 e^{\alpha l}}{1 - \beta(e^{\alpha l} - 1)} \quad (4.36)$$

The analogy of a feedback amplifier going into oscillation is useful in understanding a self-maintaining discharge. Consider Fig. 4.10, which depicts an amplifier with a forward voltage gain A and a feedback network with a voltage ratio β. The output and input voltages are related by

$$(v_i + \beta v_0)A = v_0 \quad (4.37)$$

which gives

$$v_0 = \frac{v_i A}{1 - \beta A} \quad (4.38)$$

The condition for oscillation is $\beta A = 1$, which implies that the input voltage v_i is no longer needed.

Comparison of Eqs. (4.36) and (4.38) shows that I_1 is the input and $e^{\alpha l}$ is the forward gain. The only difference in the two expressions is in the denominator. In the case of the gaseous discharge, we have one less positive ion than we do electrons, giving an $A - 1$ instead of an A in the demoninator of Eq. (4.36).

The voltage V_L at which a discharge becomes self-maintaining has several names: breakdown, ignition, sparking, striking, starting. We will call it breakdown voltage and assign it the label V_B. V_B is a function of pressure, temperature, electrode spacing, and the type of gas. To relate these quantities, we return to Eq. (4.24) and note that, for a specified gas and temperature, the number of collisions of electrons with atoms is a function of the product of the pressure and the electrode spacing. The breakdown voltage

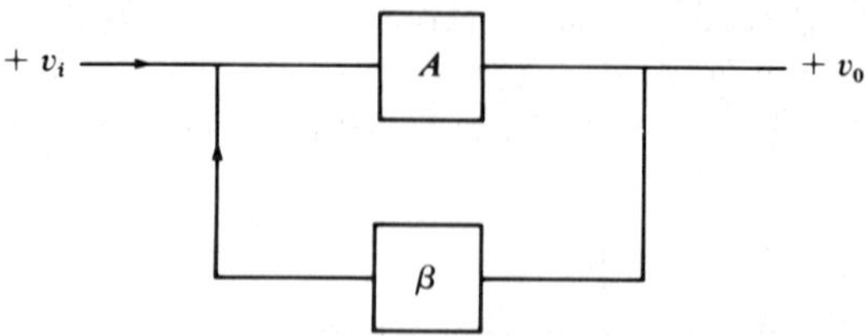

Figure 4.10 Feedback amplifier analogy.

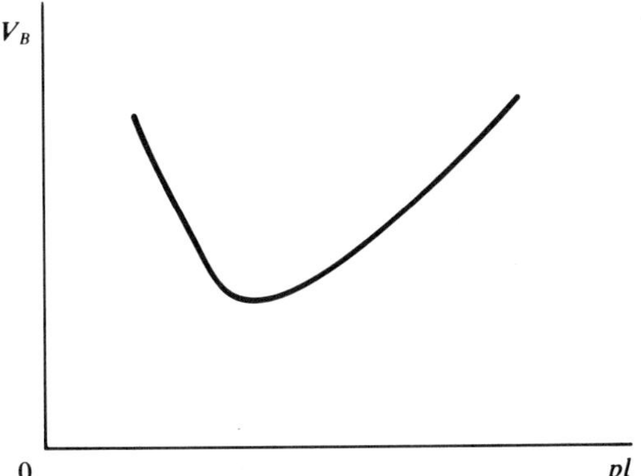

Figure 4.11 Illustration of Paschen's law.

V_B is directly related to the number of collisions. This leads to Paschen's law, which states that, for parallel-plane electrodes and a particular gas at a specified temperature, V_B is a function of the pl product only. The general relationship is shown in Fig. 4.11.

At large values of pl, either the pressure is so high that the mfp is too short or the electrode spacing is so great that the electric field is too small. In either case, insufficient ionization occurs. If the pl product is too small, either the pressure is so low that there are insufficient gas atoms present or the electrode spacing is so small that an insufficient number of collisions is possible. In either case, there is once again insufficient ionization. In between these two extremes, there is a value of pl that requires the minimum V_B for breakdown. The minimum breakdown voltage for air is about 350 V, occurring at $pl = 0.006$, where p is in millimeters of mercury and l is in meters.

4.10 GLOW DISCHARGE

The voltage V is increased so that the load line now touches the volt-ampere characteristic at point C in Fig. 4.9. This is not a stable operating point, and operation will move quickly from C to D and then to the intersection of the load line with the portion of the volt-ampere characteristic between D and E. During this movement, the current increases and the voltage across the tube decreases and then becomes constant. The discharge is now defined as a glow discharge and is self-maintained. The voltage across the tube is 100 to 300 V, slightly less than the breakdown voltage of the gas, and the current is in the milliampere range. The voltage is essentially independent of current, with the latter being limited by the external circuit.

To gain further insight into a glow discharge, it is useful to consider the potential, electric field, and charge distributions present. The governing relation is Poisson's equation

$$\frac{d^2V}{dx^2} = -\frac{\rho}{\epsilon} \tag{4.39}$$

where V is the voltage across the tube, ϵ is the permittivity, and ρ is the charge concentration. By including electric field intensity, Eq. (4.39) can be split into two first-order differential equations.

$$\mathscr{E} = -\frac{dV}{dx} \qquad \frac{\rho}{\epsilon} = \frac{d\mathscr{E}}{dx} \tag{4.40}$$

Graphs of V, $\mathscr{E}$, and ρ versus x appear in Fig. 4.12, along with a diagram showing the four regions within a glow discharge.

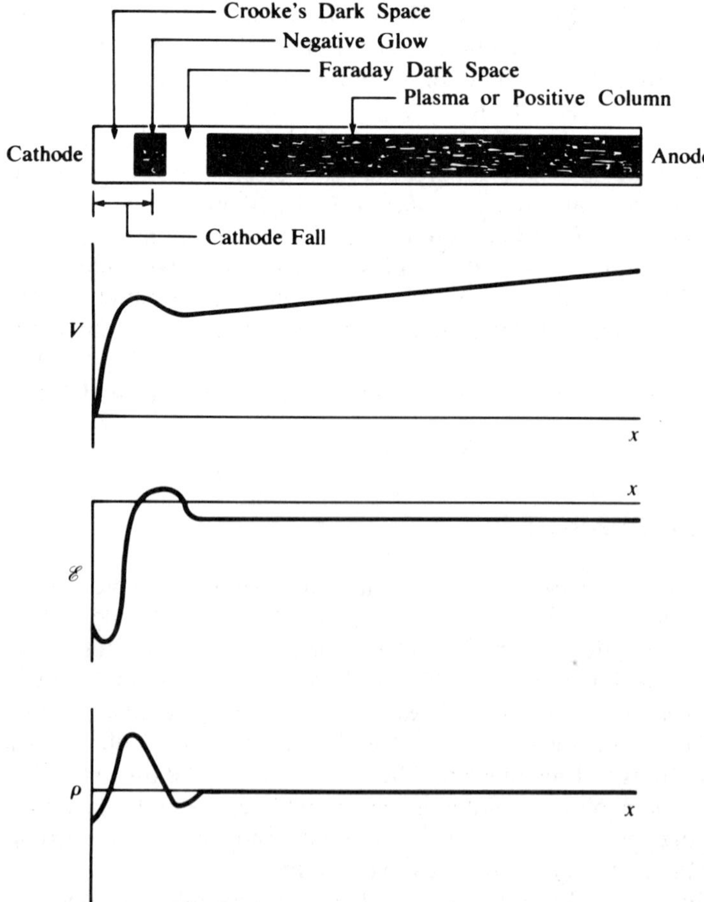

Figure 4.12 Potential, electric field, and charge distributions in a glow discharge.

There are large numbers of positive ions and free electrons in a glow discharge. Together they produce a current given by

$$I = I_p + I_e \tag{4.41}$$

where I_p is the positive ion current and I_e is the electron current. Each of these currents may be expressed in terms of charge concentration and particle velocity as

$$I_p = \rho_p v_p A \qquad I_e = \rho_e v_e A \tag{4.42}$$

where A is the cross-sectional area of the discharge. Both types of particles are accelerated by the electric field present and achieve kinetic energies given by

$$\mathrm{KE}_p = q_e V = \tfrac{1}{2}\, m_p v_p^2 \qquad \mathrm{KE}_e = q_e V = \tfrac{1}{2}\, m_e v_e^2 \tag{4.43}$$

Solving Eqs. (4.43) for the velocities and inserting into Eqs. (4.42) yields

$$\frac{I_e}{I_p} = \frac{\rho_e}{\rho_p} \sqrt{\frac{m_p}{m_e}} \tag{4.44}$$

The positive column of a discharge contains slightly more positive ions than electrons but is essentially a region of charge neutrality, as Fig. 4.12 shows. From Eq. (4.44), the current in the positive column will be predominantly electron current because $m_p >> m_e$. Thus 1 A of positive ion current can neutralize the effect of perhaps 100 A of electron current because the positive ions are traveling relative slowly and remain in the positive column for a much longer time than do the electrons.

The situation in the cathode fall region near the cathode is somewhat different. For the discharge to be self-maintained, large numbers of positive ions and a high electric field must be present to produce the required secondary electrons. Again the positive ions remain in the region much longer than do the free electrons and the result is a net positive charge (space charge) near the cathode. Nearly all of the electric field lines originating at the cathode terminate on positive ions in the cathode fall region and thus most of the voltage drop across the tube appears here. The current at the cathode is predominantly positive ion current and the ratio in Eq. (4.44) is much less than unity.

The cathode fall voltage is essentially independent of the gas pressure within the tube. What happens is that, as pressure is increased, the distance from the cathode covered by the cathode fall decreases so that the product of pressure and distance remains constant at a value to yield the minimum value of breakdown voltage, according to Paschen's law and Fig. 4.11.

In Fig. 4.12, Crooke's dark space is indicated as a region which is not luminous. The reason for this is that the predominant activity in this region is ionization. The electrons produced by secondary emission are in such a strong field that they acquire sufficient energy for ionization in a very short distance. At the anode end of Crooke's dark space, the electrons, after many ionizing collisions, are moving more slowly. They no longer can ionize gas atoms, but they can excite them. The result is a rather intensely luminous region called the negative glow. Near the anode end of the negative glow, the electrons are traveling more slowly yet and do not have enough energy even for excitation. Thus a region of essentially elastic collisions, called the Faraday dark space, results. Because of their low velocity the electrons in this region congregate, producing a slightly negative space charge and a dip in potential distribution. At the anode end of this space the electrons are accelerated once again, this time by the small constant positive column field, and they acquire sufficient energy to excite gas atoms. Thus, the entire length of the positive column is luminous.

4.11 ARC DISCHARGE

We are still in the region *DE* on the discharge curve in Fig. 4.9, experiencing a normal glow discharge where the current is essentially independent of the voltage across the tube. Let us now increase the applied voltage again so that the intersection of the load line and the volt-ampere characteristic approaches *E*. If we observe the cathode as this is going on, we note that the glow covers an increasing portion of the cathode surface until, at *E*, all the surface is covered. What this means is that, as we move from *D* to *E* the current and the cross-sectional area of the cathode that it covers increase at the same rate, keeping the current density, and thus the voltage across the tube, constant.

From *E* to *F* the discharge is an abnormal glow. The current increases further, but so now does the current density and tube voltage. The glow is no longer uniform across the cathode but rather begins to concentrate at a spot, which increases in temperature. At point *F* this cathode spot becomes hot enough (about 3300 K for tungsten) so that it begins to emit thermionically. As with point *C*, point *F* is not a stable operating point and a rapid transition occurs to the intersection of the load line with the arc discharge portion of the volt-ampere characteristic. This is a negative resistance characteristic (voltage decreases as current increases)and thus the current must be limited by an external element, such as the resistor in Fig. 4.9 or the ballast in fluorescent and high-intensity discharge lamps.

In the arc discharge region, the current is high (amperes) and the voltage across the tube is low, of the order of the ionization potential of the gas (10 to 20 V). The latter condition occurs because it is no longer necessary to have a large cathode fall voltage to accelerate positive ions to the cathode so that the required number of secondary electrons are emitted. Rather, all that is needed is to have sufficient bombardment to keep the hot spot emitting thermionically. This requires much less ionization in the cathode fall region.

In a glow discharge, the emission of electrons from the cathode is largely the result of positive ion bombardment. In an arc discharge, there are three basic mechanisms for securing these electrons.

1. The cathode is heated to a temperature at which it can thermionically emit by the discharge itself. This is the mechanism just

described. The carbon arc lamp, used for streetlighting in the late 1800s and early 1900s, is an example of this type of arc discharge. So also is the instant-start fluorescent lamp. In this lamp, the cathodes emit a few electrons per second thermionically at room temperature. These are accelerated by a high applied voltage, thus creating positive ions which bombard the cathodes, rapidly raising their temperature. The result is that the discharge starts as a thermionic arc, bypassing the glow discharge stage.

2. The cathode is externally heated. This is the procedure used in contemporary preheat and rapid-start fluorescent lamps. The cathodes in these devices are coated with an emission material which makes them abundant suppliers of electrons when heated. In this case, the discharge is not truly self-maintained. It depends on the presence of a heating current for starting but will continue if that current is removed or reduced after starting.
3. Electrons are extracted from a relatively cold cathode by application of a high field. This case differs from case (1) in that the mechanism for electron release is not bombardment by positive ions. Rather, the electrons are removed directly from the surface of the cathode by the force of the electric field. An example of this sort of arc is the cold-cathode fluorescent lamp, which is little used today.

4.12 THE LASER

The first maser (*m*icrowave *a*mplification by *s*timulated *e*mission of *r*adiation) was completed in 1954 by C. H. Townes, J. P. Gordon, and H. J. Zeiger in an effort to generate electromagnetic waves at frequencies higher than was possible with electronic tubes. In 1958, Townes and A.L. Schawlaw published a paper showing the feasibility of an optical maser, later called a laser (with the "l" standing for "light").

The first successful laser was built by T. H. Maiman in 1960. Maiman's laser used a ruby crystal. Ruby is composed largely of aluminum oxide, with a few of the aluminum atoms replaced by chromium atoms, about 4% in Maiman's crystal. The ruby is machined into a rod 0.5 cm in diameter and 4 cm long. Its ends are polished to be perfectly flat and are silvered, providing a fully reflecting mirror at one end and a partially reflecting mirror at the other end. The rod is placed inside a helically coiled electronic flash tube, which is connected to a large capacitor-charged power supply.

When the power supply is activated, the capacitors charge and the voltage builds up. Then a firing button is pushed and the flash tube strikes, providing an intense flash of luminous flux for a very short period of time. This is called "pumping" and the pulse of light raises most of the chromium atoms to one of the excitation levels shown in the left portion of Fig. 4.13. The atoms immediately fall to a lower energy level, shown in the center portion of Fig. 4.13. This level is a metastable state. While they are in this state, they are likely to be impacted by chromium photons, which "stimulate"

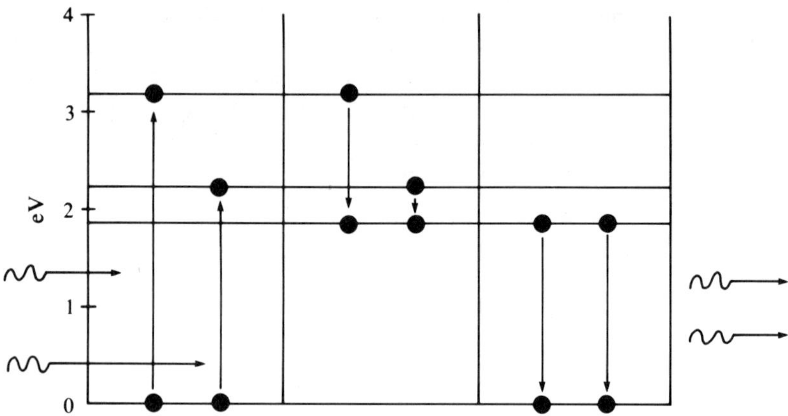

Figure 4.13 Ruby laser.

them to emit photoms of the same wavelength and revert to the ground state, as shown in the right portion of Fig. 4.13. In the case of chromium, the photon wavelength is 694.3 nm.

It is important to contrast what occurs in Fig. 4.13 with the phenomena of photon excitation and imprisonment of radiation which occur in high-intensity discharge lamps. Consider, for example, a sodium lamp. If a sodium photon at 589 nm collides with a sodium atom in its *ground* state, the atom is excited to the 2.1-eV level and the photon ceases to exist. When the atom reverts to its ground state, a new 589-nm photon is generated. In Fig. 4.13, the chromium atom is in an *excited* state when the chromium photon collides with it. The result is two chromium photons; the original photon does not cease to exist. This is how light amplification occurs.

From the foregoing, we conclude that, for the laser action to occur, a sizable number of atoms must already be in an excited state, rather than in the ground state. This situation is known as *population inversion*, and it is important that this inversion be maintained continually to make stimulation of radiation more likely than absorption of radiation.

Laser light is spatially and temporally coherent. Light from conventional sources is not. What is meant by coherence? In a conventional light source, the atoms radiate independently of one another, producing light waves that are both spatially incoherent (moving in all directions) and temporally incoherent (consisting of many wavelenths). A pinhole can be used to obtain nearly spatially coherent light from a spatially incoherent source, but most of the power is lost. A narrow band filter can be used to obtain nearly temporally coherent light, but again at a large sacrifice in power output.

When a chromium photon in the ruby laser encounters a chromium atom in the excited metastable state, not only is another photon of exactly the same frequency emitted, but the two photons are in phase. Some of these photons will travel to the side of the rod and escape, but many will proceed along the rod to its ends, where they are reflected back by the mirrored surfaces. The rod is an integral number of chromium wavelengths long and thus acts as a cavity resonator, building up the intensity of the laser beam. Ultimately an intense collimated directional monochromatic beam emerges through the partially silvered mirror at one end of the rod.

In the ruby laser, beam intensities of 10^4 W/cm^2 have been produced, lasting for about 0.5 ms. If the beam is focused by an output lens, concentrations of 10 W/cm^2 can be achieved.

The width of the red spectral line at 694 nm is about 0.002 nm, corresponding to a bandwidth of about

$$\Delta f = c \frac{\Delta\lambda}{\lambda^2} = 1200 \text{ MHz} \tag{4.45}$$

This is comparable to the bandwidth of the sharpest spectral line from a low-pressure gaseous discharge light source.

The divergence of the laser beam is less than 0.5° at high power outputs and 0.05° at low power outputs, the latter corresponding to 5 ft/mi. If the beam is sent through a telescope in reverse, almost negligible divergence can be achieved. A ruby laser beam sent to the moon in this manner illuminated a spot less than 2 mi in diameter.

All of the original lasers were operated in the pulsed mode, providing short bursts of energy. It is possible to build a solid continuous beam laser, but it is more feasible to use a gas laser. A widely used laser of this latter type is the helium-neon laser.

Consider Fig. 4.14, in which helium atoms are rings and neon atoms are solid dots. At the start, both are in the ground state. Through radio frequency excitation, a glow discharge is created and helium atoms are raised to an excited state, as shown in the left portion of Fig. 4.14. The excited helium atoms collide with neon atoms in the ground state in the center portion of Fig. 4.14, exciting each of the latter to one of four distinct energy levels. The helium atoms revert to the ground state.

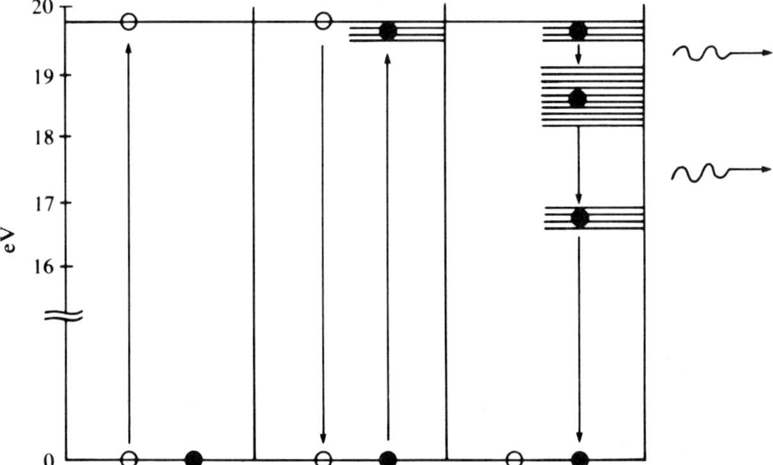

Figure 4.14 Helium-neon laser.

In the right portion of Fig. 4.14, the excited neon atoms are stimulated by neon photons, producing new photons. The neon atoms fall to one of 10 distinct energy levels, and then to the ground state in steps. The lower photon in Fig. 4.14 does not contribute to the laser beam. The upper photons reflect back and forth between end mirrors, as in the ruby laser, before emerging at one end.

The first helium-neon laser was constructed by A. Javan, W. R. Bennett, and D. R. Herriott in 1961. It operated at the very low power of 50 W and produced a beam with a wavelength of 1000 nm in the infrared and a bandwidth of 1 MHz, much less than that of the ruby laser or of the best low-pressure gaseous discharge lamp. Because of its long life, low power and low cost, it has become the most widely used laser.

Most lasers, including the two already discussed, emit about the same amount of luminous energy as does a flashlight. The difference lies in the space and time coherence. The laser's output is highly directional and its output power per nanometer is very high.

There is one continuous laser which does provide high total power output. This is the carbon dioxide laser, invented by C. K. N. Patel. Carbon dioxide lasers produce infrared wavelengths and can generate up to 9 kW of continuous power. At the other extreme are semiconductor diode lasers the size of the head of a pin, composed of gallium arsenide. A summary of types of lasers and their properties is given in Table 4.5.

The uses of lasers are many and are increasing daily. A brief summary of some of the important ones follows:

Optical Ranging

The spatial coherence of the laser makes it ideally suited for optical ranging and scanning applications. In 1962, laser echoes were received from the moon. In 1969, following the Apollo lunar landing, a laser was used to illuminate a two-mile-wide spot on the moon. This experiment provided an estimate of the moon's distance from the earth within an accuracy of 15 cm.

Recently a range finder has been developed which may be held in the hand. It includes a sighting telescope, a laser transmitter, and a laser receiver. Operation is from two small batteries and it has a range of 3 to 4 km with an accuracy within 5 m.

Searching lasers are used to scan the sky, in much the same way as radar does. At 500 mi, radar can determine distance within 100 ft. A ruby laser narrows this to 25 ft and, in addition, requires a smaller antenna and is less sensitive to interference by enemy signals.

Industrial Applications

The laser is used widely in industry. Computer-controlled CO_2 lasers spot-weld sheet metal parts on automotive production lines at the rate of 50

TABLE 4.5

Laser type	Typical operating wavelengths (μm)	Principal characteristics
Helium-neon	0.6328, 1.15, 3.39	Highly monochromatic, rugged, simple, long-life, inexpensive, low power
Helium-cadmium	0.3250, 0.4416	Highly monochromatic, visible and UV
Carbon dioxide	10.6	High power, high efficiency, IR
Argon	0.3511, 0.3638, 0.4579 0.4765, 0.4880, 0.4965 0.5017, 0.5145	CW, high power, highly monochromatic, inefficient
Krypton	0.3507, 0.3564, 0.4680, 0.4762, 0.5208, 0.5682, 0.6471, 0.6741	Similar to argon except for lines in the red portion of the spectrum
Nitrogen	0.3371	UV, inefficient, short pulses
Ruby	0.6943	High peak power, moderate spatial coherence
Glass	1.06	Low cost, excellent optical quality, high peak power, low average power, wide spectral line
Ga As	0.85–0.91	Very small, inexpensive, efficient, poor coherence, low power, temperature sensitive

ft/min. Laser "microwelders" are used to spot-weld today's microelectronic circuits.

Lasers are used in heat-treating metals, raising temperatures over 2000°F in a few milliseconds. They can aid in the critical inspection of many products, particularly those with hard-to-see crevices. They can be used to cut extremely hard and brittle materials, such as tungsten.

In the furniture and garment industries, lasers are used for the nondestructive cutting of wood and fabric. In the aircraft industry, they are used to cut epoxy sheets.

Lasers can drill holes in diamonds, which then become dies through which fine wire can be drawn. A single drilling takes only minutes with a pulsed ruby laser, as compared with 2 days for the previous process which involved driving steel pins coated with olive oil.

Surveying and Measurement

Again because of coherence, lasers are excellent surveying and aligning tools. They are widely used in laying pipe, boring tunnels, and creating plumb lines. For example, fields an acre or so in size have been leveled flat within one-fourth of an inch under laser control. Tunnels have been bored with a deviation of less than an inch in one and a half miles. In this latter application, photocells serve as a target for the laser, which is mounted on the boring machine. If the maching drifts off course, a signal is sent to the operator indicating what correction to make.

Laser interferometers have been built which can detect changes in length of 10^{-11} in. Gas lasers are used in seismometers to detect earth tremors. This is done by allowing one of the laser mirrors to move slightly during a tremor, thus altering the wavelength of the emitted laser beam.

Medical Applications

It is perhaps in the field of medicine that the laser will find its widest and most beneficial use. Already there are a myriad of medical laser applications. One of the most significant is the laser photocoagulator. This device is being widely used by eye surgeons to repair torn retinas. A weak laser beam, focused through the eye's lens, forms a set of tiny scars in the retinal tissues surrounding the tear, thus keeping it from expanding. The entire process takes only about 20 min and the patient feels essentially no pain.

In cancer treatment, the laser is also becoming prominent. It is being used to burn away skin tumors and blemishes. A CO_2 laser scalpel has shown some success in selectively destroying inoperable brain tumors. The laser scalpel is also promising in treating burn victims, where the surgeon must remove large amounts of tissue. Its major virtue is that is cauterizes as it cuts and thus reduces bleeding significantly.

In conjunction with fiber optics, lasers are also being used to treat internal problems. In the case of ulcer treatment, two fiber optic cables can be inserted into the stomach. Through one the doctor views the ulcer and through the second transmits a laser beam to close it. This, of course, avoids the major incision of conventional surgery.

The use of the laser as a tool of medical research is also important. For example, the physiology of a single cell can be studied by slicing off a portion of its nucleus with a laser. It is even possible to split off chromosomes for genetic research.

Energy Generation

Research on laser fusion is being conducted at several laboratories. In one project, a hollow glass ball, 0.1 mm in diameter and filled with hydrogen gases, is rapidly laser-heated by focusing the beam from a large pulsed glass laser on the ball. This causes the ball to implode, and the resulting fusion reaction releases energy equivalent to that of a tank of gas. It is hoped that work such as this will lead ultimately to generating plants in which small pellets of heavy hydrogen are injected rapidly into a chamber and heated to the point of fusion by lasers, producing inexpensive and environmentally clean energy.

Optical Communication

The visible spectrum extends from 400 to 700 nm, representing a frequency spread of 321×10^6 MHz. A VHF television channel has a bandwidth of 6 MHz. Thus it is theoretically possible to insert about 50 million TV channels into the visible spectrum. This should serve to indicate the vast possibilities for communication at optical frequencies.

An example of optical communications is a fiber optic telephone system being installed in a few places in the United States. The system is driven by an aluminum gallium arsenide laser operating at 45 MHz and can handle 50,000 simultaneous telephone conversations. The transmission efficiency is such that the signals can travel up to 4 mi before amplification is required.

Holography

Holography is a photographic technique in which the patterns of the wavefronts of the reflected light waves from an object are recorded, rather than recording the image of the object itself. The result is startling three-dimensional imagery with depth and perspective.

The holographic process was first conceived, quite accidentally, by Dennis Gabor, a Hungarian physicist, in 1948. However it was not until the introduction of the laser that the process became practical.

The holographic recording process is shown schematically in Fig. 4.15. A laser beam is directed at a half-silvered mirror called a beam splitter, where it becomes two beams. One beam lights the object to be photographed, with the reflected light from the object proceeding to the holographic plate, usually a high-resolution photographic plate. The other beam, termed the reference beam, is reflected by a mirror directly to the

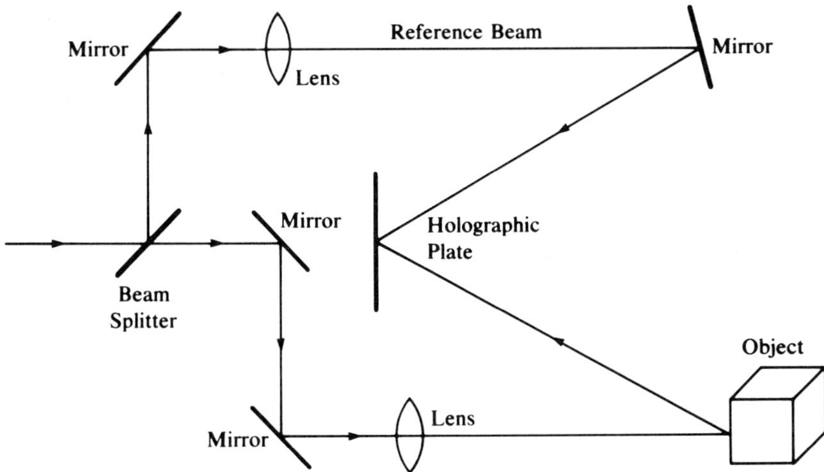

Figure 4.15 Holographic recording.

holographic plate. This beam plays the key role of providing a phase, as well as an amplitude, reference. When the two beams meet at the holographic plate the information that is recorded is in the form of interference patterns. Specifically, a pattern of concentric alternating light and dark rings is formed on the plate for each point on the object which is illuminated. If the ring is light, the two beams are reinforcing; if it is dark, they are destructively interfering.

The exposed holographic plate is developed as in normal photography. If it is then illuminated by conventional light, nothing is seen and the plate appears as would a uniformly underexposed film. To reproduce the image, it must be illuminated by the exact laser reference beam that was used to make the hologram. Only then does the full three-dimensional fidelity appear.

Holograms are currently used in photomicrography to view and analyze extremely small objects. In photographic art, holographic techniques are being used to recreate rare museum objects in three dimensions for public view. The objects themselves are stored safely away. Holograms of robots or famous performers in action are also becoming popular.

REFERENCES

Bitter, F.: *Currents, Fields and Particles*, Wiley, New York, 1957.

Dow, W. G.: *Fundamentals of Engineering Electronics*, Wiley, New York, 1952.

Elenbaas, W.: *Light Sources,* Macmillan Press Ltd., London, 1972.

Gray, H. B.: *Chemical Bonds,* Benjamin, Menlo Park, Calif., 1973.

Hughes Aircraft Co.: "What You Always Wanted to Know About Lasers," *Vectors,* Winter 1973.

Martin, T. L.: *Physical Basis for Electrical Engineering,* Prentice-Hall, Englewood Cliffs, N.J., 1957.

Meloy, T.: "The Laser's Bright Magic," *National Geographic,* December 1966.

Millman, J.: *Vacuum-Tube and Semiconductor Electronics,* McGraw-Hill, New York, 1958.

Schawlow, A. L.: "Lasers and Light," *Readings from Scientific American,* Freeman, San Francisco, 1969.

Sears, F. W., and M. W. Zemansky: *University Physics*, Addison-Wesley, Cambridge, Mass., 1955.

Ward, B.: "Laser Technology," *Sky Magazine,* August 1979.

Waymouth, J. F.: *Electric Discharge Lamps,* MIT Press, Cambridge, Mass., 1978.

Chapter 5

Vision and Color

5.1 THE FACTORS OF SEEING

In earlier chapters, we considered the process of seeing from the point of view of the functioning of the human eye. We discussed the eye's optical system, its two types of photoreceptors, the photochemical-electrical processing of light energy within the eye, and the phenomena of accommodation, adaptation, and selective response to wavelength. In short, we considered the physiological factors of seeing.

In this chapter we will discuss some of the physical factors of seeing, those factors external to the eye but of paramount importance in determining our ability to see a visual task and to do visual work. There are five major ones, namely, size, contrast, time, luminance, and color. We will discuss the first four of these together in detail now, show how they are interrelated, and use them to establish criteria for prescribing the amount and type of illumination needed to perform a visual task. Then we will close the chapter by considering the fifth factor, color, which not only affects seeing but also our perception of objects and the spaces that they occupy.

Before discussing each of the first four factors in detail, we should define more carefully what we mean by a visual task and distinguish between visual potential and visual performance. Driving a car, threading a needle, and washing a dog are examples of human tasks. Each of these contains a visual component, which we call a visual task. Success in performing a total task depends in part on the ability to perform the visual portion of the task. This, in turn, depends on the visual characteristics of the task and on the physical characteristics of the observer, particularly those of his/her seeing system. A task may have excellent visibility but may not be well performed because of a tired, aged, untrained, or distracted worker. The visual potential is present but the visual performance is not. Thus optimization of the four factors of seeing creates the potential for excellence in visual performance but does not guarantee it.

To explore the relationship of the factors of seeing to the visual task, let us borrow from the field of electrical communications. To perform a visual task we must be able to receive clearly the "signal" being sent to us from the task, in the form of reflected or transmitted lumens. Our ability to detect the signal is influenced by what other visible radiation we are receiving from the areas surrounding the task. This is the lighting system "noise" and is often as strong as the signal itself. Our ability to detect the signal is also influenced by how fast the signal is changing, for our visual receiver, like a radio or television receiver, has an upper limit on its capacity to process rapidly changing signals. The strength of the signal is certainly of significance, and this, in turn, is a function of both the density of luminous flux emitted by the task and the surface area of the task. Last, the bandwidth of the system channel, which is 5 kHz for a radio channel, 6 MHz for a TV channel and 320×10^6 MHz for the visual channel, the latter obtained from

$$\Delta f = c\left(\tfrac{1}{400} - \tfrac{1}{700}\right) \times 10^9 \text{ Hz} \tag{5.1}$$

and selectivity within that channel [Q for radio and TV and $V(\lambda)$ for visible] determine the quality (sound, picture, or color fidelity) of the the output signal.* System noise, rapidly changing signals, signal intensity, source surface area, and system bandwidth and selectivity are thus nothing more than other ways of identifying contrast, time, luminance, size, and color as the external factors of seeing.

5.2 SIZE

When we talk about the size of an object affecting our ability to see that object, what do we mean? Consider Fig. 5.1*a*, where a road is shown disappearing into the distance with a row of telephone poles at its side. We know that all the telephone poles are the same height, yet those in the distance appear much smaller than those close by and it is harder to see the details of the distant poles (their crossarms, for example). Evidently physical size of an object is not a good criterion for determining whether or not we can see that object.

Consider now Fig. 5.1*b*, where we have shown an eye observing two telephone poles, each of height h. The angles which these poles subtend at the eye are labeled θ_1 and θ_2; the distances from the eye to the poles are labeled d_1 and d_2. From the definition of radian measure, we can write

*No analogy should be stretched too far. In electronic systems, information is processed sequentially in time. In the visual system, there is simultaneous temporal and spatial processing of information.

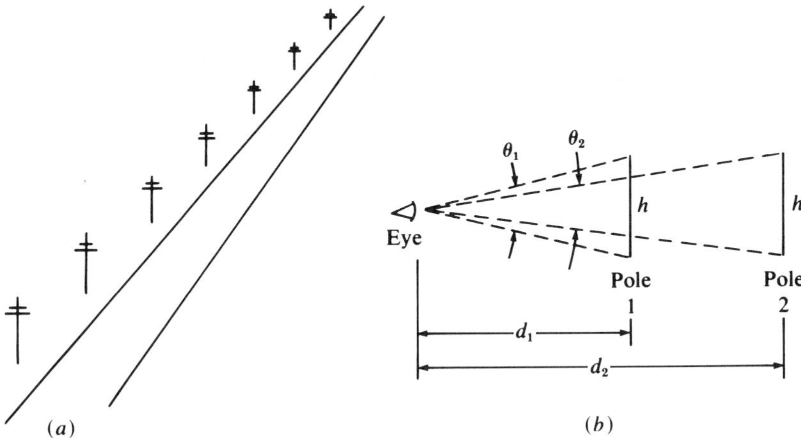

Figure 5.1

$$\theta_1 \simeq \frac{h}{d_1} \quad \text{and} \quad \theta_2 \simeq \frac{h}{d_2} \tag{5.2}$$

We call θ_1 and θ_2 the visual sizes of the objects, measured in radians, or more commonly in minutes of arc, where 1 radian = 3438 minutes of arc. Thus an object 10 ft high and 200 ft from an observer would have a visual size of $\frac{10}{200} = 0.05$ rad = 172 minutes of arc (about 3°). We observe that, even though the pole heights are the same, their visual sizes are different. It is visual (angular) size, rather than physical size, that helps to determine how well an observer can see an object.

It is not always easy to decide what dimension of a visual task to use in determining its visual size. Basically we need to determine the visual size of the smallest detail that needs to be seen in order to perform the visual task.

Consider Fig. 5.2. The first diagram is the standard parallel bar test object, a square with the center third removed. The bars are generally black and the space white. This test object was originated in the General Electric

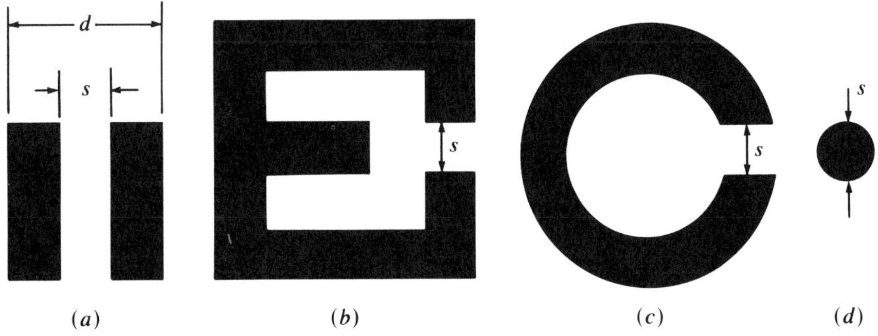

Figure 5.2 Visual test objects.

Lighting Research Laboratory in the early 1900s and served as the fundamental test object for visibility studies until after World War II. Assuming the visual task is simply to determine the presence or absence of the object in the field of view, the dimension *d* is the one to use in determining visual angle. But if, on the other hand, one is interested in resolution of detail so that the test object may be identified as parallel bars, the separation *s* should be taken. The latter situation of identifying, rather than simply detecting, is usually the case, although occasionally, such as when driving at night, we simply want to know if an object is present or not on the road ahead.

Consider the other seeing tasks in Fig. 5.2. If we are again concerned with the details of the objects, we must in each case use the dimension *s*. Thus all four objects in Fig. 5.2 have the same visual size from the point of view of identifying the object, not just detecting its presence.

Two other terms concerning size require definition. One is threshold size, which is the visual size in minutes of arc of an object which can just be seen 50% of the time in a given seeing situation. At threshold, there is zero certainty of seeing. An observer scoring 5 correct identifications out of 10 in a forced-choice, yes or no, situation could just as easily have achieved this score by chance. Threshold size is not a fixed parameter. It varies with the contrast between the object and its background, with the time available for seeing, and with the luminance of the background.

The other term requiring definition is visual acuity, which is simply the reciprocal of threshold size. Thus a visual threshold angle of 0.4 minute of arc gives a visual acuity of 2.5. It should be understood that visual size is a physical quantity independent of the lighting environment whereas threshold size and visual acuity are very dependent on the other factors of seeing.

A plot of visual acuity versus background luminance in footlamberts* for a circular black object on a white background appears in Fig. 5.3. We note that, as background luminance decreases, object size must increase if we are to be able to continue to just see the object. Also, the maximum visual acuity possible for this object is 2.46, corresponding to a threshold size of 0.41 minute of arc and a background luminance of 100,000 fL. Ninety percent of maximum visual acuity occurs at about 100 fL. Thus the final 10% gain in visual acuity requires a 1000 to 1 increase in background luminance. At the very low background luminance of 0.01 fL, visual acuity is 0.4 and threshold size is 2.5 minutes.

It would be useful to develop some "feel" for the visual size of real-world seeing tasks that we encounter every day. The most common seeing task is the reading of bookprint. Customarily we hold a book about 14 in from the eye. At that distance, the visual size of various type sizes in

*In most of the research on vision, background luminance is expressed incorrectly in footlamberts (lumens per square foot). It should be either luminous exitance in lumens per square foot or luminance in candelas per square foot, preferably the latter. However, the tasks and backgrounds involved are generally assumed to be perfectly diffusing surfaces so that one can convert to candelas per square foot if necessary by dividing by π. Thus we will use the misnomer in order to be consistent with the literature.

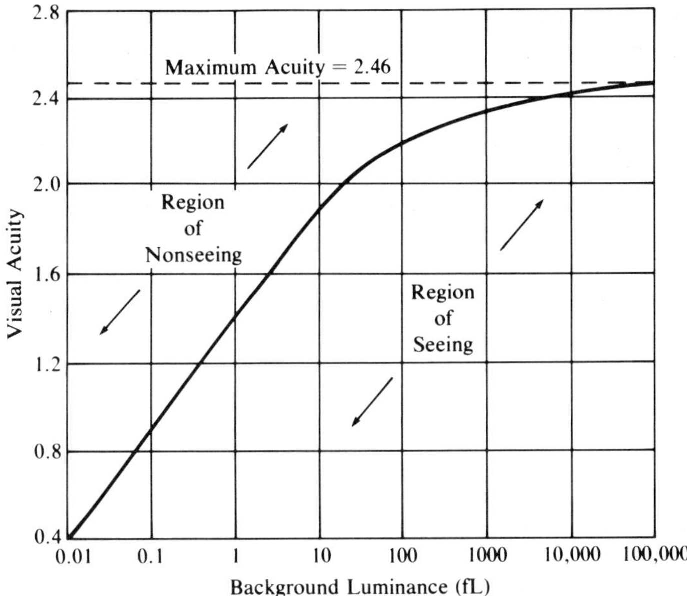

Figure 5.3 Visual acuity: Black object on white background. [*Source:* J. E. Kaufman (ed.), *IES Lighting Handbook,* 5th ed., IES, New York, 1972, fig. 3-13, with permission.]

minutes of arc, measured using the distance between the dot and the body of the letter i, is as follows: 6 point, 1.7; 8 point (paperback book print), 3.9; 10 point (normal book print), 4.9; 12 point (normal typing), 6.4; 14 point, 7.8. The smallest visual task we normally encounter is 6 point type at 14 in. For example, the width of the eye of a small needle at 14 in subtends a visual angle of 2.5 minutes of arc at the eye. As an extreme, the letters in the third from the bottom line of an eye chart, which is the lowest line a person with 20/20 vision can just read at the specified test distance, subtend an angle of 1 minute of arc at the eye.

Referring to Fig. 5.3, we find that 0.025 fL is required to just see (threshold) 6 point type, 0.01 fL to just see the needle eye, and 0.18 fL to just see the row on the eye chart. These are all quite low levels of luminance and show that, in general, we are well above threshold in contemporary seeing situations.

5.3 CONTRAST

Contrast, or more properly luminance contrast, between an object and its adjacent background is defined by the following equation:

$$C = \frac{|L_b - L_t|}{L_b} \tag{5.3}$$

where L_t is the luminance of the object (or task), L_b is the luminance of the background (also the adaptation luminance of the observer), and the vertical bars in the numerator indicate that we are to take the absolute value of the difference.* Thus contrast is positive whether the object luminance is greater than that of its background or vice versa. For example, an object whose luminance is 30 fL has the same contrast against a background of 20 fL as it does against a background of 60 fL. In each case the contrast is 0.5, or 50%.

If a visual task and its background are lighted to the same illuminance level and if both are perfectly diffusing, contrast becomes a function of only the reflectances of the two surfaces. Then Eq. (5.3) can be rewritten as:

$$C = \frac{|\rho_b - \rho_t|}{\rho_b} \tag{5.4}$$

If the task is black print and the background is good white paper, contrast can approach 1.0 (100%). But for a faded ditto page or the fifth carbon copy of a letter, contrast will be considerably less than 1.0. The diagram in Fig. 5.4 is intended to illustrate the wide differences in contrast that can exist. The black lines against the white background are relatively easy to see, but against the black background are difficult to detect. But we can't always have black objects on white backgrounds. A dark task on a dark background is not an uncommon seeing task, sewing black thread on black cloth being a prime example. It requires a high level of illumination to provide sufficient visibility for such a task.

Just as with size, there are two additional terms concerning contrast at threshold that need to be defined. The first is minimum perceptible contrast (MPC), which is the least contrast a task may have with its background and be seen 50% of the time in a given seeing situation, that is, with the other three factors specified. Its reciprocal is called contrast sensitivity (CS). Thus a minimum perceptible contrast of 0.1 yields a contrast sensitivity of 10. Minimum perceptible contrast and contrast sensitivity are each very dependent on background luminance, visual size of the object, and the time available for seeing, and, in that sense, they are similar to threshold size and visual acuity.

Contrast sensitivity is plotted versus background luminance in Fig. 5.5 for a 4-minute test object with a time for seeing of 0.1 s. Note that, as with visual acuity, a saturation effect occurs as background luminance increases. The maximum contrast sensitivity for this size object is 45, corresponding to a minimum perceptible contrast of 0.022. This occurs when object and background reflectances are nearly the same, for example, a task of 88% reflectance against a background of 90% reflectance. The value of contrast sensitivity at 0.001 fL is 0.035, with a minimum perceptible contrast of 28.6. Such a contrast requires a very dark background, for example, a task with 60% reflectance and a background with 2% reflectance. From these ex-

*Occasionally the absolute value bars are omitted in the definition, permitting negative contrasts.

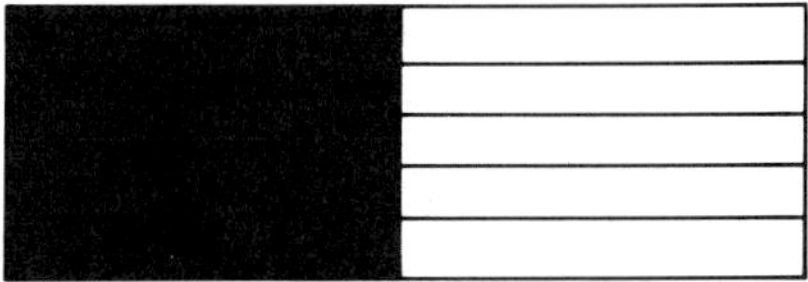

Figure 5.4 Contrast differences.

tremes, it can be seen that Fig. 5.5 covers the full range of contrast possibilities.

Size and contrast are closely interrelated. As size increases, there is a gradual transition from size being the governing factor in determining visual threshold to contrast assuming this role. The transition occurs at 1.5 to 2.5 minutes of arc. For small objects (less than 1.5 minutes), a small percent change in visual size requires a larger percent change in contrast to rereach threshold conditions with the same background luminance. For larger objects, (greater than 2.5 minutes), the reverse is true. Since, as we saw earlier, we are usually concerned with objects whose visual size is greater than 2.5 minutes, visual threshold is in general a much more sensitive function of contrast than it is of size.

5.4 TIME

It takes time to see. Those of us who are a bit older recall the Burma Shave signs along the nation's highways during the 1930s and 1940s. These were spaced so that a motorist traveling at a speed of up to 45 mi/h had time to read the message on each sign while passing by it (at interstate highway

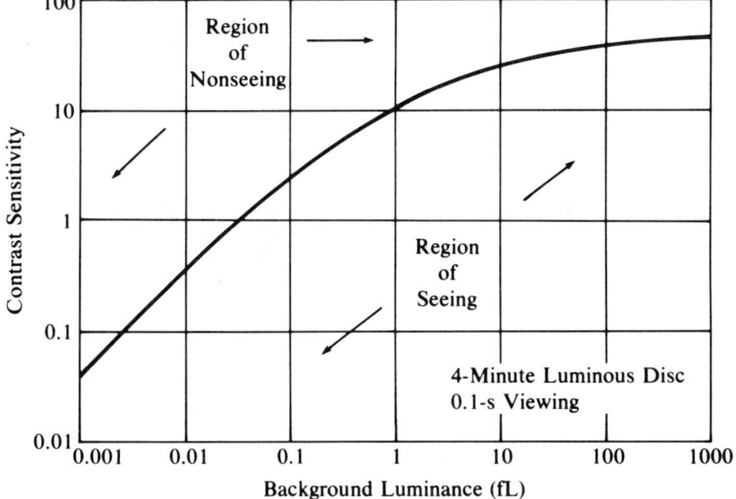

Figure 5.5 Contrast sensitivity. [*Source:* J. E. Kaufman (ed.), *IES Lighting Handbook,* 5th ed., IES, New York, 1972, fig. 3-10, with permission.]

speeds, the signs would have to be placed further apart). As another illustration, we know that we cannot see quickly enough to catch a balloon in the act of popping: we can see it fully inflated and we can see it fully deflated but we can't see it in the process of deflating right after we prick it with a pin. This is a job for high-speed photography, which is capable of catching an event that lasts for only a thousandth of a second or less.

The time required for seeing decreases as background luminance increases, if contrast and size are held constant. For example. it takes about 1 s to see a 4-minute test object having a contrast of 1.0 with its background when the background luminance is 0.015 fL. This becomes 0.1 s at 0.03 fL, 0.01 s at 0.54 fL, and 0.0015 s at 100 fL. Thus for normal lighting levels of 10 to 100 fc, and for most seeing tasks that we encounter, time is not a problem.

5.5 LUMINANCE

It is luminance and not illuminance with which we are concerned in determining task visibility. Consider Fig. 5.6. The upper three visual tasks are white, gray, and black and there is an illuminance of 100 fc on each. The tasks have different reflectances and, assuming the tasks are perfectly diffusing surfaces, we can multiply the illuminance by the reflectances to obtain task luminances in footlamberts. These are given below the tasks. The tasks will appear white, gray, and black, reading from left to right, assuming that the area surrounding the tasks is gray and that the eyes of the viewer are adapted to this surround.

Now consider the same three tasks lighted to different illumination levels so as to produce the same luminance for each, as shown in the bottom row in Fig. 5.6. Assuming continued viewer adaptation to the gray surround, each task now appears gray and we are unable to tell that the task are different. By adjusting the illumination levels on the tasks we have made them appear identical to the eye in the given environment.

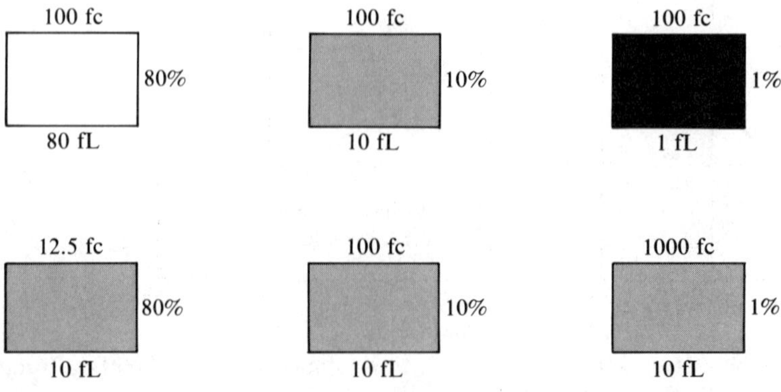

Figure 5.6

Luminance is often confused with the more popular term "brightness." Luminance is the visual excitation, whereas brightness is the visual response. Luminance is a quantitative measure. Brightness is a qualitative and subjective term referring to how a surface appears to an observer. Snow on a sunny day, neon signs at night, and movie screens are examples of surfaces that are usually considered to be bright. But brightness is a relative thing. For example, a full moon on a clear night can appear bright to an observer. But the same observer would not think the moon bright at all if he saw it in the daytime sky. The same is true of a streetlight; it seems bright at night but, even though lighted, not so bright by daylight. Brightness is not appropriate in quantitative determinations of visibility; luminance is.

5.6 RELATING THE FOUR FACTORS

Before discussing the vast amount of research that has gone on during the past 30 years to determine quantitative relationships between the four factors of seeing, let us review qualitatively what we have shown thus far.

1. *Size.* Increasing visual size increases visibility.
2. *Time.* The more time we are given to see, the better our likelihood of seeing becomes.
3. *Contrast.* Increasing contrast improves visibility. Given a dark object on a light background, if we make the object darker or the background lighter, seeing improves.
4. *Luminance.* Visibility improves with an increase in luminance.

These conclusions are very rudimentary, but they do, at least, give a feeling for the effect of each factor on visibility.

Much of the early work prior to 1940 on visibility and the four factors was done by Matthew K. Luckiesh and Frank K. Moss and is summarized in their 1937 book "The Science of Seeing." Luckiesh and Moss did extensive studies of the effects of contrast and luminance on threshold size, using the parallel bar test object shown in Fig. 5.2*a*. Their results indicated that threshold size varies approximately inversely as the logarithm of luminance over a luminance range from 1 to 100 fL, with contrast held constant. For example, with 100% contrast, threshold size was found to be approximately 0.6 minute at 100 fL and 1.1 minutes at 1 fL. When contrast was reduced to 2%, they found that threshold size at 100 fL became 5 minutes and at 1 fL, 15 minutes. Thus threshold size increased by about 8 to 1 when contrast decreased from 100% to 2% at 100 fL and about 14 to 1 at 1 fL for the same percentage contrast change.

A more comprehensive study of the four factors was undertaken by H. Richard Blackwell after World War II and is continuing to this day. Blackwell's early work was reported by him in a 1952 article entitled "Brightness Discrimination Data for the Specification of Quantity of Illumination." His basic experiment consisted of having observers detect the presence (or absence) of a lighted disc target at the center of a darker uniform background. Four target sizes were used from 1 to 64 minutes of arc, five target luminances from 0.001 to 800 fL, and seven time durations from 0.001 to 1 s. The observers, knowing the location of the target, were required to identify the time interval during which the target appeared from four possible time intervals. Each observer was given 250 target presentations at each viewing session, 50 of each of five target luminances. The observers were young, trained, and had 20/20 vision.

The results of the experiment were plotted as a family of curves, one set of which appears in Fig. 5.7. Each set presents contrast plotted against background luminance on logarithmic scales for various values of target visual size and for one value of target time duration. Seven such sets appear in the article, one for each of the seven target time durations. The vertical axis of each set of curves has four scales, one for each value of performance level (99% performance means the target is found 99 out of 100 times; 50% performance is threshold).

EXAMPLE 5.1

As an example of the use of these curves, let us suppose we wish to find the background luminance necessary to see with 50% accuracy (threshold seeing) a 4-minute disc target having a contrast with its background of 0.1 when the target is lighted for 1 s.

Solution. To enter the curves in Fig. 5.7, we move horizontally from the −1 log contrast line on the 50% accuracy scale and intersect the 4-minute size curve at a log background luminance of −0.62. This gives a background luminance of 0.24 fL. Then we can find the task luminance from Eq. (5.3) as

$$0.1 = \frac{|0.24 - L_t|}{0.24}$$

$$L_t = 0.26 \text{ fL}$$

Suppose now we wish to achieve 99% accuracy of discernment. We have two choices. We can keep size and contrast constant and increase background luminance. From the curves in Fig. 5.7, $L_b = 1.0$ fL for this situation. Solving for task luminance gives $L_t = 1.10$ fL. We can also achieve 99% accuracy of discernment by increasing contrast while keeping size and background luminance fixed. Proceeding vertically from −0.62 on the log L_b axis to the 4-minute size curve and then horizontally to the 99% scale we find $\log C = -0.65$, giving $C = 0.22$. This gives a task luminance of 0.29 fL. We note the ratio of the contrasts required for 99% and 50% visual performance, with all other factors held constant, is $0.22/0.1 = 2.2$ whereas the ratio of background luminances required, with all other factors held constant, is $1.0/0.24 = 4.2$. Thus task visibility is a more sensitive function of contrast in this instance than it is of background luminance.

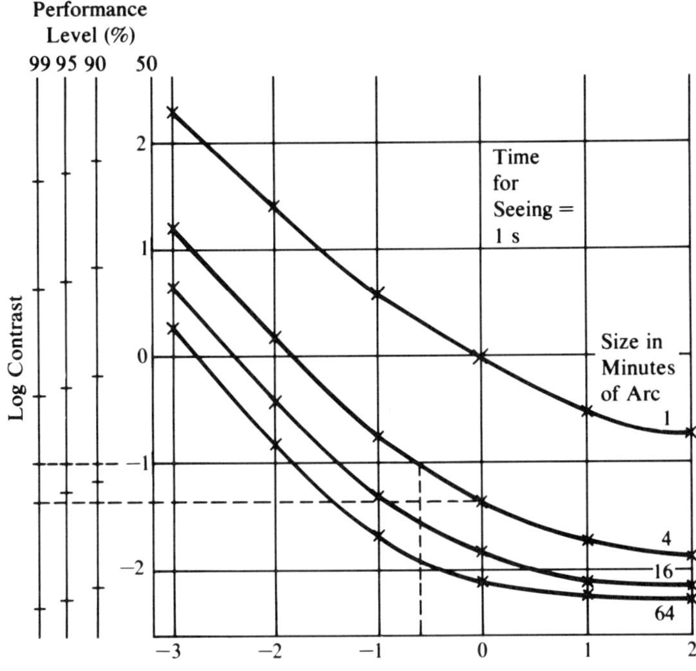

Figure 5.7 Relating the four factors of seeing. (*Source:* H. R. Blackwell, "Brightness Discrimination Data for the Specification of Quantity of Illumination," *IES Transactions,* November 1952. Reprinted with permission of the Illuminating Engineering Society of North America.)

5.7 VISIBILITY LEVELS

In Blackwell's early experiments, observers knew when and where the visual task would appear. Also their viewing was static, not dynamic; that is, the observers were not required to scan with their eyes prior to focusing on the visual task. Last, they were required to detect only the presence or absence of a task, not its details.

Blackwell postulated that each of these effects could be represented by a contrast multiplier, a number greater than one by which the contrast values from his original study could be multiplied to take into account these additional constraints. From numerous studies, these multiplying factors were determined as:

1.5	When and where factor
2.78	Dynamic seeing factor
2.51	Detection of detail factor (also called "common-sense" factor)
1.9	Accuracy factor

The accuracy factor (99%) came from Blackwell's early experiments, previously mentioned.

Blackwell's original curves (one set has been shown in Fig. 5.7) are a bit cumbersome to work with, especially if we have to shift them by the multipliers we have just presented. What was needed was a standard visibility reference function to which all other seeing situations could be related. Threshold seeing is the easiest condition to measure and it was decided to relate all seeing to this level. It was also decided to include the common-sense factor in the threshold level, the rationale being that most of the time we want to identify an object, not just detect its presence.

What standards for size and time should we use? Reading is the most common seeing task and the average duration of fixation pauses in reading is 0.2 s. Thus it was decided to standardize on a $\frac{1}{5}$-s exposure. Also a 4-minute luminous circular disc was chosen as the standard test object, largely because so much data had been gathered on the visibility of such an object and because 8 point type at 14 in subtends an angle of about 4 minutes at the eye.

Last, it was necessary to determine what characteristics of task, background, and surround luminance should be required. By "surround" we mean outside the immediate task background but still in the field of view. It was decided to stipulate a background luminance which is diffuse and to require a surround luminance which is uniform and equal in value to the background luminance. Such a lighting environment can be produced by placing the task at the center of a luminous diffusing sphere and thus can be readily achieved in the laboratory.

In Fig. 5.8, the agreed-upon visibility reference function (VRF) is shown as VL1 (visibility level 1). To repeat, this is a common-sense visibility-threshold curve for a 4-minute circular task with $\frac{1}{5}$-s time for seeing under diffuse illumination.

The next step was to decide what visibility level is reasonable to expect in "normal" seeing tasks. It was decided that normal seeing should require 99% accuracy, that it should not rely on knowing when and where an object might appear, and that it should include normal eye movement. Using the contrast multipliers previously listed, this means an overall multiplier of $1.5(2.78)(1.9) = 7.92 \simeq 8$. The second curve from the top in Fig. 5.8 is labeled VL8 (visibility level 8) and was named the visual performance criterion function (VPCF). For a given background and luminance, VL8 represents eight times the contrast of VL1. It is this visibility level which was aimed for until recently in prescribing task illuminance levels.

EXAMPLE 5.2

Assume the background of the 4-minute standard test object has a luminance of 10 fL. What luminance of the test object is required to achieve VL1? VL8?

Solution. From Fig. 5.8, to reach VL1 (threshold) requires a contrast of 0.11 at a background luminance of 10 fL. To reach VL8 requires a

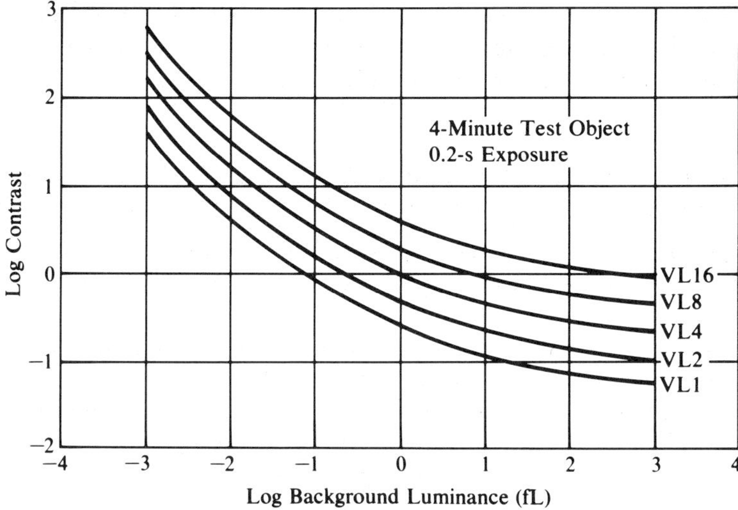

Figure 5.8 Visibility level curves.

contrast of 0.91. Using Eq. (5.3), test object luminance for VL1 is 11.1 fL and for VL8 is 19.1 fL. Note that the ratio of the two contrasts is approximately 8, as it should be.

Assume the luminous disc in a lighted environment has a background luminance L_b, a task luminance L_t, and thus, from Eq. (5.3), a contrast C. It is desired to reduce the task to threshold by changing its contrast while keeping background luminance constant. This may be done by adding a veiling luminance L_v and simultaneously reducing both L_t and L_b by a factor $K < 1$ such that

$$L_v + KL_b = L_b \tag{5.5}$$

and

$$\bar{C} = \frac{|L_v + KL_b - L_v - KL_t|}{L_v + KL_b} = KC \tag{5.6}$$

where $\bar{C}$ is the contrast of the disc task at threshold in the lighted environment. The threshold condition is reached by adding the veiling luminance while reducing uniformly the output of all luminaires in the environment so as not to alter the distribution of luminous flux within the environment. From Eq. (5.6), the visibility level present in the actual lighted environment is simply

$$\text{VL} = \frac{C}{\bar{C}} = \frac{1}{K} \tag{5.7}$$

Several visibility level curves in addition to VL1 and VL8 are shown in Fig. 5.8.

An instrument called a Visual Task Evaluator (VTE) can perform the measurement of $\bar{C}$ in a given environment without disturbing the lighting in that environment. The instrument places a veiling luminance over the task, while simultaneously, through neutral density filters, reducing the luminance of both the task and its background until threshold is reached.

EXAMPLE 5.3

Assume the 4-minute luminous disc task and its background have luminances of 5 and 7 fL, respectively. What is the visibility level?

Solution. From Eq. (5.3), the contrast is 0.29. From Fig. 5.8, with log $L_b = 0.845$, $\bar{C} = 0.13$. Thus, from Eq. (5.7), VL = 2.2.

As can be seen from the previous two examples, it is somewhat inaccurate to determine visibility levels from the curves in Fig. 5.8. Values of contrast as a function of background luminance for VL1 are given in Table 5.1. For any other VL, these contrasts are simply multiplied by that VL. The values in Table 5.1 may be determined approximately within 1% for 10

TABLE 5.1 CONTRAST VERSUS BACKGROUND LUMINANCE IN fL FOR VL1

L_b	$\bar{C}$	L_b	$\bar{C}$
0.00292	12.50	26	0.0939
0.00875	4.41	28	0.0929
0.0292	1.74	30	0.0918
0.0875	0.83	35	0.0896
0.292	0.43	40	0.0880
0.875	0.26	45	0.0866
2.92	0.1586	50	0.0855
4	0.1418	60	0.0835
5	0.1331	70	0.0818
6	0.1269	80	0.0803
7	0.1220	90	0.0790
8	0.1179	100	0.0781
9	0.1144	120	0.0764
10	0.1115	140	0.0750
11	0.1094	160	0.0738
12	0.1076	180	0.0728
13	0.1060	200	0.0719
14	0.1045	250	0.0700
15	0.1033	300	0.0688
16	0.1020	400	0.0666
17	0.1010	500	0.0652
18	0.1000	750	0.0625
19	0.0990	1000	0.0610
20	0.0981	1500	0.0591
22	0.0965	2000	0.0581
24	0.0951	2920	0.0574

fL $\leq L_b \leq$ 1600 fL by the equation

$$\bar{C} = 5.74 \times 10^{-4.94037/(2.25 + L_b^{-0.2})} \tag{5.8}$$

5.8 EQUIVALENT CONTRAST

Thus far we have dealt only with a 4-minute luminous disc task. We need to relate these results to other, more realistic, visual tasks (IES RQQ Committee, 1970).

Perhaps the most important factor affecting the visibility of a visual task in a given environment is veiling reflections. This effect results when rays from lighting units, usually on the ceiling of a room, reflect from the visual task directly into our eyes (as though from a mirror), rather than being absorbed or reflected diffusely. The net result is that the details of the visual task are partially or sometimes totally obscured.

Veiling reflections can be quantified as an effective reduction in contrast between the task and its background. To develop this concept, we define the equivalent contrast $\tilde{C}_0$ of an actual task as the contrast of the 4-minute luminous reference task which makes the reference task equal in visibility to the actual task when both are viewed for $\frac{1}{5}$ s under threshold seeing conditions at the center of a diffusing luminous sphere. The measurement procedure is as follows: The reference and actual tasks are placed in the sphere and viewed through the VTE. The background luminances of the two tasks, as seen in the VTE, are matched so as to insure that the adaptation of the operator is the same when observing each task. The contrast of the actual task is then reduced to threshold while keeping background luminance constant by using veiling luminance and neutral density filters, as described earlier. The fraction (K_a) of the actual task luminance remaining is measured. The process is repeated for the reference task and the fraction (K_r) of reference task luminance remaining is noted. Generally $K_a > K_r$. $\tilde{C}_0$ is obtained by multiplying the physical contrast of the reference task by the ratio of the two K's.

$$\tilde{C}_0 = \frac{K_r}{K_a} C_r \tag{5.9}$$

where C_r is the contrast of the reference task.

With $\tilde{C}_0$ in hand, we enter whatever VL curve is desired and determine the necessary background luminance L_r. In Fig. 5.9, we have chosen VL8. Dividing by the background reflectance ρ_b gives the amount of sphere illuminance E_r required. Thus E_r is the illuminance necessary to place the actual task at a visibility level of VL8 under reference lighting conditions in a sphere.

We turn now to the actual task in its real lighting environment and ask whether or not we can follow the same procedure to determine required illuminance. If we repeat our measurements of contrasts at threshold and at

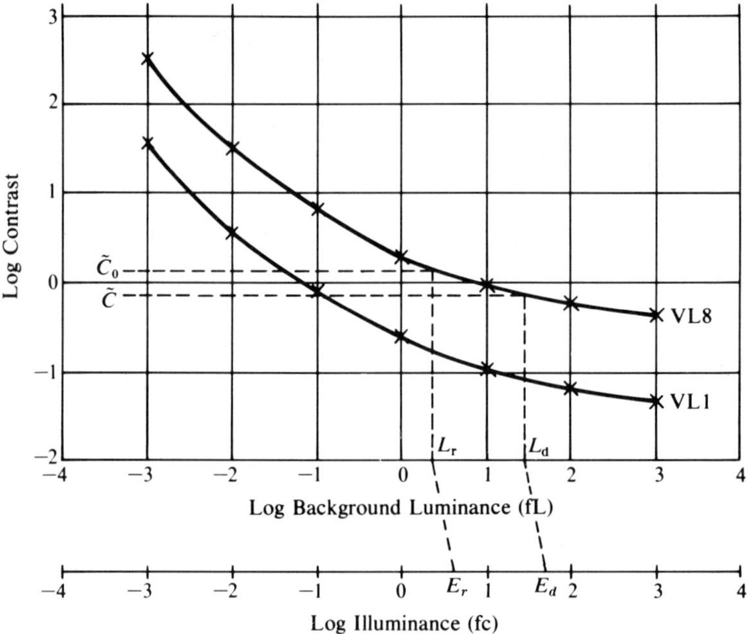

Figure 5.9 Equivalent contrasts.

the same viewing angle with the VTE, this time with the actual and reference tasks in the real lighting environment, we will usually get a lower value for equivalent contrast. The main reason for this is the veiling reflections previously mentioned. Particularly troublesome in this regard are luminaire luminances slightly in front of desks, which cause light to reflect into our eyes from printed material on the desk tops.

The new value of equivalent contrast is labeled $\tilde{C}$ and with it we can enter the VL8 curve in Fig. 5.9 and proceed as we did previously to find the required design values of task background luminance L_d and task illuminance E_d. We can then compare E_d with the actual illuminance present in the real lighting environment to see if we are meeting the VL8 criterion.

The effect of veiling reflections was to reduce the equivalent contrast of the task from $\tilde{C}_0$ to $\tilde{C}$. We define contract rendition factor (CRF) as the measure of this change in going from sphere conditions to the actual lighting environment. Thus

$$\text{CRF} = \frac{\tilde{C}}{\tilde{C}_0} \tag{5.10}$$

and is generally less than unity. We should note that a small percentage change in equivalent contrast produces a large percentage change in required illuminance if we are to maintain the same visibility level. In the example in Fig. 5.9, $\tilde{C}_0 = 1.3$ and $\tilde{C} = 0.7$. Thus CRF = 0.54, quite a low value. This gives $L_r = 2.6$ fL and $L_d = 30.9$ fL. Their ratio is 11.9. Thus a reduction in contrast of 46% resulted in an increase in luminance, and thus illuminance, of 1090% for the same task visibility. As a rough rule of thumb, a decrease of 10% in contrast requires an increase of 100% in illuminance for equal visibility.

Measurements of $\tilde{C}_0$ and $\tilde{C}$ using the contrast-reducing visibility meter and the standard reference task are awkward to perform. Experience has shown that CRF may be obtained reasonably accurately in terms of *physical* measurements and use of Eq. (5.3). Thus we place the task in the sphere, measure its luminance and that of its background in the direction of viewing and use Eq. (5.3) to obtain $\tilde{C}_0$. We repeat the process in the actual lighting environment and obtain $\tilde{C}$. Then Eq. (5.10) gives the value of CRF.

We will discuss methods of predetermining CRF and VL for a proposed lighting installation in Chap. 8.

EXAMPLE 5.4

Assume $\tilde{C}_0$ for a certain task at a certain viewing angle has been measured as 0.25. The task is placed in a lighted environment where the illuminance is 70 fc and the task and background luminances at the same viewing angle are 40 and 50 fL, respectively. Find VL, CRF, and the illuminance required in the sphere to achieve the same visibility as in the actual environment.

Solution. From Eq. (5.3), the constrast of the task in the lighted environment is

$$\tilde{C} = \frac{50 - 40}{50} = 0.20$$

From Table 5.1, the equivalent contrast of the task to achieve VLl at $L_b = 50$ fL is $\bar{C} = 0.0855$. Thus, from Eq. (5.7)

$$\text{VL} = \frac{0.20}{0.0855} = 2.3$$

is being achieved in the lighted environment.

From Eq. (5.10),

$$\text{CRF} = \frac{0.20}{0.25} = 0.80$$

If the task had threshold contrast in the lighted environment, it would have a contrast in the sphere of

$$\frac{0.0855}{0.80} = 0.1069$$

From Table 5.1, this yields a background luminance in the sphere of

$$L_b = 12.46 \text{ fL}$$

The background reflectance is

$$\rho_b = \tfrac{50}{70} = 0.714$$

and thus the required sphere illuminance is

$$\frac{12.46}{0.714} = 17.5 \text{ fc}$$

The result is that 17.5 fc of sphere illuminance provides the same visibility level for this task as 70 fc of illuminance in the actual lighted environment.

5.9 OTHER FACTORS AFFECTING VISUAL TASK PERFORMANCE

We have stated that veiling reflections are the predominant cause of reduced contrast. There are two other major factors which create contrast reduction, namely a disability glare factor (DGF) and a transient adaptation factor (TAF). Overall reduction in contrast is found from the product of CRF, DGF, and TAF.

Glare is the sensation produced by luminances within the visual field that are sufficiently greater than the luminance to which the eyes are adapted to cause annoyance, discomfort, and/or decrease in visibility and visual performance. There are two types of glare. Discomfort glare is glare which produces discomfort but does not necessarily impair visibility and visual performance. We will discuss this form of glare in Chap. 8 when we consider the quality of a lighting installation. Disability glare, on the other hand, is glare which reduces visibility and visual performance but does not necessarily produce discomfort. It comes in two forms: reflected glare, which produces the veiling reflections previously discussed, and direct glare. Direct disability glare is caused by luminances in the field of view which create stray light within the eye, producing a veil of light over the task image on the retina. Calling the veiling luminance L_v, the total luminances of task and background in the presence of direct disability glare become $L_t + L_v$ and $L_b + L_v$, respectively. Then from Eq. (5.3), contrast becomes

$$C = \frac{|L_b + L_v - L_t - L_v|}{L_b + L_v} = \frac{|L_b - L_t|}{L_b + L_v} \tag{5.11}$$

It is easily seen from Eq. (5.11) that L_v reduces contrast.

A second factor affecting visual task performance is transient adaptation, which is an effect resulting from an observer glancing about his environment. As he does so, he generally encounters a variety of luminances, producing brief transients in his eye sensitivity which reduce visibility. To obtain a TAF, it is necessary to obtain a measure of visibility under conditions of eye movement and a second measurement under conditions of eye fixation on the task, and take their ratio. Studies to determine this factor for typical situations of eye movement are underway but are incomplete and inconclusive at this time.

In addition to the factors which reduce contrast, there are other parameters which should be considered when deciding what illuminance to prescribe for a given lighting situation. One of these is the age of the person performing the visual task.

The eye undergoes several changes as a person ages. First, the lens becomes less flexible and there comes a time when the ciliary muscles are unable to alter its curvature sufficiently for it to focus on close work. It is then that reading glasses become necessary. The accommodation of the eye is expressed in diopters, which is equal to the reciprocal of the distance in meters from a plane 1.4 cm in front of the cornea (the spectacle plane) to the nearest point on which the eye can focus when using a lens that permits a distant object to be seen clearly. In a young person, an accommodation of 12 diopters (8 cm) is average. In a person 50 years old this is reduced to 2 diopters (50 cm). A second change is a continuing discoloration of the lens with age resulting in a selective absorption of certain wavelengths, particularly in the blue portion of the spectrum. Blue objects then tend to appear gray. A third change that occurs with age, and the one that we are most concerned with in prescribing illuminance, is a reduction in the size of the pupil, which requires that lighting levels be increased if good visibility is to be maintained. With a luminance of 1 fL, a person age 20 has a pupil diameter of about 5 mm. At age 50, the diameter is less than 4 mm. Other changes with age include an increased sensitivity to glare, a loss in contrast sensitivity, a decrease in speed of perception, and an increase in adaptation time.

Additional parameters to consider when prescribing illuminance are the speed and accuracy required of the worker. The processing of welfare application forms may require greater speed than the reading of the latest company sales brochure. The verification of a company's income and expenses may require greater accuracy than a third-grade spelling quiz. In determining whether or not speed and accuracy are important in the performance of a given visual task, one should ask:

1. Will speed increase productivity?
2. Will speed decrease costs?
3. Will errors produce an unsafe product?
4. Will errors affect health?
5. Will errors produce a poor public image?

If the answer to one or more of these questions in "yes," it will probably be necessary to increase the illuminance on the task.

5.10 PROBLEMS IN PRESCRIBING ILLUMINANCE

It could be inferred that the 30 or so years of visual research leading to the results presented in Fig. 5.8 would have provided the lighting community with a scientific basis for prescribing the illuminance required to perform a given visual task. Unfortunately the process through which an individual is able to carry out visual work is so complex that, despite the excellent research that has been done in the past, there is still much to do before such a scientific basis can be established and defended.

Let us reconsider VL1 and VL8 in Figs. 5.8 and 5.9. Recall that VL1 was for threshold seeing. The viewer knew when and where the object would appear and no eye movement was required prior to focusing on the task. On the other hand, VL8 included not knowing when or where the object would appear and required eye scan. It also required 99% accuracy as opposed to 50% accuracy for VL1.

Is VL8 actually eight times the contrast at threshold for a given background luminance? The answer is no. The conditions for VL8 are, with the exception of the 99% accuracy requirement, a new set of *threshold* conditions. Thus VL8 is in reality 1.9 times the contrast at threshold for this new set of threshold conditions. It is suprathreshold seeing, but not as much so as previously may have been implied.

Also, the assumption was made in developing VL8 that the human vision system was linear at a particular background luminance so that the overall effect of the contrast multipliers could be found by simply multiplying them together. There is evidence now that this assumption may be incorrect and that the multipliers are coupled nonlinearly in ways which we do not as yet know.

As a third point, we need to ask what is involved in the most common seeing task, namely reading. We know where the book is and when it's going to be there, so it does not appear that the when and where factor is needed for that task. Also, there is some eye scan involved in reading, but it does not involve moving the eye through large angular swings. Thus the dynamic seeing factor may need to be modified for reading. The integration of all these factors could mean that a level of say, VL3, would be adequate for reading. We simply don't know as yet.

An additional concern stems from measurements of $\tilde{C}_0$ which have been done for a variety of office and school tasks. These were performed in the sphere using the contrast-reducing visibility meter. The results are shown in Table 5.2, where ρ_b is the background reflectance and E is the illuminance required in the sphere to achieve VL8. For example, entering the VL8 curve in Fig. 5.9 with a contrast of 2.15 (8-point text type) yields a background luminance of 0.79 fL, which, when divided by ρ_b, gives a required sphere illuminance of 1.13 fc.

The concern with this table is twofold. First, experience throughout the world over many decades argues against 1.13 fc being sufficient illuminance for reading 8-point text type, even under quality diffuse lighting. It is possible to read at this level but one would not feel comfortable doing so nor would one's speed and comprehension be satisfactory. An analogy might be that a person could see sufficiently to drive a car along a country road on a moonlit night without the car's headlights being on, but the seeing would be difficult and could not be described as safe. The second concern with Table 5.2 is with the tremendous range of illuminance values and with the extreme

TABLE 5.2

	$\tilde{C}_0$	ρ_b	E for VL8 (fc)
6-point text type	1.39	0.7	2.98
8-point text type	2.15	0.7	1.13
10-point text type	2.36	0.7	0.94
12-point text type	3.01	0.7	0.6
ink writing	1.87	0.76	1.38
#2 pencil	0.68	0.76	63
#3 pencil	0.659	0.79	76.5
typing (good ribbon)	2.49	0.62	0.97
typing (poor ribbon)	0.459	0.62	3140
fifth carbon	0.65	0.53	133
Thermofax (poor copy)	0.527	0.63	589

sensitivity of E to small changes in $\tilde{C}_0$. It is difficult to believe that the ratio of illuminances to produce the same visibility for typing with poor and good ribbons is 3237:1.

The VL system had been used for many years to prescribe illuminances. During that time, the Illuminating Engineering Society of North America (IESNA) developed increasing concern about the validity of its use and about the wisdom of specifying a single level of illumination for a given task, irrespective of worker age, speed and accuracy requirements, and background reflectance. In 1979, a new procedure was approved which incorporates these factors, includes the essence of the many years of visual research that has been carried on, and results in illumination levels which are in better agreement with those adopted by the world community through the Commission Internationale De L'Eclairage (CIE).

5.11 SELECTION OF ILLUMINANCE VALUES FOR INTERIOR LIGHTING

The procedure to be described below was adopted by the IESNA in 1979 and is drawn in part from the *Guide on Interior Lighting* published by the CIE in 1975. The procedure will be described in two steps.

Step 1: Selection of Illuminance Category

Determine the types of visual activities to be carried out in each space to be lighted and, for each activity, consult Table 5.3(a) to determine the appropriate illuminance category. This table is adapted in part from Table 1.2 of the CIE Guide mentioned above. If values of $\tilde{C}_0$ are known for the particular tasks, these may be used instead of the activity descriptions in determining the illuminance category, except that, if $\tilde{C}_0 > 1.0$, 20 fc should be prescribed and the remaining step in the procedure ignored. Also, when using $\tilde{C}_0$, if the

TABLE 5.3(a)

Category	Illuminance range (fc)	$\tilde{C}_0$	Type of activity
A	2-3-5	—	Public areas with dark surroundings (e.g., lobbies)
B	5-7.5-10	—	Simple orientation for short temporary visits (e.g., corridors, storage rooms)
C	10-15-20	—	Working spaces where visual tasks are only occasionally performed (e.g., waiting rooms, reception desks)
D	20-30-50	0.75–1.0	Performance of visual tasks of high contrast or large size (e.g., printed material, typed originals, ink handwriting, rough industrial work)
E	50-75-100	0.62–0.75	Performance of visual tasks of medium contrast or small size (e.g., medium pencil handwriting, poorly printed or reproduced material, medium industrial work)
F	100-150-200	0.50–0.62	Performance of visual tasks of low contrast or very small size (e.g., hard pencil handwriting on poor quality paper, faded dittos, difficult industrial work)
G	200-300-500	0.40–0.50	Performance of visual tasks of low contrast and very small size over a prolonged period (e.g., fine industrial work, difficult inspection)
H	500-750-1000	0.30–0.40	Performance of very prolonged and exacting visual tasks (e.g., extra fine bench, machine, or assembly work)
I	1000-1500-2000	<0.30	Performance of very special visual tasks of extremely low contrast and small size (e.g., surgical procedures, sewing academic gowns)

Source: J. E. Kaufman (ed.), *IES Lighting Handbook,* Application volume, 1981 ed., IES, New York, 1981, fig. 2-2. Reprinted with permission.

task reflectance is between 5 and 20%, use the next higher illuminance category; if less than 5%, use two categories higher.

Step 2: Selection of Illuminance Level

For the illuminance category selected in step 1, determine which of the three values within the range of illuminance to choose. For categories D through I, this is done through Table 5.3(b). From that table, determine the appropriate weighting factor to use for each characteristic. Sum the weighting factors. If the sum is −2 or −3, choose the lowest of the three illuminances; if it is +2 or +3, choose the highest illuminance; otherwise choose the middle illuminance.

For categories A through C, a slightly different procedure is used because there are no sustained task activities in these areas. Thus only

TABLE 5.3(b)

Task and Worker Characteristics	Weighting factors		
	−1	0	1
Average worker age	<40	40–55	>55
Demand for speed and accuracy	Not important	Important	Critical
Task background reflectance	>0.70	0.30–0.70	<0.30

Source: J. E. Kaufman (ed.), *IES Lighting Handbook*, Application volume, 1981 ed., IES, New York, 1981, fig. 2-4. Reprinted with permission.

worker age and the average reflectance of the walls and floor (instead of the task background) are used in Table 5.3(b) in determining the illuminance level. If the sum of these two weighting factors is −1 or −2, choose the lowest of the three illuminances; if it is +1 or +2, use the highest value; otherwise use the middle value.

As a final point, it is recommended that 20 fc be regarded as the minimum acceptable horizontal illuminance for interior spaces where continuous work is done. Also, adjacent areas with different activities should not have illuminances differing from each other by more than a 5:1 ratio. For example, a corridor outside an office lighted to 50 fc should have an illuminance of at least 10 fc.

EXAMPLE 5.5

A conference room is being remodeled to become an engineering office where the engineers and their clients will review drawings and specifications for design work. The room is finished in dark wood paneling and dark carpeting. Task background reflectance is medium. Determine the desired illuminance on the conference table.

Solution. From Table 5.3(a), the illuminance category is estimated as E. People of all ages will be present (including those over 55), speed and accuracy are important, and the reflectances are 30 to 70%. Thus, from Table 5.3(b), a weighting factor of +1 is obtained. Therefore, from Table 5.3(a), 75 fc should be provided.

5.12 VISUAL PERFORMANCE

Vision researchers have been attempting for decades to obtain meaningful correlations between visibility, as measured by the four factors, and visual performance. Some results have been obtained. In one study of keypunch operators, productivity, as measured by the number of keypunch cards processed per minute, dropped an average of 12% when the illumination level was reduced from 150 to 50 fc. A second study involved a search and identification task in a warehouse environment. Workers in the study were required to locate price information for 15 lamps on a requisition list from 30 bin locations. The average productivity of the workers, as measured by processing time per requisition, increased 13.5% when the illuminance was raised from 5 to 50 fc. In a third study, the productivity of 500 employees in a Social Security office was monitored. The visual task was checking typed and occasionally handwritten copy of employee wage earnings against computer printout data. When the illumination level was lowered from 100 to 50 fc, productivity dropped almost 30%.

Six experiments involving visual performance have been conducted recently by the team of Dr. Stanley Smith and Dr. Mark Rea (1976, 1977). These involved the office tasks of reading, proofreading, check reading, reading-typing, and numerical verification, with performance based on speed and accuracy. Of 448 comparisons of performance under background luminance levels of 50 to 106 fL and 150 to 418 fL, performance was better in 225 cases under the higher level, in 195 cases under the lower level and there were 28 ties. According to the authors, this indicates clearly that performance improved as illuminance increased, even for these rather high luminance levels.

As a result of the above studies, and others, there seems to be rather general, although not unanimous, agreement that visual performance, in terms of speed and accuracy, improves as illuminance increases, up to background luminances of at least 1000 cd/m^2.

In addition to the work relating visual performance to illuminance (luminance), there is more recent work by Dr. Rea (1981, 1982) which strongly relates visual performance to contrast. The task was to find discrepancies in two number lists as contrast was varied. Time and errors were used as the dependent variables. Four methods of changing task contrast were used: lighting geometry, ink pigment density, ink specularity, and polarization of luminous flux. A general transfer function was obtained relating performance (score) to contrast and is presented graphically in Fig. 5.10.

With these recent results, it appears that we are nearing the time when visual performance, in terms of speed and accuracy, can be related directly to the four factors, particularly contrast and luminance. Perhaps then, with the inclusion of such other parameters as worker age, we will be able to recommend with confidence illuminance and luminance levels for a wide variety of visual tasks.

5.13 COLOR THEORIES AND CONCEPTS

Much of the foundation for our current understanding of color stems from Sir Isaac Newton and is contained in his treatise "Opticks" (Wintringham, 1951). Newton asserted that the sensation of color arose because different wavelengths produced different visual effects. He concluded that the color appearance of objects was due to selective wavelength reflection or transmission of incident light. He also pointed out that it was impossible to distinguish the color components in a color mixture and that colors could appear the same and yet have different spectral compositions.

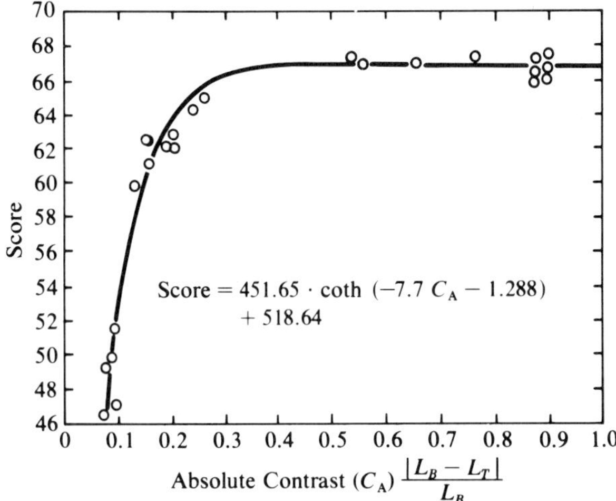

Figure 5.10 Relating performance to contrast. (*Source:* M. S. Rea, "Visual Performance with Realistic Methods of Changing Contrast." Reprinted from the April 1981 issue of the *Journal of IES* with permission of the Illuminating Engineering Society of North America.)

Newton was able to show that mixtures of adjacent spectral colors produced intermediate spectral colors but that a mixture of red and blue, which are far apart spectrally, produced purple, which is a nonspectral color. He further showed that several spectral colors could be added to produce white. He actually assumed seven primaries (red, orange, yellow, green, blue, indigo, and violet) and used these to calculate the colors of other spectral wavelengths.

In 1792, Wunsch was able to show that four of Newton's primaries could be obtained from the other three. This was the beginning of tri-color colorimetry. Then Sir Thomas Young (1773–1829, English physician and physicist) published a paper "On the Theory of Light and Colours" in 1801. This was the first serious attempt to explain color vision. Young hypothesized that the retina contained three types of "resonators" which responded selectively to different wavelengths. Initially he described these resonators as red, yellow and blue, essentially the pigment primaries. Later he changed to red, green and violet, nearly the light primaries, and concluded that all spectral colors could be produced by summations of these lights.

Young, in Lectures on Natural Philosophy in 1807, stated: "From three simple sensations, with their combinations, we obtain several primitive distinctions of colours, but the different proportions in which they may be combined afford a variety of tints beyond all calculations. The three simple sensations being red, green and violet, the three binary combinations are yellow, consisting of red and green; crimson, of red and violet; and blue of green and violet; and the seventh in order is white light, composed by all the three united."

Young's qualitative statement of the three-component color theory did not gain wide acceptance until H. L. F. von Helmholtz (1821–1894), German physicist, along with Maxwell, developed color-mixing equations. Helmholtz was one of the most brilliant scientists of the nineteenth century. He invented the opthalmoscope and clarified the mechanisms of sight in addition to developing a theory of color vision based on Young's work. In 1856 he published *Physiological Optics,* a complete study of the physics and physiology of human vision.

The Young-Helmholtz theory states that any color can be synthesized by a unique mixture of three lights of different colors. It assumes that the eye contains three kinds of receptors that react to wavelength selectively and in proportion to the light intensity at any given wavelength. Each of the three receptors is presumed to have its own set of nerves and the optic nerve and brain produce the integrating effect to give the sensation of the particular color being viewed. The theory has difficulty explaining the perception of single colors, particularly yellow. It does not adequately explain perceived shifts in color when luminance is changed, nor does it explain color blindness well. However, quantitative work by G. Wald (1964) of Harvard, among others, has lent considerable credence to the Young-Helmholtz theory.

A second theory of color vision which is widely used today is the Hering Opponent Colors theory, advanced by the German psychologist, E. Hering in 1878. The Hering theory has a response rather than a stimulus basis. It assumes six basic colors, red, yellow, green, blue, black and white, which act in three pairs of processes, rather than separately. There is a blue-yellow opponent pairing such that a color cannot appear bluish and yellowish simultaneously. In other words, blue cancels yellow and vice versa. The same is true of the red-green pair. For the white-black pair, a color may move toward the black or white from the gray, but not toward both at once. White does not cancel black, but rather blends with it to produce gray. The theory fails to explain two types of dichromatism, partial color blindness in which distinctions can be made between light and dark and either yellow and blue or green and red, but not both. Also it cannot explain the difference between brightness and lightness of a gray object. The Hering theory was quantified in the mid-1950s by L. M. Hurvich and D. Jameson through considering an excitation mechanism in the receptors and a response mechanism in the visual nerve center beyond the retina.

Many refinements of these two color theories have been advanced in recent years. However, the present state of affairs is that both theories seem to be necessary for an adequate description of the color vision process, even though there is very good evidence of late that there are at least three types of cone receptors in the eye with peak sensitivities in the violet-blue, green, and yellow-red regions of the spectrum, as shown in Fig. 5.11 (Kaufman, 1981).

It is important for the lighting designer to distinguish between color, object color, and perceived object color. The IESNA (Kaufman, 1981) defines color as "the characterisitics of light by which an observer may dis-

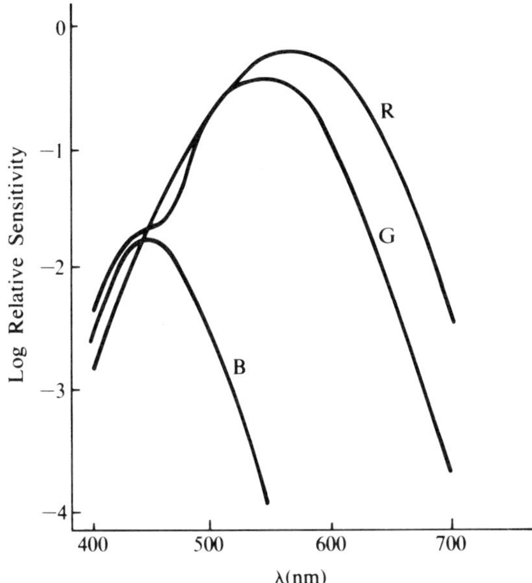

Figure 5.11 Sensitivity of cone receptors. [*Source:* J. E. Kaufman (ed.), *IES Lighting Handbook,* Reference volume, 1981 ed., IES, New York, 1981, fig. 3-5, with permission.]

tinguish between two structure-free patches of light of the same size and shape." This definition relates clearly to the spectral distribution of radiant power leaving the object in question. The IESNA goes on to define object color as "the color of the light reflected or transmitted by an object when illuminated by a standard CIE light source." This definition assumes that the radiation will be evaluated in terms of the standard CIE observer and under controlled conditions of viewing and adaptation. Thus object color does not tell us how an object will appear in a lighted space but rather categorizes its appearance when lighted by certain standard sources. Last, the IESNA defines perceived object color as "the color perceived as belonging to an object, resulting from characterisitcs of the object, of the incident light, and of the surround, the viewing direction and observer adaptation." This is an attribute that is perceived almost instantly by an observer and is influenced by past memory. It can be altered by altering the surroundings and the color last seen and by mood.

The single word "color" is often used to mean all three of the above entities, but, as noted, they can be quite different. It is perceived object color with which the lighting designer is often most concerned, because it is this attribute of color which is a major factor in determining how occupants feel about a lighted space and perceive objects in that space. Although we will quantify color in succeeding sections through the CIE system, it should be recognized that such a mathematical representation may not totally describe the overall effect that a color produces on a viewer.

5.14 COLORIMETRY

Colorimetry is defined as the science of measuring color and designating it systematically. One of the fundamental aspects of colorimetry is color mixing and matching.

Any color may be matched by proper proportions of three component colors spaced far apart in the spectrum, subject only to the constraints that no one of the components can be matched by a combination of the other two and that, in some cases, it may be necessary to add one of the components to the color being matched in order to achieve a match (equivalent to adding two of the components and subtracting the third). This is usually true when matching spectral colors.

The three component colors are called *primaries*, and are additive when adding colored lights and subtractive when adding paint pigments, as is illustrated in Fig. 5.12. In Fig. 5.12*a*, the additive system for lights is shown. The most common primaries are red, blue, and green. Two colors which, when mixed in the proper proportion, are achromatic (neutral) are called complementary colors. Addition of red and blue produces magenta, which is the complement of green. Blue and green form cyan, the complement of red. Red and green produce yellow, the complement of blue. The proper combination of all three primaries produces white.

The subtractive primaries for pigments are the complements of the additive primaries for lights, and vice versa, as shown in Fig. 5.12*b*. Here the combination of all three primaries yields black. Subtractive color mixing is commonly used for overlays in color printing and photography and with colored inks, dyes, and paints. The primaries in Fig. 5.12*a* and *b* can be considered as forming triangles, within which almost all colors lie (the exception is when a negative amount of one primary is required for a match).

The two most common color-measuring instruments in use today are the spectrophotometer, previously described in Chap. 3, and the colorimeter. The latter is a device for visually matching an unknown color with a combination of three (or more) colors of light, by varying the intensities of

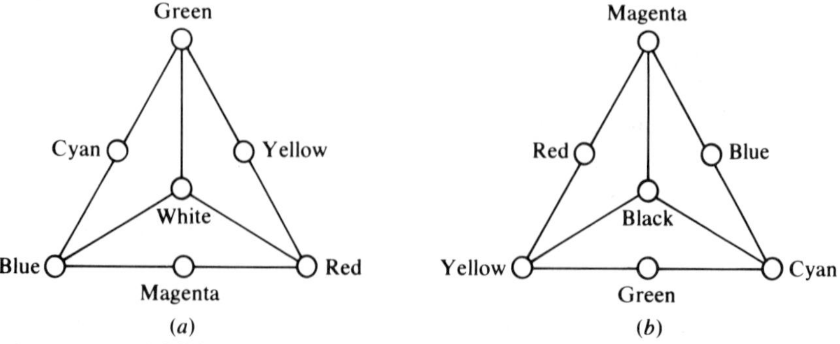

Figure 5.12 Additive and subtractive color mixtures.

each. Unlike the spectrophotometer, the colorimeter does not give the radiant power of a source in each wavelength band, but rather measures the overall radiant power for each primary component.

There are eight fundamental rules of tricolor colorimetry, which were formulated by Professor Ḥ. Grassman in "On the Theory of Compound Colors" in 1854. These may be stated as follows:

1. A color match is independent of luminance over a wide range.
2. The luminance of a mixture of colors is the sum of the individual luminances.
3. Any color can be matched by a mixture of no more than three colored lights, provided no two yield the third.
4. The components of a color mixture cannot be resolved by the eye.
5. A color match can be stated in equation form as

$$C = R(\mathrm{R}) + G(\mathrm{G}) + B(\mathrm{B}) \tag{5.12}$$

   This means that R units of (R), G units of (G), and B units of (B), when summed, yield a match with color C. The coefficients R, G, and B are called tristimulus values.
6. Color matches obey the law of addition. Thus, if C_1 matches C_2 and C_3 matches C_4, then $C_1 + C_3$ matches $C_2 + C_4$. This rule is known as Grassman's law.
7. Color matches obey the law of subtraction.
8. Color matches obey the law of transition. Thus if C_1 matches C_2 and C_2 matches C_3, then C_1 matches C_3.

5.15 COLOR-MEASURING SYSTEMS

There are several systems in use to describe the color of a source or object. All of these are based on three variables or descriptors. In the CIE system, color is expressed in terms of three tristimulus values, or, alternatively, in terms of two chromaticity coordinates and a measure of luminance. The result is a chromaticity diagram on which a given color's coordinates can be located. From these coordinates, the dominant wavelength and purity of the color may be determined. This system will be discussed in detail later in the chapter.

Other color systems are more visually related and subjective. The Plochere color system (Jerome, 1969) is an example of a colorant mixture system, in which pigments or dyes are blended in prescribed proportions. It consists of 1248 painted cards, with 26 hues, 6 shades for each hue and 8 tints for each shade.* The system is quite widely used by interior decorators.

An example of a color mixture system is the Ostwald system (GE, 1978; Jerome, 1969). This system is based on ideal reflecting objects, which

*Comparable to hue, value and chroma, as described below for the Munsell system.

have a constant spectral reflectance between two complementary visible wavelengths and a second (lower) constant spectral reflectance elsewhere in the visible spectrum. The lesser reflectance is called white content; the difference between unity and the greater reflectance is called black content; the difference between the two reflectances is called full color content. The sum of these three numbers is unity. The two complementary wavelengths determine hue and there are 24 uniformly spaced hues in the Ostwald system.

A third type of color system is a color appearance system. The two major examples are the Munsell and DIN systems (Kaufman, 1981; Jerome, 1969). The systems are similar, with the former the more widely used. In the Munsell system, a color is described in terms of hue, value, and chroma. There are 10 named hues, positioned uniformly around a hue circle, with each hue subdivided into 10 finer divisions. Each of these, in turn, is divided into 10 visually equal steps to describe value, the lightness or darkness of the color on a gray scale. Last, the deviation of a given hue from a neutral gray of the same value is called chroma. There are up to 14 chromas for each hue and value. As an example of the complete Munsell notation, a color designated a 6GY5/8 would have a named hue of green-yellow and subdivision 6, a value of 5 and a chroma of 8.

The Inter-Society Color Council and the National Bureau of Standards (Kelly and Judd, 1976) have developed a table of 267 standarized names for describing paint, based on the Munsell system. The table utilizes 28 basic hue names with up to 11 adverbs and adjectives for each hue to describe lightness and saturation (value and chroma).

The problem of color matching is a critical one for lighting designers and engineers. Two objects may appear to be of the same color under one source, for example, incandescent, but appear to be different in color under a second source, such as daylight. We define metamers as lights of the same color but of different spectral energy distributions. Two objects whose spectral reflectances differ but which appear to have the same color under one or more sources are called metameric objects. Two lamps which differ in spectral power distribution but appear to have the same color are called metameric souces.

There are several metrics that are used to describe the ability of one source or object to match the color of another source or object. Among these are correlated color temperature, (CCT), and color rendering index (R or CRI). Correlated color temperature was introduced in Chap. 3. It and color rendering index will be discussed later in this chapter, after we have developed the mathematics of the CIE system for describing color.

5.16 EARLY MATHEMATICAL FORMULATIONS OF COLOR

In the 1920s, W. D. Wright, J. Guild, and D. Judd conducted color-matching experiments which led to the "chromaticity characteristic of the standard observer" adopted by the CIE in 1931 (Boast, 1953; Wintringham, 1951).

The colorimeter used in these experiments is diagrammed in Fig. 5.13. The instrument has a divided visual field of 2 to 3 degrees. The spectral color to be matched is displayed in the left-half field, the mixture of the three primaries in the right-half field. Three calibrated attenuators are included to adjust the intensity of each primary.

The procedure used is similar to that employed by Gibson and Tyndall in obtaining the eye's spectral luminous efficiency curve. Monochromatic radiation of variable wavelength λ and contant power density impinges on the left-half field. Three monochromatic radiations of fixed wavelength and variable (through the attenuators) power density are superimposed on the right-half field. The primaries chosen were a red line at 700 nm, obtained by using a narrow filter with an incandescent source, a blue line at 435.8 nm, and a green line at 546.1 nm, obtained from a mercury spectrum. The sources were defined as standard by the CIE in 1931.

The data from these measurements of radiant fluxes of the CIE primaries which color match known amounts of radiant fluxes of spectral colors are shown in Fig. 5.14. These data are based on 17 observers and represent the average results of many observations by each. For each spectral color, each observer adjusted the attenuators in Fig. 5.13 until the two halves of the field appeared alike in both luminance and color. The watts of the spectral color and of the attenuator outputs of each primary were then measured. For many of the spectral colors, it was necessary to transfer one of the primaries to the left-half field to obtain a match, corresponding to a negative amount of that primary.

As an example of the use of the curves in Fig. 5.14, to match 1 W of pure spectral color at 500 nm requires approximately −20 W of the red primary, 0.35 W of the green primary, and 0.19 W of the blue primary. In Fig. 5.13, this particular match is achieved by adding 20 W of red to the spectral color in the left-half field.

It might appear that we are done, that the data in Fig. 5.14 give us a sound basis for the specification of color. Such is not the case for several reasons. First of all, the curves in Fig. 5.14 are difficult to use with any degree of accuracy. The watts of red far exceed the watts of green and blue

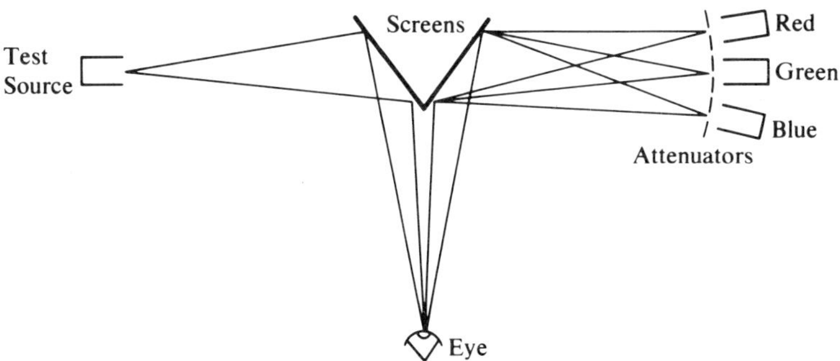

Figure 5.13 Colorimeter for basic color experiment.

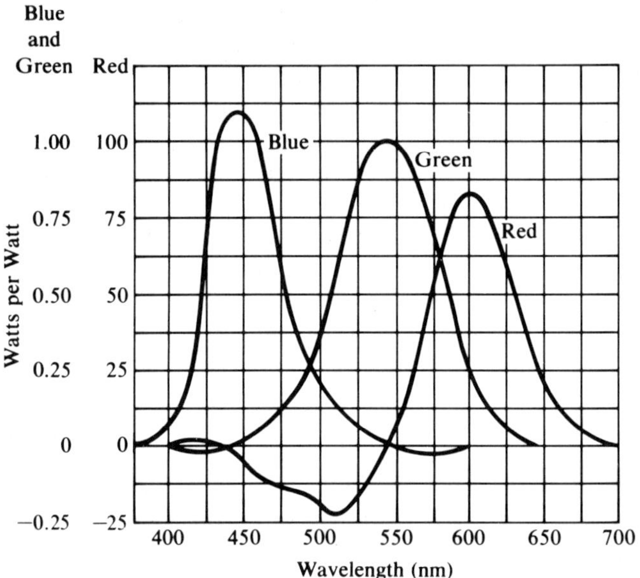

Figure 5.14 Wattages of CIE primaries to match spectral colors. (*Source:* G. M. Glasford, *Fundamentals of Television Engineering*, McGraw-Hill, New York, 1955.)

and negative amounts of red are required at several wavelengths. We could try to remedy the first difficulty by replotting the curves in Fig. 5.14 on a lumens per watt basis. This would entail multiplying each curve by 680* times its $V(\lambda)$ value, 0.0041 for red, 0.9842 for green, and 0.0177 for blue. However, we would now find that the luminous flux of the green primary far exceeds those of the red and blue primaries. Thus, for either plot, if we are to use the same number of significant figures for each primary, different numbers of decimal places will be required and this will make summations inaccurate.

An additional deficiency of Fig. 5.14 is that it is a plot of absolute, rather than relative, measurements and thus is quite sensitive to observer error. There is no reference base with which the observer is required to become calibrated at the outset, so that one can be assured that variations between observers are minimized and that sequential measurements by the same observer are consistent.

A final problem with Fig. 5.14 is that luminance and color are locked together in the matching process. Observers tend to disagree more on luminance matches than they do on color matches.

As an alternative to Fig. 5.14, two additional sets of color specification curves have been developed and these form the basis of modern colorimetry.

*At the time of this work, the luminous efficacy constant was 680, rather than the current 683.

The first is a set of tristimulus value curves. The tristimulus values are proportional to the number of units of the spectral primaries which will match a given quantity of spectral radiance (the radiometric equivalent of luminance) when the primaries are mixed.

A second basic way to specify color is to use chromaticity coordinates and luminance. Again three parameters are required, two normalized chromaticity values to describe the color of the object and a separate measurement or calculation to obtain its luminance. We will describe each of these specification systems in some detail.

Both types of color specification are based on calibrating the curves in Fig. 5.14 in terms of a "reference white." Equal energy white was chosen, with, by definition, the same radiance in watts per square meter per steradian per nanometer at each wavelength within the visible spectrum. Calculations were made to determine the luminance required of each primary which, when summed, would produce a match with unit radiance per nanometer of equal energy white.

The luminance of equal energy white with unit radiance per nanometer is

$$L_w = 680(1)\int_{380}^{770} V(\lambda)\, d\lambda \tag{5.13}$$

The area under the $V(\lambda)$ curve was obtained in Sec. 3.6 and has the value of 106.858. Thus

$$L_w = 680(1)(106.858) = 72{,}660 \text{ cd/m}^2 \tag{5.14}$$

This luminance must be equated to the luminances provided by the three primaries. Thus

$$L_w = 680\left[V_r(\lambda)\int_{380}^{770} S_r(\lambda)\, d\lambda + V_g(\lambda)\int_{380}^{770} S_g(\lambda)\, d\lambda + V_b(\lambda)\int_{380}^{770} S_b(\lambda)\, d\lambda\right] \tag{5.15}$$

where the S's inside the integrals are spectral radiances.*

When the three integrals in Eq. (5.15) are evaluated, using the data from Fig. 5.14, the ratios of the luminances of the primaries are:

$$L_r : L_g : L_b = 1 : 4.5907 : 0.0601 \tag{5.16}$$

Thus, from Eq. (5.15),

$$72{,}660 = L_r + 4.5907L_r + 0.0601L_r \tag{5.17}$$

giving the sizes of the luminances as

$$L_r = 12{,}860 \text{ cd/m}^2 \qquad L_g = 59{,}030 \text{ cd/m}^2 \qquad L_b = 722.8 \text{ cd/m}^2 \tag{5.18}$$

*We will use the symbol S to represent radiant flux, irradiance, radiant emittance, or radiance. This is to avoid confusion between radiometric and photometric quantities, which have the same letter designations.

We are now ready to scale the curves in Fig. 5.14. The wattage peaks of these curves are at

$$\begin{aligned} \text{Red} &= 85.4\ \text{W/W} \\ \text{Green} &= 1.00\ \text{W/W} \\ \text{Blue} &= 1.09\ \text{W/W} \end{aligned} \tag{5.19}$$

To visually evaluate these curves requires scaling each by the appropriate $V(\lambda)$ factor.

$$\begin{aligned} &\text{Red:} && 0.0041(85.4) = 0.350 \\ &\text{Green:} && 0.9842(1.00) = 0.984 \\ &\text{Blue:} && 0.0177(1.09) = 0.0193 \end{aligned} \tag{5.20}$$

Finally, we scale each curve by the luminance ratios in Eq. (5.16)

$$\begin{aligned} &\text{Red:} && \frac{0.350}{1} = 0.350 \\ &\text{Green:} && \frac{0.984}{4.5907} = 0.214 \\ &\text{Blue:} && \frac{0.0193}{0.0601} = 0.321 \end{aligned} \tag{5.21}$$

The quantities in Eqs. (5.21) are the peaks of the tristimulus value curves. These curves, with the labels $\bar{r}(\lambda)$, $\bar{g}(\lambda)$ and $\bar{b}(\lambda)$, are plotted in Fig. 5.15.

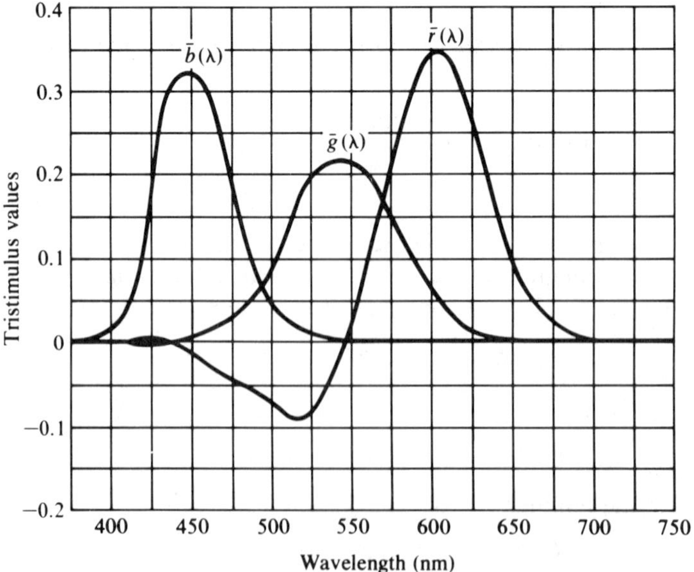

Figure 5.15 Tristimulus values of spectral stimuli of equal radiance. (*Source:* G. M. Glasford, *Fundamentals of Television Engineering*, McGraw-Hill, New York, 1955.)

Each of the tristimulus curves is simply a scaled version of its counterpart in Fig. 5.14. The scaling was carried out so that equal energy white would require the same number of units of each primary for a match. This requires that the areas under the tristimulus curves be equal, which is guaranteed by the luminance scaling we did using Eq. (5.16). Thus

$$\int_{380}^{770} \bar{r}(\lambda)\, d\lambda = \int_{380}^{770} \bar{g}(\lambda)\, d\lambda = \int_{380}^{770} \bar{b}(\lambda)\, d\lambda \tag{5.22}$$

For any color of radiance $S(\lambda)$, we can write

$$\begin{aligned} R &= K \int_{380}^{770} \bar{r}(\lambda) S(\lambda)\, d\lambda \\ G &= K \int_{380}^{770} \bar{g}(\lambda) S(\lambda)\, d\lambda \\ B &= K \int_{380}^{770} \bar{b}(\lambda) S(\lambda)\, d\lambda \end{aligned} \tag{5.23}$$

where R, G, and B are the numbers of units of the primaries required for a match. The areas in Eqs. (5.22) can be calculated as 18.91 each. Then, if unit radiance of equal energy white is to be matched by one unit of each primary, the constant K must be assigned the value $1/18.91 = 0.05288$. Thus, for equal energy white of 1 W/(sr · m² · nm), R, G, and B in Eqs. (5.23) are each unity.

EXAMPLE 5.3

Find R, G, and B for unit radiance of a spectral color at 500 nm. Find the luminance of this color.

Solution. From Fig. 5.15, the approximate tristimulus values are:

$$\bar{r}(\lambda) = -0.075 \qquad \bar{g}(\lambda) = 0.085 \qquad \bar{b}(\lambda) = 0.050$$

Then, from Eqs. (5.23),

$$\begin{aligned} R &= 0.05288(-0.075) = -0.00397 \\ G &= 0.05288(0.085) = 0.00449 \\ B &= 0.05288(0.050) = 0.00264 \end{aligned}$$

The luminance of the color is obtained using Eqs. (5.18).

$$\begin{aligned} L &= 12{,}860(-0.00397) + 59{,}030(0.00449) + 722.8(0.00264) \\ &= -51.1 + 265.0 + 1.9 \\ &= 216 \text{ cd/m}^2 \end{aligned}$$

As a check, this luminance can also be found from $680\, V(500) = 680(0.323) = 220$ cd/m².

The calculation of luminance in Example 5.3 can be generalized for any spectral color as

$$680\ V(\lambda) = L_r R + L_g G + L_b B$$
$$= K[L_r\,\bar{r}(\lambda) + L_g\,\bar{g}(\lambda) + L_b\,\bar{b}(\lambda)] \tag{5.24}$$

The second basic system for specifying color is through chromaticity coordinates. For each pure spectral color, we calculate R, G, and B from Eqs. (5.23). We then form the ratios

$$r = \frac{R}{R+G+B}$$
$$g = \frac{G}{R+G+B}$$
$$b = \frac{B}{R+G+B} \tag{5.25}$$

These are the chromaticity coordinates and values of these at 5-nm intervals were standardized by the CIE in 1931 as the definition of the chromaticity characteristics of the standard observer. They are plotted in Fig. 5.16. In this system, reference white has coordinates $(\frac{1}{3}, \frac{1}{3}, \frac{1}{3})$. The coordinates of the primaries are (1,0,0), (0,1,0), and (0,0,1). Since, from Eqs. (5.25),

$$r + g + b = 1 \tag{5.26}$$

there are only two independent chromaticity coordinates, and we have com-

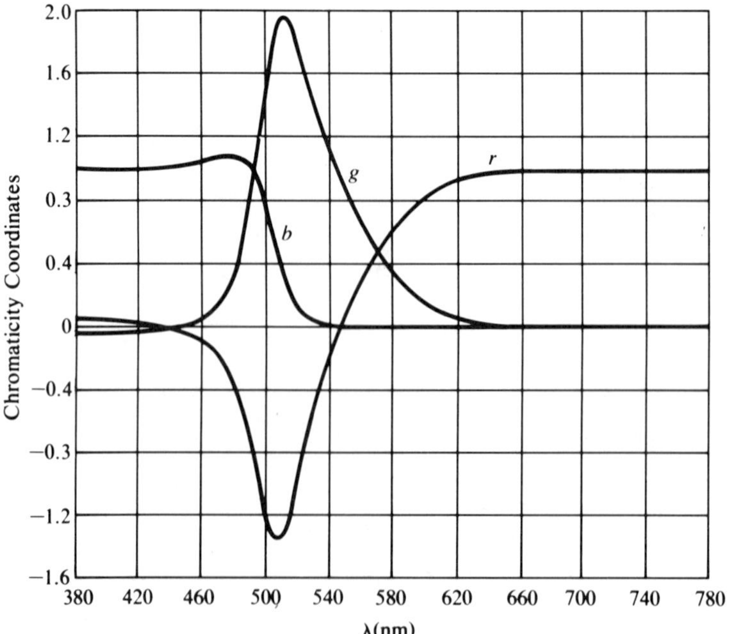

Figure 5.16 Chromaticity coordinates of spectral colors.

pletely lost luminance information in Eqs. (5.25). It may be regained from a separate calculation using Eq. (5.24).

EXAMPLE 5.4

Find the chromaticity coordinates of the radiation in Example 5.3.

Solution. From Example 5.3 and Eqs. (5.25),

$$R + G + B = 0.00316$$

$$r = \frac{-0.00397}{0.00316} = -1.26$$

$$g = \frac{0.00449}{0.00316} = 1.42$$

$$b = \frac{0.00264}{0.00316} = 0.84$$

These values could also be read directly from Fig. 5.16.

5.17 THE CIE COLOR SPECIFICATION SYSTEM

There are two major deficiencies in the color system which was presented in the preceding section. First, it contains both positive and negative tristimulus values. This makes summations awkward and subject to error. Second, the luminance of a color mixture depends on all three tristimulus values, again complicating the calculations.

As was stated at the outset, there is nothing unique about the set of primaries that was chosen, except that, because they are physical primaries, actual experiments can be carried out using them. Other than this, a set of nonphysical primaries would have done just as well.

The system ultimately adopted by the CIE in 1931 uses nonphysical primaries, denoted by X, Y, and Z. These are derived from the R, G, B physical primaries through a linear transformation, under the following conditions:

1. The primaries are chosen so that all real colors have positive tristimulus values.
2. Calculations of luminance should involve only one of the primaries. This primary (Y) is chosen equal to $V(\lambda)$, making luminance calculations identical to those for a black and white system.
3. The scales of the new primaries are chosen so that one unit of each primary yields equal-energy white.
4. Two of the primaries (X and Y) are chosen such that the line connecting them is tangent to the spectral locus at 770 nm. This makes the value of the third primary (Z) negligible over a considerable range at the red end of the spectrum, simplifying calculations.

The transformation from one set of primaries to the other is straightforward but messy mathematically (see Wintringham, 1951). The final result is

$$\begin{aligned} X &= 2.7690R + 1.7518G + 1.1300B \\ Y &= 1.0000R + 4.5907G + 0.0601B \\ Z &= 0.0000R + 0.0565G + 5.5943B \end{aligned} \tag{5.27}$$

Note that since Y contains all of the luminance information, its coefficients are those originally derived in Eq. (5.16).

These equations are true for any color. Thus they apply to the pure spectral colors. If we insert the R, G, and B values for the spectral colors, obtained as in Example 5.3, into Eqs. (5.27), we will obtain the X, Y, and Z of the spectral colors. These are the spectral tristimulus values for the new system of primaries and are labeled $\bar{x}$, $\bar{y}$, and $\bar{z}$. The chromaticity coordinates for the new system are labeled x and y and are obtained from the tristimulus values using the form of Eqs. (5.25). The spectral tristimulus values and chromaticity coordinates for the standard CIE (1931) observer are presented in Table 5.4. These are for a field of view of from 1 to 4°. Additional data have been obtained for the CIE (1964) supplementary observer when the field of view exceeds 4° (Kaufman, 1981).

Just as in the case of the physical primaries, the tristimulus values of any color can be written in terms of the tristimulus values of the spectral colors as

$$X = \int S(\lambda)\bar{x}(\lambda)\,d\lambda \qquad Y = \int S(\lambda)\bar{y}(\lambda)\,d\lambda \qquad Z = \int S(\lambda)\bar{z}(\lambda)\,d\lambda \tag{5.28}$$

where $S(\lambda)$ is spectral irradiance, radiant flux, radiance, or radiant exitance of the source or object. Then the chromaticity coordinates are obtained from

$$\begin{gathered} D = X + Y + Z \\ x = \frac{X}{D} \qquad y = \frac{Y}{D} \qquad z = \frac{Z}{D} \\ x + y + z = 1 \end{gathered} \tag{5.29}$$

The x and y chromaticity coordinates for the spectral colors are plotted in Fig. 5.17. All real colors are included within this locus and the line connecting its ends. The physical primaries of Sec. 5.16 are located on the spectral locus. Their coordinates are, from Table 5.4:

		x	y
(R)	700 nm	0.7347	0.2653
(G)	546.1 nm	0.2757	0.7147
(B)	435.8 nm	0.1661	0.0094

If these three points are connected by straight lines to form a triangle, as shown in Fig. 5.17, then all colors within the triangle can be matched by positive amounts of these three physical primaries. Colors outside the triangle

TABLE 5.4 CIE TRISTIMULUS VALUES AND CHROMATICITY COORDINATES

	Spectral tristimulus values on equal-energy basis			Chromaticity coordinates of pure spectral colors	
λ, μ	$\bar{x}$	$\bar{y}$	$\bar{z}$	x	y
0.38	0.0014	0.0000	0.0065	0.1741	0.0050
0.39	0.0042	0.0001	0.0201	0.1738	0.0049
0.40	0.0143	0.0004	0.0679	0.1733	0.0048
0.41	0.0435	0.0012	0.2074	0.1726	0.0048
0.42	0.1344	0.0040	0.6456	0.1714	0.0051
0.43	0.2839	0.0116	1.3856	0.1689	0.0069
0.44	0.3483	0.0230	1.7471	0.1644	0.0109
0.45	0.3362	0.0380	1.7721	0.1566	0.0177
0.46	0.2908	0.0600	1.6692	0.1440	0.0297
0.47	0.1954	0.0910	1.2876	0.1241	0.0578
0.48	0.0956	0.1390	0.8130	0.0913	0.1327
0.49	0.0320	0.2080	0.4652	0.0454	0.2950
0.50	0.0049	0.3230	0.2720	0.0082	0.5384
0.51	0.0093	0.5030	0.1582	0.0139	0.7502
0.52	0.0633	0.7100	0.0782	0.0743	0.8338
0.53	0.1655	0.8620	0.0422	0.1547	0.8059
0.54	0.2904	0.9540	0.0203	0.2296	0.7543
0.55	0.4334	0.9950	0.0087	0.3016	0.6923
0.56	0.5945	0.9950	0.0039	0.3731	0.6245
0.57	0.7621	0.9520	0.0021	0.4441	0.5547
0.58	0.9163	0.8700	0.0017	0.5125	0.4866
0.59	1.0263	0.7570	0.0011	0.5752	0.4242
0.60	1.0622	0.6310	0.0008	0.6270	0.3725
0.61	1.0026	0.5030	0.0003	0.6658	0.3340
0.62	0.8544	0.3810	0.0002	0.6915	0.3083
0.63	0.6424	0.2650	0.0000	0.7079	0.2920
0.64	0.4479	0.1750	0.0000	0.7190	0.2809
0.65	0.2835	0.1070	0.0000	0.7260	0.2740
0.66	0.1649	0.0610	0.0000	0.7300	0.2700
0.67	0.0874	0.0320	0.0000	0.7320	0.2680
0.68	0.0468	0.0170	0.0000	0.7334	0.2666
0.69	0.0227	0.0082	0.0000	0.7344	0.2656
0.70	0.0114	0.0041	0.0000	0.7347	0.2653
0.71	0.0058	0.0021	0.0000	0.7347	0.2653
0.72	0.0029	0.0010	0.0000	0.7347	0.2653
0.73	0.0014	0.0005	0.0000	0.7347	0.2653
0.74	0.0007	0.0003	0.0000	0.7347	0.2653
0.75	0.0003	0.0001	0.0000	0.7347	0.2653

require a negative amount of one primary, as we found in Sec. 5.16. In particular, note that many of the more saturated greens and cyans require negative amounts of the red primary.

In the early development of color television, it was thought that mov-

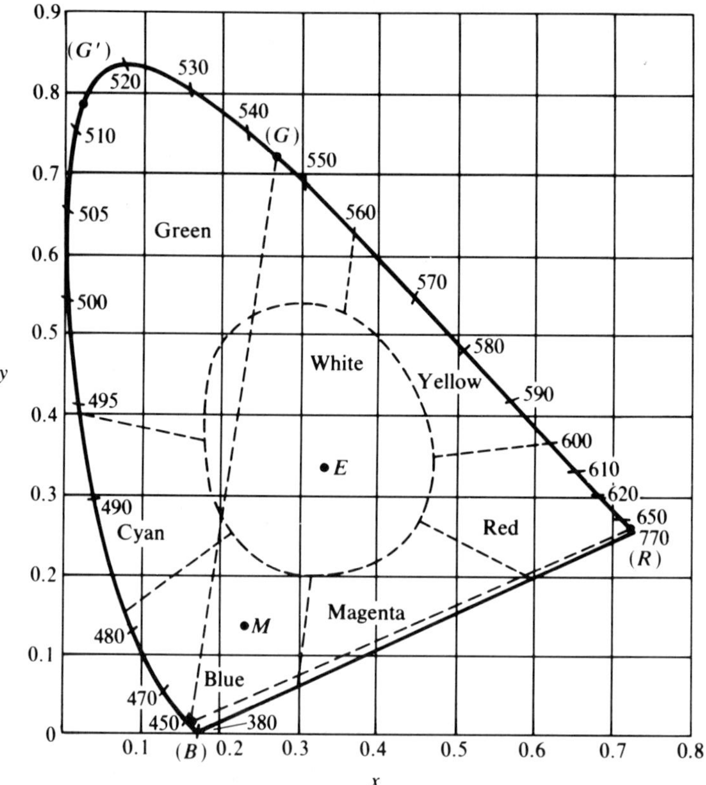

Figure 5.17 CIE chromaticity diagram.

ing the green primary to a new location, such as (G') in Fig. 5.17, might produce a better color picture. This change would include more of the saturated blues and greens but would exclude more of the yellows and oranges. However experience showed that the latter are more important for pleasing color reproduction than the former, particularly with regard to skin tones. Thus the CIE primaries are a better choice.

Equal-energy white (E) is shown in Fig. 5.17, with coordinates $x = y = 0.3333$. This point is close to the center of the region labeled "white," where the unsaturated colors lie. Actually the color and white regions are not sharply defined, but rather blend, one into another. Also, "perceived white" is a function of the current adaptation of the eyes and thus the "white point" can lie anywhere in the central region of the diagram. Nevertheless, it is possible to show roughly the regions of white, the three primary colors of light, and their three complements, and this has been done in Fig. 5.17 with dashed lines radiating outward from point E.

The nonphysical primaries used to create Fig. 5.17 do not appear on the diagram. Their coordinates are at (1, 0), (0, 1), and (0, 0). As expected, these are located outside the region of real colors defined by the horseshoe locus.

EXAMPLE 5.5

Verify that the chromaticity coordinates of the physical green primary of Sec. 5.16 are at $x = 0.2757$ and $y = 0.7147$.

Solution. From Table 5.4 and Eqs. (5.29), after interpolation,

$$x = \frac{0.3776}{0.3776 + 0.9790 + 0.0132} = 0.2757$$

$$y = \frac{0.9790}{0.3776 + 0.9790 + 0.0132} = 0.7147$$

EXAMPLE 5.6

Find the chromaticity coordinates of the mercury lamp in Example 3.5.

Solution. The calculations will be done in tabular form, using Eqs. (5.28) and Table 5.4.

λ (nm)	M_i(mW/cm²)	$\bar{x}_i$	$\bar{y}_i$	$\bar{z}_i$	$\bar{x}_i M_i$	$\bar{y}_i M_i$	$\bar{z}_i M_i$
404.7	1.0	0.0280	0.0008	0.1335	0.0280	0.0008	0.1335
435.8	0.8	0.3213	0.0182	1.5953	0.2570	0.0146	1.2762
546.1	0.1	0.3776	0.9790	0.0132	0.0378	0.0979	0.0013
578.0	0.2	0.8855	0.8864	0.0018	0.1771	0.1773	0.0004
					0.4999	0.2906	1.4114

Thus $X = 0.4999$, $Y = 0.2906$, and $Z = 1.4114$. Then, from Eqs. (5.29)

$$D = 2.2019 \qquad x = 0.227 \qquad y = 0.132 \qquad z = 0.641$$

The point is labeled M (for mercury lamp) on Fig. 5.17, and lies in the blue region of the chromaticity diagram.

5.18 COLOR SPECIFICATIONS OF SOURCES

Consider first a blackbody radiator. For each temperature, we can use Eqs. (5.28), with $S(\lambda)$ replaced by the M_λ of a blackbody from Eq. (3.3), and calculate X, Y, and Z. Then the chromaticity coordinates of the blackbody for each temperature may be obtained from Eqs. (5.29).

The chromaticities from such calculations are plotted on the CIE diagram in Fig. 5.18 as a function of temperature. The resulting curve is called the Planckian locus. We see that near 1000 K, the blackbody is red. It is yellow near 3000 K, white near 5000 K, bluish-white near 10,000 K, and pale blue near 30,000 K. It must be pointed out, as was done earlier, that the perceived color of a blackbody at each of these temperatures may be quite different from the color indicated by its chromaticity coordinates because of viewer adaptation.

The chromaticity of a source lying on the Planckian locus may be specified in terms of color temperature, whereas the chromaticity of a source

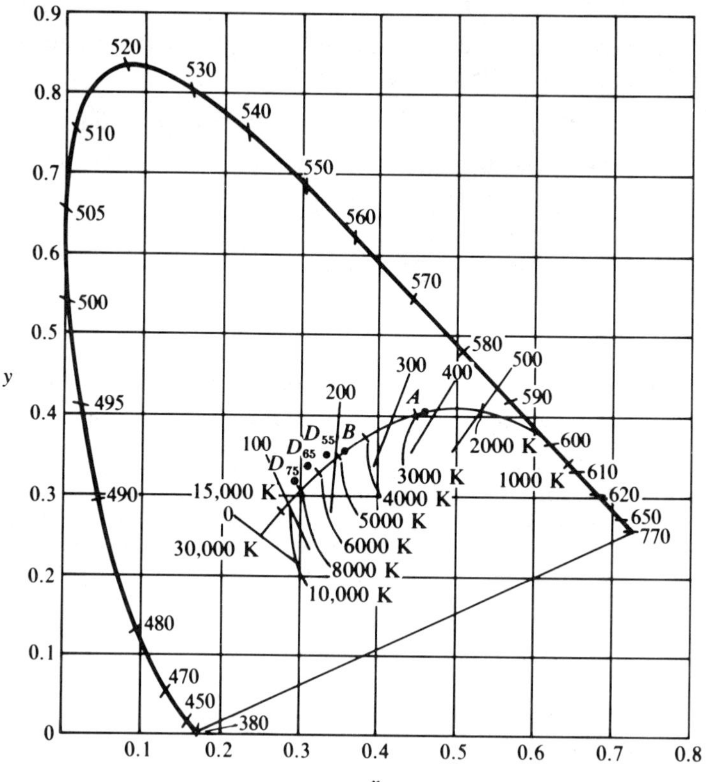

Figure 5.18 Source loci.

lying near to the Planckian locus may be specified in terms of correlated color temperature (CCT). Recall that in Chap. 3 we defined CCT as the absolute temperature at which a blackbody must operate if its output is to be the closest possible color match to the output of the light source under consideration. Stated another way, CCT is the absolute temperature of a blackbody whose chromaticity most nearly approximates that of the light source.

CCTs of sources whose chromaticity coordinates lie near the Planckian locus can be determined approximately by plotting lines of constant CCT on the chromaticity diagram near the locus. Equal color differences on the Planckian locus are nearly in direct proportion to reciprocal color temperature, rather than color temperature itself. The customary unit is reciprocal megakelvins obtained by dividing 10^6 by the color temperature. Lines of constant CCT are shown in 100-$(\text{MK})^{-1}$ intervals in Fig. 5.18.

We turn now to a discussion of standard sources and illuminants.* In addition to specifying the standard observer and the chromaticity coordinate

*The former is a specified physical emitter of light; the latter is a specified spectral power distribution.

system, the CIE has also defined three standard sources and three standard illuminants, designated A, B, C, D_{55}, D_{65}, and D_{75}. Source A is a tungsten filament incandescent lamp operating at 2856 K. It matches very closely a blackbody operating at that temperature. Sources B and C are obtained by combining source A with two liquid filters. Source B is designed to approximate noon sunlight and has a CCT of 4874 K; source C has a CCT of 6774 K and represents a combination of direct sunlight and clear skylight.

The three standard illuminants represent daylight distributions with CCTs of 5500 K, 6500 K, and 7500 K. These illuminants are nonphysical sources specified in terms of their spectral power distributions. They were reconstituted from 622 spectral irradiance distributions of daylight measured by D. B. Judd, D. L. MacAdam, and G. Wysecki at three widely separated locations (Jerome, 1969). They found that all of the data taken could be represented closely by chromaticities defined by

$$y = 2.870x - 3.000x^2 - 0.275 \qquad (5.30)$$

Observers have found since that sources B and C do not represent common daylight distributions adequately. Thus the current CIE recommendation is that source A and illuminant D_{65} be used in evaluating the spectral reflectance or transmittance characteristics of colored objects.

Relative spectral power distributions of the CIE standard sources and illuminants are given in Table 5.5. These have been normalized so that each has the value 1000 at $\lambda = 560$ nm. The chromaticity coordinates for these emitters are given in Table 5.6 and are shown on the chromaticity diagram in Fig. 5.18. The chromaticity coordinates and CCTs for several common lamp types are given in Table 5.7 and the spectral power distribution curves for common fluorescent and HID lamps are shown in Fig. 5.19.

EXAMPLE 5.7

Verify the numbers given in Table 5.6 for the chromaticity coordinates of source C.

Solution. We must evaluate the integrals in Eqs. (5.28) when $S(\lambda)$ is the spectral power distribution of source C. This has been done (Kaufman, 1981) for all six of the standard emitters and the results are given in Table 5.8, where values have been normalized so that

$$\int_0^\infty S(\lambda)\bar{y}(\lambda)\, d\lambda = 100 \qquad (5.31)$$

From that table, and using Eqs. (5.29),

$$D = 98.041 + 100.000 + 118.103 = 316.144$$

$$x = \frac{98.041}{316.144} = 0.3101$$

$$y = \frac{100.000}{316.144} = 0.3163$$

TABLE 5.5 RELATIVE SPECTRAL POWER DISTRIBUTIONS OF CIE STANDARD EMITTERS

λ(nm)	*A*	*B*	*C*	D_{55}	D_{65}	D_{75}
380	98	218	313	326	500	667
390	121	304	450	381	546	700
400	147	402	601	610	828	1019
410	177	507	765	686	915	1119
420	210	615	932	716	934	1128
430	247	711	1067	679	867	1031
440	287	786	1154	856	1049	1212
450	331	831	1178	980	1170	1330
460	378	859	1169	1005	1178	1324
470	429	895	1176	999	1149	1273
480	482	926	1177	1027	1159	1268
490	539	939	1146	981	1088	1178
500	599	916	1065	1007	1094	1066
510	661	882	972	1007	1078	1137
520	725	871	920	1000	1048	1087
530	791	897	931	1042	1077	1104
540	859	943	970	1021	1044	1063
550	929	982	999	1030	1040	1049
560	1000	1000	1000	1000	1000	1000
570	1072	998	972	972	963	956
580	1144	982	929	977	958	942
590	1217	965	885	914	887	870
600	1290	953	852	944	900	872
610	1363	958	840	951	896	861
620	1436	970	837	942	877	836
630	1508	982	836	904	833	787
640	1580	994	834	923	837	784
650	1650	1011	838	889	800	748
660	1720	1021	835	903	802	743
670	1788	1020	820	939	823	754
680	1854	1011	798	900	783	716
690	1919	988	762	797	697	639
700	1983	964	725	828	716	651
710	2044	936	688	848	743	681
720	2104	904	649	702	616	564
730	2161	870	612	793	699	642
740	2217	845	584	850	751	692
750	2270	829	562	719	636	586
760	2321	824	552	528	464	426
770	2370	831	553	759	668	614

Source: J. E. Kaufman (ed.), *IES Lighting Handbook,* Reference volume, 1981 ed., IES, New York, 1981, fig. 5-2, with permission.

TABLE 5.6 CHROMATICITY COORDINATES OF CIE STANDARD EMITTERS

	x	y
A	0.4476	0.4074
B	0.3484	0.3516
C	0.3101	0.3163
D_{55}	0.3324	0.3475
D_{65}	0.3127	0.3290
D_{75}	0.2990	0.3150

5.19 COLOR SPECIFICATIONS OF OBJECTS

Consider now that radiant flux from one of the standard emitters or from a common light source impinges on a colored object and is selectively reflected and/or transmitted. We desire to calculate the chromaticity coordinates of the reflected or transmitted light.

For this situation, Eqs. (5.28) are modified so that $S(\lambda)$ is either $\rho(\lambda)E(\lambda)$ or $\tau(\lambda)E(\lambda)$, where $E(\lambda)$ is the spectral irradiance of the surface and $\rho(\lambda)$ and $\tau(\lambda)$ are the spectral reflectance and transmittance. The procedure will be illustrated by an example.

EXAMPLE 5.8

The spectral reflectance characteristic of a sample of green paint is known. The sample is to be lighted by illuminant A. Determine the chromaticity coordinates of the light reflected from the sample.

TABLE 5.7 CIE CHROMATICITY COORDINATES AND CCTs OF COMMON LAMPS

Lamp	x	y	CCT (K)
Tungsten halogen	0.424	0.399	3190
Cool white fluorescent	0.373	0.385	4250
Warm white fluorescent	0.436	0.406	3020
Deluxe cool white fluorescent	0.376	0.368	4050
Deluxe warm white fluorescent	0.440	0.403	2940
Daylight fluorescent	0.316	0.345	6250
Clear mercury	0.326	0.390	5710
Color-improved mercury	0.373	0.415	4430
Clear metal halide	0.396	0.390	3720
High-pressure sodium	0.519	0.418	2100

Source: J. E. Kaufman (ed.), *IES Lighting Handbook*, Reference volume, 1981 ed., IES, New York, 1981, fig. 5-26, with permission.

TABLE 5.8 PRODUCTS OF TRISTIMULUS VALUES AND SOURCE RADIANCES

Wave-length (nm)	A			B		
	$\bar{x}(\lambda)S(\lambda)$	$\bar{y}(\lambda)S(\lambda)$	$\bar{z}(\lambda)S(\lambda)$	$\bar{x}(\lambda)S(\lambda)$	$\bar{y}(\lambda)S(\lambda)$	$\bar{z}(\lambda)S(\lambda)$
380	0.001	0.000	0.006	0.003	0.000	0.014
390	0.005	0.000	0.023	0.013	0.000	0.060
400	0.019	0.001	0.093	0.056	0.002	0.268
410	0.071	0.002	0.340	0.217	0.006	1.033
420	0.262	0.008	1.256	0.812	0.024	3.890
430	0.649	0.027	3.167	1.983	0.081	9.678
440	0.926	0.061	4.647	2.689	0.178	13.489
450	1.031	0.117	5.435	2.744	0.310	14.462
460	1.019	0.210	5.851	2.454	0.506	14.085
470	0.776	0.362	5.116	1.718	0.800	11.319
480	0.428	0.622	3.636	0.870	1.265	7.396
490	0.160	1.039	2.324	0.295	1.918	4.290
500	0.027	1.792	1.509	0.044	2.908	2.449
510	0.057	3.080	0.969	0.081	4.360	1.371
520	0.425	4.771	0.525	0.541	6.072	0.669
530	1.214	6.322	0.309	1.458	7.594	0.372
540	2.313	7.600	0.162	2.689	8.834	0.188
550	3.732	8.568	0.075	4.183	9.603	0.084
560	5.510	9.222	0.036	5.840	9.774	0.038
570	7.571	9.457	0.021	7.472	9.334	0.021
580	9.719	9.228	0.018	8.843	8.396	0.016
590	11.579	8.540	0.012	9.728	7.176	0.010
600	12.704	7.547	0.010	9.948	5.909	0.007
610	12.669	6.356	0.004	9.436	4.734	0.003
620	11.373	5.071	0.003	8.140	3.630	0.002
630	8.980	3.704	0.000	6.200	2.558	0.000
640	6.558	2.562	0.000	4.374	1.709	0.000
650	4.336	1.637	0.000	2.815	1.062	0.000
660	2.628	0.972	0.000	1.655	0.612	0.000
670	1.448	0.530	0.000	0.876	0.321	0.000
680	0.804	0.292	0.000	0.465	0.169	0.000
690	0.404	0.146	0.000	0.220	0.080	0.000
700	0.209	0.075	0.000	0.108	0.039	0.000
710	0.110	0.040	0.000	0.053	0.019	0.000
720	0.057	0.019	0.000	0.026	0.009	0.000
730	0.028	0.010	0.000	0.012	0.004	0.000
740	0.011	0.006	0.000	0.006	0.002	0.000
750	0.006	0.002	0.000	0.002	0.001	0.000
760	0.004	0.002	0.000	0.002	0.001	0.000
770	0.002	0.000	0.000	0.001	0.000	0.000
Total	109.828	100.000	35.547	99.072	100.000	85.223

C			D_{55}		
$\bar{x}(\lambda)S(\lambda)$	$\bar{y}(\lambda)S(\lambda)$	$\bar{z}(\lambda)S(\lambda)$	$\bar{x}(\lambda)S(\lambda)$	$\bar{y}(\lambda)S(\lambda)$	$\bar{z}(\lambda)S(\lambda)$
0.004	0.000	0.020	0.004	0.000	0.020
0.019	0.000	0.089	0.015	0.000	0.073
0.085	0.002	0.404	0.083	0.002	0.394
0.329	0.009	1.570	0.284	0.008	1.354
1.238	0.037	5.949	0.916	0.027	4.400
2.997	0.122	14.628	1.836	0.075	8.959
3.975	0.262	19.938	2.838	0.187	14.235
3.915	0.443	20.638	3.136	0.354	16.528
3.362	0.694	19.299	2.780	0.574	15.959
2.272	1.058	14.972	1.858	0.865	12.243
1.112	1.618	9.461	0.933	1.357	7.936
0.363	2.358	5.274	0.299	1.941	4.341
0.052	3.401	2.864	0.047	3.094	2.606
0.089	4.833	1.520	0.089	4.819	1.516
0.576	6.462	0.712	0.602	6.754	0.744
1.523	7.934	0.338	1.641	8.546	0.418
2.785	9.149	0.195	2.821	9.267	0.197
4.282	9.832	0.086	4.245	9.747	0.085
5.880	9.841	0.039	5.656	9.466	0.037
7.322	9.147	0.020	7.048	8.804	0.019
8.417	7.992	0.016	8.520	8.089	0.016
8.984	6.627	0.010	8.927	6.584	0.010
8.949	5.316	0.007	9.541	5.668	0.007
8.325	4.176	0.002	9.074	4.553	0.003
7.070	3.153	0.002	7.658	3.415	0.002
5.309	2.190	0.000	5.528	2.280	0.000
3.693	1.443	0.000	3.934	1.537	0.000
2.349	0.886	0.000	2.397	0.905	0.000
1.361	0.504	0.000	1.417	0.524	0.000
0.708	0.259	0.000	0.781	0.286	0.000
0.369	0.134	0.000	0.401	0.145	0.000
0.171	0.062	0.000	0.172	0.062	0.000
0.082	0.029	0.000	0.090	0.032	0.000
0.039	0.014	0.000	0.047	0.017	0.000
0.019	0.006	0.000	0.019	0.007	0.000
0.008	0.003	0.000	0.011	0.004	0.000
0.004	0.002	0.000	0.006	0.002	0.000
0.002	0.001	0.000	0.002	0.001	0.000
0.001	0.001	0.000	0.001	0.001	0.000
0.001	0.000	0.000	0.001	0.000	0.000
98.041	100.000	118.103	95.655	100.000	92.102

TABLE 5.8 PRODUCTS OF TRISTIMULUS VALUES AND SOURCE RADIANCES (*Continued*)

Wavelength (nm)	D_{65}			D_{75}		
	$\bar{x}(\lambda)S(\lambda)$	$\bar{y}(\lambda)S(\lambda)$	$\bar{z}(\lambda)S(\lambda)$	$\bar{x}(\lambda)S(\lambda)$	$\bar{y}(\lambda)S(\lambda)$	$\bar{z}(\lambda)S(\lambda)$
380	0.007	0.000	0.031	0.009	0.000	0.041
390	0.022	0.001	0.104	0.028	0.001	0.132
400	0.112	0.003	0.532	0.137	0.004	0.650
410	0.377	0.010	1.796	0.457	0.013	2.179
420	1.189	0.035	5.711	1.424	0.042	7.839
430	2.330	0.095	11.370	2.749	0.112	13.414
440	3.458	0.228	17.343	3.964	0.262	19.883
450	3.724	0.421	19.627	4.199	0.475	22.133
460	3.243	0.669	18.614	3.614	0.746	20.746
470	2.124	0.989	13.998	2.336	1.088	15.395
480	1.048	1.524	8.915	1.139	1.655	9.682
490	0.330	2.142	4.791	0.354	2.301	5.146
500	0.051	3.343	2.815	0.054	3.536	2.978
510	0.095	5.132	1.614	0.099	5.371	1.689
520	0.628	7.041	0.775	0.646	7.245	0.798
530	1.687	8.785	0.430	1.717	8.941	0.438
540	2.869	9.425	0.201	2.899	9.523	0.203
550	4.267	9.796	0.086	4.270	9.803	0.086
560	5.625	9.415	0.037	5.584	9.345	0.037
570	6.947	8.678	0.019	6.844	8.550	0.019
580	8.304	7.885	0.015	8.108	7.699	0.015
590	8.612	6.352	0.009	8.387	6.186	0.009
600	9.046	5.347	0.007	8.704	5.171	0.007
610	8.499	4.264	0.003	8.114	4.071	0.002
620	7.089	3.101	0.002	6.710	2.992	0.002
630	5.062	2.088	0.000	4.754	1.961	0.000
640	3.547	1.386	0.000	3.301	1.290	0.000
650	2.147	0.810	0.000	1.993	0.752	0.000
660	1.252	0.463	0.000	1.152	0.426	0.000
670	0.680	0.249	0.000	0.620	0.227	0.000
680	0.347	0.126	0.000	0.315	0.114	0.000
690	0.150	0.054	0.000	0.136	0.049	0.000
700	0.077	0.028	0.000	0.070	0.025	0.000
710	0.041	0.015	0.000	0.037	0.013	0.000
720	0.017	0.006	0.000	0.015	0.005	0.000
730	0.009	0.003	0.000	0.008	0.003	0.000
740	0.005	0.002	0.000	0.005	0.002	0.000
750	0.002	0.001	0.000	0.002	0.001	0.000
760	0.001	0.000	0.000	0.001	0.000	0.000
770	0.001	0.000	0.000	0.001	0.001	0.000
Total	95.018	100.000	108.845	94.954	100.000	122.520

Source: J. E. Kaufman (ed.), *IES Lighting Handbook,* Reference volume, 1981 ed., IES, New York, 1981, fig. 5-3, with permission.

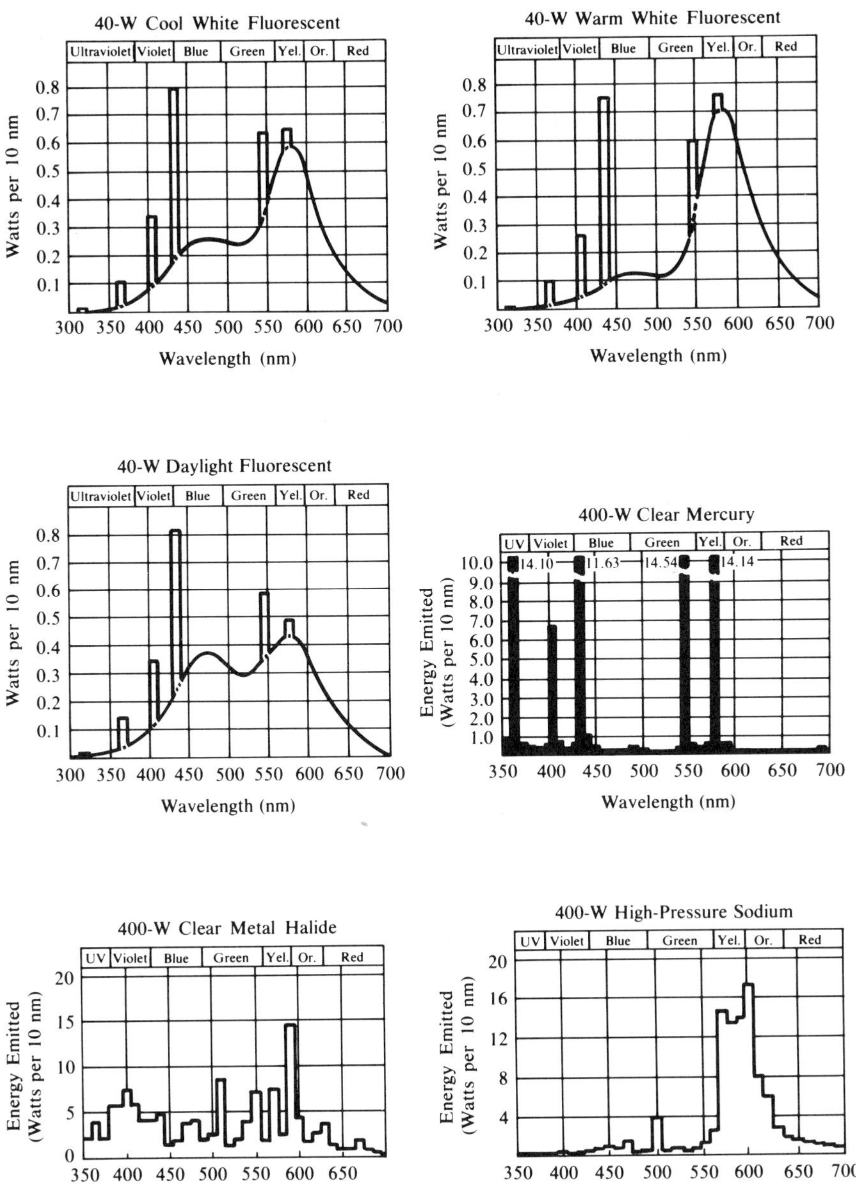

Figure 5.19 Fluorescent and HID SPD curves. (*Source:* GTE Sylvania, Engineering Bulletins 0-341, 0-344, 0-346, 0-348, with permission.)

Solution. From Eqs. (5.28), Table 5.8, and the known spectral reflectance, we can prepare the following tabulation.

λ(nm)	$\rho(\lambda)$	$\bar{x}(\lambda)S(\lambda)$	ΔX	$\bar{y}(\lambda)S(\lambda)$	ΔY	$\bar{z}(\lambda)S(\lambda)$	ΔZ
400	0.200	0.019	0.004	0.001	—	0.093	0.019
410	0.150	0.071	0.011	0.002	—	0.340	0.051
420	0.125	0.262	0.033	0.008	0.001	1.256	0.157
430	0.100	0.649	0.065	0.027	0.003	3.167	0.317
440	0.085	0.926	0.079	0.061	0.005	4.647	0.395
450	0.075	1.031	0.077	0.117	0.009	5.435	0.408
460	0.080	1.019	0.082	0.210	0.017	5.851	0.468
470	0.130	0.776	0.101	0.362	0.047	5.116	0.665
480	0.330	0.428	0.141	0.622	0.205	3.636	1.200
490	0.465	0.160	0.074	1.039	0.483	2.324	1.081
500	0.550	0.027	0.015	1.792	0.986	1.509	0.830
510	0.590	0.057	0.034	3.080	1.817	0.969	0.572
520	0.600	0.425	0.255	4.771	2.863	0.525	0.315
530	0.580	1.214	0.704	6.322	3.667	0.309	0.179
540	0.530	2.313	1.226	7.600	4.028	0.162	0.086
550	0.470	3.732	1.754	8.568	4.027	0.075	0.035
560	0.410	5.510	2.259	9.222	3.781	0.036	0.015
570	0.340	7.571	2.574	9.457	3.215	0.021	0.007
580	0.280	9.719	2.721	9.228	2.584	0.018	0.005
590	0.230	11.579	2.683	8.540	1.964	0.012	0.003
600	0.180	12.704	2.287	7.547	1.358	0.010	0.002
610	0.140	12.669	1.774	6.356	0.890	0.004	0.001
620	0.110	11.373	1.251	5.071	0.558	0.003	—
630	0.080	8.980	0.718	3.704	0.296	—	—
640	0.070	6.558	0.459	2.562	0.179	—	—
650	0.055	4.336	0.238	1.637	0.090	—	—
660	0.053	2.268	0.120	0.972	0.052	—	—
670	0.050	1.448	0.072	0.530	0.027	—	—
680	0.040	0.804	0.032	0.292	0.012	—	—
			21.843		33.164		6.811

Then, from Eqs. (5.29)

$$D = 61.818 \qquad x = 0.353 \qquad y = 0.536$$

The coordinates are located at point S on Fig. 5.20. The coordinates of the source are shown at point A. We observe that the sample's chromaticity falls in the greenish-yellow region of the diagram when illuminated by source A.

The coordinates on the chromaticity diagram of the radiation incident on and reflected (transmitted) from an object can be used to determine dominant wavelength and excitation purity. Dominant wavelength (λ_d) is defined as the wavelength of radiant energy of a single frequency that, when combined in suitable proportion with the radiant energy of a reference standard,

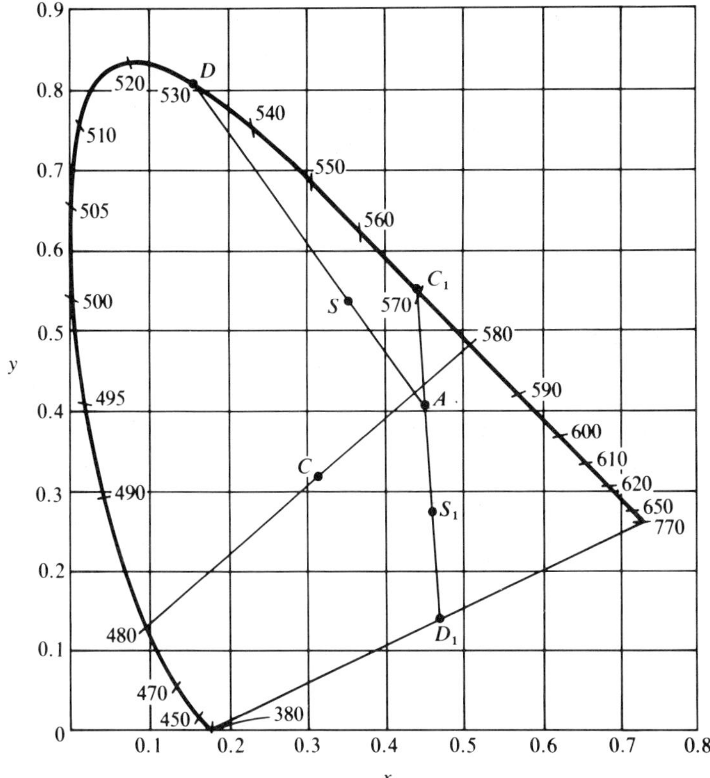

Figure 5.20 Object loci.

matches the color of the light reflected (transmitted) from an object. Excitation purity is defined as the ratio on the CIE chromaticity diagram of the distance between the reference point and the light point to the distance in the same direction between the reference point and the spectrum locus (including the straight line closing the horseshoe).

On the chromaticity diagram, a line drawn from the chromaticity coordinates of the source through the chromaticity coordinates of the sample will intersect the spectral locus at the dominant wavelength. For the case in Example 5.8, $\lambda_d = 530$ nm, as shown in Fig. 5.20. The spectral locus represents 100% purity; the source point 0% purity. From Fig. 5.20,

$$\% \text{ excitation purity} = 100 \frac{AS}{AD} \tag{5.32}$$

giving 32% for the green sample of Example 5.8.

Assume a sample S_1 when illuminated by source A gives chromaticity coordinates as shown in Fig. 5.20. What are its dominant wavelength and excitation purity? The intersection with the straight line closing the spectral locus is at point D_1, a nonspectral color. In this case we require a new quantity, the complementary wavelength (λ_c), defined as the wavelength of radiant energy of a single frequency that, when combined in suitable proportion with the light reflected (transmitted) matches the color of the reference standard. For S_1, the complementary wavelength is 570 nm in Fig. 5.20. The % excitation purity is $100(AS_1/AD_1)$, in this case 51%.

5.20 COLOR MIXING

According to Grassman's law, stated in Sec. 5.14, color mixtures obey the law of addition. Thus if we know the tristimulus values of two colors, we may add them to obtain the tristimulus values of the resultant.

Usually a color is specified in terms of its chromaticity coordinates (x, y) and its luminance (L), which is proportional to the tristimulus value Y. Before we can add the two colors, we must calculate their tristimulus values.

Let two colors have tristimulus values X_1, Y_1, Z_1 and X_2, Y_2, Z_2. The chromaticity coordinates of the mixture of these two colors are

$$x = \frac{X_1 + X_2}{D_1 + D_2} \qquad y = \frac{Y_1 + Y_2}{D_1 + D_2} \tag{5.33}$$

Noting that, for any color,

$$X = xD \qquad Y = yD \tag{5.34}$$

we can express Eqs. (5.33) as

$$x = \frac{x_1 D_1 + x_2 D_2}{D_1 + D_2} = \frac{x_1(Y_1/y_1) + x_2(Y_2/y_2)}{(Y_1/y_1) + (Y_2/y_2)} \tag{5.35}$$

$$y = \frac{y_1 D_1 + y_2 D_2}{D_1 + D_2} = \frac{Y_1 + Y_2}{(Y_1/y_1) + (Y_2/y_2)} \tag{5.36}$$

The results in Eqs. (5.35) and (5.36) can be directly extended to more than two colors by writing

$$x = \frac{\sum_1^n \left(x_k \dfrac{Y_k}{y_k}\right)}{\sum_1^n \dfrac{Y_k}{y_k}} \qquad y = \frac{\sum_1^n \left(y_k \dfrac{Y_k}{y_k}\right)}{\sum_1^n \dfrac{Y_k}{y_k}} \tag{5.37}$$

The chromaticity coordinates of a mixture of two colors will be on a straight line connecting them on the chromaticity diagram. To prove this feature, it is necessary to show that

$$\frac{x - x_1}{y - y_1} = \frac{x_2 - x_1}{y_2 - y_1} \tag{5.38}$$

which is the equal-slope condition. If Eqs. (5.35) and (5.36) are inserted into the left side of Eq. (5.38), the simplified result equals the right side of Eq. (5.38), proving the relation.

Sometimes a source is specified in radiometric, rather than photometric, terms. Let P_k be the watts emitted by a source. Then

$$Y_k = \int \bar{y}(\lambda) P_k(\lambda)\, d\lambda \tag{5.39}$$

and

$$\phi_k = 683 \int V(\lambda) P_k(\lambda)\, d\lambda \tag{5.40}$$

But $V(\lambda) = \bar{y}(\lambda)$, giving $\phi_k = 683 Y_k$. Therefore, the tristimulus value Y_k may be replaced by its equivalent luminous value in Eqs. (5.35) to (5.37) without altering the resulting chromaticity coordinates.

EXAMPLE 5.9

Two colors are specified as follows:

x	y	M (lm/ft²)
0.25	0.45	10
0.30	0.20	15

Find the chromaticity coordinates and luminance exitance of the mixture of the two colors.

Solution. From Eqs. (5.35) and (5.36)

$$x = \frac{0.25(10/0.45) + 0.30(15/0.20)}{(10/0.45) + (15/0.20)} = 0.29$$

$$y = \frac{10 + 15}{(10/0.45) + (15/0.20)} = 0.26$$

$$M = 10 + 15 = 25 \text{ lm/ft}^2$$

EXAMPLE 5.10

Determine the ratio of radiant fluxes from two spectral (single-wavelength) sources that will produce a match with source C, if one of the spectral radiances has a wavelength of 480 nm.

Solution. We must first find the second wavelength. The chromaticity coordinates are

	x	y
Source C	0.3101	0.3163
480 nm	0.0913	0.1327

These two points are located on Fig. 5.20 and a line is drawn between them and extended to the locus. The intersection is at $\lambda \simeq 580$ nm with coordinates $x = 0.5125$ and $y = 0.4866$.

For a spectral line, $Y_k = \bar{y}_k P_k$ where P_k is the radiated watts of source k. Then, from Eqs. (5.35) and (5.36)

$$x = \frac{(x_1 \bar{y}_1 P_1/y_1) + (x_2 \bar{y}_2 P_2/y_2)}{(\bar{y}_1 P_1/y_1) + (\bar{y}_2 P_2/y_2)} \qquad y = \frac{\bar{y}_1 P_1 + \bar{y}_2 P_2}{(\bar{y}_1 P_1/y_1) + (\bar{y}_2 P_2/y_2)} \tag{5.41}$$

Everything is known in these two equation except P_1 and P_2. Thus either equation may be solved to yield $P_1/P_2 = 1.7$.

5.21 COLOR RENDERING

In Sec. 5.15, we stated that various metrics are used to describe the ability of one source or object to match the color of another source or object. One of these, correlated color temperature, was discussed earlier in the chapter and is a means of specifying chromaticities of sources lying near the Planckian locus. The other metric, to be discussed now, is color rendering index (CRI or R). The CRI of a light source is a measure of the shift in chromaticity of an object when it is lighted by the source as compared to being lighted by a reference source of comparable color temperature.

The resultant perceived color shift seen by an observer when an object is lighted first by one source and then by another is a combination of the actual colormetric shift, as measured by changes in luminance and chromaticity, and an adaptive color shift, caused by changes in chromatic adaptation. CRI permits us to calculate the former and to partially account for the latter, about which we still do not know a great deal. Thus CRI provides a useful, but as yet incomplete, measure of the color-rendering properties of light sources.

CRI is based on a test color method (Kaufman, 1981; Jerome, 1969) which was proposed to the CIE by the IES Color Committee in 1961. In 1965, CIE Publication No. 13, "Method of Measuring and Specifying Color Rendering Properties of Light Sources," based on the U.S. proposal, was approved. With minor improvements (1971), which partially take into account chromatic adaptation of the observer, this is the accepted contemporary method for rating the ability of a lamp to give a true representation of the color of an object.

The determination of CRI is carried out on a uniform *uv* chromaticity diagram, rather than on the *xy* chromaticity diagram previously presented. It was found that the *xy* diagram does not have a uniform chromaticity scale. Thus the perceived color difference between two points on the diagram separated by a given distance varies with the location of the two points.

The *uv* diagram has more uniform perceptible color differences than the *xy* diagram. It is called "the CIE 1960 UCS Diagram" (uniform color space) and is based on a tranformation described in 1937 by D. L. MacAdam. The diagram is derived from the *xy* diagram through the coordinate transformations:

$$u = \frac{4x}{-2x + 12y + 3} \quad \text{or} \quad u = \frac{4X}{X + 15Y + 3Z}$$
$$v = \frac{6y}{-2x + 12y + 3} \quad \text{or} \quad v = \frac{6Y}{X + 15Y + 3Z} \tag{5.42}$$

and is displayed in Fig. 5.21.

The uv diagram is two-dimensional and contains no luminance information. In 1964, the CIE extended this diagram to three dimensions with its "CIE 1964 UCS Diagram." The coordinates of this latter diagram are derived from

$$W^* = 25Y^{1/3} - 17 \tag{5.43a}$$

$$U^* = 13W^*(u - u_0) \tag{5.43b}$$

$$V^* = 13W^*(v - v_0) \tag{5.43c}$$

where u_0 and v_0 are the uv coordinates of the light source and are placed at the origin of the $U^*V^*W^*$ system.

This new system, developed by G. Wyszecki, converts Y into what is called a lightness index (W^*) by Eq. (5.43a). W^* approximates the value function in the Munsell color system in the range $1 \leq Y \leq 100$. U^* and V^* are called chromaticness indexes and the system is designed so that a lightness difference (ΔW^*) of 1 corresponds to a chromaticness difference $(\Delta U^{*2} + \Delta V^{*2})^{1/2}$ of 13. Color differences are calculated from

$$\Delta E = (\Delta W^{*2} + \Delta U^{*2} + \Delta V^{*2})^{1/2} \tag{5.44}$$

based on one unit of ΔE corresponding to one National Bureau of Standards unit of color difference.

The procedure for determining the CRI of a light source involves several steps. First reference light sources are selected. To minimize effects due

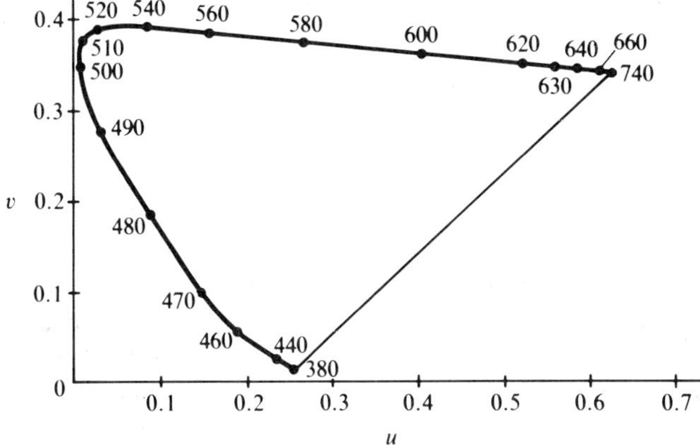

Figure 5.21 uv Chromaticity diagram.

to chromatic adaptation, it was decided that the reference source should have nearly the same chromaticity coordinates as the source being evaluated. Thus the reference source for any light source whose CCT is less than 5000 K is a Planckian radiator of that color temperature. For a source whose CCT exceeds 5000 K, the reference source is the nearest in CCT of a series of mathematically derived daylight spectral power distributions.

Next the spectral power distribution curve for the light source being evaluated is determined and its CCT calculated. This determines the reference source to be used and its spectral power distribution.

The third step is to select color samples. The CIE has chosen eight Munsell chip color samples, distributed uniformly around the Munsell hue circle, with colors (R), yellow (Y), greenish-yellow (GY), green (G), bluish-green (BG), purplish-blue (PB), purple (P), and reddish-purple (RP). Each of these samples has a Munsell value of 6 and a medium chroma (4 to 8) so that there is some reflectance from each sample at each wavelength. Each of the samples has a specified and known spectral reflectance (Jerome, 1969).

In addition to the eight standard samples, there are six special samples suggested by the CIE. Four of these have high chroma (8 to 13) so that the effect of the test source on saturated colors can be estimated. The other two are colors which represent a Caucasian complexion and green foliage. The CRI calculation procedure includes only the eight standard samples. A CRI for each sample is calculated. Then a general CRI is found as the average of the eight samples. This general index is the single number most often quoted.

The fourth step is to calculate the tristimulus values and chromaticity coordinates for the eight colored samples when lighted by the reference source and the source being evaluated. The procedure is identical to that used in Example 5.8. These sets of coordinates are then transformed to uv space using Eqs. (5.42). Then a perturbation is made of the uv coordinates of the color samples when lighted by the source being evaluated to partially account for chromatic adaptation by correcting for any difference in the chromaticity coordinates of the source being evaluated and the reference source.

The uv coordinates for the samples under the reference source and the altered uv coordinates for the samples under the source being evaluated are transformed into $U^*V^*W^*$ coordinates by Eqs. (5.43). Then the resultant color shift of each test color is determined from Eq. (5.44).

The final step involves deciding on a CRI scale and calibrating the results obtained from Eq. (5.44). It was decided to have an index in which a better performance was represented by a larger number [in Eq. (5.44) a larger number would represent a poorer performance]. Also it was decided to place the CRI of the reference source at 100 and to make the CRIs of all other sources be less than 100. Finally it was agreed that a standard fluorescent white lamp at 3000 K would have a CRI of 50. With these decisions made, the CRI for each color sample can be written as

$$R_i = 100 - 4.6\,\Delta E_i \tag{5.45}$$

and the general CRI for all samples as

$$R_a = \frac{1}{8} \sum_{1}^{8} R_i \tag{5.46}$$

The 4.6 in Eq. (5.45) arises from the calibration to a CRI of 50 for the 3000 K white fluorescent lamp. Thus

$$50 = \frac{1}{8} \sum_{1}^{8} (100 - k \, \Delta E_i) \tag{5.47}$$

yields $k = 4.6$ when ΔE_i are the vectors for the 3000 K lamp.

The vector shifts in position of the sample colors when illuminated by a test source of $R_a = 63$ are shown in the U^*V^* diagram in Fig. 5.22. CRIs for common lamp types are given in Table 5.9.

The color rendering index is not a panacea for insuring color quality of lighting. It has several deficiencies. First, since each light source has a specific CCT reference, two light sources of differing color temperatures cannot be compared directly. Second, since CRI is an average of eight responses, two lamps with the same CCT and CRI may differ drastically in their ability to render one or several colors. For example, a lamp giving six indexes of 80 and two of 40 has the same CRI as a lamp with eight indexes of 70. Third, some of the indexes are negative for some lamps, for example, low-pressure sodium and clear mercury. Finally, CRI is based on Planckian radiators and sources of reconstituted daylight. There is nothing inherent in these sources that says they are perfect in a color-rendering sense.

There are two additional color coordinate systems, the $L^*u^*v^*$ and L^*a^*b systems, which were adopted by the CIE in 1976 to replace the 1964 system (Kaufman, 1981; Jerome, 1969). They arose because color differences calculated by the 1964 system did not correlate well with the small

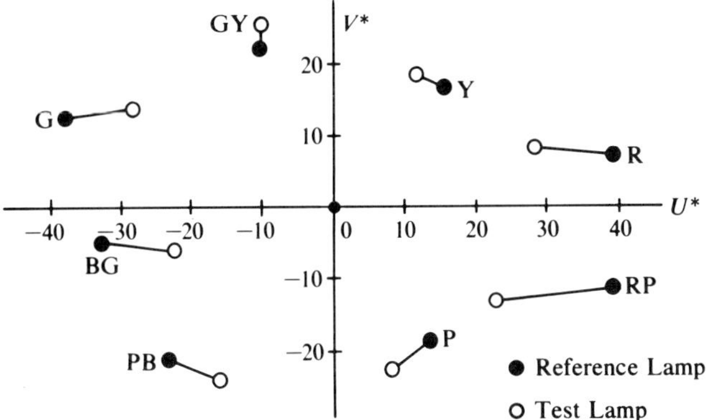

Figure 5.22 Graphic representation of CRI. (*Source:* C. W. Jerome, "Basic Colorimetry for the Lighting Industry," GTE Lighting Products, Salem, Mass., 1969, fig. 5.19, with permission.)

TABLE 5.9 COLOR RENDERING DATA FOR COMMON LAMPS

Lamp	CCT (K)	CRI
Tungsten halogen	3190	100
Cool white fluorescent	4250	62
Warm white fluorescent	3020	52
Deluxe cool white fluorescent	4050	89
Deluxe warm white fluorescent	2940	73
Daylight fluorescent	6250	74
Clear mercury	5710	15
Color-improved mercury	4430	32
Clear metal halide	3720	60
High-pressure sodium	2100	21

Source: J. E. Kaufman (ed.), *IES Lighting Handbook,* Reference volume, 1981 ed., IES, New York, 1981, fig. 5-26, with permission.

color differences often encountered in industry. Although these new systems supersede the $U^*V^*W^*$ system, the latter is still used for the calculation of CRI.

REFERENCES

Vision

Blackwell, H. R.: "Brightness Discrimination Data for the Specification of Quantity of Illumination," *IES Transactions,* November 1952.

Flynn, J. E.: "The IES Approach to Recommendations Regarding Levels of Illumination," *Lighting Design and Application,* September 1979.

IES RQQ Committee: "RQQ Report No. 4," *Journal of IES,* August 1970.

———: "Selection of Illuminance Values for Interior Lighting," *Journal of IES,* April 1980.

Kaufman, J. E.: *IES Lighting Handbook,* 1981, Illuminating Engineering Society, New York, 1981 (Vision, chap. 3; Color, chap. 5).

Luckiesh, M., and F. K. Moss: *The Science of Seeing,* Van Nostrand, New York, 1937.

Rea, M. S.: "Visual Performance with Realistic Methods of Changing Contrast," *Journal of IES,* April 1981.

———: "An Overview of Visual Performance," *Lighting Design and Application,* November 1982.

Rubin, A. I., and J. Elder: "Building for People," National Bureau of Standards Special Publication 974, U.S. Government Printing Office, Washington, D.C., 1980.

Smith, S. W.: "Is There an Optimum Level for Performing Office Tasts," *IERI International Symposium on Light and Vision and Biological Effects,* Munich, 1977.

———, and M. S. Rea: "Performance of Office-Type Tasks under Different Types and Levels of Illumination," IERI Report, October 1976.

Color

Boast, W. B.: *Illumination Engineering,* McGraw-Hill, New York, 1953.

Committee on Colorimetry: *The Science of Color,* Optical Society of America, Washington, D.C., 1953.

General Electric Co.: "Light and Color," TP119, Large Lamp Department, Nela Park, Cleveland, 1978.

Glasford, G. M.: *Fundamentals of Television Engineering*, McGraw-Hill, New York, 1955.

GTE Sylvania: "The Color Rendering Index," GTE Products Corp., Danvers, Mass., 1972.

Jerome, C. W.: "Basic Colorimetry for the Lighting Industry," GTE Lighting Products, Salem, Mass. 1969.

Kelly, K. L., and D. B. Judd: "Color-Universal Language and Dictionary of Names," National Bureau of Standards, Washington, D.C., 1976.

Michaels, E. L.: "Basic Colorimetry for Color Television," *Electrical Engineering,* November 1956.

Wald, G.: "The Receptors of Human Color Vision," *Science,* vol. 145, September 1964.

Waymouth, J. F., and R. E. Levin: "Designers Handbook—Light Source Applications," GTE Products Corp., Danvers, Mass., 1980.

Wintringham, W. T.: "Color Television and Colorimetry," *Proceedings of IRE,* vol. 39, no. 10, October 1951.

Wyszecki, G., and W. S. Stiles: *Color Science,* Wiley, New York, 1967.

Chapter 6

Lamps

6.1 INTRODUCTION

There are nearly 6000 different lamps being manufactured today, most of which can be placed in the following six categories: incandescent, tungsten halogen, fluorescent, mercury, metal halide, and sodium. In this chapter, we will consider each of these categories in detail, tracing the history of development of the various lamp types, explaining their engineering design, and delineating their electrical and illumination parameters.

The major characteristics to be considered when choosing a lamp are its luminous efficacy, life, lumen depreciation, and color. Luminous efficacy is the measure of the lamp's ability to convert input electric power, in watts, into output luminous power, in lumens, and is measured in lumens per watt (lm/W). If, in order to operate the lamp properly, auxiliary equipment is required, the watts dissipated in that equipment must be charged to the lighting system. This is always true in the case of fluorescent, mercury, metal halide, and sodium lamps, which require a ballast (inductor, capacitor, auto transformer, or electronic) for starting and operating the lamp. If a fluorescent lamp, for example, emits 3000 lm and consumes 40 W, and its ballast consumes 10 W, the lamp-ballast efficacy would be listed as 60 lm/W, not 75 lm/W.

The life of a lamp is the number of hours for approximately 50% of a large group of lamps of the same kind to fail. Failure generally implies that the lamp will no longer light, although in a few cases it means that the light output has dropped to a specific percentage value. It would be ideal if all lamps of the same kind burned out at the same time, but this is not the case. In a group of, say, 100 lamps all rated at 1000 h, it is likely that 10 of them will have failed by 700 h, and 5 of them may last as long as 1400 h. However at 1000 h, we can expect that about 50 of them will have failed.

Lumen depreciation during life is also a characteristic of all lamps. This process of "aging" is caused by many factors and can result in a lamp emitting 70% or less of its initial lumens when it reaches rated life. It is important in lighting design to include lumen depreciation. If a lighting task requires, for example, 500 lux and the designer installs just enough lighting to provide that amount initially, the lighting level may fall to 350 lux as the lamps approach rated life. Often a compromise is sought which provides the design illumination at, say, 70% of life, which means that the installation will provide more than 500 lux initially and less than 500 lux at 100% life.

Finally, there is the matter of color, or more properly, color rendering. Each of the lamp types does not provide the same nominal "white." Their differences in spectral distribution can produce two effects within a lighted space. First, some of the colors of objects within that space can appear unnatural or faded—reds can appear brown, violets nearly black, etc. This was discussed in Chap. 5 and is measured in terms of color rendering index. Second, the entire space many "feel" warm or cool. For example, a mercury lamp, because it is lacking in reds and oranges, makes a space seem cool, whereas an incandescent lamp, with deficiencies in the blue and violets, makes a space feel warm. Although personal preference and intended use of a space often govern the choice of warm versus cool, as a general guideline, a room which has a northern exposure should have "warmer" lamps—those richer in reds and yellows—whereas a room with a southern exposure should be lit with "cooler" sources—those richer in blues and greens. The "warmness" or "coolness" of a light source was quantified in Chap. 2 in terms of its correlated color temperature.

6.2 INCANDESCENT LAMP DEVELOPMENT

The incandescent lamp preceded all others in development and, aside from the carbon arc lamp and the Cooper Hewitt mercury pool lamp, was the only type available until fluorescent and mercury lamps appeared in the 1930s. Edison's group at his Menlo Park, N.J., laboratory began working on the development of an incandescent lamp in the fall of 1877. Edison was a relatively late starter in this work, as there were at least six other individuals in the 1870s working along similar lines—Joseph Swan and St. George Lane-Fox in England and Moses Farmer, Hiram Maxim, William Sawyer, and Albon Man in the United States. Edison won the race in developing a commercially practical lamp for two main reasons:

1. He realized the need for an electrical system, not just a lamp. He rejected the series-circuit distribution system commonly used in arc lighting, reasoning that each lamp must be capable of operating independently of the others. This led him to develop a new type of dy-

namo, a shunt machine providing a constant voltage output of 110 V at close to 90% efficiency. The series dynamos used in arc lighting provided constant current and were roughly 50% efficient.

2. All incandescent lamps invented prior to 1879 had used thick low-resistance radiating elements. Edison, displaying considerably greater understanding of Ohm's law than any of his contemporaries, reasoned that a high-resistance filamentary element was necessary. This was his greatest contribution to lamp design, the use of a thin, threadlike filament of high resistance which could be heated to incandescence.

With these decisions made, Edison turned to a consideration of what filament material to use. Before deciding that he needed a high-resistance element, Edison had rejected carbon and begun working with platinum. But platinum was very expensive and the temperature at which it became incandescent was very close to its melting temperature. Thus, in 1879, he returned to carbon, finally settling on carbonized cotton sewing thread as the filament material.

As is so often the case, one technological breakthrough must occur before another can transpire. Early lamp developers were plagued by being unable to produce a high vacuum inside the bulb. With some gases remaining, their filaments expired in a matter of minutes. The development of the vacuum pump by Herman Sprengel, a German chemist working in England, solved this problem in the late 1860s. Edison used Sprengel's pump to evacuate his lamps. He noticed that not only was it necessary to evacuate the lamp when the filament was cold, but it was also necessary to continue the evacuating process as the filament was raised to its operating temperature. This is because residual gases are released from both filament and glass bulb as temperature rises.

Edison's first practical carbonized sewing thread filament lamp was lit on October 19, 1879. It had an efficacy of 1.4 lm/W and burned for 40 h, after which Edison raised the applied voltage slightly and it expired. Public announcement of Edison's success occupied the entire front page of the *New York Herald* on December 21, 1879, as a means of announcing a public "open house" at Menlo Park on New Year's Eve. Several thousand people visited the park that evening to see 60 of the new lamps lighting the buildings and pathways.

Since Edison's first lamp, there have been many improvements in lamp design leading to the incandescent lamps of today. We will discuss a few of the most important of these.

Filament Materials Although carbon has a very high melting point (3870 K), it was necessary to operate carbon filaments much below this temperature because of rapid evaporation and resulting short life. A number of other materials were tried but ultimately tungsten was chosen. Tungsten's melting point (3650 K) was lower than that of carbon but it could be operated at a higher temperature for the same life. The early tungsten filaments (1907) provided 7.8 lm/W and had lives of 800 h, a vast improvement over Edison's first lamp.

Ductile Tungsten The problem with tungsten was that the early filaments were fragile and costly. They were made of tungsten powder, mixed with a binder and squirted through a die to form a thread. The lamps were packed in cotton wadding for shipment but, even so, there was much filament breakage.

During the period from 1906 to 1910, Dr. William Coolidge, General Electric research scientist noted also for his development of the x-ray tube, developed a process for converting crystalline tungsten into fibrous tungsten, with the latter being very ductile and having five times the tensile strength of steel. The process involves converting tungsten powder into ingots. These are sintered and swaged to produce tungsten rods. The rods are then drawn through a series of diamond dies, each slightly smaller in diameter than its predecessor, to create tungsten wire.

Gas Filling Even the new tungsten filaments evaporated more rapidly than was desired, and a black tungsten deposit appeared on the inside of the bulb, which reduced light output. At first researchers felt the problem could be solved by creating an even better vacuum inside the bulb, but improvement was minimal.

Dr. Irving Langmuir, a General Electric chemist, began working on this problem in 1909. He first discovered that the presence of even minute amounts of water vapor (10 parts per million) inside the lamp greatly increased the amount of tungsten carried from the filament to the bulb wall. Next he examined the effect of adding inert gases, such as nitrogen, and found that these reduced evaporation significantly. The nitrogen formed a blanket around the filament, retarding evaporation and reducing the ability of the water vapor to pick up those tungsten particles which were evaporated. However the nitrogen also took heat away from the filament, reducing its temperature. Langmuir found that the loss in light output was greater than the gain from reduced blackening.

Langmuir continued his work and found that the increase in heat loss was proportionately less for larger filaments. This led him to coil the filament to make it more compact and resulted in lamps with much improved efficacy for the same life. Today double-, and even triple-, coiled filaments are used.

Inside Frosting Early incandescent lamps used clear glass bulbs and produced disturbing glare. Several methods of treating the outside of the bulb were tried, but reduction in light output was too severe. An attempt was made to acid-etch the inside of the bulb but this produced many small jagged corners which weakened the bulb and caused early breakage.

In 1925 Marvin Pipkin of the General Electric Lamp Development laboratory decided to re-etch the inside of a previously etched bulb with a dilute solution of acid. The effect was to round off the sharp corners and fill in the crevices, producing a strong bulb which diffused the light to such an extent that the glowing filament could barely be seen. Such lamps are called *inside frosted* and are still commonly available today.

Reflectorized Lamps The idea of one-piece, built-in reflectorized lamps occurred in the mid-1930s. Prior to that time, automobile head lamps, for example, consisted of a small lamp bulb mounted in an auxiliary reflector. Then Daniel Wright, a General Electric researcher, inserted two terminals into a common custard cup, aluminized the inner surface, mounted a filament and sealed a glass cover to the cup. The result was the first sealed-beam lamp and has led to the R and PAR lamps, discussed in Sec. 2.6, and to the one-piece auto headlight, which was introduced first on 1939 cars.

These are only a few of the many incandescent lamp developments over the past 100 years. Although certainly not all-inclusive, they do illustrate the high degree of technology in every contemporary incandescent lamp.

6.3 CONSTRUCTION OF AN INCANDESCENT LAMP

An incandescent lamp is shown diagrammatically in Fig. 6.1. Let us consider each of its parts in some detail.

Bulbs

Bulbs are manufactured in two, quite different, ways. General service lamps have blown bulbs and are generally of soft, or lime, glass, with a maximum bulb temperature rating of about 650 K. They are mass produced by a ribbon machine, first introduced by Corning Glass Works in 1927. This unique machine takes molten glass from a furnace, presses it into a ribbon, and automatically blows it into molds as the ribbon moves along a conveyor belt. Current ribbon machines can produce up to 50,000 bulbs per hour.

The second manufacturing method is a pressed glass technique, used in the manufacture of sealed-beam lamps. These bulbs are made of hard, heat-resistant lead and borosilicate glasses. The glass is heated to soften it, cut into chunks, and placed between the two halves of a metal die. When the halves are closed, the glass is pressed into the shape of the die cavity. The process allows the accurate forming of reflectors and lenses, which are later fused together to complete the lamp.

Lamp bulbs come in a a variety of shapes and sizes. The lamp manufacturers use a letter-number code to distinguish between bulb types. The letter denotes the bulb shape and the number gives its diameter in eighths of an inch. As examples, a PS30 lamp has a pear-shaped bulb, 3.75 in in diame-

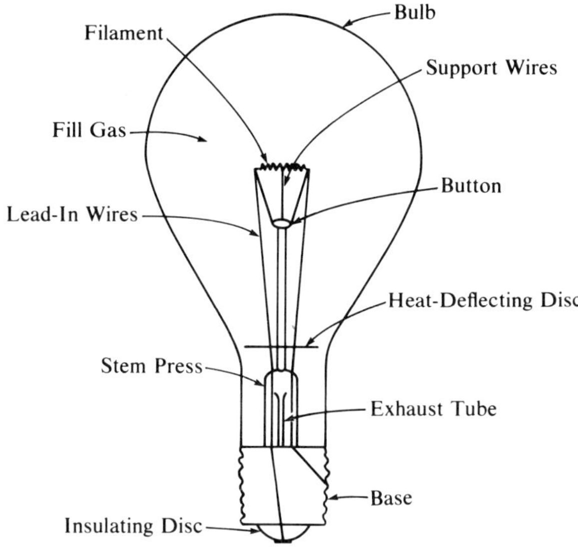

Figure 6.1 Parts of an incandescent lamp.

ter; a T12 bulb is tubular with a diameter of 1.5 in; an R40 bulb is a reflector lamp envelope with a diameter of 5 in. The common household incandescent lamp, available in 25-, 40-, 60-, 75-, 100-, 150- and 200-W sizes, has a tear-drop-shaped bulb, given the letter A for "arbitrary" shape. It has efficacies from 9 to 20 lm/W and lives of 750 to 2500 h.

Fill Gas

In his early work, Langmuir used pure nitrogen as the filling gas for incandescent lamps, primarily to prevent arcing. Later he found that the heat loss from the filament could be further reduced by the use of an inert gas such as argon. But pure argon produced a glow around the filament which increased bulb blackening significantly.

Today's lamps use various mixtures such as 85% argon and 15% nitrogen. Occasionally krypton, with lower heat conductivity but greater cost, is used instead of argon. The gases are introduced into the lamp at about 80% of atmospheric pressure.

Filling gas is used in lamps of 40 W and above. In lamps of less than 40 W, the increased heat loss due to the addition of the gas more than offsets the gain in light output from operating the filament at a somewhat higher temperature.

It is mind-boggling to note that one drop of water, entering with the fill gas and distributed uniformly among 50,000 lamps, will produce severe bulb blackening in all of them. This illustrates the extremely close control required in incandescent lamp manufacture.

Filaments

The almost universally used filament material in tungsten. It is obtained from wolframite ore, which is a compound of tungsten, manganese, and iron. Although tungsten melts at 3650 K, tungsten filaments in general service lamps are operated much below this point (2800 to 3000 K) to obtain reasonable life.

Most incandescent lamp filaments today are singly or doubly coiled. For example, the filament for a 120-V, 60-W lamp has a diameter of 0.0045 cm and an uncoiled length of 53 cm. After its first coiling (1130 turns), the filament length is about 8 cm. The second coiling reduces the length to about 2 cm. If filament diameter at any point is 1% less than specified, filament life may be reduced by as much as 25%. This again indicates the need for close manufacturing controls.

Supports

The filament must be mounted and supported so that is does not sag or otherwise distort in shape. A glass rod, called a button, extends upward from the glass stem press in the neck of the lamp to the filament support wires. The latter are of molybdenum, to reduce heat loss, and are imbedded in the button during manufacture.* The lead-in wires, which carry current to and from the filament, also serve as supports. These are copper from base to stem press and nickel from stem press to filament. Within the stem press, the lead-in wires are Dumet wire, a nickel-iron alloy core with a copper sleeve. This unusual material is needed to match the coefficient of expansion of glass and lead-in wires so that the vacuum seal will not be broken when the lamp is turned on and the temperature rises.

There are many filaments shapes and varying numbers of supports, depending on the ultimate use to which the lamp is to be put. For example, coils for projection lamp filaments are planar, to create the greatest intensity in a particular direction (perpendicular to the plane). As a second example, rough-service lamp filaments have many more supports than do those for general service lamps.

Neck and Base

A heat-deflecting disc is placed just above the neck and base of the bulb in some higher-wattage lamps to reduce the circulation of hot gases into the stem press and base regions.

The exhaust tube, which projects beyond the base during manufacture, terminates in the stem press. Exhausting was done originally through the top of the lamp, leaving a rather dangerous glass tip. In contemporary lamps, the

*There has been some trend recently to alternative forms of construction to reduce the number of supports.

exhausting and adding of gases is done through the bottom of the lamp, after which the exhaust tube is sealed and cut off inside the base.

The base itself is of brass or aluminum. It may be cemented in place or mechanically anchored. One lead-in wire is attached to the side of the base, the other to a contact in the center of the base, which is insulated from the rest of the base by a glass disc. There are four standard screw-type bases and several bayonet and prefocus bases, the latter to position filaments accurately.

We have mentioned the automatic ribbon machine in connection with bulb production. The rest of the incandescent lamp manufacturing process is also highly automated and consists of three complex machines. The first is a mount machine, which prepares the stem press unit consisting of stem press, button rod, exhaust tube, filament, supports, and lead-in wires. Stem press units are transferred by conveyor to a sealing and exhaust machine, where each one is inserted into a bulb and sealed. Next the lamp is exhausted, the fill gases are added, and then the exhaust tube is sealed. Each lamp is then sent to a third machine which connects the lead-in wires to the base, attaches the base, and lights the completed lamp.

6.4 LIGHT, LIFE, AND LOSSES

The goals of any lamp design are that the lamp produce light most economically for the service intended and that lamps having the same ratings be uniform from one lamp to another with respect to lumen output, power input, and life.

In the design of an incandescent radiator, there is always a tradeoff between light and life. At one extreme, a photoflash lamp emits thousands of lumens but lasts only for a few milliseconds. At the other extreme, a stovetop heating unit glows a dull red and has a life measured in years. Between these extremes are the myriad of incandescent lamps in today's marketplace.

The maximum luminous efficacy of a blackbody radiator is 95 lm/W and occurs at an operating temperature of 6800 K. Tungsten melts at 3650 K and is a selective radiator. Thus the maximum efficacy of an uncoiled tungsten filament, if it had zero life, would be 52 lm/W. To gain life, the filament temperature must be reduced well below tungsten's melting point. For example, a standard 100-W incandescent lamp operates at a filament temperature of 2860 K, giving it a maximum efficacy of about 21 lm/W if all of the input power is radiated beyond the bulb. But an incandescent lamp has losses in the form of heat conducted away from the filament by the lead-in and support wires (end loss), heat carried away from the filament by the gas (gas loss), and direct radiation absorbed by the bulb and base (bulb and base loss). In the 100-W lamp, these losses amount to 18% of the power input, leaving a realized initial efficacy of $17\frac{1}{2}$ lm/W.

The energy distributions of several 120-V, inside-frosted, general service incandescent lamps are shown in Table 6.1 and the operating data

TABLE 6.1 GENERAL SERVICE INCANDESCENT LAMP ENERGY DISTRIBUTIONS

	% Radiation beyond bulb			% Energy losses			
Watts	Light	Infrared	Total	End	Gas	Bulb and base	Total
25	8.7	85.3	94.0	1.5	—	4.5	6.0
40	7.4	63.9	71.3	1.6	20.0	7.1	28.7
60	7.5	73.3	80.8	1.2	13.5	4.5	19.2
75	8.4	72.8	81.2	1.2	12.8	4.8	18.8
100	10.0	72.0	82.0	1.3	11.5	5.2	18.0
200	10.2	67.2	77.4	1.7	13.7	7.2	22.6
300	11.1	68.7	79.8	1.8	11.6	6.8	20.2
500	12.0	70.3	82.3	1.8	8.8	7.1	17.7

Source: General Electric Co., "Incandescent Lamps," Bulletin TP-110R1, 1980. Reprinted with permission.

for these same lamps are shown in Table 6.2. All lamps in these tables have single-coiled filaments except the 60-, 75-, 100-, and 500-W size, whose filaments are double coiled. The bulb sizes are A19 for the first five lamps, PS35 for the last two, and PS30 for the 200-W size.

An analysis of the losses in Table 6.1, as a function of power rating, is instructive. Consider end loss first. This is the percent of the input power conducted away from the filament by the various wires connected to it. End loss increases as wattage increases, except for a decrease for the three double-coiled filaments. We should expect this. The larger the wattage, the longer the filament and the greater the number of supports required for a given filament configuration. Double-coiling reduces the number of supports needed, thus reducing end loss.

The percentage of input energy conducted away from the filament by the gas decreases as the wattage increases, for a given filament configuration. This is because the filament, regardless of its size, is surrounded by a sheath of hot gas of essentially constant thickness. Larger-wattage lamps have thicker filaments and thus their percentage heat loss to the gas is less.

TABLE 6.2 GENERAL SERVICE INCANDESCENT LAMP CHARACTERISTICS

Watts	Filament temperature (K)	Initial lumens	Initial lm/W	Life (h)	Uncoiled filament length (cm)	Filament diameter (μm)	% Lumen depreciation
25	2560	235	9.4	2500	56.4	30.5	79
40	2740	455	11.4	1500	38.1	33.0	88
60	2770	870	14.5	1000	53.3	45.7	93
75	2840	1,190	15.9	750	54.9	53.3	92
100	2860	1,750	17.5	750	57.9	63.5	91
200	2890	3,710	18.5	750	63.5	96.5	85
300	2920	5,820	19.4	1000	76.2	129.5	86
500	3000	10,850	21.7	1000	87.4	180.3	89

Source: General Electric Co., "Incandescent Lamps," Bulletin TP-110R1, 1980. Reprinted with permission.

Also, double-coiled filaments have less percent gas loss than do single-coiled filaments, because of less surface area exposed to the gas sheath.

Bulb and base losses result from direct filament radiation absorbed by the bulb and base. A portion of this energy is reradiated at longer wavelengths and the remainder is conducted and convected to the surrounding air. Percent bulb and base loss increases slightly with wattage because higher-wattage lamps operate at higher temperatures and thus emit more radiation, some of which is absorbed by the bulb and base.

Initial lumens and lumens per watt are given in Table 6.2. The light output of an incandescent lamp decreases over the life of the lamp because of tungsten evaporation, which lowers filament temperature, and, more important, because of bulb blackening. The decrease is approximately straight line, with those lamps having double-coiled filaments showing the least depreciation.

The effect of bulb blackening depends on bulb orientation. In base-up operation, the tungsten is deposited mainly in the neck and the light loss is less than for other orientations.

The power consumed also decreases as the lamp ages, because the filament evaporation narrows the filament, thus increasing its resistance. The wattage decrease is less than the lumen decrease, about 5% at 100% life. Thus the luminous efficacy decreases as the lamp ages.

The correlated color temperature (CCT) of an incandescent lamp is related to its luminous efficacy, over quite a wide range of wattages. This approximate relationship is illustrated in Table 6.3 for gas-filled lamps. Such data are useful in photographic and theatre applications. From the published lumens per watt figure, the designer can use Table 6.3 to estimate the CCT of a particular incandescent lamp.

6.5 FILAMENT DESIGN PARAMETERS

In 1927, Jones and Langmuir developed a set of curves describing the properties of a straight tungsten filament in a vacuum as a function of filament temperature. These curves are shown in Fig. 6.2, where

TABLE 6.3 CCT VERSUS EFFICACY FOR INCANDESCENT LAMPS

CCT (K)	Approximate lumens per watt
2170	4
2480	8
2680	12
2860	16
3000	20
3130	24
3250	28

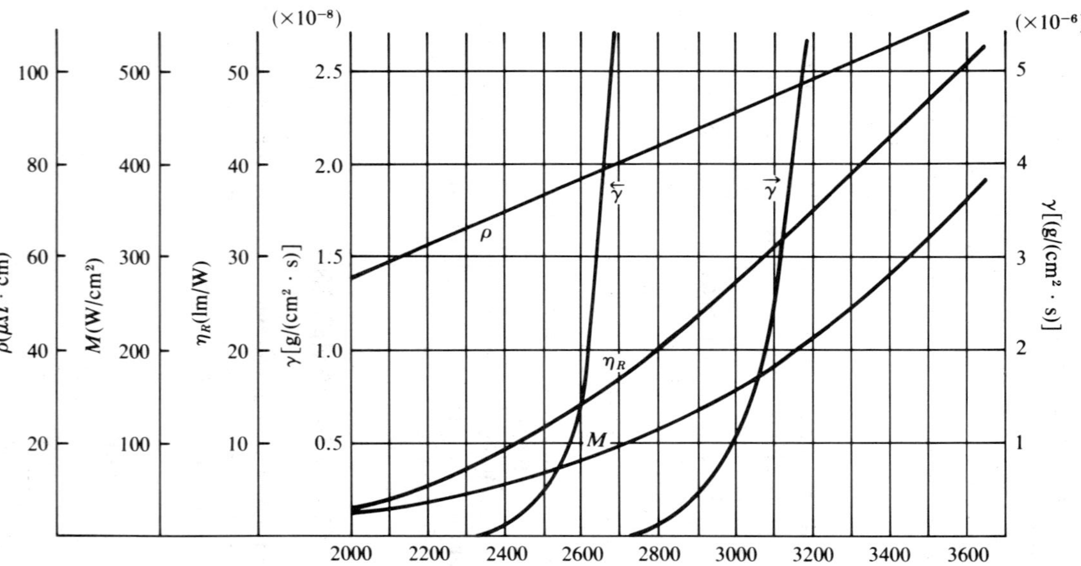

Figure 6.2 Properties of a straight tungsten filament in a vacuum. (*Source*: Jones and Langmuir, "The Characteristics of Tungsten Filaments as Functions of Temperature," GE Review, 30, 1927, pp. 310, 354, 408.)

1. γ is the evaporation rate of tungsten in a vacuum in grams of filament material per square centimeter of filament surface area per second. For gas-filled lamps, the evaporation rate is 2% of the values given in Fig. 6.2.
2. ρ is the resistivity of tungsten in microohm-centimeters. The resistance of the filament in ohms is given by

$$R = \frac{\rho l}{A_c} \times 10^{-6} \tag{6.1}$$

   where l is filament length in centimeters and A_c is filament cross-sectional area in square centimeters.
3. M is the emitted power density in watts per square centimeter of filament surface area.
4. η_R is the luminous conversion efficiency in initial lumens emitted per radiated watt.

We will illustrate the use of these curves by working two examples.

EXAMPLE 6.1

Assume a vacuum incandescent lamp has an uncoiled filament operating at a temperature of 2500 K. Find the following: life, watts consumed, lumens emitted, luminous efficacy, and operating voltage.

The density of tungsten is 19.35 g/cm³. The filament is 44 cm long and 0.03 mm in diameter at this temperature. Assume losses amount to 7%.

The life of a filament is defined as the arithmetic average of the burning hours of the lamps tested, which is approximately the number of hours for 25% of the filament material to be evaporated.

Solution. From the curves in Fig. 6.2, we obtain, for $T = 2500$ K,

$$M = 70 \text{ W/cm}^2 \qquad \rho = 73.8 \ \mu\Omega \cdot \text{cm}$$
$$\gamma = 0.2 \times 10^{-8} \text{ g/(cm}^2 \cdot \text{s)} \qquad \eta_R = 11.5 \text{ lm/radiated watt}$$

The volume, mass, surface area, and cross-sectional area of the filament are

$$V = \pi \frac{d^2}{4} l = \pi\left(\frac{9 \times 10^{-6}}{4}\right)(44) = 3.11 \times 10^{-4} \text{ cm}^3$$

$$m = (3.11 \times 10^{-4})(19.35) = 6.02 \times 10^{-3} \text{ g}$$

$$A_s = \pi dl = \pi(3 \times 10^{-3})(44) = 0.415 \text{ cm}^2$$

$$A_c = \frac{\pi d^2}{4} = 7.07 \times 10^{-6} \text{ cm}^2$$

The evaporation rate of the filament is

$$r = \gamma A_s = (0.2 \times 10^{-8})(0.415)(3600) = 2.99 \times 10^{-6} \text{ g/h}$$

Thus the life of the filament is

$$L = 0.25\,\frac{m}{r} = \frac{0.25(6.02 \times 10^{-3})}{2.99 \times 10^{-6}} = 503 \text{ h}$$

The filament radiates

$$P_R = MA_s = 70(0.415) = 29.1 \text{ W}$$

With the losses included, the input power is

$$P = \frac{29.1}{0.93} = 31.3 \text{ W}$$

The luminous efficacy is

$$\eta = 11.5(0.93) = 10.7 \text{ lm/W}$$

giving a lumen output of

$$\Phi = \eta P = 10.7(31.3) = 335 \text{ lm}$$
$$\text{or } \Phi = \eta_R P_R = 11.5(29.1) = 335 \text{ lm}$$

From Eq. (6.1), the filament resistance is

$$R = \frac{\rho l}{A_c} = \frac{(73.8 \times 10^{-6})(44)}{7.07 \times 10^{-6}} = 459\ \Omega$$

Thus

$$V = \sqrt{RP} = \sqrt{459(31.3)} = 120 \text{ V}$$

EXAMPLE 6.2

Repeat Example 6.1 if the lamp is gas filled. Assume the losses increase to 17% and that the filament is replaced by one that has the same dimensions at a new operating temperaute of 2800 K.

Solution. At 2800 K,

$$M = 117 \text{ W/cm}^2 \qquad \rho = 84.5\ \mu\Omega \cdot \text{cm}$$
$$\gamma = 0.2 \times 10^{-8} \text{ g/(cm}^2 \cdot \text{s)} \qquad \eta_R = 20.3 \text{ lm/radiated watt}$$

The filament evaporation rate is the same as before because γ is unchanged, the increase in filament temperature being exactly offset by the introduction of filling gas. Thus the life is unchanged at 503 h.

The filament radiates

$$P_R = 117(0.415) = 48.6 \text{ W}$$

giving an input power of

$$P = \frac{48.6}{0.83} = 58.6 \text{ W}$$

The lumen output and luminous efficacy are

$$\Phi = 20.3(48.6) = 987 \text{ lm} \qquad \eta_R = 20.3(0.83) = 16.8 \text{ lm/W}$$

The resistance of the filament is

$$R = \frac{84.5(44)}{7.07} = 526\ \Omega$$

requiring an applied voltage of

$$V = \sqrt{526(58.6)} = 176\ \text{V}$$

Note that this lamp is much more efficient, for the same life, than the lamp in Example 6.1, but requires a much higher voltage.

Examples 6.1 and 6.2 have dealt with the analysis of an already existing filament. An equally interesting problem is that of filament design. Assume we have available the curves in Fig. 6.2 and are asked to design a filament to operate at a voltage V, a power input P, a life L, and a filament temperature T. We wish to find its length and diameter.

Two design equations are needed

$$\pi dl = \frac{P_R}{M} \tag{6.2}$$

$$\frac{V^2}{P} = \frac{4\rho l}{\pi d^2} \tag{6.3}$$

The first equation is simply two expressions for the surface area of the filament; the second, for the resistance of the filament. If we choose a temperature, all terms in these two equations are known except d and l. Thus our design solution must be one of trial and error. We will illustrate the procedure through an example.

EXAMPLE 6.3

Design a tungsten filament to operate on 120 V ac, consume 500 W, have at least 1000 h life, and provide at least 9000 initial lumens when operated in a gas-filled incandescent lamp.*

Solution. Try a temperature of 2800 K. Then, from Fig. 6.2,

$$M = 117\ \text{W/cm}^2 \qquad \gamma = 0.02 \times 10^{-7}\ \text{g(cm}^2 \cdot \text{s)}$$
$$\eta_R = 20.3\ \text{lm/radiated watt} \qquad \rho = 84.5\ \mu\Omega \cdot \text{cm}$$

From Table 6.1, the losses for a 500-W lamp will be 17.7%. Thus

$$P_R = 500(0.823) = 411.5\ \text{W}$$

From Eq. (6.2),

$$\pi dl = \frac{411.5}{117} \qquad dl = 1.12$$

and from Eq. (6.3)

*Power, life, and lumens are not independent variables. Thus it is possible that a set of specifications of this sort cannot be satisfied.

$$\frac{(120)^2}{500} = \frac{4(84.5 \times 10^{-6})}{\pi} \frac{l}{d^2} \qquad \frac{l}{d^2} = 2.68 \times 10^5$$

Thus,

$$\frac{dl}{l/d^2} = d^3 = \frac{1.12}{2.68 \times 10^5}$$

giving $d = 0.0161$ cm and $l = 1.12/0.0161 = 69.6$ cm.

We must now check the life and lumens that this d and l provide. Proceeding as in Examples 6.1 and 6.2,

$$V = 1.42 \times 10^{-2}\ \text{cm}^3 \qquad m = 0.274\ \text{g} \qquad A_s = 3.52\ \text{cm}^2$$

The evaporation rate is

$$r = (0.2 \times 10^{-8})(3.52)(3600) = 2.53 \times 10^{-5}\ \text{g/h}$$

giving a life of

$$L = \frac{0.274(0.25)}{2.53 \times 10^{-5}} = 2700\ \text{h}$$

The lumen output is

$$\Phi = 20.3(411.5) = 8350\ \text{lm}$$

Our first trial has given us too long a life and too small a lumen output. Choosing a higher temperature of 2950 K gives $L = 445$ h and $\Phi = 10{,}500$ lm. A proper temperature lies between these two values at approximately 2875 K.

The preceding three examples should be considered as simplifications of the very complex problem of filament design. In practical coiled or coiled-coil filaments, each part of the filament radiates to other parts. Some of this radiation is absorbed; the rest is multiply reflected. Also the filament temperature varies from coil to coil due to conduction to support wires. As a result, if the filament is coiled, the curves in Fig. 6.2 theoretically do not apply. But if we use coil diameter and coil length in calculating surface area and wire diameter and wire length in calculating volume, our answers will still be reasonable approximations.

6.6 VOLTAGE EFFECTS

From the time of the first incandescent lamp, it was known that lamp performance, particularly lamp life, was seriously affected by changes in applied voltage. In 1932, Barbow and Meyer at the National Bureau of Standards used curve-fitting techniques to quantify this effect for variations in lumen output and power input with applied voltage in the following equations:

$$\log \frac{\Phi}{\Phi_0} = k_1 \log \frac{V}{V_0} + k_2 \log \left(\frac{V}{V_0}\right)^2$$
$$\log \frac{P}{P_0} = k_3 \log \frac{V}{V_0} + k_4 \log \left(\frac{V}{V_0}\right)^2 \tag{6.4}$$

where the subscript 0 refers to rated values. Barbow and Meyer determined values for the constants experimentally, for certain types of both vacuum and gas-filled lamps. Also they showed that for applied voltages within 77 to 130% of rated value, the squared terms in Eqs. (5.4) and (6.5) could be neglected if errors in Φ and P of up to 5% could be tolerated. Later, similar equations were developed relating life and voltage and CCT and voltage.

The result is a set of four equations from which variations in all incandescent lamp parameters with voltage can be approximately determined:

$$\frac{\Phi}{\Phi_0} = \left(\frac{V}{V_0}\right)^k \tag{6.5}$$

$$\frac{P}{P_0} = \left(\frac{V}{V_0}\right)^n \tag{6.6}$$

$$\frac{L}{L_0} = \left(\frac{V_0}{V}\right)^d \tag{6.7}$$

$$\frac{\text{CCT}}{(\text{CCT})_0} = \left(\frac{V}{V_0}\right)^m \tag{6.8}$$

The exponents in these equations are functions of lamp type, wattage, and voltage. Values are given in Table 6.4 and should be considered as averages over a typical operating range.

To illustrate how other relationships may be derived from these basic equations, suppose it is desired to know how lamp current varies with voltage. We can write

$$\frac{I}{I_0} = \frac{P/V}{P_0/V_0} = \frac{P}{P_0}\left(\frac{V_0}{V}\right) = \left(\frac{V}{V_0}\right)^n \left(\frac{V_0}{V}\right) = \left(\frac{V}{V_0}\right)^{n-1} \tag{6.9}$$

Thus the constants to use in expressing the current ratio are 0.54 for gas-filled lamps and 0.58 for vacuum lamps.

From Table 6.4, it can be seen that life is the most sensitive parameter to voltage variation. Five percent overvoltage yields roughly 50% life. This represents 126 V for lamps rated at 120 V and it is not at all uncommon for utility voltages to be this high to some residential customers.

EXAMPLE 6.4

A baseball field is to be lighted with incandescent lamps rated at 120 V, 1000 W, and 1000 h life. Fifty night games, each requiring 4 h of lamp burning, are played each season. At what voltage should the lamps be operated so that they will be at 70% of their rated life at the end of the season (which would mean about 90% survivors)? What percentage of

TABLE 6.4 INCANDESCENT LAMPS—VOLTAGE VARIATIONS

Constant	Gas-filled	Vacuum
k	3.38	3.51
n	1.54	1.58
d	13.1	13.5
m	0.42	0.42

% Rated volts	% Rated lumens	% Rated watts	% Rated life	% CCT
80	47	71	1860	91
85	58	78	840	93
90	70	85	398	96
92	75	88	298	97
95	84	92	196	98
98	93	97	130	99
100	100	100	100	100
102	107	103	77	101
105	118	108	53	102
108	130	113	36	103
110	138	116	29	104
115	160	124	16	106
120	185	132	9.2	108
125	213	141	5.4	110

the lamps required when 120 V is used would the owner use at this new voltage, while still maintaining the same initial illumination level? What percentage of the energy consumption at 120 V will be saved over the season?

Solution. The required lamp life is 200/0.7 = 286 h. From Table 6.4, this requires 110% of rated voltage, or 132 V. The lumen output at this voltage is 38% more than at 120 V. Thus only 100/1.38 = 72.5% as many lamps are required for the same initial illumination level. The wattage consumption per lamp will be 16% higher than at 120 V, but, with only 72.5% of the lamps, the net effect is a reduction in total initial power to $1.16(0.725)(100) = 84\%$. The average percentage of lamps burning at any time during the season is 95%. Thus the percentage energy saved over the season is $100[1 - 0.84(0.95)] = 20\%$.

6.7 REFLECTORIZED LAMPS

Reflectorized lamps were introduced in Sec. 2.6, but they are so widely used that we should address them further. There are three types, R, PAR and ER. In R and PAR lamps, the reflectors are parabolic with the filament at the focal point. Thus the reflected rays are roughly parallel. The R lamp consists of a one-piece blown bulb in the shape of a parabola. The reflector portion of the inner bulb surface is coated with vaporized silver or aluminum and the flat face is lightly frosted if a narrow (spot) distribution of light is desired and more heavily frosted if a wider (flood) distribution is wanted. R lamps are generally made of "soft" glass for use indoors.

PAR lamps are made of hard, heat-resistant glass and thus are suitable for outdoor as well as indoor use. These lamps are of two-piece construction, one piece for the reflector and the other for the lens cover. Prior to fusing the lens to the reflector, the latter is coated with a thin layer of vaporized aluminum and the filament is mounted accurately at the parabola's focal point, allowing for more precise beam control than in R lamps. Beam spreads available range from a very narrow spot to a very wide flood, with the amount of "stippling" of the lens determining the spread of the beam.

The ER lamp is a one-piece, blown-bulb elliptical reflector lamp. An ellipse has two focal points. In the ER lamp, the filament is placed at one focal point, which causes an appreciable fraction of the lumens to pass near to the second focal point, 2 in in front of the face of the lamp. When used in downlights, this design minimizes the amount of light trapped inside the fixture and can provide more light on the work surface than a larger-size R lamp.

Typical characteristics of R, PAR, and ER lamps are shown in Table 6.5.* The beam spread is the cone angle out to the point where the intensity

*A line of 12-V PAR lamps, with heavy compact filaments for precise beam control, is also available.

TABLE 6.5 120-V REFLECTOR LAMPS

Watts	Bulb	Type	Central cone candelas	Beam spread (degrees)	Beam lumens	Total lumens	Lumens per watt	Life (h)
50	R20	Flood	530	95	350	440	8.8	2000
75	R30	Flood	400	136	730	900	12.0	2000
75	R30	Spot	1,540	78	760	900	12.0	2000
150	R40	Flood	1,040	124	1650	1900	12.7	2000
150	R40	Spot	5,400	49	1550	1900	12.7	2000
300	R40	Flood	1,950	123	3200	3650	12.2	2000
300	R40	Spot	8,900	60	3100	3650	12.2	2000
50	ER30	—	—	—	—	525	10.5	2000
75	ER30	—	—	—	—	850	11.3	2000
120	ER40	—	—	—	—	1475	12.3	2000
75	PAR38	Flood	1,750	57	690	765	10.2	2000
75	PAR38	Spot	4,500	27	540	765	10.2	2000
150	PAR38	Flood	4,000	60	1690	1740	11.6	2000
150	PAR38	Spot	11,500	28	1200	1740	11.6	2000
200	PAR46	Flood	11,500	23 × 38	1300	2300	11.5	2000
200	PAR46	Spot	31,000	19 × 23	1100	2300	11.5	2000
300	PAR56	Flood	24,000	19 × 34	2100	3840	12.8	2000
300	PAR56	Spot	68,000	14 × 20	1800	3840	12.8	2000
500	PAR64	Flood	37,000	19 × 35	3300	6500	13.0	2000
500	PAR64	Spot	11,000	14 × 19	3000	6500	13.0	2000

Source: General Electric Co., "PAR and R Lamps," Bulletin 260-399R4. 1976. Reprinted with permission.

is 10% of its maximum value. The beam lumens are the initial lumens present in this cone. If the beam is oval, the maximum and minimum beam spreads in two planes separated by 90° are given. These are called *vertical* and *horizontal* spreads. The central cone candelas are the average intensity in the central 10° cone for floods and 5° cone for spots.

Generally each wattage size of R and PAR lamp is made in both flood and spot distributions. For a given wattage, the spot has greater maximum intensity than the flood and a narrower beam spread, but the total initial lumens from each lamp are the same.

Reflectorized lamps are the "gas guzzlers" of the incandescent lamp field. Because they are often located in hard-to-service places and are more costly than general service lamps, their lives are set at 2000 h, in contrast with the 750- and 1000-h life ratings of most standard incandescent lamps. But, as we have seen, greater life generally means less light and reduced lumens per watt. Thus, in these times of energy conservation, reflectorized lamps should be used sparingly in those applications for which other lamps are not suitable.

Recently, PAR lamps have appeared with improved luminous efficacy. These incorporate modifications in the optics to increase beam efficiency by reducing stray light at large angles from the beam axis. Also, higher-efficiency tungsten halogen capsules (see Sec. 6.8) are replacing the filament as the internal source.

6.8 TUNGSTEN-HALOGEN LAMPS

In 1958, E. G. Fridrich and E. H. Wiley discovered that introducing a halogen gas, originally iodine, into an incandescent lamp could increase the lumens per watt and improve the lumen maintenance significantly. The reason for this lies in the halogen regenerative cycle, shown diagrammatically in Fig. 6.3.

In an ordinary incandescent lamp, evaporated tungsten particles are transported to the inside of the bulb by gas convection currents. The bulb wall is relatively cool and the particles adhere to it.

When iodine is introduced and if the temperature near the filament is maintained at about 2000 K, the evaporated tungsten atoms combine with the iodine vapor to form tungsten iodide. The tungsten iodide is carried by the convection currents to the bulb wall, but if the bulb temperature is kept between 500 and 1500 K, it does not adhere to the bulb wall but rather circulates back toward the filament. Very near the filament, where the temperature exceeds 2800 K, the tungsten iodide is reduced to tungsten and iodine vapor. The former redeposits on the filament and the latter is free to acquire another tungsten atom and repeat the cycle. The chemical equations for these relations are:

$$\mathrm{W} + 2\mathrm{I} \xrightarrow{2000\ \mathrm{K}} \mathrm{WI_2} \qquad \mathrm{WI_2} \xrightarrow{2800\ \mathrm{K}} \mathrm{W} + 2\mathrm{I} \tag{6.10}$$

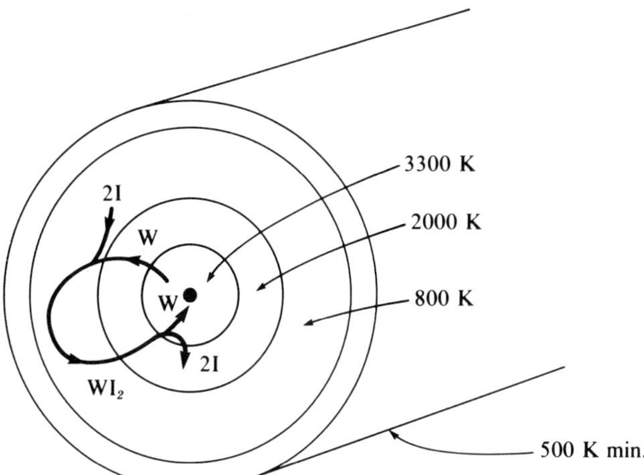

Figure 6.3 Halogen cycle.

Whereas an incandescent lamp can be expected to provide only about 80% of its initial lumens at the end of life, the figure for a tungsten-halogen lamp is above 95%.

Because of the high bulb-wall temperatures required, it is necessary to use quartz for tungsten-halogen lamps. Quartz has a softening point of 1900 K, far above the temperature requirements of the halogen cycle.

The stronger bulb wall and smaller volume mean that the lamp can be operated up to several atmospheres of internal gas pressure. For a given filament temperature, this higher pressure reduces the rate of filament evaporation, thus lengthening life. Or, the lamp can be operated at a higher filament temperature for the same life, with resulting higher luminous efficacy. Parenthetically it should be noted that if each tungsten particle could be guided back to the spot on the filament from which it originated, the life would be infinite. Such is, of course, not the case and thus life can be increased but remains finite.

In addition to the halogen gas, tungsten-halogen lamps are filled with a mixture of argon and a small amount of nitrogen. Most lamps today use bromine instead of iodine as the halogen gas. It is colorless, whereas iodine can impart a purplish tint. Also it permits about 50 K lower bulb-wall temperature. Correlated color temperatures of tungsten-halogen lamps range from 2800 K to 3400 K and their lives are 2000 h or more in the 3000 K range.

Tungsten-halogen lamps can have a number of shapes but are most often tubular with the filament oriented axially. They are available in both double-ended and single-ended types, with the latter having a small screw base. The two types are illustrated in Fig. 6.4 and a summary of characteristics of common tungsten-halogen lamps is given in Table 6.6.

Because they are compact sources providing good correlated color

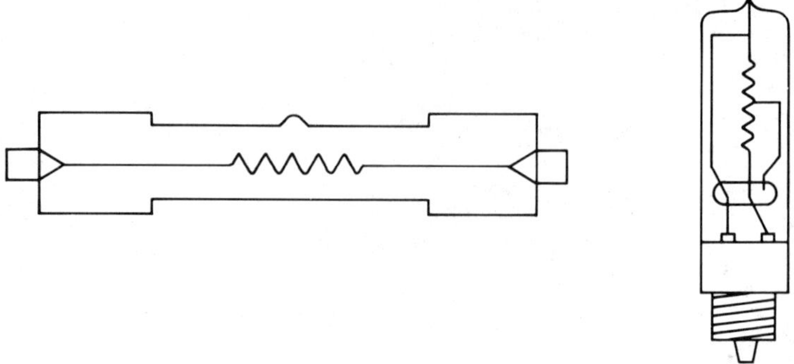

Figure 6.4 Double- and single-ended tungsten-halogen lamps.

temperature, excellent lumen maintenance, and reasonable life, tungsten-halogen sources are well-suited for outdoor lighting applications, particularly sports lighting, and for studio, theater, and television lighting. Their filaments are generally mechanically stable and can be positioned with high precision. Their small envelopes permit optical components to be located close to the filament, permitting precise optical control. Because of the higher filament temperature, high filament luminances can be obtained. Thus the tungsten-halogen lamp is widely used in optical systems, for example, in spotlights, film projectors, and scientific instruments.

Recently a new line of low-voltage tungsten-halogen lamps has been introduced by several manufacturers. These are MR-16 (multifaceted reflec-

TABLE 6.6 TUNGSTEN-HALOGEN LAMPS

Watts	Bulb	Lighted length (in)	Life	Lumens	Lumens per watt
Double-ended					
300	T2½	2⅛	2000	6,000	20.0
300	T4	⅝	2500	5,000	16.7
400	T4	⅝	2000	7,500	18.8
500	T3	2	2600	10,000	20.0
1000	T3	6¾	2000	21,500	21.5
1000	T6	1⅛	2000	22,000	22.0
1500	T3	6¾	3000	29,000	19.3
Single-ended					
100	T4	1⅜	1000	1,900	19.0
150	T4	1½	2000	2,600	17.3
250	T4	1⅝	2000	5,000	20.0
500	T4	2	2000	9,500	19.0

tor, 2-inch diameter) lamps, operating, through a transformer, at 12 volts. They are available in 20, 42, 50, and 75 watts, operate at 3000 K and have lives ranging from 2000 to 3500 hours.

The lamps were originally developed for optical projection equipment but are now being widely used in display lighting. They are available in flood, narrow flood, and narrow spot distributions and feature precise beam control with very little spill light.

The heart of the lamp is a small tungsten-halogen capsule, cemented into a one-piece, all-glass reflector with facets for optically controlling the beam. The reflector has a dichroic coating (see Chap. 10) which transmits two-thirds of the infrared energy generated by the capsule out the back of the reflector.

MR-16 lamps have slightly higher luminous efficacy than standard-voltage incandescent lamps, but, because of their precise optics, can replace standard lamps of twice their wattage. Also, their small size (lamp and socket can fit into a 2-inch cube) permits compact fixtures that can be easily concealed.

6.9 FLUORESCENT LAMP DEVELOPMENT

As noted in Sec. 1.2, the basic principle of transforming ultraviolet radiation into visible radiation was discovered in 1852 by Sir George Stokes. In the interim, from then until the 1920s, much work was done in the development of low- and high-pressure electric discharges in mercury and sodium but very little work was done on fluorescence. Also the electric discharge devices that were developed were relatively inefficient in transforming the electrical input to the lamp into visible radiation. The electrodes emitted electrons inefficiently into the arc stream, many of the collisions of electrons with gas atoms were elastic, and the excitations that did result were not concentrated in one special line.

A major breakthrough occurred in the 1920s when it was discovered almost simultaneously in several locations that a mixture of mercury vapor at low pressure and an inert gas at higher pressure was approximately 60% efficient in converting electrical input power into a single spectral line at 253.7 nm. This, along with the development of efficiently emitting and long-lived electrodes and appropriate fluorescent materials (phosphors), paved the way for the introduction of the first fluorescent lamp in the late 1930s.

In 1934, Dr. W. L. Enfield, manager of General Electric's Lamp Development Laboratory, received a report from England from Dr. A. H. Compton outlining the progress being made by the British in developing lamps using tubes coated with fluorescent materials. Enfield immediately created a research team composed of G. E. Inman, R. N. Thayer, A. E. Lemmers, and W. A. Roberts and charged them with the task of producing a commercially practical fluorescent lamp.

By the summer of 1935, Inman's team had produced a prototype green fluorescent lamp with an efficacy of 60 lm/W. From their work emerged a set of guidelines for future fluorescent lamp development:

A. The basic discharge process is always in three steps:
  1. Free electrons are derived from the electrodes and accelerated by the applied field.
  2. The kinetic energy of the free electrons is converted into excitation energy of the gas atoms.
  3. The excitation energy of the gas atoms is converted into radiation.

B. High efficiency is obtained if mercury vapor at low current density is used to produce a single ultraviolet spectral line at 253.7 nm.

C. Phosphors are needed whose peak spectral sensitivity is at 253.7 nm, and these must be applied to the inside of the tube because 253.7 nm does not pass through ordinary glass.

D. To provide optimum arc temperature and pressure for generation of 253.7 nm, a bulb temperature of 105 to 115°F, a mercury pressure of 6 to 10 μm, and an inert gas pressure of 2 to 3 mm are required.

E. A tubular lamp of appreciable length-to-diameter ratio is needed so that the fixed wattage losses at each end, in the cathode and anode fall regions, are a small percentage of the total watts input.

F. Oxide-coated hot cathodes should be used to provide an abundance of free electrons.

Late in 1935, a 2-ft-long green fluorescent lamp was demonstrated at the annual IES Technical Conference as "a laboratory development of great promise." It was 2½ years later, in April 1938, that the fluorescent lamp was introduced commercially in white and six colors, obtained by using various mixtures of phosphor powders. The new lamps were displayed publicly in 1939 at the New York World's Fair and the Golden Gate Exposition in San Francisco.

The first lamps were 15, 20, and 30 W in 18-, 25-, and 36-in lengths, respectively. Soon after the 40-W T12, 4-ft-long lamp was introduced and took the lead rapidly for general lighting in offices, schools, industry, and commerce. As better phosphor combinations were found, the early yellowish-white lamp gave way to a 3500 K lamp, which is essentially today's standard warm white, and a 6500 K daylight lamp, designed to simulate an average north skylight on an overcast day. Later a standard cool white, 4500 K lamp was introduced. From early skepticism that the lamp was too long and bulky, sales increased until, in 1952, fluorescent lamps passed incandescent lamps as the major source of general lighting in the United States.

Since the introduction of the standard 4-ft, 1½-in-diameter, 40-W lamp in 1940, there has been very little change in the basic arc discharge portion of this lamp. Argon is still used, although its pressure is somewhat less than it was originally; the mercury vapor is maintained at the same pressure it was initially; the lamp still draws 425 mA and has a voltage drop of 100 to

105 V. There have been dramatic improvements in phosphors and in electrode design since 1940, and a number of basic improvements in lamp circuitry have occurred. The more important of these will be discussed in later sections of the chapter.

6.10 FLUORESCENT LAMP CONSTRUCTION AND TYPES

A cutaway view of a fluorescent lamp is shown in Fig. 6.5*a*. It consists of a lime glass tube, which serves as an airtight enclosure for the drop of mercury, the argon filling gas, the phosphor coating and the electrodes and their mount assemblies, and also includes two bases, which have two contacts (bipin) for preheat and rapid-start lamps and one contact (single pin) for instant start lamps.

The electrode mount assembly is very similar to the stem press unit in incandescent lamps. The filament often functions as both cathode and anode, because the applied voltage is alternating. Sometimes separate anodes are used, in the form of wires or small plates attached to the filament ends. These protect the filaments from undue electron bombardment and reduce the wattage loss at the ends of the lamp.

Typical cathodes are shown in Fig. 6.5*b*. They are similar to incandescent lamp filaments except that they are dipped in a waffle-batter-consistency mixture of barium, strontium, and calcium carbonates, which are baked during manufacture to become oxides, providing an active cathode with an abundance of easily obtainable free electrons.

Fluorescent lamps are available in a variety of lengths, diameters, and wattages. Lengths range from 6 in to 8 ft; diameters from T5 to T17($\frac{5}{8}$ to $2\frac{1}{8}$ in)

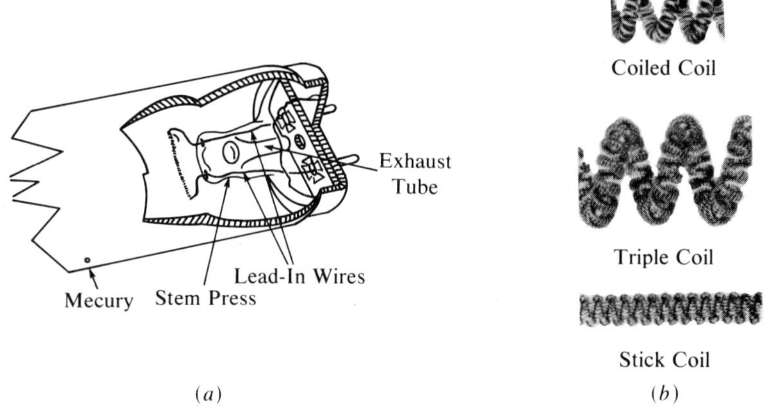

Figure 6.5 Fluorescent lamp construction. (*Source*: General Electric Co., "Fluorescent Lamps," Bulletin TP-111R, Nela Park, Cleveland, Ohio, 1978. Reprinted with permission.)

and wattages from 4 to 215 W. They are available in many varieties of white and five colors—blue, green, pink, red, and gold.

A summary of the electrical and lighting characteristics of several of the more common fluorescent lamps is given in Table 6.7. The life figures are based on operating the lamp for 3 h each time it is started. Each start is deleterious to the electrodes; thus life is a function of the number of burning hours per start. At 8 h per start, the life values should be multiplied by 1.3; for 12 h per start by 1.5 and for continuous burning by 1.7.

The final column in Table 6.7 indicates lamp lumen depreciation. In fluorescent lamps, this may be as much as 10% during the first 100 h of operation. For rating purposes, the 100-h lumen value is called *initial lumens* and the lumen depreciation is calculated from that point onward. Quite often two depreciation factors are given. The first is called the mean lumen factor and is the percent of the initial lumens to be expected at 40% of rated life. The second is called the lamp lumen depreciation factor and is the percent of the initial lumens to be expected at 70% of rated life. Lamp manufacturers generally publish the former figure, but the latter figure is more relevant in lighting design. The two factors will be discussed further in Sec. 7.3.

The symbols used in the lamp number designations in Table 6.7 mean the following:

CW	Standard cool white
WW	Standard warm white
CWX	Deluxe cool white
WWX	Deluxe warm white
ES	Energy saving
HO	High output

The lamp companies have standardized on four basic white lamps, two standard and two deluxe. The standard lamps have high efficacy and average color rendition. The deluxe lamps have lower efficacy and improved color rendition, particularly of complexions. Recently several lamp companies have developed high CRI lamps with higher efficacy than deluxe lamps. These lamps have an altered phosphor response so that output power is concentrated in carefully selected wavelength bands.

The ES lamps in Table 6.7 merit further discussion. In response to concerns about conserving energy and reducing energy costs, all of the major companies have developed reduced-wattage fluorescent lamps for operation on standard fluorescent ballasts. In these lamps, the argon filling gas is replaced by an argon-krypton mixture which decreases lamp operating voltage and thus lamp wattage. For the standard 40-W, 4-ft-long fluorescent lamp, lamp power is reduced to 34 W (15%) but initial light output is also reduced to 2750 lm (9.5%).

TABLE 6.7 FLUORESCENT LAMP DATA

Type	Lamp no.	Watts	Length (in)	Life (at 3 h per start)	Initial lumens	Lamp current (mA)	Initial lamp lumens per watt	Lumens at 70% life
Preheat	F14T8CW	14	15	7,500	650	420	46	520
	F14T12CW	14	15	7,500	675	390	48	555
	F15T8CW	15	18	7,500	870	300	58	690
	F15T12CW	15	18	9,000	800	330	53	650
	F20T12CW	20	24	9,000	1,250	380	63	1,060
Rapid start and preheat	F40CW	40	48	20,000	3,150	425	79	2,650
	F40WW	40	48	20,000	3,200	425	80	2,690
	F40CWX	40	48	20,000	2,200	425	55	1,670
	F40WWX	40	48	20,000	2,150	425	54	1,635
	F40ES	34	48	20,000	2,750	450	81	2,310
Instant start	F48T12CW	39	48	9,000	3,000	425	77	2,490
	F72T12CW	55	72	12,000	4,600	425	84	4,090
	F96T12CW	75	96	12,000	6,300	425	84	5,610
	F64T6CW	40	64	7,500	2,800	200	70	2,160
	F72T8CW	35	72	7,500	3,000	200	86	2,490
	F96T8CW	50	96	7,500	4,200	200	84	3,740
	F96T12ES	60	96	12,000	5,600	440	93	4,985
High output rapid start	F48T12CWHO	60	48	12,000	4,300	800	72	3,525
	F72T12CWHO	85	72	12,000	6,650	800	78	5,455
	F96T12CWHO	110	96	12,000	9,200	800	84	7,545
	F96T12ESHO	95	96	12,000	8,300	810	87	6,805
Very high output rapid start	F48T12CW1500	115	48	10,000	6,800	1,500	59	4,690
	F72T12CW1500	160	72	10,000	10,900	1,500	68	7,850
	F96T12CW1500	215	96	10,000	15,000	1,500	70	12,300
	F96T12ES1500	195	96	10,000	14,000	1,580	72	10,100

Later versions of these lamps employ a new white phosphor, which is similar to standard cool white but gives somewhat greater luminous efficacy and somewhat poorer color rendition. These lamps operate at 34 W and produce 2925 lm. Energy saving lamps are also now available in instant start, high output, and very high output types.

6.11 FLUORESCENT LAMP BASICS

The basic low-pressure gas discharge process begins when electrons are emitted from the electrodes at the ends of the tube, either by thermionic emission or by high-field emission. The former mechanism is used in all preheat fluorescent lamps (the ones introduced in 1938), where the external circuitry is arranged so that current flows through each electrode, heating them to incandescence, prior to application of voltage across the lamp. The latter mechanism is used in instant-start fluorescent lamps, which first appeared in 1944. Voltage is applied across the lamp immediately and is sufficiently high to overcome the work function at the surface of the electrode and remove electrons directly, causing the electrode to heat instantly and emit thermionically. A combination of these two mechanisms is used in rapid-start fluorescent lamps, which were introduced in 1952 and dominate the market today. In these lamps, medium voltage and circulating filament current are applied simultaneously and the lamps light within 1 s.

As the free electrons leave the electrodes, by whatever mechanism, they are accelerated by the applied electric field. Their drift velocity down the tube is only 1 to 2% of their random velocity in all directions within the tube. Thus they may be more properly thought of as a swarm of bees rather than as a line of fast-moving cars.

The free electrons encounter argon atoms and excite and ionize them. The first excitation level of argon is metastable and thus argon forms what is known as a Penning mixture with mercury. A Penning mixture of two gases occurs when the major gas (argon) has a metastable state as its lowest excited state and the energy of this state is slightly greater than the ionization energy of the minor gas (mercury). A Penning mixture converts a high probability that every electron will ultimately gain enough energy for an excitation collision into a high probability that such collisions will ultimately result in ionization. In the argon-mercury situation, metastable argon at 11.56 eV can easily ionize mercury (10.39 eV). At low temperatures most of the mercury is in the form of small liquid droplets on the inside of the tube, but the heat from the argon discharge quickly vaporizes the mercury, which then is ionized. At room temperatures, there is enough vaporized mercury so that the ionization of mercury begins immediately. In both cases, mercury is the predominant ionized constituent of the arc.

Of course it is not the ionized mercury atoms that produce the desired radiation; it is the excited ones. It requires 10.39 eV to ionize mercury and only 4.88 eV to reach the excited state of interest, so it is to be expected that considerable excitation will occur. The 4.88-eV level in mercury is not metastable, so the excited atoms revert immediately to their ground states and radiate 253.7-nm photons. These photons impinge on the phosphor on the inside of the tube and are converted into visible photons.

To explain further the underlying mechanisms in a fluorescent lamp, let us ask and answer several questions:

(a) Why Is It So Desirable to Use Mercury in Fluorescent Lamps?

The answer to this question is threefold. First, for high luminous efficacy, it is necessary to have a high quantum ratio, defined as

$$QR = \frac{\text{average energy of visible photons}}{\text{energy of UV photon}} \tag{6.11}$$

From Eq. (4.6), Eq. (6.11) can be rewritten as

$$QR = \frac{\lambda \text{ of UV photon}}{\text{average } \lambda \text{ of visible photons}} \tag{6.12}$$

For mercury, the wavelength of the ultraviolet photon is 253.7 nm and the average wavelength of the visible photons, assuming the phosphor radiation is distributed reasonably uniformly throughout the visible spectrum, is 555 nm. This gives a relatively high quantum ratio of 0.46.

Second, we desire a single ultraviolet line with a high probability of reaching its excitation level. The first three excitation levels of mercury are 4.66 eV, 4.88 eV, and 5.43 eV. The first and last of these are metastable and thus do not produce radiation directly. Also 4.88 eV is less than $\frac{1}{2} \times 10.39$ eV, which indicates a high probability of excitation. Thus we have a situation in mercury where there are three easily reached excitation levels but only one is capable of radiation. The result is that 60% of the input watts appear as 253.7-nm radiation.

Third, it is essential to have a material with a vapor pressure at room temperature such that the lamp temperature for vaporization will not be excessive. By vapor pressure is meant the pressure exerted by a vapor when it is in equilibrium with its liquid, that is, when molecules are returned to the liquid at the same rate as they are leaving it. For mercury the vapor pressure at room temperature is 1.8 μm.

Each fluorescent lamp has an excess of liquid mercury to insure an adequate supply during the life of the lamp. The mercury vapor pressure within the lamp is determined by the temperature of the coolest spot on the bulb and the optimum situation occurs when this spot is about 40°C (104°F). At this temperature, the mercury vapor pressure is about 10 μm.

(b) Why Is an Inert Gas Necessary for Operation of a Fluorescent Lamp?

Again the answer is threefold. First, since the mercury vapor pressure at room temperature is relatively low, it is desirable to have an auxiliary agent to help establish the arc. Inert gases ionize at lower tube voltages than do other gases because the ratio of the first excitation potential and the ionization potential in an inert gas is relatively high, increasing the likelihood of ionization. For example, this ratio for argon is 11.56/15.68 = 0.74.

Second, it is necessary to have a means of ionizing vaporized mercury. As we have mentioned, the ability of some inert gases to form Penning mixtures with mercury greatly enhances the efficiency of this process.

Third, it is necessary to contain the mercury arc. Without the pressure of the inert gas, the diffusion of positive mercury ions and electrons out of the arc stream to the bulb wall, and the resultant recombination of these particles, would be excessive and the mercury arc would be inefficient.

(c) Why Should the Inert Gas Be Argon?

The answer to this question is in two parts. First we desire the ratio of masses of the electron and the gas atom to be as small as possible so that electrons lose very little energy in elastic collisions with gas atoms. The inert gases, in order of increasing mass, are helium, neon, argon, krypton, and xenon. Based on the mass ratio criterion, xenon would be best and helium worst.

Second, it is necessary that the inert gas form a Penning mixture with mercury. It turns out that helium, neon, and argon do, but krypton and xenon do not. Also the metastable energy of argon (11.56 eV) is much nearer to the ionization energy of mercury than is that of helium (19.80 eV) or neon (16.62 eV). Thus our first choice for the inert gas is argon.

6.12 ENERGY DISTRIBUTION AND EFFICIENCY

The energy distribution of a typcial standard cool white fluorescent lamp is shown in Fig. 6.6. We note that 2% of the input energy is converted directly into light. This represents transitions from higher states on the mercury energy level diagram which results in emission of the four visible lines in the mercury spectrum at 404.7 nm, 435.8 nm, 546.1 nm, and 577 to 579 nm.

As mentioned earlier, 60% of the input energy goes directly to producing 253.7-nm ultraviolet radiation. One-third of this energy (20% of the total input energy) results ultimately in the emission of visible energy by the phosphor. Two-thirds of the 253.7-nm radiation and 38% of the input energy appear as heat. The former is the fraction of the 253.7-nm energy which is converted to heat by the phosphor, whereas the latter is the percentage of the input energy that provides electrode heating and bulb warmth. The heat is ultimately distributed externally in two forms: 46% of it (36% of the input energy) manifests itself as long-wavelength infrared radiation, largely by the bulb; the other 54% (42% of the input energy) is dissipated by conduction and convection to the surrounding air.

We can also look at the distribution of the input energy from the point of view of realized efficacy, much as we did for the incandescent lamp. If all of the input energy were converted into a single spectral line at 555 nm, the luminous efficacy would be 683 lm/W. But phosphors emit over a range of

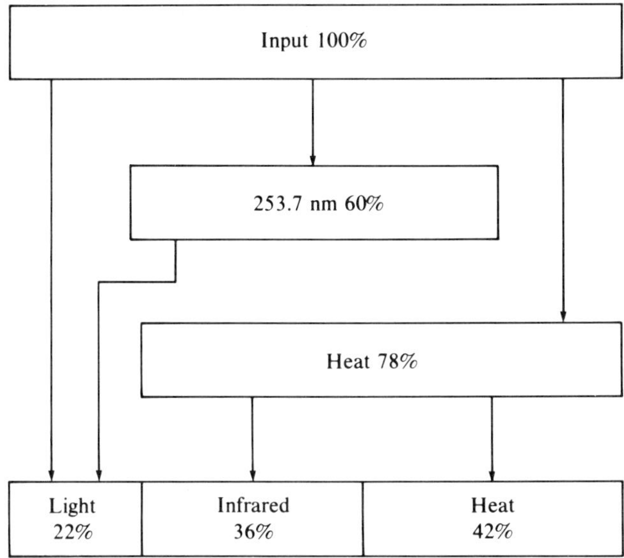

Figure 6.6 Fluorescent lamp energy distribution.

wavelengths, including some infrared energy. This reduces the maximum luminous efficacy by 49% to 348 lm/W. The quantum ratio is 46%, reducing the efficacy further to 160 lm/W. Next, only 60% of the input energy becomes 253.7-nm radiation, causing a further reduction to 96 lm/W maximum. Last, losses due to phosphor absorbtion, bulb absorbtion, end loss in luminance, and nonutilization of 253.7 nm result in a realized lamp efficacy of 70 to 80 lm/W.

6.13 FACTORS AFFECTING FLUORESCENT LAMP PERFORMANCE

There are five major factors affecting fluorescent lamp performance, all of which are interrelated. These are

1. Lamp loading
2. Argon pressure
3. Bulb-wall temperature (and associated mercury vapor pressure)
4. Tube length
5. Tube diameter

The loading of a fluorescent lamp may be expressed as the arc watts per square inch of bulb-wall surface area according to the following equation:

$$\text{Loading} = \frac{P - 16I}{\pi d l} \tag{6.13}$$

where P is the input power, 16 is the average electrode fall voltage (both ends added), I is the lamp current, and d and l are the lamp's diameter and length. For the standard 40-W lamp, Eq. (6.13) yields

$$\text{Loading} = \frac{40 - 16(0.425)}{226} = 0.15 \text{ W/in}^2$$

Increasing the current for a given length and diameter of bulb tends to decrease efficacy. This is borne out by comparing the efficacies of the various 4-ft T12 lamps of different current ratings in Table 6.7. The results appear in Table 6.8, where it is seen that the efficacy of the standard lamp is higher than for the other two and its loading is less. It turns out that maximum luminous efficacy results for a loading of from 0.12 to 0.15 W/in² for T12 lamps.

EXAMPLE 6.5

The question might be legitimately raised as to why higher wattage fluorescent lamps are not available. What length and diameter would be required for a 1000-W fluorescent lamp of reasonable efficacy?

Solution. Assume a loading of 0.15 W/in² is sought. Then

$$A_s \simeq \frac{1000}{0.15} = 6700 \text{ in}^2$$

neglecting the cathode fall loss. If d were 5 in, l would have to be 355 in $\simeq$ 30 ft, a totally unreasonable size for a lamp.

One may explain the fact that there is an optimum loading in the following way. As current increases, the probability of multiple collisions of electrons and mercury atoms increases, decreasing the probability of generation of the 253.7-nm line. Also, an increase in current tends to increase mercury vapor pressure, which increases the imprisonment of 253.7-nm radiation. On the other hand, as current decreases, mercury vapor pressure decreases and this reduces the number of collisions of electrons and mercury atoms, reducing the number of 253.7-nm photons.

The effect of inert gas pressure is also an important factor. If it is too high, the mean free path in reduced and the number of elastic collisions increases. If it is too low, there is excessive migration of mercury ions and electrons out of the arc stream.

TABLE 6.8 LOADING EFFECTS OF 4-FT T12 LAMPS

Lamp	Milliamperes	Watts	Lumens	Lumens per watt	Loading (W/in²)
Standard	425	40	3150	79	0.15
High output	800	60	4300	72	0.21
Very high output	1500	115	6800	59	0.40

Fluorescent lamps are extremely sensitive to temperature changes. It was pointed out in Sec. 6.11 that optimum performance results when the coolest spot on the bulb is maintained at about 104°F. The situation when bulb temperature varies is shown in Fig. 6.7. As the temperature of the coolest part of the bulb decreases, mercury vapor pressure decreases. The result is a 1% decrease in efficacy for each 1.5°F decrease in bulb-wall temperature. As temperature increases, more mercury vaporizes, the mean free path decreases, and the proportion of elastic collisions increases. Also there is more imprisonment of radiation. The net effect is less severe, representing a 1% decrease in efficacy for each 2.5°F increase in bulb-wall temperature.

Lamp voltage is a linear function of positive column length for a given lamp loading. Thus luminous efficacy increases as tube length increases because the constant wattage loss represented by the electrode fall potential becomes a smaller fraction of the total watts input. Also the end loss in light output becomes a smaller portion of the total as tube length increases. For a T12 lamp operating at 425 mA, the theoretical maximum efficacy is about 100 lm per lamp watt when the length is infinite. This reduces to 75 lm per lamp watt when the length is 48 in.

Last, luminous efficacy is a function of tube diameter. As diameter is increased, luminous efficacy increases, then decreases. With increasing diameter, there is more imprisonment of 253.7-nm radiation. This means that there are more opportunities for collisions in which a mercury atom already excited to the 4.88-eV level is excited to a higher state or ionized before it can release its 253.7-nm photon. This effect decreases luminous efficacy, but there is a counter effect. As tube diameter increases, the rate of loss of electrons and ions to the bulb wall decreases. This means that less new ionization is required to sustain the arc and thus more electron energy can be utilized in achieving excitation to the 4.88-eV level. The result is an increase in luminous efficacy. As diameter is increased, the second factor dominates until a peak efficacy is reached, after which the first factor dominates.

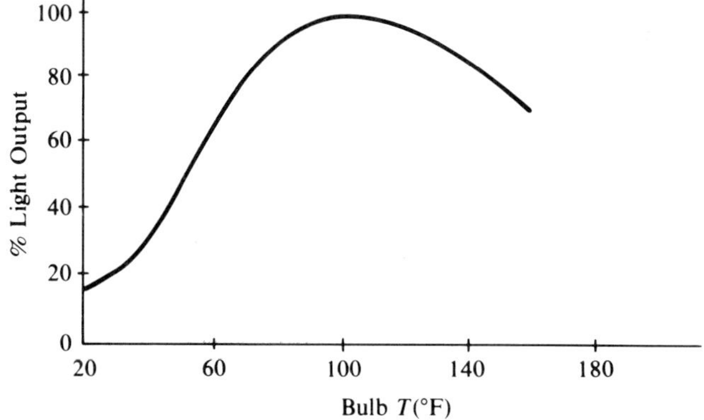

Figure 6.7 Bulb-wall temperature effects.

6.14 PHOSPHORS

As we have noted, the inside of the tube of a fluorescent lamp is coated with a mixture of phosphors, each of which has its peak spectral sensitivity near 253.7 nm. The early fluorescent lamps used zinc beryllium silicate as the basic phosphor. However beryllium turned out to be toxic and was replaced by calcium halophosphate. Contemporary lamps contain mixtures of calcium halophosphate and magnesium tungstate to produce the various whites. The former produces yellowish-white light; the latter, bluish-white light.

A phosphor is basically a semiconductor material to which activators have been added. The band theory model for a typical phosphor is shown in Fig. 6.8. It contains three zones, namely a valance band, a forbidden zone, and a conduction band.

In a pure semiconductor, the valance electrons are tied to their respective atoms and there are very few free electrons and ions present. In the model this is portrayed by having no permitted energy levels in the forbidden zone between the valance band and the conduction band. Thus to free an electron from an atom, we must apply enough energy (electronvolts) to cause an electron to leap directly from the valance band to the conduction band. In the case of a phosphor, activators are added which create permitted energy levels in the forbidden zone, as indicated in Fig. 6.8. For ZnS, an activator is silver.

The process of converting ultraviolet energy to visible energy in the phosphor may be explained in terms of Fig. 6.8. A 253.7-nm photon impinges on the phosphor. It causes an electron to jump from divalent sulfur to divalent zinc, creating a univalent zinc ion (with an excess electron) in the conduction band and a univalent sulfur ion (deficient one electron) in the valance band. The excess electron in the conduction band migrates through the crystal lattice from one zinc ion to another. At the same time, the deficiency, or positive hole, in the valance band migrates from one sulfur ion to another. When the positive hole arrives near an activator, it captures an electron from it. This traps the positive hole in the activator. Then a migrating electron in the valance band falls into the positive hole and a photon is released.

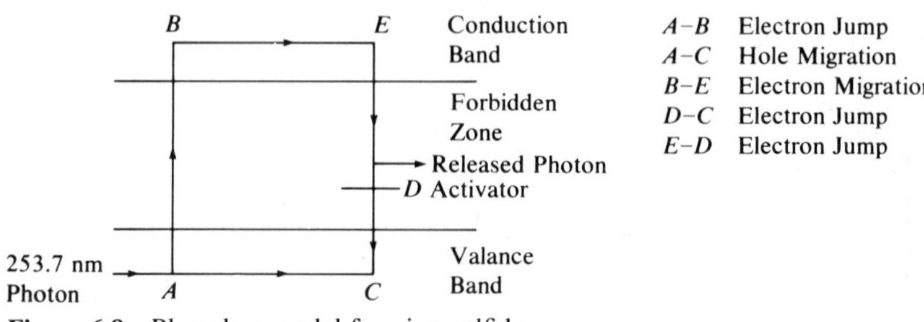

Figure 6.8 Phosphor model for zinc sulfide.

The diagram in Fig. 6.8 helps to explain why the conversion is always from ultraviolet to visible, rather than vice-versa. The photon released when an electron falls to the activator level always has less energy than the incident photon. Therefore its wavelength must be longer.

6.15 BALLASTS

Arc discharges are negative resistance phenomena, which means that if lamp current increases, lamp voltage decreases. Thus it is essential to place some form of current-limiting device in series with the lamp. The simplest control element is a series resistor, but this leads to significant power loss. Thus resistive ballasting is used only when lamps are operated on dc; otherwise reactive ballasting is employed, or more recently, electronic ballasting.

The ballast has six functions in addition to its fundamental one of limiting current in the negative resistance arc:

1. Provide sufficient open-circuit secondary voltage to initiate the arc
2. Regulate the lamp current against line voltage changes
3. Relight the lamp on each half cycle of the applied ac voltage
4. Minimize power loss
5. Permit cathode heating, for preheat and rapid-start lamps
6. Provide high power factor

Let us discuss each of these functions in turn. The peak of the applied voltage wave governs the starting of the lamp because, at 60 Hz, the time required for significant ionization to occur is small compared with the time of half a cycle, that is, the ionization growth rate is much greater than the applied voltage growth rate. However, size, weight, cost, and power loss of the ballast are a function of rms open-circuit secondary voltage and short-circuit secondary current. What is desired then is a ballast which provides a relatively small rms voltage for a given peak voltage.

The ratio of peak-to-rms voltages is called the voltage crest factor:

$$\mathrm{VCF} = \frac{V_{\mathrm{PK}}}{V_{\mathrm{rms}}} \tag{6.14}$$

For a sinusoidal voltage, VCF = 1.4; a typical VCF for a fluorescent lamp ballast is 2.0 and the resulting secondary voltage waveform is as shown in Fig. 6.9. Through Fourier analysis, it can be shown that this wave contains a significant third harmonic component, resulting from the fact that the magnetic circuit of the ballast is driven into its saturation region by the applied voltage.

The ballast also performs the function of regulating against line voltage changes. A general guideline is that the lamp operating voltage should be about one-half the required starting voltage if good regulation is to be

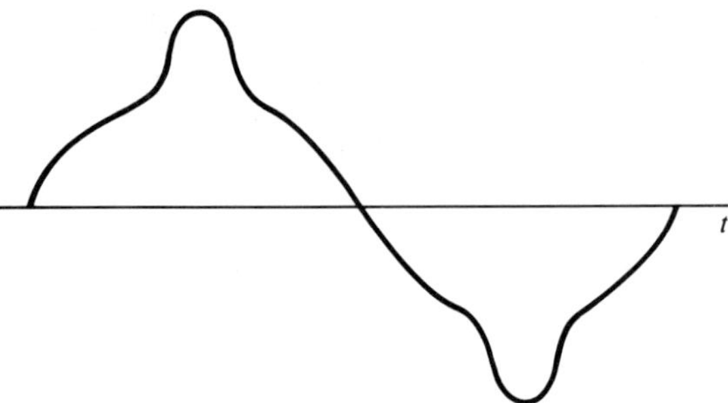

Figure 6.9 Ballast voltage waveform.

achieved. This will yield a situation where a 1% change in line voltage will change lumen output by only about 1%.

The relighting of the lamp each $\frac{1}{120}$ s requires that the open-circuit secondary voltage of the ballast be near its maximum when the lamp current passes through zero. This is indeed what the ballast does, and thus reignition is not a serious problem in most fluorescent lamp circuits.

Any power loss in a ballast reduces the overall luminous efficacy. Ballast power loss results from hysteresis and eddy current losses in the transformer core and copper losses in the windings. A summary of power losses for fluorescent lamp inductive ballasts is given in Table 6.9.

The last two functions of the ballast, namely cathode heating and provision for high power factor, will be discussed in subsequent sections.

The development of electronic (solid-state) ballasts over the past 5 years has been rapid and it is likely that these will replace reactive ballasts in a great many lighting circuits within the next decade.

Electronic ballasts are basically frequency converters, changing the 60-Hz input voltage to dc and then to 20 to 25 kHz. They are available in two types, nondimming and dimming. Each type is packaged in the same size envelope as the reactive ballast it is designed to replace, although its weight is about 50% less.

The nondimming type is now manufactured by several companies for use with either the standard 4-ft rapid-start fluorescent lamp or its energy-saving equivalent. It is designed to provide the same lumen output as its reactive counterpart and the same lamp life. Its PF $\simeq$ 0.95 and its draws about 25% less line current. Because of the frequency conversion, flicker effects are absent. As seen in Table 6.10, the ballast provides a considerable savings in power, about 22% with the standard lamp and 32% with the energy-saving lamp.

The dimming type electronic ballast has most of the features of the nondimming type and, in addition, permits smooth control of lumen output between 30 and 100% of rated value. It also saves power, about 11% with standard lamps and 26% with energy-saving lamps. It is available for use

TABLE 6.9 FLUORESCENT LAMP BALLAST LOSSES

Lamp type	Lamp no.	Number of lamps	Ballast watts	Total watts
Preheat	F14T12	1	5	19
	F15T12	1	5	20
	F20T12	1	5	25
Rapid start	F40T12	1	12	52
	F40T12	2	12	92
	F40T12	3	20	140
Instant start	F48T12	1	20	59
	F48T12	2	32	110
	F72T12	1	27	82
	F72T12	2	34	144
	F96T12	1	27	102
	F96T12	2	34	184
High-output rapid start	F48T12HO	1	25	85
	F48T12HO	2	30	150
	F72T12HO	1	25	110
	F72T12HO	2	31	201
	F96T12HO	1	30	140
	F96T12HO	2	21	241
Very high output rapid start	F48T121500	1	28	143
	F48T121500	2	15	245
	F72T121500	1	20	180
	F72T121500	2	15	335
	F96T121500	1	15	230
	F96T121500	2	15	455

with high-output, very high output, and instant-start fluorescent lamps, in addition to the standard 4-ft lamp. Three dimming options are included: local dimming through an adjustment on the ballast, remote dimming through a potentiometer circuit, and automatic photocell dimming, which is useful in applications involving daylight.

TABLE 6.10 COMPARISON OF ELECTRONIC AND REACTIVE BALLASTS FOR TWO-LAMP 4-FT RAPID-START FLUORESCENT SYSTEMS

2 F40CW	Inductor ballast	92–94 W
2 F40CW	Electronic ballast	66–70 W
2 F40ES	Electronic ballast	58–60 W
2 F40CW	Electronic dimming ballast	77 W
2 F40ES	Electronic dimming ballast	64 W

6.16 PREHEAT FLUORESCENT LAMPS

The original fluorescent lamp circuit consisted of a lamp, an inductor, and a starter button. One of the first improvements was to replace the starter button by an automatic starting device, as shown in Fig. 6.10. The starter is a small neon glow lamp, containing a fixed contact, a bimetallic strip, and a small capacitor. When the line switch is closed, the full line voltage appears across the neon glow tube, ionizing the gas and establishing a glow discharge which then heats the bimetallic strip, causing it to bend and connect to the fixed contact. This creates a circuit for current to flow and preheats the electrodes. The current causes a voltage drop across the inductor, reducing the voltage across the glow tube, which causes the latter to deionize. The strip cools and breaks away from the fixed contact, producing a large $L\,di/dt$ voltage surge across the inductor, which also appears across the lamp, causing the arc to strike. During lamp operation there is insufficient voltage across the glow tube to reestablish the glow discharge. The purpose of the capacitor is to short out any high-frequency signals generated by the glow switch which could produce radio interference.

The voltage and current waveforms in a fluorescent lamp circuit are not sinusoidal, due to nonlinearities in the ballast and lamp. However, for the simple series circuit of Fig. 6.10 (and the two-lamp circuit in Fig. 6.13), a conventional ac circuit analysis based on sinusoidal waveforms gives a reasonable approximation. This will be illustrated by two examples.

EXAMPLE 6.6

A fluorescent lamp and ballast are connected in series across a 200-V rms, 60-Hz supply and the following rms measurements are obtained: $V_{\text{lamp}} = 90$ V, $I = 500$ mA, $P_{\text{lamp}} = 40$ W, $P_{\text{ballast}} = 10$ W. Find the resistance and inductance of both the lamp and ballast.

Solution. A phasor diagram of voltages is shown in Fig. 6.11. The subscripts R and I refer to real and imaginary parts. The subscripts S, L, and B refer to supply, lamp, and ballast.

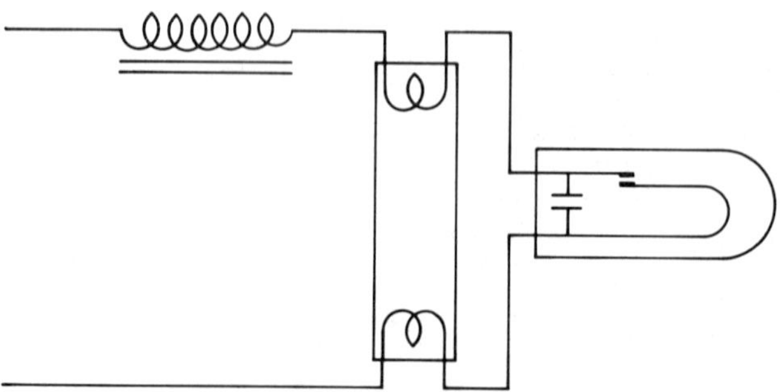

Figure 6.10 Glow switch starter.

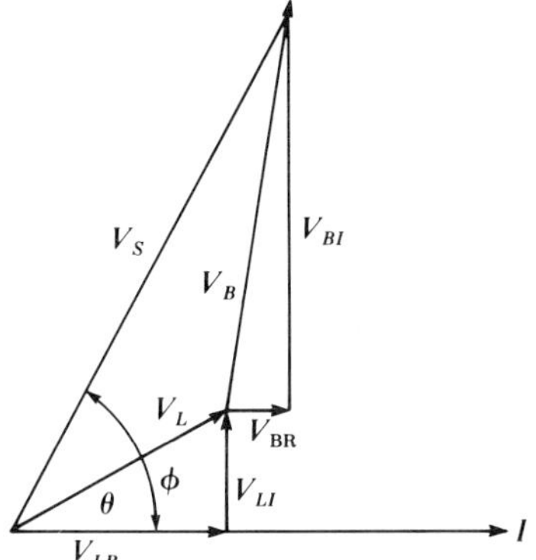

Figure 6.11

The calculations are as follows:

Lamp	Supply	Ballast
$\cos\theta = \dfrac{40}{90(0.5)} = 0.889$	$\cos\Phi = \dfrac{50}{200(0.5)} = 0.500$	$V_{BR} = 100.0 - 80.0 = 20.0$ V
$\theta = 27.3°$, $\sin\theta = 0.458$	$\phi = 60°$, $\sin\phi = 0.866$	$V_{BI} = 173.2 - 41.2 = 132.0$ V
$V_{LR} = 90(0.889) = 80.0$ V	$V_{SR} = 200(0.500) = 100.0$ V	$R_{\text{ballast}} = \dfrac{20.0}{0.5} = 40\ \Omega$
$V_{LI} = 90(0.458) = 41.2$ V	$V_{SI} = 200(0.866) = 173.2$ V	$X_{\text{ballast}} = \dfrac{132.0}{0.5} = 264\ \Omega$
$R_{\text{lamp}} = \dfrac{80}{0.5} = 160\ \Omega$		$L_{\text{ballast}} = \dfrac{264}{377} = 0.70$ h
$X_{\text{lamp}} = \dfrac{41.2}{0.5} = 82.4\ \Omega$		
$L_{\text{lamp}} = \dfrac{82.4}{377} = 0.22$ h		

EXAMPLE 6.7

What value of capacitor, placed directly across the line in Fig. 6.10, will cause the overall circuit in Example 6.6 to operate at unity power factor?

Solution. A phasor diagram of the currents involved is shown in Fig. 6.12.

The calculations are

$$I_c = 0.5 \sin\Phi = 0.5(0.866) = 0.433 \text{ A}$$

$$X_c = \frac{200}{0.433} = 462\ \Omega$$

$$C = \frac{1}{377(462)} = 5.7\ \mu\text{F}$$

These examples illustrate a problem inherent in fluorescent lamp circuits,

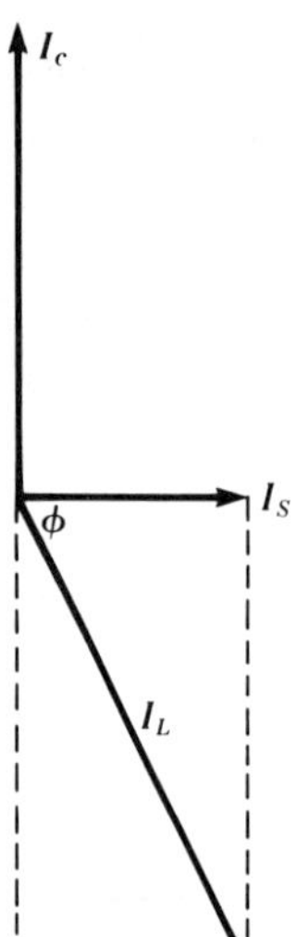

Figure 6.12

namely poor power factor. One solution was shown in Example 6.7. An alternative solution is shown in Fig. 6.13. The autotransformer steps up the 120-V supply voltage. This is necessary on all fluorescent lamps of greater than 20 W in order to provide sufficient voltage to start the lamp. Lamp *A* operates with its current lagging the secondary voltage, whereas lamp *B*'s current leads. The circuit elements are chosen so that a resulting power factor of 90% or more is achieved. It should be noted that the lead-ciruit portion of Fig. 6.13 contains an inductor and capacitor in series, rather than simply a capacitor. At 60 Hz, pure capacitive ballasting produces a peaked lamp current which is deleterious to the electrodes, resulting in shortened lamp life. This problem is not present at higher frequencies because the period of the applied voltage is now comparable to the time for ionization growth. The result is that there is insufficient time during each cycle for the lamp current

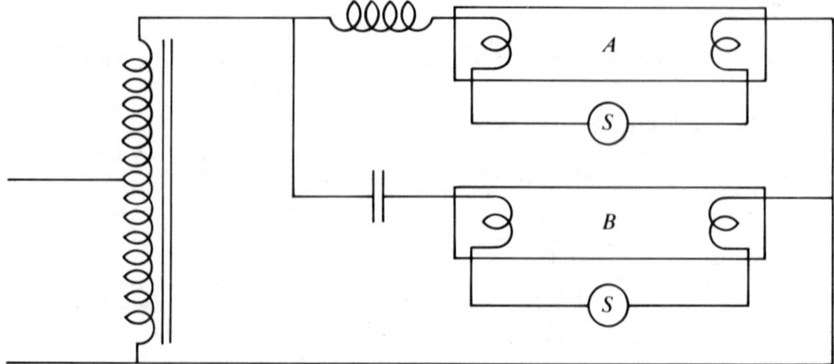

Figure 6.13 Two-lamp, lag-lead circuit.

to increase excessively. Thus capacitor ballasts, with their advantages of small size, weight and cost, are quite acceptable at higher frequencies.

6.17 INSTANT-START FLUORESCENT LAMPS

Dating from its introduction in 1938, the preheat fluorescent lamp was criticized for taking too long to start. It was not until 1944 that instant-start fluorescent lamps appeared. These new lamps required no starters, had single-pin bases which were stronger than the bipin bases of preheat lamps, were available in longer lengths of up to 8 ft, and could be operated over a range of currents, offering a 2:1 range of lumen output in a given lamp. The major disadvantage was that, for a given lamp size, they required a much larger ballast. For example, a two-lamp ballast for 40-W T12 preheat lamps weighs 4 lb and consumes 11 W. Its instant-start counterpart weighs $9\frac{3}{4}$ lb and consumes 20 W.

There were three major problems to overcome in designing an instant-start lamp, namely reduced life, early end blackening, and nonreliable starting. The first two are governed largely by electrode design. Preheat filaments are double coiled (Fig. 6.5*b*). If sufficient voltage is applied to start them when cold, severe sputtering of emission material results. This blackens the lamp ends and reduces life to about one-third its normal value.

Attention was focused on designing an electrode that would deliver reasonable life and minimal end blackening despite some sputtering during each start. The result was the triple-coiled filament, also shown in Fig. 6.5*b*. It consists of the normal double-coiled filament with a loose overwind of fine tungsten wire. The overwind reduces sputtering time by causing the filament to reach its emitting temperature more rapidly and it acts as a basket to hold a larger quantity of emission material.

The problem of reliable starting was a more baffling one. An experiment was performed in which the voltage across a 4-ft-long T12 lamp without cathode preheating was gradually increased until the lamp started. The result was a wide range of starting voltages from a minimum of 250 V to a maximum of 750 V (a preheat lamp of this size requires 200 V to start). Even with the same lamp, the starting voltages on two different days could differ widely.

It was found that humidity was responsible for this large variation in starting voltages. As shown in Fig. 6.14, starting voltage increases as relative humidity increases, which corresponds to a reduction in the resistance of the bulb-wall surface. The reason for this effect is that the starting of discharges by combined field and thermionic emission requires a high electric field near each electrode. When the bulb resistance is very high, the metal fixture provides a low-resistance path between electrodes and thus concentrates most of the applied voltage near the lamp ends. When the bulb resistance is very low, the bulb surface performs this function of concentrating the voltage. At intermediate values of bulb resistance, which occur when

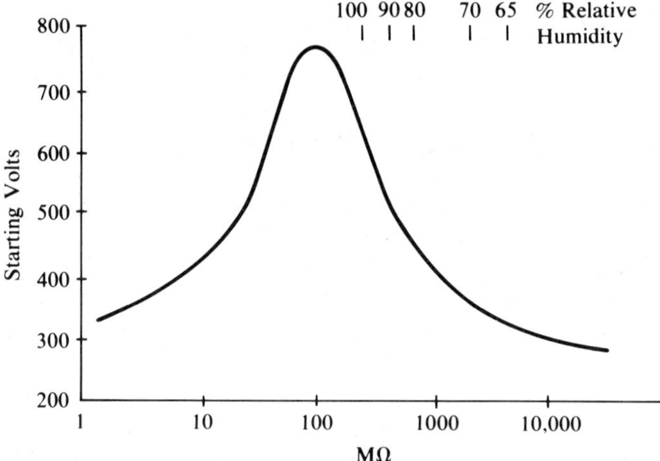

Figure 6.14 Starting voltage variation with humidity and bulb resistance.

relative humidity is 65 to 100%, neither the fixture nor the bulb can concentrate the voltage and starting is difficult.

Several remedies for this situation were tried, but the most successful was a silicone coating of the exterior bulb wall, producing an effect much like that of a car wax, in that any moisture present tended to form droplets, rather than distributing itself uniformly over the surface. This caused the bulb resistance to remain at a high value independent of humidity, thus allowing the metal fixture to concentrate the voltage at the electrodes. The result was that reliable starting of the 4-ft T12 lamp could be achieved with 430 V.

A single instant-start lamp can be operated in a circuit similar to that in Fig. 6.10, except that the starter is omitted and the filament leads at each end of the lamp are shorted together. Two instant-start lamps can be operated in the lag-lead circuit of Fig. 6.13, which places the lamps in parallel. However it was found that ballast size, weight, and voltampere product are reduced if the lamps are operated in the series arrangement shown in Fig. 6.15. In this circuit, winding A is the primary and winding B the secondary. Winding C is a high-voltage low-current winding. When the line switch is closed, the high voltage of winding C is impressed directly across lamp 1 and that lamp starts. The voltage across lamp 2 is the phasor sum of the voltages of the three windings. The capacitor introduces a phase shift which causes the voltage of winding C to lag that of windings A and B by about 120°. The resultant voltage across lamp 2 is sufficient to start it a few thousandths of a second after lamp 1.

6.18 RAPID-START FLUORESCENT LAMPS

Although the advantages of instant-start lamps over preheat lamps were considerable, the disadvantages of large inefficient ballasts and relatively

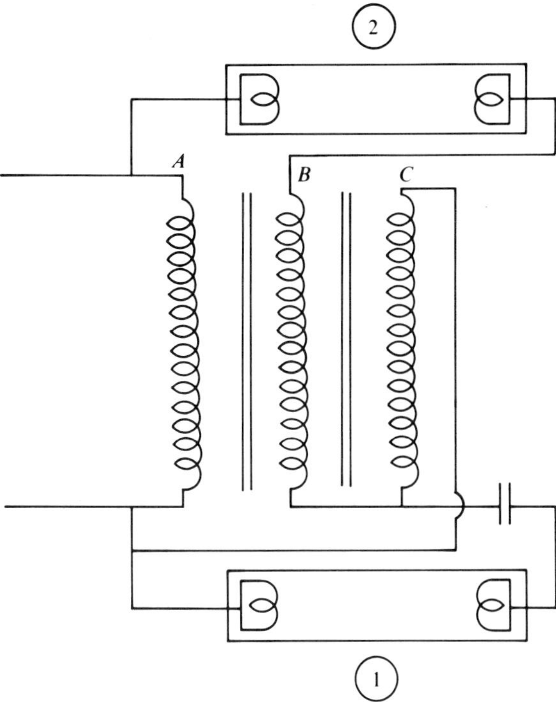

Figure 6.15 Sequence-start circuit.

modest lives caused lamp researchers to seek a new lamp which provided the best features of both preheating and instant starting. The result was the introduction of the rapid-start lamp in 1952. Rapid-start lamps require no starters. They start in less than a second. Their ballasts are smaller and more efficient than those of instant-start lamps. They can be dimmed and flashed relatively easily. Their lives are long, in excess of 20,000 h at 3 or more burning hours per start.

In the development of the rapid-start lamp, it was decided first to investigate the effects on lamp performance when electrode heating and transformer secondary voltage were applied to a lamp simultaneously. In Fig. 6.16, the starting voltage for 40-W T12 fluorescent lamps is plotted against electrode heating current for several conditions. Curve *A* is for isolated lamps with no fixture present to concentrate the applied voltage at the electrodes. Curve *B* is for a lamp with a conductive stripe running lengthwise of the lamp. Curve *C* is for a lamp with a silicone coating in a metal fixture grounded to one end of the lamp.

Based on these curves, it was determined that a ballast circuit which could provide about 380 mA of heating current simultaneously with the application of applied voltage would start a lamp in about $\frac{1}{2}$ s. This presumes that the lamp is silicone coated and that it is housed in a metal fixture. In addition, it was decided to use shorter electrodes, having a voltage drop of 2 to 4 volts, to reduce the time required for heating and to decrease the wattage loss at the electrodes to about 1.3 W at each end.

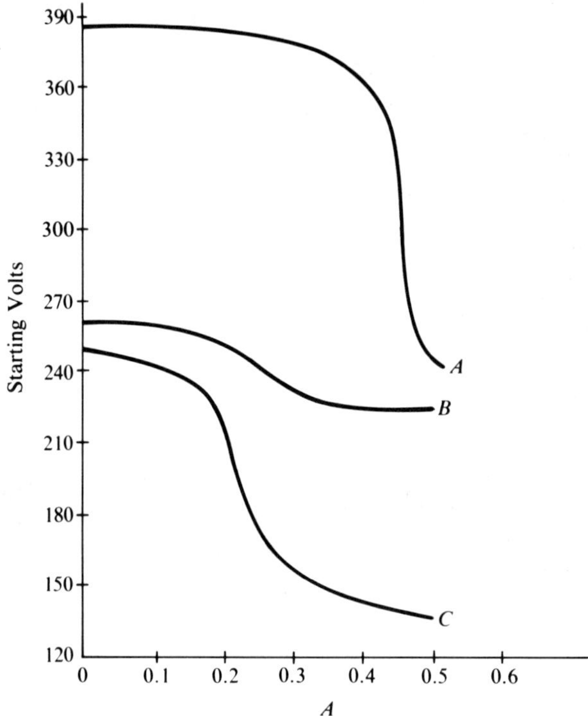

Figure 6.16 Starting volts versus heating current for 40-W T12 lamps.

The basic rapid-start circuits are shown in Fig. 6.17. The one-lamp circuit is the same as that in Fig. 6.11 for the preheat lamp except for the replacement of the starter by the small auxiliary windings on the transformer. The two-lamp circuit is a sequence-start circuit similar to that for instant-start lamps in Fig. 6.15. The capacitor across the upper lamp aids in starting the lower lamp first. After starting, the voltage across the lower lamp is relatively low, placing a large voltage across the upper lamp, which then starts. The three auxiliary windings on the transformer provide the necessary electrode preheating during starting.

For reliable starting, all rapid-start installations require that each lamp be mounted in a fixture which has a grounded metal plate extending the length of the lamps. Optimally, the strip should be 1 in wide and from $\frac{1}{2}$ to 1 in from the lamp.

6.19 HIGH-INTENSITY DISCHARGE LAMP DEVELOPMENTS

As described in Sec. 1.2, Peter Cooper Hewitt developed the first practical mercury lamp in 1901, a 385-W, 4-ft-long tubular lamp giving 12.4 lm/W. The first commercial use of this source was in the composing room of the *New York Evening Post* in 1903.

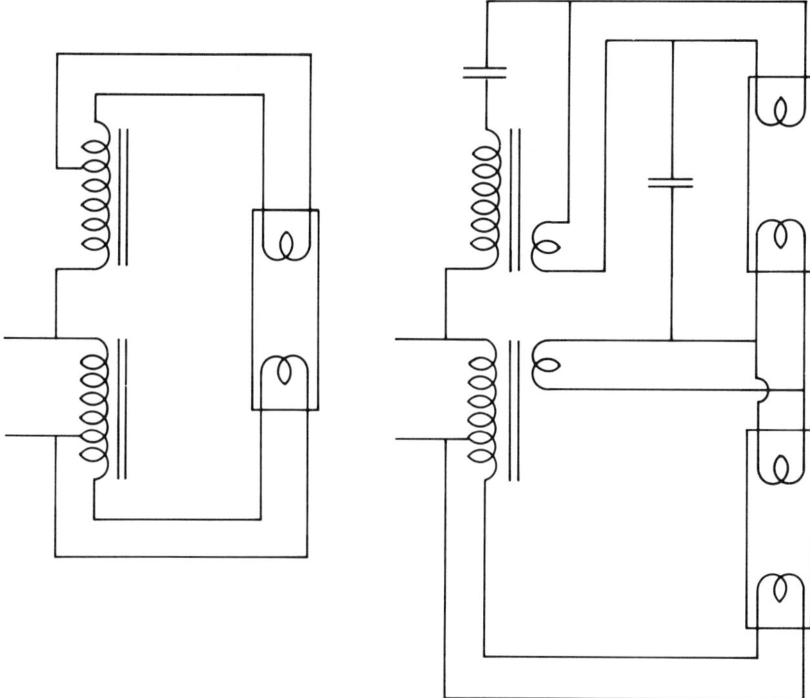

Figure 6.17 Rapid-start circuits.

So-called Moore tubes were introduced in 1904. These were 7- to 9-ft-long tubes filled first with nitrogen, later with carbon dioxide, and still later with air freed of oxygen. Those filled with carbon dioxide, although inefficient, produced a white light whose correlated color temperature matched daylight and they were widely used in color matching and grading applications.

Although some development work on high-intensity discharge lamps was carried on in Europe in the early 1900s, it was not until the 1930s that such lamps began to have an impact commercially. In 1932, Dr. Giles Holst in Holland developed a low-voltage, low-pressure sodium lamp, utilizing a special glass which could withstand the highly alkaline effects of vaporized sodium. The lamp is still used extensively in Europe in street lighting installations. It was first introduced in the United States in 1933, but has received limited acceptance over the years, largely because of its low intensity, difficulty of optical control due to the size of the arc, and monochromatic spectrum (the double yellow line at 589.0 and 589.6 nm which was discussed in Sec. 4.9).

Interest in mercury vapor sources also quickened during the 1930s. Lamps at nearly atmospheric pressure (called high pressure) and rated at 400 W were introduced in Europe in the early 1930s and in the United States by the middle of the decade. These lamps consisted of an arc tube enclosed within an outer glass bulb and provided a lumen output about equal to that of a 750-W incandescent lamp. A 250-W mercury lamp was introduced in the late 1930s and 1000- and 3000-Watt mercury lamps appeared in the early 1940s to satisfy a need for high-output, efficient lighting for industry during the war years.

Mercury arc discharges had two inherent problems. The first involved the bowing of the arc when the lamps were burned horizontally. This caused the arc to touch the glass tube, often melting it. All mercury lamps for street lighting were equipped with magnets to pull the arc away from the arc tube. This was the case until the early 1950s, when synthetic quartz was introduced. Contemporary mercury lamps have smaller quartz arc tubes, which can withstand the high temperatures generated by the discharge, and they do not require magnets.

The second problem with mercury lamps was their color. In Sec. 4.9, it was pointed out that a mercury discharge provides four visible lines at 404.7, 435.8, 546.1, and 577 to 579 nm. This produces a bluish-green white light with no reds or oranges and makes human complexions appear as though rigor mortis has set in. Mercury vapor lamps were reluctantly accepted for street lighting and a few other applications but could not be used in any application where color rendition was important.

A first solution to this problem was simply to use a 400-W mercury lamp side-by-side with a 750-W incandescent lamp in industrial settings. The excess of reds and oranges in the incandescent lamp counterbalanced their absence in the mercury lamp. Then, in the 1950s, color-corrected mercury lamps were introduced. These consisted of a conventional mercury lamp with a phosphor added to the inside of the outer glass bulb which converted ultraviolet mercury radiation into visible radiation in the red portion of the spectrum. These phosphors improved the overall color properties of the lamp somewhat, but not enough to make the lamps acceptable for a wide range of interior applications. Also, the phosphors increased the source size, that is, the source was no longer the inner arc tube but rather the outer glass bulb. This made optical control difficult. Improvements in phosphors have been made over the years, particularly the development of a deep red phosphor, which have made color-corrected mercury lamps more acceptable indoors.

In the early 1960s work proceeded in a different direction to try to improve the color of mercury lamps, as well as their luminous efficacy. Dr. Gilbert Reiling found that the addition of certain metals, in the form of their halide salts (usually iodides), would produce sufficient concentration of metal atoms in the vapor to provide additional spectral lines in the discharge and enhance the overall spectral characteristics of the lamp. At the same time, the halides were able to recombine with the metals leaving the arc stream thus preventing these metals from coming in contact with the wall of the arc tube, which would seriously degrade the latter. The result was the metal-halide lamp, introduced in the mid-1960s.

Metal-halide lamps are now available in a variety of sizes ranging from

100 to 1500 W. The metallic additives include indium, thallium, sodium, and scandium. The lamps have excellent color characteristics when compared with mercury. Also they are relatively easy to control optically compared with color-corrected mercury lamps because of their small source size. They suffer in that their life is about half that of a mercury lamp and also in that it is difficult to control their color from lamp to lamp, because of the sensitivity of the additives to temperature variations.

The most recent major milestone in lamp development was the in-introduction of the high-pressure sodium lamp, also in the mid-1960s. The lamp was developed by Dr. Kurt Schmidt and William Louden and was introduced commercially in 1965 in the 400-W size, with ratings of 6000 h and 42,000 initial lumens. The key to the introduction of the high-pressure sodium lamp was the development of a special ceramic material for the arc tube, a polycrystalline translucent alumina which permits 92% light transmission and yet is free of minute pores that would allow the active sodium to pass through. The material is also chemically resistant to sodium and can withstand a central arc temperature of 1500 K.

High-pressure sodium lamps are available in wattages ranging from 35 to 1000 with efficacies of well over 100 lm/W. Their spectra are deficient in reds, greens, and blues and they have a correlated color temperature of 2100 K, quite a bit lower than for incandescent lamps. Their lives are long, 12,000 to 20,000 h.

6.20 MERCURY LAMPS

The question might be asked, Why not just remove the phosphor coating from the inside of a fluorescent lamp to create a mercury lamp? The answer, of course, lies in the relative efficiency of generation of lines in the mercury spectrum. At low mercury pressure, such as exists in fluorescent lamps, conditions are optimum for a large amount (60%) of the input watts to be converted into a single line at 253.7 nm. As Fig. 4.5 shows, this transition requires the least amount of input energy from a colliding electron.

As pressure is increased, the likelihood of multiple collisions is enhanced, particularly those in which an outer-shell electron in either the 4.66- or 5.43-eV metastable levels is raised to a higher energy state. For exanple, an outer-shell electron at 4.66 eV could receive a 3.07-eV boost from a colliding electron, which would place it at 7.73 eV. From there, it might fall to 5.43 eV and emit a 546.1-nm photon or it might return to 4.66 eV, emitting a photon at 404.7 nm, or it might fall to 4.88 eV, in which case it would emit a 435.8-nm photon. In the last instance, it would then likely fall to the ground state and emit a 253.7-nm photon as well. It is also possible for an outer-shell electron at the 4.66- or 5.43-eV level to move to the 8.84 to 8.85-eV levels and then return to the 6.70-eV level, emitting a photon in the 577 to 579-nm range. Thus for optimum generation of visible lines in mercury, the pressure should be high enough (1 to 2 atmospheres) so that many outer-shell electrons reside in metastable states and can take part in multiple collisions which raise them to higher levels.

A cutaway view of a mercury lamp is shown in Fig. 6.18. As mentioned earlier, the lamp consists of an inner quartz arc tube and an outer borosilicate glass envelope. The quartz tube must withstand an arc temperature of 1300 K, whereas the outer tube operates at a maximum of 700 K.

Nitrogen is used as a filling gas between the inner and outer bulbs to thermally insulate the arc tube and to protect metal parts from oxidation. The arc tube contains mercury and argon gas, the latter serving the same function as in a fluorescent lamp.

Within the arc tube are two main electrodes and a starting electrode. Each main electrode consists of a tungsten rod upon which a double layer of coiled tungsten wire is wound. The electrode is dipped into a mixture of thorium, calcium, and barium carbonates and then heated to convert these compounds to oxides. The electrodes are connected through the quartz tube by molybdenum foil leads.

When voltage is applied to the lamp, it appears between the starting electrode and the adjacent main electrode, as well as between the two main electrodes. This creates a local argon arc, with current limited by the starting resistor. The heat generated by the small argon discharge vaporizes some of the mercury present, and soon the main arc strikes. The resistance of the main arc circuit is much less than that of the starting arc circuit, and so the starting arc ceases. Following initiation of the main arc, mercury vaporization continues and, after 5 to 7 min, all of the mercury is vaporized and the lamp reaches steady state.

Characteristics of some typical clear-bulb mercury lamps are shown in Table 6.11. These five lamps are also available with phosphor coatings to

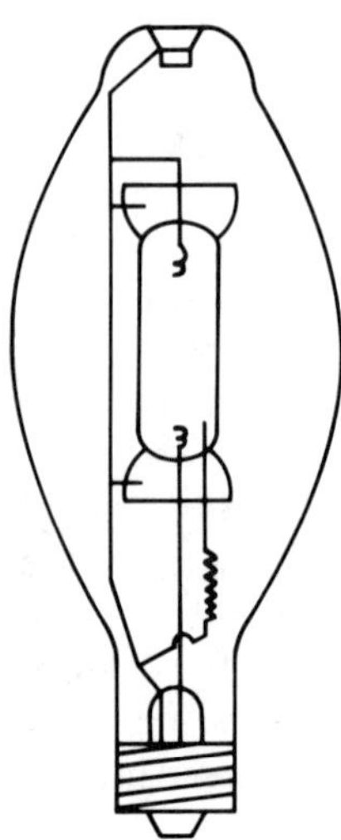

Figure 6.18 Mercury lamp construction.

TABLE 6.11 HID LAMP DATA

Watts	Bulb	Clear lamp initial lumens	Initial lamp lumens per watt	Life (h)	Mean lumens
		Mercury			
100	E23½	3,850	38	24,000	3,120
175	E28	7,950	43	24,000	7,470
250	E28	11,200	45	24,000	10,300
400	E37	21,000	52	24,000	19,100
1000	BT56	57,000	57	24,000	48,400
		Metal halide			
175	BT28	14,000	80	7,500	10,800
250	BT28	20,500	82	7,500	17,000
400	B37	34,000	85	15,000	20,400
1000	BT56	115,000	115	10,000	92,000
1500	BT56	155,000	103	1,500	140,000
		High-pressure sodium			
100	E23½	9,500	95	24,000	8,550
150	E23½	16,000	106	24,000	14,400
150	E28	12,000	80	12,000	10,800
215	E28	19,000	88	12,000	17,100
250	E18	27,500	110	24,000	24,750
360	E18	37,000	103	16,000	33,300
400	E18	50,000	125	24,000	45,000
1000	T18	140,000	140	24,000	126,000

provide improved color performance. The initial lumen ratings for the phosphor-coated lamps are 4200, 8600, 12,100, 22,500, and 63,000, in order of increasing wattage. The initial lumens of HID lamps, as in the case of fluorescent lamps, are taken at 100 h, because the "cleaning-up" of inpurities reduces light output significantly during the first 100 h of operation. The mean lumens are taken at 50 to 70% of life at 10 h per start, or about 12,000 to 16,000 h in the case of mercury lamps.

The mercury lamp life rating of 24,000 h is not the 50% survivor point. Rather it is the point at which the lumen output has dropped so low that the lamp is no longer economically viable. At 24,000 h, the number of survivors is about 67%.

6.21 METAL-HALIDE LAMPS

There had been interest since the 1930s in using metals other than mercury in arc discharge light sources to improve color. But all efforts to use them in their elemental form had failed because of their low vapor pressures (less than 0.1 mm) at normal arc tube temperatures, which prevented the efficient production of visible radiation.

The discovery in 1960 by Dr. Reiling that such metals could be introduced into the arc stream via their iodide salts, which have vapor pressures of several millimeters at arc tube temperatures of 1000 K, made metal-halide lamps feasible. When the lamp is cold, the metal iodides reside on the bulb wall. As arc temperature rises, they vaporize, diffuse from the wall into the arc stream, and dissociate, yielding free metal and iodine atoms. The former are then excited and ionized in the same fashion as are the mercury atoms.

In a multiple-metal-halide lamp, the iodides do not all vaporize at the same time. In the lamp containing indium, thallium and sodium iodides, the indium vaporizes first and forms a blue sheath around the mercury arc. Next, the thallium vaporizes and forms a green sheath around the indium. Last, the sodium vaporizes and forms a yellow sheath around the thallium. Such a lamp is very sensitive to changes in lamp wattage, largely because of sodium, the final iodide to vaporize. If the lamp watts are lower than rated, too little sodium will be vaporized and the lamp will be deficient in yellow and red. Too high a lamp wattage causes a color shift toward pink.

This lamp has been replaced throughout the industry by the sodium-scandium metal-halide lamp. In this lamp, the scandium vaporizes first, followed by the sodium. The lamp is much less sensitive to lamp wattage changes than its three-iodide counterpart and thus gives much more stable color performance.

The basic requirements for a metal additive are

1. The iodide of the metal must be stable at the bulb-wall operating temperature.
2. The iodide vapor pressure must be relatively high.
3. The metal atom must have excitation levels lower than the average of those of mercury (about 7.8 eV).
4. The metal must have an energy level configuration which encourages a high percentage of visible radiation.

Thallium is an example of a metal which satisfies these constraints (see Fig. 4.7). It has a strong line at 535 nm requiring only 3.3 eV for its generation, and the vapor pressure of its iodide salts is about 10 mm when the coolest spot on the arc tube is at about 800 K. Scandium, sodium, and indium also fulfill these requirements.

A cutaway view of a metal-halide lamp is shown in Fig. 6.19. The lamp is very similar in construction to a mercury lamp with the following exceptions:

1. The arc tube ends are treated with a reflective white coating to redirect energy back into the tube so as to maintain a more uniform temperature over the entire arc tube length.

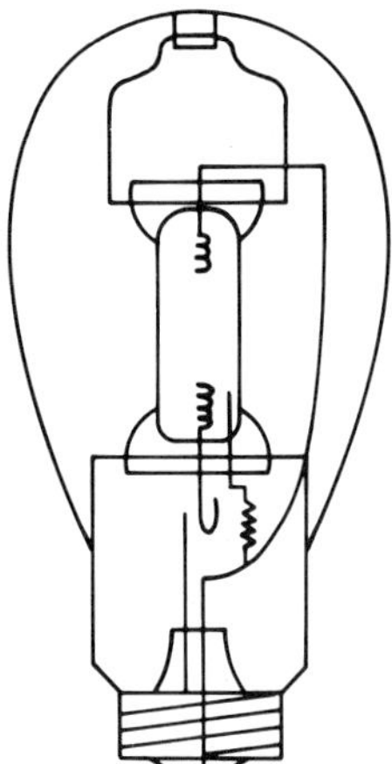

Figure 6.19 Metal-halide lamp.

2. A bimetal switch is included to short the starter electrode to the main electrode after starting. This prevents the build up of small electric fields which can destroy the seal between the lead-in wires and the glass.
3. The construction is frameless and the outer electric wire is of non-magnetic molybdenum. The reason for this is that ac currents would flow through the frame and create pulsating magnetic fields. These, in turn, would cause electrons to leave the frame and deposit on the outside of the arc tube. The electrons would attract and combine with sodium ions that would migrate through the arc tube wall. The result would be a serious depletion of sodium from the arc stream.

The starting of a metal-halide lamp is identical with that of a mercury lamp up until the point where the additives begin to vaporize. Thus the discharge is first in argon, then in mercury, and later in mercury plus the additives. The entire starting process to full light output takes about 5 min.

Data on selected metal-halide lamps are given in Table 6.11. Note that for a given wattage size, because of the halide additives, the lumen output and luminous efficacy of a metal-halide lamp are significantly greater than those of a mercury lamp. However lamp life is shorter. The addition of the metallic iodides necessitates a higher voltage to start the arc at a given temperature and a higher voltage for reignition each half cycle than is required in mercury lamps. This increases electrode deterioration and has an adverse effect on lamp life. Metal-halide lamps show greater lumen depreciation than the other HID types. This is caused by a shift in the chemical balance of the halide additives as the lamps age.

Metal-halide lamps are much like fluorescent lamps in that the life is dependent on the number of burning hours per start. The life ratings given in Table 6.11 are for 10 h per start. Life is reduced by 25% at 5 h per start and increased by 35% for continuous burning. The 400-W metal-halide lamp has the longest life (20,000 h); the 1500-W lamp the shortest life (3000 h). In this latter lamp, which is used in sports lighting, life is sacrificed in order to get greater lumen output per lamp, thus reducing the number of poles and fixtures required.

6.22 HIGH-PRESSURE SODIUM LAMPS

We have pointed out that the development of polycrystalline alumina (PCA) in 1959 paved the way for the introduction of the high-pressure sodium (HPS) lamp. The first commercial lamp appeared in 1965 and was rated at 400 W, 42,000 initial lumens, and 6000 h. Improvements since then have permitted these ratings to be raised to 50,000 initial lumens and 24,000 h at 10 h per start. Thus we have a lamp which gives 2.4 times the lumen output of its mercury counterpart and yet has the same rated life.

A diagram of a high-pressure sodium lamp appears in Fig. 6.20. The inner PCA arc tube is filled with xenon gas, for starting, and a sodium-mercury amalgam. Backwound and coated tungsten electrodes are mounted in each end. Originally these were sealed in place with niobium end caps but now a monolithic seal is used. The arc tube is inserted into a heat-resistant outer bulb and supported by a floating end clamp which permits the entire structure to expand and contract without distorting. The space between the tube and bulb is evacuated.

There is no starting electrode in a high-pressure sodium lamp because the small diameter ($\frac{3}{8}$ in) prohibits it. Thus the ballast must provide a much higher starting voltage than in the case of mercury and metal-halide lamps. This is commonly done by superimposing a low-energy high-voltage pulse on the ballast open-circuit voltage waveform. A typical pulse has a peak voltage of 2500 V and lasts for 1 μs. This pulse ionizes the xenon gas sufficiently for the open-circuit ballast voltage to then initiate and maintain the xenon arc. Initially the arc is bluish-white, representing the effect of excited xenon and mercury. As temperature rises, sodium becomes excited and a low-pressure monochromatic yellow sodium spectrum results. The spectrum broadens as temperature rises further and becomes an absorption spectrum in that the sodium resonance line at 589 nm is self-absorbed. During this period sodium pressure increases from 0.02 atm in the monochromatic discharge to over 1 atm in the final steady-state, broad-spectrum condition.

A high-pressure sodium lamp, like a metal-halide lamp, is an "excess"

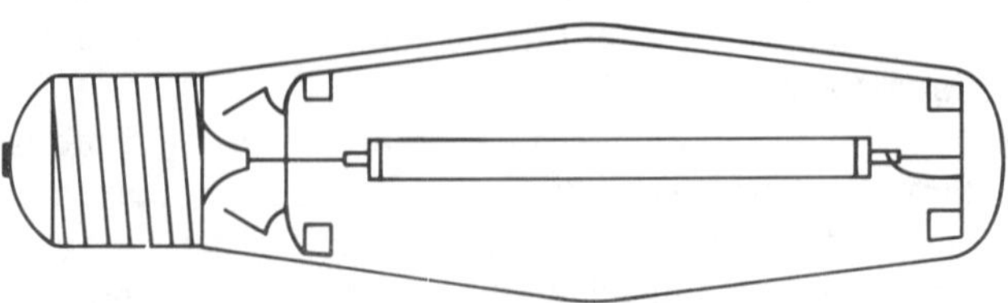

Figure 6.20 High-pressure sodium lamp.

discharge source. There is more amalgam in the reservoir formed inside the arc tube in back of one of the electrodes than is normally vaporized during lamp operation. During the first 100 h of burning, lamp voltage may rise or fall considerably as greater or lesser amounts of amalgam enter the arc stream. During the remainder of the lamp's life, the lamp voltage gradually increases until a point is reached at which the ballast voltage is no longer sufficient to support the arc. This is the end of the useful life of the lamp and is manifested by the lamp "cycling," that is, repeatedly starting and extinguishing.

Data for several common high-pressure sodium lamps appear in Table 6.11. These lamps have the highest luminous efficacies of the HID family, have life ratings of 24,000 h when used with the ballasts specifically designed for them, and give excellent lumen maintenance. There is a group of high-pressure sodium lamps which have been designed to directly replace mercury lamps of comparable wattage size. Each uses the same ballast and fixture as does the comparable mercury lamp, gives a higher lumen output than its mercury counterpart, and has a life rating of 12,000 h. In Table 6.11, these are the second 150-W, the 215-W, and the 360-W lamps, which can operate directly on mercury ballasts designed for 175-, 250-, and 400-W mercury lamps, respectively.

6.23 HID BALLASTS

Ballasts for HID lamps can be placed in four categories, as shown in the circuit diagrams of Fig. 6.21. The simplest is the reactor ballast in Fig. 6.21*a*, which is essentially a wire coil on an iron core placed in series with the lamp to limit lamp current. A power-factor correcting capacitor across the line is often included. A reactor ballast provides about a 5% change in lamp watts for an 18% change in lamp volts or a 5% change in line volts. Thus it regulates well against lamp voltage variations but poorly against variations in line voltage. It provides a relatively low current crest factor of about 1.5 and the amount of starting voltage it can provide to the lamp is limited to the line voltage.

A lag ballast, shown in Fig. 6.21*b*, is a combination autotransformer and reactor. It has the same regulation characteristics as the reactor ballast but overcomes the starting voltage limitation. However it is larger, more costly, and has greater losses.

A regulator ballast is shown in Fig. 6.21*c*. It has isolated primary and secondary windings and achieves current limiting through a series capacitor, which causes the current to lead the secondary voltage. Excellent regulation is obtained with a regulator ballast. Line voltage changes of ±13% cause only about ±3% changes in lamp wattage. The power factor is about 95% and the isolation of the lamp circuit from the supply line minimizes grounding and fusing problems. Its only disadvantages are cost and a higher current crest factor of between 1.65 and 2.0.

The autoregulator ballast in Fig. 6.21*d* includes features of both the lag

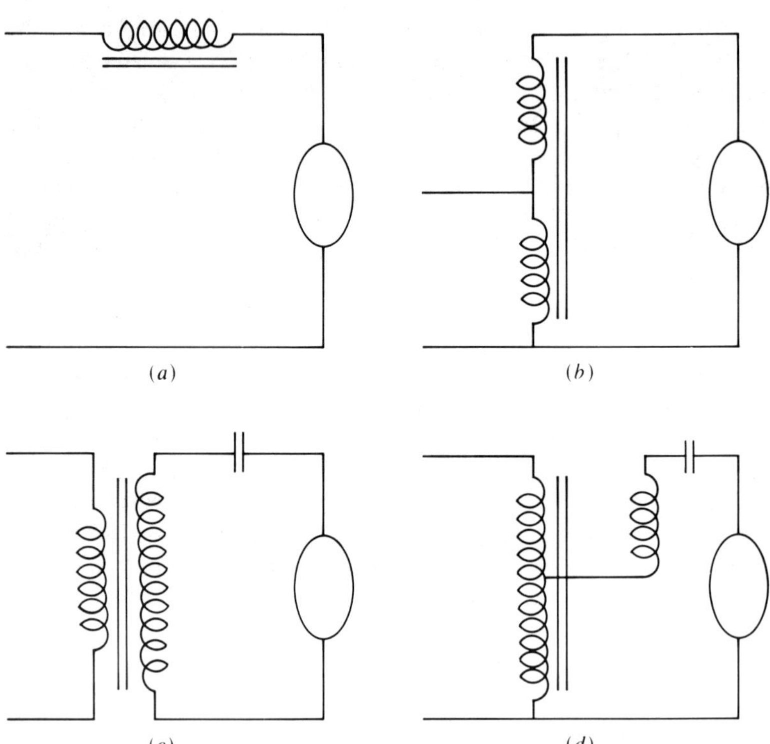

Figure 6.21 HID ballast circuits.

and regulator ballasts. This is the most popular ballast in new installations and is intended as a tradeoff. It is smaller and less costly than a regulator ballast, but its regulation is not as good in that line voltage variations of $\pm 10\%$ produce lamp wattage changes of $\pm 5\%$. Also it does not provide isolation between primary and secondary.

High-pressure sodium lamps require ballasts which are different from those for other HID sources. Crest factors should not exceed 1.8 for proper lamp operation. Required starting voltages are much higher and an auxiliary circuit is necessary to provide the high-voltage spike to initiate the ionization of xenon. Also, lamp wattage must be closely controlled so as to control the amount of amalgam vaporized.

A high-pressure sodium lamp ballast has an operating locus, rather than an operating point, as shown in Fig. 6.22. This is because the lamp wattage and voltage change as the lamp ages. As the lamp is burned, the arc tube darkens because of evaporated electrode material and the outer bulb darkens because of sodium leakage. This darkening causes arc tube temperature to increase, vaporizing more amalgam and raising lamp voltage by about 1 V for every 1000 h of burning.

In Fig. 6.22, curve *A* illustrates a proper design. Lamp wattage and lumen output start below rated values and increase as the lamp ages, finally

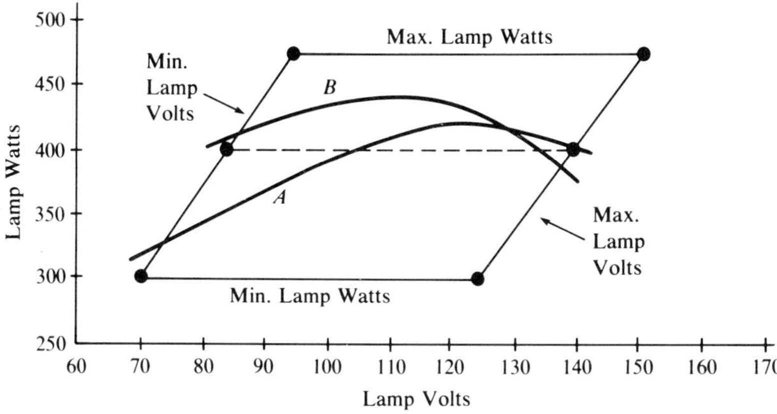

Figure 6.22 400-W HPS ballast locus.

decreasing at the end of life. With curve *B*, light output is decreasing for a significant portion of lamp life. The trapezoid boundary in Fig. 6.22 indicates the limits specified by the lamp manufacturer at 100 h, in this case between 84 and 140 lamp volts and between 300 and 475 lamp watts. The ballast designer must provide a ballast that causes the lamp to operate within these limits and to give peak lumens at approximately mid-life.

A short summary of HID ballast wattage losses is given in Table 6.12.

TABLE 6.12 HID BALLAST LOSSES

Lamp watts	Ballast watts	Total watts
	Mercury	
100	20	120
175	30	205
250	40	290
400	50	450
1000	75	1075
	Metal halide	
175	35	210
250	50	300
400	60	460
1000	100	1100
1500	125	1625
	High-pressure sodium	
100	35	135
150	50	200
250	60	310
400	75	475
1000	100	1100

REFERENCES

Anderson, J. M., and J. S. Saby: "The Electric Lamp: 100 Years of Applied Physics," *Physics Today*, October 1979.

Cox, J. A.: *A Century of Light*, Benjamin/Rutledge, New York, 1979.

Elenbaas, W.: *Light Sources*, Macmillan, London, 1972.

Frier, J. P., and M. E. G. Frier: *Industrial Lighting System*, McGraw-Hill, New York, 1980.

General Electric Co.: "Fluorescent Lamps," Bulletin TP-111, Nela Park, Cleveland, Ohio, 1973.

———: "High Intensity Discharge Lamps," Bulletin TP-109R, Nela Park, Cleveland, Ohio, 1975.

———: "Incandescent Lamps," Bulletin TP-110R, Nela Park, Cleveland, Ohio, 1977.

GTE Sylvania: "Fluorescent Lamps," Bulletin 0-341, Lighting Center, Danvers, Mass.

———: "Tungsten Halogen Lamps," Bulletin 0-349, Lighting Center, Danvers, Mass.

Inman, G. E.: "Fluorescent Lamps: Past, Present, Future," *G.E. Review*, July 1954.

Kaufman, J. E.: *IES Lighting Handbook*, Reference volume, IESNA, New York, 1981, chap. 8.

Lemmers, A. E., and W. W. Brooks: New Fluorescent Lamp and Ballast Design for Rapid Starting," IES Technical Conference, September 1952.

Moon, P.: *Scientific Basis of Illuminating Engineering*, McGraw-Hill, New York, 1936.

Oetting, R. L.: "Electric Lighting in the First Century of Engineering," *AIEE Transactions*, November 1952.

Shelby, B. L.: "High Pressure Sodium Lighting: An Overview," *Electrical Consultant Magazine*, September 1974.

Thayer, R. N., and A. C. Barr: "Instant Start Slimline Lamps," *G.E. Review*, March 1952.

Waymouth, J. F.: *Electric Discharge Lamps*, M.I.T. Press, Cambridge, Mass. 1971.

———, W. C. Gungle, J. M. Harris, and F. Koury: "A New Metal Halide Arc Lamp," IES Technical Conference, 1964.

Chapter 7

Interior Lighting Design: Average Illuminance

7.1 INTRODUCTION

This chapter is the first on lighting design. It opens with a brief review of interior lighting techniques. Next, characteristics of lamps used in interior lighting are discussed. Following this, the requirements and properties of luminaires used in interior lighting are presented. Last, a procedure is developed for calculating the average maintained illuminance level on horizontal and vertical surfaces in an interior space and the concept of flux transfer is introduced. Then in the next chapter, methods will be presented for determining the illuminances at various points throughout the space, the task visibilities produced by these illuminances, and the visual comfort of the occupants of the space.

The procedure used to determine the average illuminances within a space is called the zonal cavity method of interior lighting design. In Chap. 2, we calculated the illuminance at a point in a plane from both point and area sources of light. Our attention was restricted to light coming directly from the source to the point in question (direct component) and we did not consider reflections from room surfaces (reflected component), which also result ultimately in light reaching the point. The zonal cavity method includes both direct and reflected components of luminous flux and deals with average illuminance on a horizontal plane, rather than with illuminance at a point.

7.2 GENERAL INTERIOR LIGHTING: PAST AND PRESENT

In the early days of electric lighting, school rooms, offices, and other general work areas were lighted by prismatic or translucent globes, suspended from the ceiling and housing incandescent lamps. Such units provided lumens both directly and indirectly (through reflections from room surfaces) to the work plane. Glass-enclosing globes tended to have high luminances and thus produced considerable glare in workers' eyes.

Totally indirect incandescent lighting appeared in the 1930s. Pan-shaped or concentric-ring luminaires, with a half-silvered lamp mounted base up in a hole in the center of the unit, redirected the lamp lumens to the ceiling. Thus the ceiling, in essence, became the light source. These indirect units produced high-quality glare-free lighting but were inherently very inefficient, both from their use of incandescent lamps and from the fact that no lumens traveled directly to the work plane. Also, because many, rather large, lamps were required in a given space to provide adequate work-plane illuminance, much heat (infrared) was generated which often caused the space to be thermally uncomfortable.

The advent of fluorescent lamps in the late 1930s initiated a renaissance in interior lighting. These lamps had much lower luminances than their incandescent counterparts and thus it was no longer necessary to send all of the lamp lumens up to the ceiling for redirection downward. Instead, with suitable louvers, lenses and the like, most of the lumens could be sent directly downward. And, of course, the fluorescent lamp had about five times the efficacy of the incandescent lamp. As a result, 70 fc of fluorescent lighting could be provided more efficiently than 30 fc of incandescent lighting. Fluorescent gradually replaced incandescent until, by the 1960s, almost all general interior lighting in offices and schools was fluorescent.

The advent of the metal-halide and high-pressure sodium lamps in the 1960s, coupled with the energy crisis in the early 1970s, caused several additional changes in interior lighting. These lamps, although concentrated and of high luminance like incandescent, had efficacies seven or more times as great. Thus it became economically feasible once again to use totally indirect lighting in interior spaces. Concurrently, in an effort to reduce energy consumption, illuminance levels were lowered and interest in so-called task-ambient lighting systems began. Such systems, instead of providing a reasonably uniform illuminance over the entire work-plane area, sought rather to provide a base level of illuminance throughout a space, supplemented by additional illuminance at the task locations.

To summarize the current state of affairs, we note that incandescent lighting is not recommended for the general lighting of interior spaces. Fluorescent lighting continues to dominate, and, in particular, the 4-ft, 40-W rapid-start lamp is far and away the most commonly used fluorescent lamp in interior lighting.* Metal-halide lamps are appearing more each year in indi-

*A wide variety of energy-efficient 4-ft fluorescent lamps are beginning to replace the standard 40-W lamp in popularity.

rect lighting, both in luminaires suspended from the ceiling and in units built into office furniture. The most popular lamp for these uses is the 400-W phosphor-coated metal-halide lamp. High-pressure sodium lamps in carefully designed luminaires are gaining some acceptance in interior lighting but are generally recommended only for rooms with high ceilings and where good color rendition is not important, such as gymnasiums.

7.3 LAMPS FOR INTERIOR LIGHTING

Before proceeding to describe the zonal cavity method, we will consider further the lamps that are used to light interior spaces. The interior lighting designer generally chooses from among the following lamp types:

Incandescent
Fluorescent
Metal halide
High-pressure sodium

Each of these types has its own particular set of strengths and weaknesses, which were discussed in detail in Chap. 6. The ultimate choice of lamp should be specified in terms of requirements of the overall lighting system and usually involves an "engineering compromise."

The factors which the designer should consider in choosing a lamp are:

1. What is its luminous efficacy? The lamp, in conjunction with the lighting system, must provide the required illuminances economically.
2. What is the life of the lamp? How difficult is it to replace burned-out lamps? Is group replacement of lamps the better choice economically?
3. What is the lumen maintenance of the lamp? Is it important to have a certain minimum level of illuminance at all times?
4. Is color a factor? Although all the lamps listed produce "white" light, their CCTs and CRIs differ. How important is it for the colors of the seeing task and its surroundings to be faithfully reproduced?
5. What auxiliary equipment is required? As we have seen, all gas discharge light sources require ballasts, whereas incandescent sources do not. The type of ballast used can affect lamp output, life, starting reliability, system efficiency, and occupant comfort.
6. What other miscellaneous factors are present in the particular environment? Is temperature a problem? Must the area be free from stroboscopic effects? Will electromagnetic interference disturb the activities going on in the space? Are fumes present which could produce corrosion or an explosive atmosphere?

Table 7.1 shows a comparison of the first three factors for the four common lamp types. Let us consider lamp efficacies first. For incandescent lamps, these range from slightly under 12 lm/W for the 40-W standard lamp to just under 22 lm/W for the 500-W standard lamp. For incandescent lamps of essentially the same design, lamp efficacy increases with lamp wattage, largely because the thicker filaments of the higher-wattage lamps may be operated at higher temperatures for the same life. Efficacies of PAR and R lamps are generally lower than those of their equal-wattage standard lamp counterparts. This is because they are designed to have longer lives.

Even including ballast losses, fluorescent lamps provide much higher efficacies than do incandescent lamps. For example, the 40-W standard cool-white fluorescent lamp emits 3150 lm initially. Its ballast consumes 12 W. Thus its efficacies are $\frac{3150}{40} = 79$ initial lumens per lamp watt and $\frac{3150}{52} = 61$ initial lumens per overall watt. It is the latter figure, of course, that is used as an overall efficacy rating. Fluorescent lamps are usually operated in pairs from a single ballast to improve overall efficacy. For example, two of the above 40-W lamps and their ballast consume 92 W, giving an initial efficacy of 68 lm/W overall. Preheat lamps have the lowest lamp efficacies in Table 7.1; the energy-saving lamps have the highest.

Metal-halide lamps, because of the addition of halide salts, generally have higher efficacies than their mercury lamp counterparts, as was noted in Table 6.11. For example, a 400-W metal-halide lamp emits 34,000 lm initially and it and its ballast consume 460 W, giving it an initial overall efficacy of 74 lm/W. Lower wattage sizes give lower efficacies. The high-pressure sodium lamp provides the highest efficacy of the types presented here (the low-pressure sodium lamp yields higher efficacy but is not suitable for interior lighting because of its poor color-rendering properties). The 400-W sodium lamp emits 50,000 initial lumens and, with its ballast, consumes 475 W. Its initial luminous efficacy is 105 lm/W. By comparison, the 100-W sodium lamp emits 9500 lm, consumes 135 W, and has an initial efficacy of 70 lm/W.

The second column in Table 7.1 gives lamp life in hours. Operation at rated voltage and normal temperature is assumed. The lives of the lamp types in Table 7.1 differ considerably. Standard incandescent lamps have life ratings of 750 or 1000 h, whereas PAR and R lamps are rated at 2000 h. Fluorescent lamp life is a function of the average number of hours the lamp is burned each time it is started. The ranges given in Table 7.1 are based on 3

TABLE 7.1 COMPARISON OF LAMP RATINGS

Lamp type	Lamp efficacy (lm/W)	Life (h)	% Lumen depreciation
Incandescent	9–22	750–2,500	10–22
Fluorescent	45–95	7,500–20,000	11–28
Metal halide	80–115	7,500–15,000	13–22
High-pressure sodium	80–140	12,000–24,000	8–10

burning hours per start. Preheat fluorescent lamps have life ratings at the low end of the range, namely 7500 or 9000 h. Instant-start lamp life is 12,000 h and the life of rapid-start lamps is 18,000 or 20,000 h.

Metal-halide lamps are much like fluorescent in that life is dependent on the number of burning hours per start. The life ratings given in Table 7.1 are for 10 h per start. The 400-W metal-halide lamp has the longest life (20,000 h); the 1500-W lamp the shortest life (3000 h). All high-pressure soidum lamps have a life rating of 24,000 h when used with the ballasts specifically designed for them. For those high-pressure sodium lamps which have been designed to directly replace mercury lamps of comparable wattage size and use the same ballast and fixture as the mercury lamp, the life rating is 12,000 h.

The third column in Table 7.1 is percent lumen depreciation. Standard incandescent lamps depreciate in lumen output by 10 to 22% during lamp life. In fluorescent lamps, the 100-h lumen value is called *initial lumens* and the lumen depreciation is calculated from that point onward and is based on 3 h per start. Quite often two lumen depreciation factors are given. The mean lumen factor is the percent of the initial lumens to be expected at 40% of rated life. Lamp lumen depreciation factor is the percent of the initial lumens to be expected at 70% of rated life. The first factor gives a measure of the average lumens between relampings and is used in economic studies; the second factor relates more to the minimum lumens between relampings and is used in determining the number of luminaires required. A range of values for percent lumen depreciation for fluorescent lamps, based on lumen depreciation factor, is given in Table 7.1. For example, the 40-W standard cool-white fluorescent lamp gives 3150 initial lumens (at 100 h) and 2650 lm at 70% of rated life (14,000 h). Thus its lumen depreciation factor is 0.84, or 16% depreciation in lumen output.

High-intensity discharge lamps also have their initial lumen ratings at 100 h. Lumen depreciation for these lamps is given in terms of *mean lumens,* which is the lumen output to be expected at about 70% of rated life. As seen from Table 7.1, metal-halide lamps show a greater lumen depreciation than do high-pressure sodium lamps.

The fourth factor to be considered by the designer is color. Two measures of this are CCT and CRI and these quantities were presented for the common lamp types in Table 5.9.

There are five common "white" fluorescent lamps.* Warm-white, cool-white, and daylight lamps are high-efficacy sources providing reasonable color rendition. The two deluxe lamps have only 70% of the efficacy of their standard counterparts but provide improved color rendition, especially of complexions. The words *warm*, *cool* and *daylight* are quite well chosen in that a warm-white lamp emits yellowish white light and does indeed make a space feel warmer. On the other hand, a cool-white lamp emits a bluish-

*Recently, triphosphor lamps have been introduced which have higher luminous efficacy and better red content than the standard warm- and cool-white lamps.

white light and tends to create a cooler atmosphere. The daylight lamp is a very cool appearing source and is a close match in CCT to an overcast day.

The inclusion of halide salts with mercury to create a metal-halide lamp adds content to the blue and yellow-red ends of the spectrum and reduces the CCT. The addition of phosphors also strengthens the red end of the spectrum, further reducing the CCT. The spectrum of a high-pressure sodium lamp is centered around its resonance radiation at 589 nm, giving a noticeably yellowish-white light. Its CCT is below that of incandescent sources. The lamp is deficient at the violet and red ends of the spectrum.

Ballasts are the fifth factor the designer should consider, and, with the advent of electronic ballasts, the choices are broader today than in the past. Inductor ballasts are reliable and long-lived but they are heavy, consume watts, produce heat, create noise, and permit lamp flicker. Electronic ballasts are lighter, more efficient, and do not have hum and flicker effects. However their track record of life and reliability is not yet established. Also there is some concern over the possible effect on nearby equipment of the high frequencies they generate.

If the designer chooses inductor-type ballasts, they should have the CBM/ETL label. This means that they have met the Certified Ballast Manufacturers Association (CBM) requirements, including specifications set by the American Natural Standards Institute (ANSI), and that they have been tested and certified by the Electrical Testing Laboratories (ETL). In addition, they should be of high power factor (greater than 90%) and have a high sound level rating. This latter factor is denoted by a letter rating, A through F. A rating of A should be specified, which means that the ballast is suitable for spaces where the ambient noise level is less than 25 dB.

Additional ballast safety standards have been set by the Underwriters Laboratories (UL) and ballasts meeting these requirements have a UL label. These standards limit ballast case temperature inside the luminaire to 90°C when the room ambient temperature is 25°C. The National Electric Code requires that ballasts be protected so that they will automatically open within 2 h if ballast case temperature reaches 110°C. Protected ballasts are called *Class P* and are so listed by the UL.

The use of an approved ballast is very important. Ballasts not meeting ANSI and UL standards can cause lamp life and lumen output to be severely reduced, by as much as 40%. Use of CBM ballasts insures at least 92½% of rated lumen output.

7.4 LUMINAIRES

A luminaire, by definition, is a complete lighting unit, consisting of a lamp or lamps together with the parts designed to electrically operate and control the lamp, to distribute and control the light, to position and physically protect the lamps, and to connect the lamps to the power source.

What is it that the designer needs to now about a luminaire in order to decide intelligently which luminaire to choose for a given interior lighting application?

Luminaire Flux Distribution

Recall that in Sec. 2.6 we explained the use of a goniophotometer in obtaining the intensity distribution curve of a lamp or luminaire. We also showed how Eqs. (2.32) and (2.33) can be used to calculate the overall lumens emitted and the lumens emitted in certain angular zones.

Generally, in luminaire selection, we are interested in the percentage of the total lumen output that is directed upward toward the ceiling, the percentage that is directed downward toward the work plane (an imaginary horizontal plane usually 30 in above the floor where the desk tops, etc., are located), and the directivity of the downward component. The latter factor is generally quantified in terms of the percentage of the luminaire lumens in the 0 to 40° zone.

The CIE has devised a six-category classification system for interior luminaires, which is presented in Table 7.2. This system classifies luminaires according to the percent of the luminaire ouput emitted above and below the horizontal. The shape of the upward and downward distributions is not considered, except in the cases of general diffuse and direct-indirect, where the latter classification implies very little luminous flux at angles near the horizontal. With this one exception, the intensity distribution curves of luminaires within a given category may take many forms, as long as the overall upward and downward lumen percentages fall within the category's bounds.

Luminaire Efficiency

This is the ratio of the lumens emitted by the luminaire to those emitted by the lamps. The lower this number, the more luminous flux is being trapped and absorbed by the luminaire. Efficiencies of less than 70% generally indicate poor luminaire design. The distinction between luminaire efficiency

TABLE 7.2 LUMINAIRE CLASSIFICATION

Category	% Upward lumens	% Downward lumens
Direct	0–10	100–90
Semidirect	10–40	90–60
General diffuse	40–60	60–40
Direct-indirect	40–60	60–40
Semi-indirect	60–90	40–10
Indirect	90–100	10–0

and the percentages in Table 7.2 should be understood. The percent upward and downward luminaire lumens always sum to 100, regardless of the luminaire efficiency.

Luminaire Luminances

Bare lamps are usually too bright and produce too much glare. Thus one of the functions of a luminaire is to filter and shield the bare lamp from the viewer so as to provide a visually comfortable seeing environment.

It is customary for a luminaire manufacturer to include maximum and average luminances in the data provided to the customer. Maximum luminance is a measured quantity and is the luminance of the brightest square inch of a luminaire at a particular viewing angle. Generally several values of maximum luminance are given within the 45 to 90° zone, where direct glare is likely to be bothersome, for both crosswise (perpendicular to the luminaire) and lengthwise (along the luminaire) viewing. The direct and indirect glare zones are shown in Fig. 7.1.

Average luminance in a particular viewing direction is a calculated quantity. It is obtained by dividing the intensity of the luminaire in the given direction by the projected area of the lighted portion of the luminaire in that direction.

Four basic types of luminaires are shown in Fig. 7.2. Luminaire *(a)* is surface mounted and has dark side panels. Luminaire *(b)* is suspended but has no upward component. It does have luminous side panels. Luminaire *(c)* is also suspended but it has an upward component and thus its effective lighted width is greater than that of the luminaire itself. Luminaire *(d)* is like luminaire *(c)* except that is has dark side panels. Thus its effective lighted width is split into two parts, a luminaire portion and a ceiling portion.

Assume each luminaire in Fig. 7.2 has a length l and a projected lighted

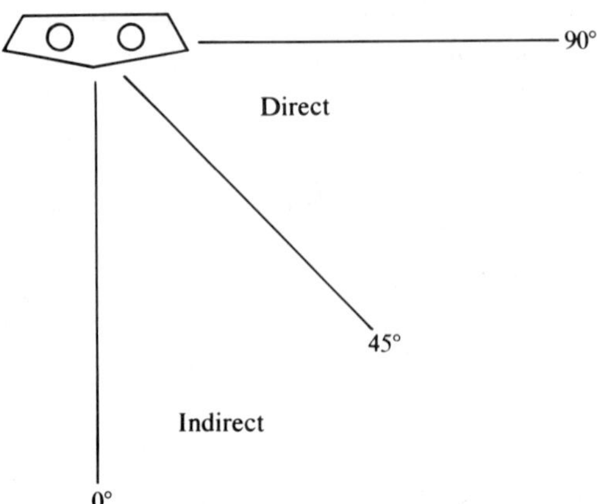

Figure 7.1 Glare zones.

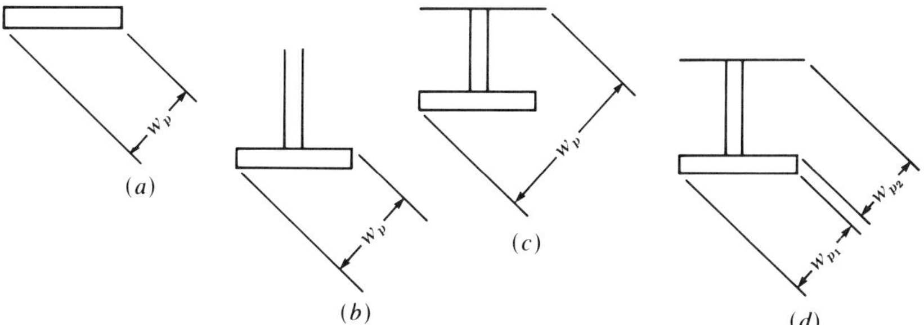

Figure 7.2 Projected areas of basic luminaires.

width w_p in the direction of an observer who is viewing the luminaires crosswise. If the average intensity of the luminaire is I in the direction of the observer, then the average luminance presented in that direction is

$$L_{\text{ave}} = \frac{I}{w_p(l)} \equiv \frac{I}{A_{p0}} \tag{7.1}$$

where A_{p0} is the crosswise projected area.

A more general case of luminaire viewing is shown in Fig. 7.3. The viewer is at point O. Parallel rows of luminaires, one of which is indicated by the line from S_0 to S, are located on the ceiling. Consider a single luminaire centered at point S_0 and having a projected area A_{p0} in the direction of O. This is the same situation as in Fig. 7.2 and yields the luminance given by Eq. (7.1).

Now let the luminaire slide down the row from S_0 to S. The width of A_{p0} does not change, but its length does. The new projected area in the direction of O is

$$A_p = A_{p0} \cos \phi \tag{7.2}$$

It is possible to express ϕ in terms of the polar angles at S_0 and S. These are

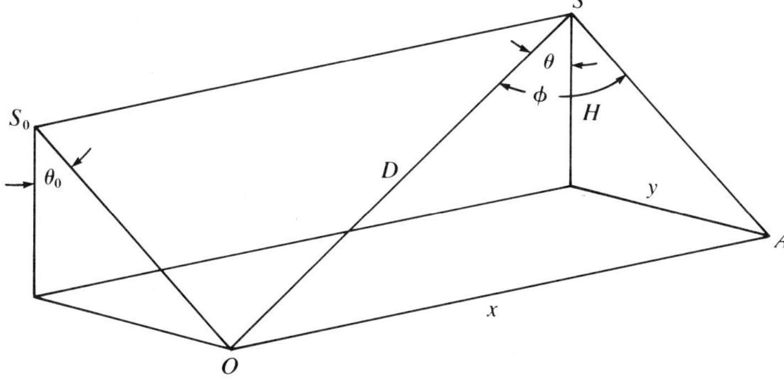

Figure 7.3 General case of luminaire viewing.

given the labels θ_0 and θ and we have

$$\cos\phi = \frac{\sqrt{Y^2 + H^2}}{D} \qquad \cos\theta_0 = \frac{H}{\sqrt{Y^2 + H^2}} \qquad \cos\theta = \frac{H}{D} \tag{7.3}$$

from which we obtain

$$\cos\phi = \frac{\cos\theta}{\cos\theta_0} \tag{7.4}$$

Assuming the luminaire at S has an average intensity I in the direction of O, the average luminance of that luminaire as seen at O is*

$$L_{\text{ave}} = \frac{I \cos\theta_0}{A_{p0} \cos\theta} \tag{7.5}$$

It is usually considered that direct glare will not be a problem if the ratio of the maximum to average luminaire luminance does not exceed 5:1 within the 45 to 90° glare zone, for both crosswise and lengthwise viewing, and if the maximum luminaire luminance is less than 5 cd/in^2 at 45°, 2.5 cd/in^2 at 65°, and 1 cd/in^2 at 85°.

Luminaire Light Control

We have stated that bare-lamp lighting installations are undesirable. They are sources of glare and they do not distribute the luminous flux efficiently and effectively.

There are many methods of light control used in contemporary luminaires. It is our purpose here to simply list these with a brief comment about each. Then optical control will be discussed in detail in Chap. 10.

Diffusers These are clear prismatic lenses, flat translucent sheets, or combinations thereof, either of glass or plastic. They are used on the bottom and sides of luminaires to redirect or scatter the light and to reduce luminances within the 45 to 90° glare zone.

Clear prismatic lenses use the principle of refraction to produce desired downward intensity distributions. A variety of patterns are used, one of the most common being a series of conical or pyramidal-shaped prisms with a density of 25 to 64 per square inch.

Translucent diffusers scatter the light in all downward directions, rather than directing it to particular zones. Transmittance is a function of pigment density and thickness. Transmittances of $\frac{1}{8}$-in sheets are 45 to 70%. For the same pigment density, these will drop 20 percentage points if thickness is doubled to $\frac{1}{4}$ in.

Diffusers generally have low absorbtion. Thus most of the luminous

*This is strictly true for the bottom luminous portion of the luminaire but is also used, by convention, for other cases.

flux that is not transmitted is reflected back into the luminaire and much of that returns to the diffuser, increasing transmission.

Baffles and Louvers A baffle is a single, usually V-shaped, shielding element, often placed parallel to and between lamps in a two-lamp luminaire, as shown in Fig. 7.4*a*. A louver is a group of baffles in an egg-crate arragement, as diagrammed in Fig. 7.4*b*. The louvers may be either straight or parabolic, as shown in Fig. 7.4*c*. The latter provide control of the directivity of the downward lumens and low luminances in the field of view.

The shielding angle of a baffle or louver is the angle between the horizontal and that line of sight for which all objects above are concealed. It is shown as θ_s in Fig. 7.4*d*. The actual lamp-shielding angle may be somewhat greater than θ_s if there is space between the lamps and the tops of the louvers.

Louvers come in a variety of sizes, shapes, materials, and finishes. In areas where low ceiling brightness is required, such as with visual display units, small hexagonal or square cell louvers are often used. The cells are typically $\frac{1}{2}$ in deep and $\frac{1}{2}$ to $\frac{3}{4}$ in wide. Each vertical cell wall is parabolic and the shielding angle is typically 45°. The louver finish can be specular silver, gold, or bronze; translucent white, or opaque black. A disadvantage of these units is that the resulting luminaire efficiency is low, often less than 50%.

At the other extreme are deep, open cell parabolic louvers. These are generally made of specular or semispecular aluminum, with a silver, gold, copper, or bronze finish. Cells are typically 5 to 7 in square by 3 to 4 in deep and provide 30 to 35° shielding. Luminaire efficiencies are higher than for small cell louvers, of the order of 55 to 65%.

Reflectors Reflectors are of two types, specular or semispecular, and diffuse, and are mounted above the lamps to redirect the upward component of luminous flux. Basic specular contours are circular, parabolic, elliptical, or combinations thereof. General specular contours may be tailor made to produce a particular overall luminous intensity distribution and are frequently designed graphically, rather than mathematically. Both types will be discussed in detail in Chap. 10.

Specular and semispecular reflectors are generally used with incandes-

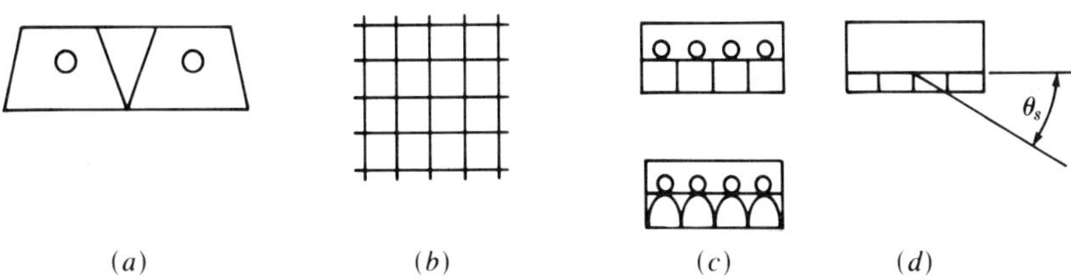

Figure 7.4 Baffle and louver shielding.

cent and HID sources. They are normally made of aluminum, which is etched, polished, brushed, plated, or anodized to produce the reflecting surface. The anodizing procedure is an electrochemical process which deposits a uniform, permanent, and high-reflectance surface of aluminum oxide on the aluminum.

Diffuse reflectors, commonly used in fluorescent luminaires, serve simply to gather and redirect as much of the upward luminous flux as possible downward to the lens, diffuser, or louver. Because of the diffusion, the contour of the reflector is relatively unimportant and the reflected intensity distribution curve will be approximately a cosine distribution, except that at angles near the horizontal it will decrease more rapidly than does the cosine of the angle because the reflecting surface is further from the light sources than for angles near the vertical.

The use of reflectors in luminaires requires several cautions. First the redirecting of reflected power back through the lamp may cause the lamp to operate at higher than normal temperature with the possible consequences of reduced lumen output and lamp life. Also concentration of reflected power on a lens or diffuser can cause thermal stress and failure. Finally, in specular reflectors, the location of the lamp with respect to the reflector is critical if the desired intensity distribution pattern is to be achieved. If the sockets are slightly out of position or the lamp base is slightly askew, the center of the light source can be displaced from the reflector focal point.

Other Considerations

There are several additional factors which the lighting designer should consider when choosing a luminaire.

Appearance The lighting of interior spaces should be both functional and aesthetically pleasing. These goals need not be incompatible, and the designer should make every effort to use luminaires which are attractive and which complement the architecture of the space being lighted.

Maintenance Luminaires should be easy to service, both for lamp replacement and for periodic cleaning. In choosing a luminaire, the designer should also be concerned about degradation of luminaire materials over time. In adverse environments, luminaires with dissimilar metals may cause problems because of the possibility of electrolytic action. Susceptibility of the luminaire to dust and dirt accumulation should also be considered.

Ventilation The ability of the luminaire to properly dissipate the build-up of heat should be considered. In some situations, it may be desirable to integrate the luminaire with the air-handling equipment of the building.

Acoustics Arc discharge light sources require ballasts, which generate noise. The luminaire can transmit and, in some cases, amplify this noise.

Luminaires should be designed so as not to resonate with the vibration frequencies of the ballast. Occasionally, when noise levels are critical, ballasts should be located remotely from their lamps and luminaires.

Mounting Luminaires may be suspended, surface-mounted, recessed (flush-mounted), connected in rows, or individually mounted. The choice affects both the functional and aesthetic aspects of the lighting. Spacing-to-mounting height guidelines for luminaires will be presented later in the chapter.

Radiation Electromagnetic radiation from arc discharge lamps can cause interference with nearby electronic equipment. The interference is of two types: direct radiation from the luminaire and feedback through the electric supply. The former may be minimized by the use of an all-metal enclosed luminaire; the latter can be controlled by electrical line filters located at each luminaire.

Hazardous Locations Luminaires for hazardous locations should have the UL label for the particular environment in question, indicating that they comply with certain safety requirements stated in the National Electric Code. If corrosive atmospheres are present, the designer should give careful consideration to the luminaire materials used.

It is impossible to review all the types of commercial, industrial, and institutional luminaires which are available for interior lighting. But, since fluorescent lighting is the dominant lighting form in building interiors, it is useful to discuss the general types of fluorescent luminaires.

Fluorescent luminaires are generally recessed or surface-mounted, rather than suspended. The troffer is the most common recessed unit. It is a metal trough with a trapezoidal cross section, recessed in the ceiling in a grid arrangement with its light opening flush with the ceiling, as shown in Fig. 7.5*a*. Troffers are generally available in 1×4-, 2×4-, 4×4-, and 2×2-ft sizes, housing two, three, or four fluorescent lamps. The 2×2-ft unit uses U-shaped lamps; the other units use standard 4-ft lamps. Troffer frames are usually of white-enameled steel. The light opening contains a prismatic diffuser, a translucent diffuser, or louvers. The latter are generally $\frac{1}{2} \times \frac{1}{2} \times \frac{1}{2}$ in, either straight or parabolic.

The most common surface-mounted unit is the wrap-around luminaire shown in Fig. 7.5*b*. This unit is a shallow rectangular box with its diffuser extending around the long sides. It is available in 2-, 4-, and 8-ft lengths to house two or four lamps. The diffuser is usually one-piece, injection-molded white translucent or clear plastic with pyramidal bottom prisms and linear horizontal side prisms. The end plates are white enameled and opaque. These luminaires often have top slits to provide a small amount (10 to 20%) of upward light to "wash" the ceiling near the luminaire and thus avoid excessive contrast between ceiling and luminaire.

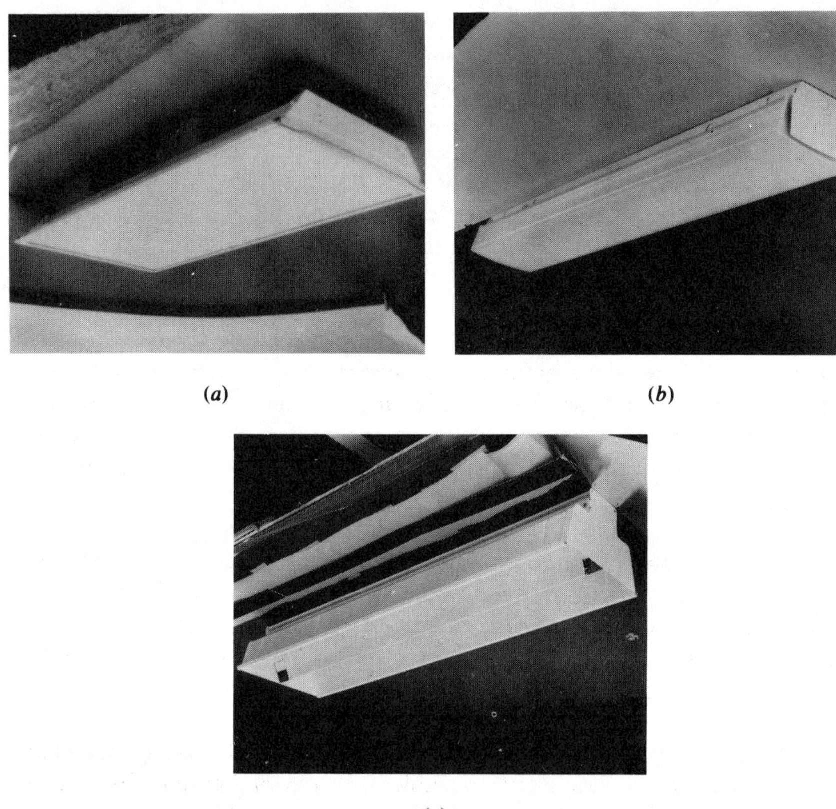

(*a*) (*b*)

(*c*)

Figure 7.5 Types of fluorescent luminaires. (*Source*: Lithonia Lighting, Fluorescent Lighting, Bulletin F-80.)

Surface-mounted units are also available with louvers instead of diffusers. The louvers may be either small $\frac{1}{2}$-in square plastic or 2-in square metal. These units have either plastic (luminous) or metal (opaque) side panels and provide a significant uplight component.

A typical fluorescent industrial luminaire is shown in Fig. 7.5*c*. It is troughlike, is generally suspended from the ceiling and mounted more than 15 ft above the floor, is all metal, and has no diffuser. It is available in 4- and 8-ft lengths housing one to four standard, high-output or very high output lamps. Most units have a V-shaped baffle running lengthwise between lamps to provide adequate crosswise shielding. The inside of the trough is a semidiffuse white porcelain enameled reflecting surface. The luminaire is often slotted in the top to provide 20 to 30% uplight.

A typical luminaire manufacturer's specification sheet, including many of the items discussed in this section, is shown in Fig. 7.6. The luminaire is a surface-mounted fluorescent unit for use in classrooms and offices. Its overall dimensions are $1\frac{1}{2}$ ft × 4 ft and it contains two 40-W 4-ft-long rapid-start fluorescent lamps. The luminaire has a plastic wrap-around prismatic

CFC—Comfort for Classrooms

Lamp Information	Size	Shipping Wt.	Catalog No.
2 Lamp, 4 ft. Rapid-Start Unit	18⅜" x 48"	25 lbs.	A88240-4
2 Row, Tandem Rapid-Start Unit	18⅜" x 96"	51 lbs.	A88240-8

Maintained Illumination Table*

Footcandles	Square Feet/Fixture		
	Room Size		
	Small**	Medium**	Large**
50 Footcandles			
2-Lamp Unit	44	55	67
70 Footcandles			
2-Lamp Unit	32	39	48
100 Footcandles			
2-Lamp Unit	22	28	34
150 Footcandles			
2-Lamp Unit	16	20	24

***40-Watt LLF = .77**

**Small RCR = 5 Width = Height Medium RCR = 3 Width = 2 x Height Large RCR = 1 Width = 5 x Height

16" 4" 18⅜"

Candlepower Distribution Curve

2 Lamp

Zone	End	45°	Cross
5	1682	1681	1688
15	1636	1669	1696
25	1540	1630	1687
35	1317	1448	1507
45	859	1043	1130
55	494	471	631
65	222	198	319
75	85	130	143
85	31	78	91
90	0	44	84
95	1	31	78
105	4	41	90
115	11	47	98
125	21	38	87
135	28	31	66
145	30	20	43
155	30	19	21
165	31	36	21
175	33	30	37
180	29	29	29

Coefficients of Utilization

Zonal Cavity Method

2-Lamp

pfc	20					
pcc	80		70		50	
pw	50	30	50	30	50	30
RCR						
1	.71	.69	.70	.67	.66	.64
2	.64	.60	.63	.59	.60	.57
3	.58	.53	.57	.52	.54	.51
4	.52	.47	.51	.46	.49	.45
5	.47	.42	.46	.41	.44	.40
6	.43	.37	.42	.37	.40	.36
7	.39	.33	.38	.33	.36	.32
8	.35	.29	.34	.29	.33	.28
9	.31	.26	.31	.26	.30	.25
10	.28	.23	.28	.23	.27	.23

Maximum recommended spacing to mounting height ratio 1.38

Zonal Interflectance Method

2-Lamp

floor		30			
ceiling		80		50	
wall		50	30	50	30
RI	**RR**				
j	**0.6**	.32	.27	.31	.26
i	**0.8**	.41	.36	.39	.35
h	**1.0**	.47	.41	.44	.40
g	**1.25**	.53	.47	.49	.45
f	**1.5**	.57	.52	.53	.49
e	**2.0**	.63	.58	.58	.54
d	**2.5**	.67	.62	.61	.58
c	**3.0**	.70	.65	.63	.60
b	**4.0**	.74	.70	.66	.64
a	**5.0**	.76	.73	.69	.66

For complete photometric information request test number #14789

***Light Loss Factor Data**
LLF = Light Loss Factor
LDD = Dirt Depreciation
IES Category V Clean Annually
LLD = Lamp Lumen Depreciation
Light Loss Factor (LLF) = LDD x LLD
LDD = Very Clean 0.93
Clean 0.88 Medium 0.82
LLD = 0.87 (Relamp @ 14,000 hrs.)

Average brightness (footlamberts) with 3100 Lumen Lamps

2-Lamp

Angle	End	Cross
45	640	707
55	454	455
65	276	282
75	173	170
85	186	173
90	0	0

Figure 7.6 Luminaire specification sheet. (*Source*: Day-Brite Lighting Division, Emerson Electric Co.)

lens. The bottom half of Fig. 7.6 contains information that will be discussed in succeeding sections.

The lighting designer should have only cautious confidence in data such as that provided in Fig. 7.6. The photometric testing of luminaires is, at best, an inexact science (see Levin, 1982). The actual luminaire, even though it has been tested by an independent laboratory, will probably not provide the values stated on the catalog sheet, for a variety of reasons. First, usually only one sample of a luminaire is sent to a testing laboratory for photometering. The unit to be tested is chosen by the manufacturer, not by the testing laboratory; thus usually a "good" sample is chosen. Second, there are bound to be statistical variations in a large number of units of the same kind. The extent of these variations is a function of the quality control exercised by the manufacturer. Third, there is no guarantee that subsequent production runs of the same unit will yield the same photometric data. Small changes in materials, fabrication methods, finishes, and assembly techniques can have a significant effect on photometric parameters. Fourth, the luminaire and lamps may be installed improperly. For example, a slight misorientation of a lamp with respect to a reflector can change a luminaire intensity distribution drastically. Finally, there is no guarantee that the parameters of the lamps and ballasts in the luminaire tested will be replicated in the commerical product. Nor is there any guarantee that the luminaire will be installed in the same type of environment as it was tested.

7.5 ZONAL CAVITY METHOD

Early in this century, calculations of horizontal illuminance in lighted interiors were done point-by-point using the inverse-square law. Such calculations were, of course, tedious and they dealt only with the direct component of illuminance, ignoring the interreflections from wall, floor, and ceiling surfaces.

The idea of a lumen method of interior lighting design had been proposed earlier, but it was not until 1920 that W. Harrison and E. A. Anderson introduced a standardized procedure which included both direct and reflected lumens. Their method involved separating the luminaire flux into three componenets (direct, indirect, and horizontal) which were then weighted by experimentally determined utilization factors that took into account room dimensions and room surface reflectances.

The Harrison-Anderson method served as the standard IES procedure until it was replaced in 1956 by the zonal factor interflectance method, an analytical procedure based on the calculation of luminous flux transfer between the surfaces in an interior space. This, in turn, was refined to the zonal cavity method, which is more versatile than its predecessor and has served as the standard IES procedure for determining average horizontal illuminance since 1964.

The analytical basis for the coefficients of utilization, cavity reflectances, and luminous exitance coefficients of the zonal cavity method will be developed later in the chapter. Our purpose in the next several sections is to become familiar with the use of the method in predicting the average illuminance on the work plane in an interior space.

To obtain this average illuminance, we need to determine the total luminous flux reaching the work plane. This flux is composed of two components: flux which comes directly from the luminaires to the work plane (direct component) and flux which is (multiply) reflected from room surfaces and then reaches the work plane (reflected component). In the zonal cavity method, the fraction of the initial lamp lumens which ultimately reaches the work plane, both directly and from reflection, is called the coefficient of utilization (CU), and it is this number that is published in tables of luminaire data for various room dimensions and room surface reflectances (see Fig. 7.6). It should be realized that CU includes both the efficiency of the luminaire in releasing lamp lumens into the interior space and the efficiency of the room surfaces in redirecting these lumens to the work plane.

In the definition of CU, initial lamp lumens are used. We have noted previously that lumen output decreases as lamps age. Also luminaire and room surfaces become less clean with time and some lamps may burn out and not be immediately replaced. The net effect of all these factors is lumped into what is known as a light loss factor (LLF). LLF is included in the calculations because we are usually interested in providing an average illumination level, maintained over time, on the work plane, not an average initial illumination level.

With CU and LLF defined, we can write an expression for the average maintained work-plane illuminance as

$$E = \frac{\Phi_i(\mathrm{CU})(\mathrm{LLF})}{A} \tag{7.6}$$

where Φ_i is the total initial lamp lumens and A is the work-plane area. The numerator of Eq. (7.6) gives the maintained lumens reaching the work plane. Our problem is now reduced to finding CU and LLF for a given interior space and luminaire.

7.6 CAVITY RATIOS

The zonal cavity method subdivides an interior space into three cavities as shown in Fig. 7.7*a*. The space beneath the work plane is termed the *floor cavity*. In the rare event that the work plane is the floor, the floor cavity ceases to exist. The space above the luminaires is called the *ceiling cavity*. This space disappears if the luminaires are mounted on (surface-mounted) or in (flush-mounted) the ceiling. The third space is named the *room cavity* and extends from the work plane to the luminaires.

Each of the three cavities is assigned a cavity ratio, which is a single number describing the proportions of the cavity, and is given by

$$\mathrm{CR} = \frac{5h(l + w)}{l(w)} \tag{7.7}$$

where l is room length, w is room width, and h is cavity height. Thus $h = h_c$ for ceiling cavity, h_r for room cavity, and h_f for the floor cavity.

Each CR is essentially the ratio of the amount of vertical surface area

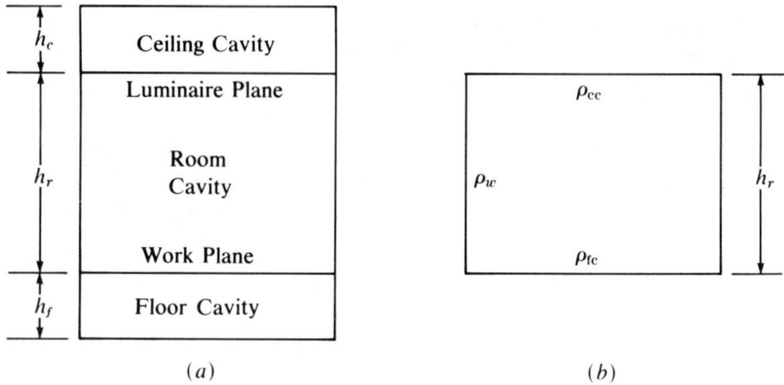

Figure 7.7 The three cavities used in the zonal cavity method.

in the cavity to the amount of horizontal surface area. In each cavity, the horizontal surface area is $2lw$ and the vertical surface area is $2h(l+w)$. Their ratio is $h(l+w)/lw$ and is generally less than 1, except for tall, narrow rooms. It was decided to use a multiplier of 5 in defining cavity ratio so as to obtain whole numbers greater than one for most rooms. For example, a $30 \times 20 \times 8$ ft school room with flush-mounted luminaires has a room cavity ratio of 2.3 ($h_r = 8 - 2.5 = 5.5$ ft).

The smaller the value of h, the smaller the cavity ratio; the smaller the values of l and w, the larger the cavity ratio. A long, wide room with a low ceiling height would be expected to utilize lumens better than a smaller room with a high ceiling. Thus as CR decreases, we will expect CU to increase, assuming all other variables remain the same.

The zonal cavity method was developed for spaces which are rectangular parallelepipeds, six-surface spaces with all sides rectangular. We will show later that the method has been extended to spaces with nonrectangular floors and sloped ceilings, but we will also point out that, in these cases, it can be questioned how accurate the resulting values of horizontal illuminance are.

7.7 CAVITY REFLECTANCE

The next step in the zonal cavity procedure is to replace the original three-cavity space by its room cavity only. The reasoning here is that if we can find effective reflectances for the two boundary planes between the room cavity and the ceiling and floor cavities, there is no need to retain these latter cavities in our calculations. The process is illustrated in Fig. 7.7*b* where ρ_{cc} and ρ_{fc} are the effective reflectances of the ceiling cavity and floor cavity, respectively.

The effective reflectance of a cavity is the ratio of the total flux out of the cavity opening to the total flux into the cavity opening. The resulting expression, involving multiple reflections and flux transfers within the cavity, is complex. Calculated values of effective cavity reflectance are given in Table 7.3.

Alternatively, a simple approximating expression, useful in computer programs and for hand calculations, is

$$\rho_{\text{eff}} = \frac{1}{1 + \left(\dfrac{A_s}{A_0}\right)\left(\dfrac{1-\rho}{\rho}\right)} \tag{7.8}$$

where A_s is the area of the cavity surfaces, A_0 is the area of the cavity opening, and ρ is the reflectance of the cavity surfaces. If the cavity is made up of surfaces having different reflectances, ρ must be the weighted-average reflectance of these surfaces.

For a rectangular ceiling cavity with walls of area A_w and reflectance ρ_w and a base of area A_b and reflectance ρ_b, the weighted average ρ is given by

$$\rho_{\text{ave}} = \frac{\rho_w A_w + \rho_b A_b}{A_w + A_b} \tag{7.9}$$

Insertion of Eq. (7.9) into Eq. (7.8) yields, following some manipulation,

$$\rho_{\text{eff}} = \frac{1}{\dfrac{[1 + (A_w/A_b)]^2}{\rho_b + \rho_w(A_w/A_b)} - \dfrac{A_w}{A_b}} \tag{7.10}$$

For a ceiling cavity $A_b = A_c$ and $\rho_b = \rho_c$. For a floor cavity $A_b = A_f$ and $\rho_b = \rho_f$.

Equation (7.10) gives exact values for ρ_{eff} when $\rho_w = \rho_c$ (or $\rho_w = \rho_f$). For the more general case when $\rho_w < \rho_c (\rho_w > \rho_f)$, the equation gives values which are slightly low (high), but these small errors do not affect seriously the ultimate values of horizontal work-plane illuminance which the zonal cavity method yields.

EXAMPLE 7.1

Calculate ρ_{eff} for a ceiling cavity $40 \times 40 \times 4$ ft whose $\rho_c = 80\%$ and $\rho_w = 30\%$. Repeat for a floor cavity of the same size with $\rho_f = 20\%$ and $\rho_w = 50\%$.

Solution. Inserting values into Eq. (7.10) yields effective cavity reflectances of

$$\rho_{\text{cc}} = \frac{1}{\dfrac{1.4^2}{0.8 + (0.3)(0.4)} - 0.4} = 0.58$$

and

$$\rho_{\text{fc}} = \frac{1}{\dfrac{1.4^2}{0.2 + (0.5)(0.4)} - 0.4} = 0.22$$

TABLE 7.3 PERCENT EFFECTIVE CEILING OR FLOOR CAVITY REFLECTANCES FOR VARIOUS REFLECTANCE COMBINATIONS

Percent base† reflectance: 90

Cavity ratio \ Percent wall reflectance	90	80	70	60	50	40	30	20	10	0
0.2	89	88	88	87	86	85	85	84	84	82
0.4	88	87	86	85	84	83	81	80	79	76
0.6	87	86	84	82	80	79	77	76	74	73
0.8	87	85	82	80	77	75	73	71	69	67
1.0	86	83	80	77	75	72	69	66	64	62
1.2	85	82	78	75	72	69	66	63	60	57
1.4	85	80	77	73	69	65	62	59	57	52
1.6	84	79	75	71	67	63	59	56	53	50
1.8	83	78	73	69	64	60	56	53	50	48
2.0	83	77	72	67	62	56	53	50	47	43
2.2	82	76	70	65	59	54	50	47	44	40
2.4	82	75	69	64	58	53	48	45	41	37
2.6	81	74	67	62	56	51	46	42	38	35
2.8	81	73	66	60	54	49	44	40	36	34
3.0	80	72	64	58	52	47	42	38	34	30
3.2	79	71	63	56	50	45	40	36	32	28
3.4	79	70	62	54	48	43	38	34	30	27
3.6	78	69	61	53	47	42	36	32	28	25
3.8	78	69	60	51	45	40	35	31	27	23
4.0	77	69	58	51	44	39	33	29	25	22
4.2	77	62	57	50	43	37	32	28	24	21
4.4	76	61	56	49	42	36	31	27	23	20
4.6	76	60	55	47	40	35	30	26	22	19
4.8	75	59	54	46	39	34	28	25	21	18
5.0	75	59	53	45	38	33	28	24	20	16
6.0	73	61	49	41	34	29	24	20	16	11
7.0	70	58	45	38	30	27	21	18	14	08
8.0	68	55	42	35	27	23	18	15	12	06
9.0	66	52	38	31	25	21	16	14	11	05
10.0	65	51	36	29	22	19	15	11	09	04

Percent base† reflectance: 80

Cavity ratio \ Percent wall reflectance	90	80	70	60	50	40	30	20	10	0
0.2	79	78	78	77	77	76	76	75	74	72
0.4	79	77	76	75	74	73	72	71	70	68
0.6	78	76	75	73	71	70	68	66	65	63
0.8	78	75	73	71	69	67	65	63	61	57
1.0	77	74	72	69	67	65	62	60	57	55
1.2	76	73	70	67	64	61	58	55	53	51
1.4	76	72	68	65	62	59	55	53	50	48
1.6	75	71	67	63	60	57	53	50	47	44
1.8	75	70	66	62	58	54	50	47	44	41
2.0	74	69	64	60	56	52	48	45	41	38
2.2	74	68	63	58	54	49	45	42	38	35
2.4	73	67	61	56	52	47	43	40	36	33
2.6	73	66	60	55	50	45	41	38	34	31
2.8	73	65	59	53	48	43	39	36	32	29
3.0	72	65	58	52	47	42	37	34	30	27
3.2	72	65	57	51	45	40	35	33	28	25
3.4	71	64	56	49	44	39	34	32	27	24
3.6	71	63	54	48	43	38	32	30	25	23
3.8	70	62	53	47	41	36	31	28	24	22
4.0	70	61	53	46	40	35	30	26	22	20
4.2	69	60	52	45	39	34	29	25	21	18
4.4	69	60	51	44	38	33	28	24	20	17
4.6	69	59	50	43	37	32	27	23	19	15
4.8	68	58	49	42	36	31	26	22	18	14
5.0	68	58	48	41	35	30	25	21	18	14
6.0	66	55	44	38	31	27	22	19	15	10
7.0	64	53	41	35	28	24	19	16	12	07
8.0	62	50	38	32	25	21	17	14	11	05
9.0	61	49	36	30	23	19	15	13	10	04
10.0	59	46	33	27	21	18	14	11	08	03

Percent base† reflectance: 70

Cavity ratio \ Percent wall reflectance	90	80	70	60	50	40	30	20	10	0
0.2	70	69	68	68	67	67	66	66	65	64
0.4	69	68	67	66	65	64	63	62	61	58
0.6	69	67	65	64	63	61	59	58	57	54
0.8	68	66	64	62	60	58	56	55	53	50
1.0	68	65	62	60	58	55	53	52	50	47
1.2	67	64	61	59	57	54	50	48	46	44
1.4	67	63	60	58	55	51	47	45	44	41
1.6	67	62	59	56	53	47	45	43	41	38
1.8	66	61	58	54	51	46	42	40	38	35
2.0	66	60	56	52	49	45	40	38	36	33
2.2	66	60	55	51	48	43	38	36	34	32
2.4	65	60	54	50	46	41	37	35	32	30
2.6	65	59	54	49	45	40	35	33	30	28
2.8	65	59	53	48	43	38	33	30	28	26
3.0	64	58	52	47	42	37	32	29	27	24
3.2	64	58	51	46	40	36	31	28	25	23
3.4	64	57	50	45	39	35	29	27	24	22
3.6	63	56	49	44	38	33	28	25	22	20
3.8	63	56	49	43	37	32	27	24	21	19
4.0	63	55	48	42	36	31	26	23	20	17
4.2	62	55	47	41	35	30	25	22	19	16
4.4	62	54	46	40	34	29	24	21	18	15
4.6	62	53	45	39	33	28	24	21	17	14
4.8	62	53	45	38	32	27	23	20	16	13
5.0	61	52	44	36	31	26	22	19	16	12
6.0	60	51	41	35	28	24	19	16	13	09
7.0	58	48	38	32	26	22	17	14	11	06
8.0	57	46	35	29	23	19	15	13	10	05
9.0	56	45	33	27	21	18	14	12	09	04
10.0	55	43	31	25	19	16	12	10	08	03

Percent base† reflectance: 60

Cavity ratio \ Percent wall reflectance	90	80	70	60	50	40	30	20	10	0
0.2	60	59	59	59	58	57	56	56	55	53
0.4	60	59	59	58	57	55	54	53	52	50
0.6	60	58	57	56	55	53	51	51	50	46
0.8	59	57	56	55	54	51	48	47	46	43
1.0	59	57	55	53	51	48	45	44	43	41
1.2	59	56	54	51	49	46	44	42	40	38
1.4	59	56	53	49	47	44	41	39	38	36
1.6	59	55	52	48	45	42	39	37	35	33
1.8	58	55	51	47	44	40	37	35	33	31
2.0	58	54	50	46	43	39	35	33	31	29
2.2	58	53	49	45	42	37	34	31	29	28
2.4	58	53	48	44	41	36	32	30	27	26
2.6	58	53	48	43	39	35	31	28	26	24
2.8	58	53	47	43	38	34	29	27	24	22
3.0	57	52	46	42	37	32	28	25	23	20
3.2	57	51	45	41	36	31	27	23	22	18
3.4	57	51	45	40	35	30	26	23	20	17
3.6	57	50	44	39	34	29	25	22	19	16
3.8	57	50	43	38	33	29	24	21	19	15
4.0	57	49	42	37	32	28	23	20	18	14
4.2	56	49	42	37	32	27	22	19	17	14
4.4	56	49	42	36	31	27	22	19	16	13
4.6	56	49	41	35	30	26	21	18	16	13
4.8	56	48	41	34	29	25	21	18	15	12
5.0	56	48	40	34	28	24	20	17	14	11
6.0	55	45	37	31	25	21	17	14	11	07
7.0	54	43	35	30	24	20	15	12	09	05
8.0	53	42	33	28	22	18	14	11	08	04
9.0	52	40	31	26	20	16	12	10	07	03
10.0	51	39	29	24	18	15	11	09	07	02

Percent base† reflectance: 50

Cavity ratio \ Percent wall reflectance	90	80	70	60	50	40	30	20	10	0
0.2	50	50	49	49	48	48	47	46	46	44
0.4	50	49	48	48	47	46	45	45	44	42
0.6	50	48	47	46	45	44	43	42	41	38
0.8	50	48	47	45	44	42	40	39	38	36
1.0	50	48	46	44	43	41	38	37	36	34
1.2	50	47	45	43	41	39	36	35	34	29
1.4	50	47	45	42	40	38	35	34	32	27
1.6	50	47	44	41	39	36	33	32	30	26
1.8	50	46	43	40	38	35	31	30	28	25
2.0	50	46	43	40	37	34	30	28	26	24
2.2	50	46	42	38	36	33	29	27	24	22
2.4	50	46	42	37	35	31	27	25	23	21
2.6	50	46	41	37	34	30	26	23	21	20
2.8	50	46	41	36	33	29	25	22	20	19
3.0	50	45	40	36	32	28	24	21	19	17
3.2	50	44	39	35	31	27	23	20	18	16
3.4	50	44	39	35	30	26	22	19	17	15
3.6	50	44	39	34	29	25	21	18	16	14
3.8	50	44	38	34	29	25	21	17	15	13
4.0	50	44	38	33	28	24	20	17	15	12
4.2	50	43	37	32	28	24	20	17	14	12
4.4	50	43	37	32	27	23	19	16	13	11
4.6	50	43	36	31	26	22	18	15	13	10
4.8	50	43	36	31	26	22	18	15	12	09
5.0	50	42	35	30	25	21	17	14	12	09
6.0	50	42	34	29	23	19	15	13	10	06
7.0	49	41	32	27	21	18	14	11	08	05
8.0	49	40	30	25	19	16	12	10	07	03
9.0	48	39	29	24	18	15	11	09	07	03
10.0	47	37	27	22	17	14	10	08	06	02

Percent base† reflectance	40										30										20										10										0									
Percent wall reflectance	90	80	70	60	50	40	30	20	10	0	90	80	70	60	50	40	30	20	10	0	90	80	70	60	50	40	30	20	10	0	90	80	70	60	50	40	30	20	10	0	90	80	70	60	50	40	30	20	10	0
0.2	40	40	39	39	39	38	38	37	36	36	31	31	30	30	29	29	29	28	28	27	21	20	20	20	20	20	19	19	19	17	11	11	11	10	10	10	10	09	09	09	02	02	02	01	01	01	01	00	00	0
0.4	41	40	39	39	38	37	36	35	34	34	31	31	30	30	29	28	28	27	26	25	22	21	20	20	20	19	19	18	18	16	12	11	11	11	11	10	10	09	09	08	04	03	03	02	02	02	01	01	00	0
0.6	41	40	39	38	37	36	34	33	32	31	32	31	30	29	28	27	26	26	25	23	23	21	21	20	19	19	18	18	17	15	13	13	12	11	11	10	10	09	08	08	05	05	04	03	03	02	02	01	01	0
0.8	41	40	38	37	36	35	33	32	31	29	32	31	30	29	28	26	25	25	23	22	24	22	21	20	19	19	18	17	16	14	15	14	13	12	11	10	10	09	08	07	07	06	05	04	04	03	02	02	01	0
1.0	42	40	38	37	35	33	32	31	29	27	33	32	30	29	27	25	24	23	22	20	25	23	22	20	19	18	17	16	15	13	16	14	13	12	12	11	10	09	08	07	08	07	06	05	04	03	02	02	01	0
1.2	42	40	38	36	34	32	30	29	27	25	33	32	30	28	27	25	23	22	21	19	25	23	22	20	19	17	17	16	14	12	17	15	14	13	12	11	10	09	07	06	10	08	07	06	05	04	03	02	01	0
1.4	42	39	37	35	33	31	29	27	25	23	34	32	30	28	26	24	22	21	19	18	26	24	22	20	18	17	16	15	13	12	18	16	14	13	12	11	10	09	07	06	11	09	08	07	06	04	03	02	01	0
1.6	42	39	37	35	32	30	27	25	23	22	34	33	29	27	25	23	22	20	18	17	26	24	22	20	18	17	16	15	13	11	19	17	15	14	12	11	09	08	07	06	12	10	09	07	06	05	03	02	01	0
1.8	42	39	36	34	31	29	26	24	22	21	35	33	29	27	25	23	21	19	17	16	27	25	23	20	18	17	15	14	12	10	19	17	15	14	13	11	09	08	06	05	13	11	09	08	07	05	04	03	01	0
2.0	42	39	36	34	31	28	25	23	21	19	35	33	29	26	24	22	20	18	16	14	28	25	23	20	18	16	15	13	11	09	20	18	16	14	13	11	09	08	06	05	14	12	10	09	07	05	04	03	01	0
2.2	42	39	36	33	30	27	24	22	19	18	36	32	29	26	24	22	19	17	15	13	28	25	23	20	18	16	14	12	10	09	21	19	16	14	13	11	09	07	06	05	15	13	11	09	07	06	04	03	01	0
2.4	43	39	35	33	29	27	24	21	18	17	36	32	29	26	24	22	19	16	14	12	29	26	23	20	18	16	14	12	10	08	22	19	17	15	13	11	09	07	06	05	16	13	11	09	08	06	04	03	01	0
2.6	43	39	35	32	29	26	23	20	17	15	36	32	29	25	23	21	18	16	14	12	29	26	23	20	18	16	14	11	09	08	23	20	17	15	13	11	09	07	06	04	17	14	12	10	08	06	05	03	02	0
2.8	43	39	35	32	28	25	22	19	16	14	37	33	29	25	23	21	17	15	13	11	30	27	23	20	18	15	13	11	09	07	23	20	18	16	13	11	09	07	05	03	17	15	13	10	08	07	05	03	02	0
3.0	43	39	35	31	27	24	21	18	16	13	37	33	29	25	22	20	17	15	12	10	30	27	23	20	17	15	13	11	09	07	24	21	18	16	13	11	09	07	05	03	18	16	13	11	09	07	05	03	02	0
3.2	43	39	35	31	27	23	20	71	15	13	37	33	29	25	22	19	16	14	12	10	31	27	23	20	17	15	12	11	09	06	25	21	18	16	13	11	09	07	05	03	19	16	14	11	09	07	05	03	02	0
3.4	43	39	34	30	26	23	20	17	14	12	37	33	29	25	22	19	16	14	11	09	31	27	23	20	17	15	12	10	08	06	26	22	18	16	13	11	09	07	05	03	20	17	14	12	09	07	05	03	02	0
3.6	44	39	34	30	26	22	19	16	14	11	38	33	29	24	21	18	15	13	10	09	32	27	23	20	17	15	12	10	08	05	26	22	19	16	13	11	09	06	04	03	20	17	15	12	10	08	05	04	02	0
3.8	44	38	33	29	25	22	18	16	13	10	38	33	28	24	21	18	15	13	10	08	32	28	23	20	17	15	12	10	07	05	27	23	19	17	14	11	09	06	04	02	21	18	15	12	10	08	05	04	02	0
4.0	44	38	33	29	25	21	18	15	12	10	38	33	28	24	21	18	14	12	09	07	33	28	23	20	17	14	11	09	07	05	27	23	20	17	14	11	09	06	04	02	22	18	15	15	13	10	08	04	02	0
4.2	44	38	33	29	24	21	17	15	12	10	38	33	28	24	20	17	14	12	09	07	33	28	23	20	17	14	11	09	07	04	28	24	20	17	14	11	09	06	04	02	22	19	16	13	10	08	06	04	02	0
4.4	44	38	33	28	24	20	17	14	11	09	39	33	28	24	20	17	14	11	09	06	34	28	24	20	17	14	11	09	07	04	28	24	20	17	14	11	08	06	04	02	23	19	16	13	10	08	06	04	02	0
4.6	44	38	32	28	23	19	16	14	11	08	39	33	28	24	20	17	13	10	08	06	34	29	24	20	17	14	11	09	07	04	29	25	20	17	14	11	08	06	04	02	23	20	17	13	11	08	06	04	02	0
4.8	44	38	32	27	22	19	16	13	10	08	39	33	28	24	20	17	13	10	08	05	35	29	24	20	17	13	10	08	06	04	29	25	20	17	14	11	08	06	04	02	24	20	17	14	11	08	06	04	02	0
5.0	45	38	31	27	22	19	15	13	10	07	39	33	28	24	19	16	13	10	08	05	35	29	24	20	16	13	10	08	06	04	30	25	20	17	14	11	08	06	04	02	25	21	17	14	11	08	06	04	02	0
6.0	44	37	30	25	20	17	13	11	08	05	39	33	27	23	18	15	11	09	06	04	36	30	24	20	16	13	10	08	05	02	31	26	21	18	14	11	08	06	03	01	27	23	18	15	12	09	06	04	02	0
7.0	44	36	29	24	19	16	12	10	07	04	40	33	26	22	17	14	10	08	05	03	36	30	24	20	15	12	09	07	04	02	32	27	21	17	13	11	08	06	03	01	28	24	19	15	12	09	06	04	02	0
8.0	44	35	28	23	18	15	11	09	06	03	40	33	26	21	16	13	09	07	04	02	37	30	23	19	15	12	08	06	03	01	33	27	21	17	13	10	07	05	03	01	30	25	20	15	12	09	06	04	02	0
9.0	44	35	26	21	16	13	10	08	05	02	40	33	25	20	15	12	09	07	04	02	37	29	23	19	14	11	08	06	03	01	34	28	21	17	13	10	07	05	02	01	31	25	20	15	12	09	06	04	02	0
10.0	43	34	25	20	15	12	08	07	05	02	40	32	24	19	14	11	08	06	03	01	37	29	22	18	13	10	07	05	03	01	34	28	21	17	12	10	07	05	02	01	31	25	20	15	12	09	06	04	02	0

†Ceiling, floor, or floor of cavity.

Source: J. E. Kaufman (ed.), *IES Lighting Handbook*, Reference volume, 1981 ed., IES, New York, 1981, fig. 9-11, with permission.

where ρ_{cc} refers to the ceiling cavity and ρ_{fc} to the floor cavity. Noting that the cavity ratio is 1.0, the exact values of these reflectances from flux transfer analysis and Table 7.3 are 0.62 and 0.19. This illustrates the close approximation provided by Eq. (7.10).

It should be noted that the cavity reflectances in Table 7.3 are for empty spaces. For the floor cavity, even though $\rho_{fc} > \rho_f$, the designer may choose to use ρ_f in subsequent calculations because of obstructions, such as furniture, which trap and absorb light within the floor cavity space.

7.8 COEFFICIENTS OF UTILIZATION

As described earlier, the coefficient of utilization (CU) is the fraction of the lamp lumens which ultimately reaches the work plane. Values of CU for a large number of luminaires used in interior spaces have been calculated using flux transfer theory (see Kaufman, 1981). The results for 10 typical ones are given in Table 7.4.*

To use Table 7.4, one first chooses a luminaire and then calculates the cavity ratios for the given room dimensions from Eq. (7.7). Next the effective reflectances of the ceiling and floor cavities (if these exist) are calculated from Eq. (7.10) or obtained from Table 7.3. Then Table 7.4 is entered with the particular room cavity ratio (RCR), ρ_{cc}, and ρ_w and the CU is determined, based on $\rho_{fc} = 20\%$ (often interpolation between adjacent rows and columns in Table 7.4 is required to obtain the CU). If ρ_{fc} is not 20%, a correction factor for the CU may be obtained from Table 7.5.

EXAMPLE 7.2

A 28 × 32 × 10 ft school room is to be lighted by luminaire 8 in Table 7.4. Forty-watt standard cool white rapid start fluorescent lamps are prescribed. Reflectances for the ceiling, walls, and floor are 80%, 50%, and 30%, respectively. An average initial illuminance of 50 fc is desired. How many luminaires are required?

Solution. From Eq. (7.7),

$$\text{RCR} = \frac{5(7.5)(60)}{28(32)} = 2.51$$

$$\text{FCR} = 2.51\left(\frac{2.5}{7.5}\right) = 0.84$$

$$\text{CCR} = 0$$

Since there is no ceiling cavity, we can enter Table 7.4 directly and obtain CU = 0.43 for $\rho_{fc} = 20\%$.

*The CU values are derived from the average luminaire intensity at each polar angle. They are based on spacing to mounting height ratios of 0.4 for fluorescent and 0.7 for incandescent and HID.

We must now check whether or not the effective floor cavity reflectance was 20%. From Table 7.3, $\rho_{fc} = 0.28$. Then from Table 7.5, a multiplier of 1.06 is obtained for a 30% floor cavity. For 28%, we will use 1.05. Thus the final CU value is 1.05(0.43) = 0.45.

Inserting this CU into Eq. (7.6), with LLF = 1, and solving for Φ_i gives

$$\Phi_i = \frac{50(896)}{0.45} = 99{,}600 \text{ lm}$$

Each luminaire houses four lamps. From Table 6.7, each lamp emits 3150 initial lumens. Thus the number of luminaires required is

$$n = \frac{99{,}600}{4(3150)} = 8$$

EXAMPLE 7.3

Repeat Example 7.2 for luminaire 5 suspended 1.5 ft from the ceiling. All other parameters are unchanged.

Solution. Because the luminaire is suspended, some of the cavity ratios change. From Eq. (7.7),

$$\text{RCR} = \frac{5(6)(60)}{28(32)} = 2.0$$

$$\text{CCR} = 2.0\left(\frac{1.5}{6}\right) = 0.5$$

$$\text{FCR} = 0.84 \qquad \text{(as before)}$$

In this case $\rho_{cc} \neq \rho_c$. Table 7.3 gives 0.73. Entering Table 7.4 gives a CU of about 0.55 for a 20% floor cavity reflectance. The value of ρ_{fc} is unchanged but ρ_{cc} and RCR have changed. From Table 7.5, the multiplying factor for a 30% floor is slightly greater than 1.06. Thus, for 28%, its value is approximately 1.05 and the final CU = 0.58.

Insertion of this CU into Eq. (7.6) yields

$$\Phi_i = \frac{50(896)}{0.58} = 77{,}200 \text{ lm}$$

which requires

$$n = \frac{77{,}200}{6300} = 12 \text{ luminaires}$$

EXAMPLE 7.4

In the previous two examples, interpolation in Table 7.4 was not required. Consider now the case when $\rho_w = 40\%$, $\rho_{cc} = 73\%$, $\rho_{fc} = 20\%$, and RCR = 1.6. Find the CU for luminaire 5.

TABLE 7.4

Typical Luminaire	Typical Intensity Distribution and Per Cent Lamp Lumens			ρ_{CC} →	80			70			50			30			10			0	WDRC
	Maint. Cat.	SC		ρ_W →	50	30	10	50	30	10	50	30	10	50	30	10	50	30	10	0	
				RCR ↓	Coefficients of Utilization for 20 Per Cent Effective Floor Cavity Reflectance (ρ_{FC} = 20)																
1 Pendant diffusing sphere with incandescent lamp	V	1.5	35½% ↑ 45% ↓	0	.87	.87	.87	.81	.81	.81	.70	.70	.70	.59	.59	.59	.49	.49	.49	.45	—
				1	.71	.67	.63	.66	.62	.59	.56	.53	.50	.47	.45	.42	.38	.37	.35	.31	.348
				2	.60	.54	.49	.56	.50	.45	.47	.43	.39	.39	.36	.33	.32	.29	.27	.23	.269
				3	.52	.45	.39	.48	.42	.37	.41	.36	.31	.34	.30	.26	.27	.24	.22	.18	.221
				4	.46	.38	.33	.42	.36	.30	.36	.30	.26	.30	.26	.22	.24	.21	.18	.15	.186
				5	.40	.33	.27	.37	.30	.25	.31	.26	.22	.26	.22	.18	.21	.18	.15	.12	.162
				6	.36	.28	.23	.33	.26	.21	.28	.23	.19	.23	.19	.16	.19	.15	.13	.10	.144
				7	.32	.25	.20	.29	.23	.18	.25	.20	.16	.21	.16	.13	.17	.13	.11	.09	.130
				8	.29	.22	.17	.26	.20	.16	.23	.17	.14	.19	.15	.12	.15	.12	.09	.07	.117
				9	.26	.19	.15	.24	.18	.14	.20	.15	.12	.17	.13	.10	.14	.11	.08	.06	.107
				10	.23	.17	.13	.22	.16	.12	.19	.14	.10	.16	.12	.09	.13	.09	.07	.05	.099
2 Prismatic square surface drum	V	1.3	18½% ↑ 60½% ↓	0	.89	.89	.89	.85	.85	.85	.77	.77	.77	.70	.70	.70	.63	.63	.63	.60	—
				1	.78	.75	.72	.74	.72	.69	.68	.66	.64	.62	.60	.58	.56	.55	.54	.51	.241
				2	.69	.65	.61	.66	.62	.58	.61	.57	.54	.56	.53	.50	.51	.49	.47	.44	.202
				3	.62	.57	.52	.60	.55	.50	.55	.51	.47	.50	.47	.44	.46	.43	.41	.39	.178
				4	.56	.50	.46	.54	.49	.44	.50	.45	.42	.46	.42	.39	.42	.39	.37	.35	.159
				5	.51	.45	.40	.49	.43	.39	.45	.41	.37	.42	.38	.35	.39	.36	.33	.31	.146
				6	.46	.40	.36	.45	.39	.35	.42	.37	.33	.39	.35	.31	.36	.32	.30	.28	.135
				7	.42	.36	.32	.41	.35	.31	.38	.33	.29	.35	.31	.28	.33	.29	.27	.25	.126
				8	.38	.32	.28	.37	.32	.28	.35	.30	.26	.32	.28	.25	.30	.27	.24	.22	.118
				9	.35	.29	.25	.34	.29	.25	.32	.27	.24	.30	.26	.23	.28	.24	.22	.20	.112
				10	.32	.27	.23	.31	.26	.22	.29	.25	.21	.27	.23	.20	.26	.22	.20	.18	.105
3 "High bay" wide distribution ventilated reflector with clear HID lamp	III	1.5	½% ↑ 77½% ↓	0	.93	.93	.93	.91	.91	.91	.87	.87	.87	.83	.83	.83	.79	.79	.79	.78	—
				1	.85	.82	.80	.83	.81	.79	.79	.78	.76	.76	.75	.74	.74	.72	.71	.70	.194
				2	.77	.73	.70	.76	.72	.69	.73	.70	.67	.70	.68	.66	.68	.66	.64	.63	.187
				3	.70	.65	.61	.68	.64	.60	.66	.62	.59	.64	.61	.58	.62	.59	.57	.56	.184
				4	.63	.58	.53	.62	.57	.53	.60	.56	.52	.58	.55	.52	.57	.54	.51	.49	.176
				5	.57	.51	.47	.56	.51	.47	.55	.50	.46	.53	.49	.46	.52	.48	.45	.44	.170
				6	.51	.45	.41	.51	.45	.41	.49	.44	.40	.48	.43	.40	.47	.43	.40	.38	.164
				7	.46	.40	.35	.45	.39	.35	.44	.39	.35	.43	.38	.35	.42	.38	.34	.33	.159
				8	.41	.35	.31	.41	.35	.31	.40	.34	.31	.39	.34	.30	.38	.33	.30	.29	.152
				9	.37	.31	.27	.37	.31	.27	.36	.30	.27	.35	.30	.27	.34	.30	.26	.25	.146
				10	.33	.28	.24	.33	.27	.23	.32	.27	.23	.31	.27	.23	.31	.26	.23	.22	.140
4 Porcelain-enameled reflector with 35°CW shielding	II	1.3	22½% ↑ 65% ↓	0	.99	.99	.99	.94	.94	.94	.85	.85	.85	.77	.77	.77	.69	.69	.69	.65	—
				1	.88	.85	.82	.84	.81	.78	.76	.74	.72	.69	.67	.66	.62	.61	.60	.57	.214
				2	.78	.73	.69	.74	.70	.66	.68	.64	.61	.62	.59	.56	.56	.54	.52	.49	.201
				3	.70	.63	.58	.67	.61	.57	.61	.56	.53	.56	.52	.49	.51	.48	.46	.43	.186
				4	.62	.55	.50	.60	.53	.49	.55	.50	.46	.50	.46	.43	.46	.43	.40	.37	.173
				5	.55	.48	.43	.53	.47	.42	.49	.44	.39	.45	.41	.37	.41	.38	.35	.32	.164
				6	.50	.43	.38	.48	.41	.37	.44	.39	.35	.41	.36	.33	.37	.34	.31	.29	.153
				7	.45	.38	.33	.43	.37	.32	.40	.34	.30	.37	.32	.29	.34	.30	.27	.25	.143
				8	.40	.34	.29	.39	.32	.28	.36	.30	.27	.33	.28	.25	.31	.27	.24	.22	.136
				9	.36	.30	.25	.35	.29	.24	.32	.27	.23	.30	.25	.22	.28	.24	.21	.19	.129
				10	.33	.27	.22	.32	.26	.22	.29	.24	.20	.27	.23	.19	.25	.21	.18	.17	.121
5 Metal or dense diffusing sides with 45°CW × 45°LW shielding	II	1.1	39% ↑ 32% ↓	0	.75	.75	.75	.69	.69	.69	.57	.57	.57	.46	.46	.46	.37	.37	.37	.32	—
				1	.67	.64	.62	.61	.59	.57	.51	.50	.49	.42	.41	.40	.34	.33	.32	.29	.084
				2	.59	.55	.52	.55	.51	.49	.46	.44	.42	.38	.36	.35	.31	.30	.29	.25	.082
				3	.53	.48	.45	.49	.45	.42	.41	.39	.36	.35	.32	.31	.28	.27	.26	.23	.077
				4	.47	.42	.39	.44	.40	.36	.37	.34	.32	.31	.29	.27	.26	.24	.23	.20	.073
				5	.43	.37	.33	.40	.35	.31	.34	.30	.28	.28	.26	.24	.23	.22	.20	.18	.070
				6	.39	.33	.29	.36	.31	.28	.31	.27	.25	.26	.23	.21	.22	.20	.18	.16	.065
				7	.35	.30	.26	.33	.28	.25	.28	.24	.22	.24	.21	.19	.20	.18	.16	.15	.062
				8	.32	.26	.23	.30	.25	.22	.25	.22	.19	.22	.19	.17	.18	.16	.15	.13	.059
				9	.29	.24	.20	.27	.22	.19	.23	.20	.17	.20	.17	.15	.16	.15	.13	.12	.057
				10	.26	.21	.18	.25	.20	.17	.21	.18	.15	.18	.15	.14	.15	.13	.12	.10	.054

Source: J. E. Kaufman (ed.), *IES Lighting Handbook*, Reference volume, 1981 ed., IES, New York, 1981, fig. 9-12, with permission.

Solution. When interpolation is required, it is desirable to set up a small table showing the quantities involved:

ρ_{cc}→	80		70	
ρ_w→	50	30	50	30
RCR				
1	0.67	0.64	0.61	0.59
2	0.59	0.55	0.55	0.51

Typical Luminaire	Typical Intensity Distribution and Per Cent Lamp Lumens		ρ_{CC} →	80			70			50			30			10			0	WDRC
			ρ_W →	50	30	10	50	30	10	50	30	10	50	30	10	50	30	10	0	
	Maint. Cat.	SC	RCR ↓	Coefficients of Utilization for 20 Per Cent Effective Floor Cavity Reflectance (ρ_{FC} = 20)																
6	V	1.5/1.2	0	.81	.81	.81	.78	.78	.78	.72	.72	.72	.66	.66	.66	.61	.61	.61	.59	—
			1	.71	.69	.66	.69	.66	.64	.64	.62	.60	.59	.58	.56	.55	.54	.53	.50	.204
	$11\frac{1}{2}$% ↑		2	.64	.59	.56	.61	.58	.54	.57	.54	.51	.53	.51	.49	.49	.48	.46	.44	.184
			3	.57	.52	.48	.55	.50	.47	.51	.48	.45	.48	.45	.42	.45	.42	.40	.38	.168
			4	.51	.46	.41	.49	.44	.41	.46	.42	.39	.43	.40	.37	.41	.38	.35	.34	.156
	$58\frac{1}{2}$% ↓		5	.46	.40	.36	.44	.39	.35	.41	.37	.34	.39	.35	.32	.37	.33	.31	.29	.147
			6	.41	.35	.31	.40	.35	.31	.38	.33	.30	.35	.31	.28	.33	.30	.27	.26	.137
			7	.37	.31	.27	.36	.31	.27	.34	.29	.26	.32	.28	.25	.30	.27	.24	.23	.129
			8	.33	.28	.24	.32	.27	.23	.30	.26	.22	.29	.25	.22	.27	.24	.21	.19	.122
2 lamp prismatic wraparound			9	.30	.24	.20	.29	.24	.20	.27	.23	.19	.26	.22	.19	.24	.21	.18	.17	.116
			10	.27	.22	.18	.26	.21	.18	.25	.20	.17	.23	.19	.16	.22	.18	.16	.15	.110
7	V	1.3	0	.52	.52	.52	.50	.50	.50	.46	.46	.46	.43	.43	.43	.39	.39	.39	.38	—
			1	.45	.43	.41	.43	.41	.39	.40	.38	.37	.36	.35	.34	.34	.33	.32	.30	.183
	8% ↑		2	.39	.35	.33	.37	.34	.32	.34	.32	.30	.32	.30	.28	.29	.28	.26	.25	.160
			3	.34	.30	.27	.33	.29	.26	.30	.27	.25	.28	.26	.24	.26	.24	.22	.21	.140
			4	.30	.26	.23	.29	.25	.22	.27	.24	.21	.25	.22	.20	.23	.21	.19	.18	.125
	$37\frac{1}{2}$% ↓		5	.26	.22	.19	.25	.21	.19	.23	.20	.18	.22	.19	.17	.20	.18	.16	.15	.114
			6	.23	.19	.16	.23	.19	.16	.21	.18	.15	.19	.17	.14	.18	.16	.14	.13	.104
			7	.21	.17	.14	.20	.16	.14	.19	.16	.13	.18	.15	.13	.16	.14	.12	.11	.096
			8	.19	.15	.12	.18	.14	.12	.17	.14	.11	.16	.13	.11	.15	.12	.10	.09	.088
2 lamp diffuse wraparound—see note 7			9	.17	.13	.10	.16	.13	.10	.15	.12	.10	.14	.11	.09	.13	.11	.09	.08	.082
			10	.15	.12	.09	.15	.11	.09	.14	.11	.09	.13	.10	.08	.12	.10	.08	.07	.077
8	IV	0.9	0	.55	.55	.55	.54	.54	.54	.51	.51	.51	.49	.49	.49	.47	.47	.47	.46	—
	0% ↑		1	.50	.48	.47	.49	.47	.46	.47	.46	.45	.45	.44	.43	.43	.43	.42	.41	.122
			2	.45	.43	.41	.44	.42	.40	.43	.41	.39	.41	.40	.38	.40	.39	.37	.37	.118
			3	.41	.38	.36	.40	.38	.35	.39	.37	.35	.38	.36	.34	.37	.35	.34	.33	.111
			4	.37	.34	.32	.37	.34	.31	.36	.33	.31	.35	.32	.31	.34	.32	.30	.29	.105
			5	.34	.30	.28	.33	.30	.28	.32	.30	.27	.32	.29	.27	.31	.29	.27	.26	.100
	46% ↓		6	.31	.28	.25	.31	.27	.25	.30	.27	.25	.29	.27	.25	.29	.26	.24	.24	.094
			7	.29	.25	.23	.28	.25	.23	.28	.25	.22	.27	.24	.22	.26	.24	.22	.21	.088
			8	.26	.23	.20	.26	.23	.20	.25	.22	.20	.25	.22	.20	.24	.22	.20	.19	.084
4 lamp, 610 mm (2') wide troffer with 45° white metal louver			9	.24	.20	.18	.24	.20	.18	.23	.20	.18	.23	.20	.18	.22	.20	.18	.17	.081
			10	.22	.19	.16	.22	.19	.16	.21	.18	.16	.21	.18	.16	.20	.18	.16	.15	.077
9	V	N.A.	0	.87	.87	.87	.84	.84	.84	.77	.77	.77	.72	.72	.72	.66	.66	.66	.64	—
			1	.76	.73	.70	.73	.70	.67	.67	.65	.63	.63	.61	.59	.58	.57	.55	.53	.272
	12% ↑		2	.66	.61	.57	.64	.59	.56	.59	.56	.52	.55	.52	.49	.51	.49	.47	.44	.241
			3	.59	.53	.48	.56	.51	.47	.53	.48	.44	.49	.45	.42	.46	.43	.40	.38	.216
			4	.52	.45	.40	.50	.44	.40	.47	.42	.38	.44	.39	.36	.41	.37	.34	.32	.196
	$63\frac{1}{2}$% ↓		5	.46	.39	.34	.44	.38	.33	.41	.36	.32	.38	.34	.31	.36	.32	.29	.27	.182
			6	.41	.34	.29	.39	.33	.29	.37	.31	.27	.34	.30	.26	.32	.28	.25	.23	.168
			7	.36	.30	.25	.35	.29	.24	.33	.27	.23	.31	.26	.23	.29	.25	.22	.20	.157
Bilateral batwing distribution—one lamp, surface mounted fluorescent with prismatic wraparound lens			8	.32	.26	.21	.31	.25	.21	.29	.24	.20	.27	.23	.19	.26	.21	.18	.17	.147
			9	.29	.22	.18	.28	.22	.18	.26	.21	.17	.24	.20	.16	.23	.19	.15	.14	.138
			10	.26	.20	.16	.25	.19	.15	.23	.18	.15	.22	.17	.14	.20	.16	.13	.12	.129
10	V	1.4/1.2	0	.75	.75	.75	.73	.73	.73	.70	.70	.70	.67	.67	.67	.64	.64	.64	.63	—
			1	.67	.65	.63	.66	.64	.62	.63	.62	.60	.61	.60	.58	.59	.58	.57	.55	.189
	0 ↑		2	.60	.57	.54	.59	.56	.53	.57	.54	.52	.55	.53	.51	.53	.51	.50	.49	.180
			3	.54	.50	.47	.53	.49	.46	.52	.48	.45	.50	.47	.45	.48	.46	.44	.43	.169
			4	.49	.44	.40	.48	.44	.40	.47	.43	.40	.45	.42	.39	.44	.41	.39	.37	.160
	63% ↓		5	.44	.39	.35	.43	.38	.35	.42	.38	.34	.41	.37	.34	.40	.36	.34	.33	.152
			6	.40	.34	.31	.39	.34	.31	.38	.34	.30	.37	.33	.30	.36	.32	.30	.29	.143
			7	.36	.30	.27	.35	.30	.27	.34	.30	.27	.33	.29	.26	.32	.29	.26	.25	.135
Fluorescent unit with flat prismatic lens, 4 lamp 610 mm (2') wide			8	.32	.27	.23	.32	.27	.23	.31	.26	.23	.30	.26	.23	.29	.26	.23	.22	.129
			9	.29	.24	.20	.28	.23	.20	.28	.23	.20	.27	.23	.20	.26	.23	.20	.19	.123
			10	.26	.21	.18	.26	.21	.18	.25	.21	.18	.24	.20	.18	.24	.20	.18	.16	.116

The ρ_w interpolation is cutomarily done first, yielding

ρ_{cc}→	80	70
RCR		
1	0.655	0.600
2	0.570	0.530

Each of these numbers is simply halfway between the coefficients for $\rho_w = 30\%$ and 50% in the first chart. Now, from the second chart, we

TABLE 7.5 MULTIPLYING FACTORS FOR FLOOR CAVITY REFLECTANCES OTHER THAN 20%
(For 30% ρ_{fc}, multiply by the factor; for 10% ρ_{fc}, divide by the factor)

ρ_{cc}	80			70			50		
ρ_w	50	30	10	50	30	10	50	30	10
RCR									
1	1.08	1.08	1.07	1.07	1.06	1.06	1.05	1.04	1.04
2	1.07	1.06	1.05	1.06	1.05	1.04	1.04	1.03	1.03
3	1.05	1.04	1.03	1.05	1.04	1.03	1.03	1.03	1.02
5	1.04	1.03	1.02	1.03	1.02	1.02	1.02	1.02	1.01
7	1.03	1.02	1.01	1.03	1.02	1.01	1.02	1.01	1.01
10	1.02	1.01	1.01	1.02	1.01	1.01	1.02	1.01	1.01

can write

$$\text{CU} = 0.530 + 0.4(0.070) + 0.3(0.055) = 0.57$$

or

$$\text{CU} = 0.530 + 0.3(0.040) + 0.4(0.085) = 0.58$$

In each case we have started with the smallest number in the second chart (0.530). In the first instance, the RCR interpolation was done first; then the ρ_{cc} interpolation. The order was reversed in the second instance. The closeness of the results indicates that the order of interpolation is relatively unimportant (the slight difference occurs because the CUs are not exactly linear functions of RCR and ρ_{cc} over the ranges taken). Also, CU values are usually rounded to two significant figures based on the accuracy of the zonal cavity procedure.

7.9 LIGHT LOSS FACTOR

It was mentioned earlier that, if we are interested in average maintained illuminance, rather than average initial illuminance, we must include a light loss factor in our calculation. Total light loss is divided conventionally into unrecoverable and recoverable losses, and the light loss factor should include as many of each of these as are quantifiable. Unrecoverable losses are those not generally subject to correction by lighting maintenance procedures. They will be discussed later in this section.

There are four recoverable losses:

1. Lamp burnouts (LBO)
2. Lamp lumen depreciation (LLD)
3. Luminaire dirt depreciation (LDD)
4. Room surface dirt depreciation (RSDD)

Lamp Burnouts

It may not be feasible to replace a lamp every time one burns out. If burned-out lamps are not replaced immediately, the average illuminance level will decrease. A decision should be made as to how many burned-out lamps will be tolerated in a given installation before they are replaced. Then the LBO factor is the ratio of the number of lamps remaining lighted to the total number of lamps, when the maximum number of burnouts is reached.

Lamp Lumen Depreciation

As discussed earlier and shown in Table 7.1, the lumen output of a lamp decreases as the lamp ages. The LLD factor is generally taken at the 70% life point for fluorescent and high-intensity discharge lamps, using the lamp lumens at 100 h as the base. For incandescent lamps, the lumens at 100% life are divided by those at 0 h to obtain the LLD.

Recommended illuminance levels, as determined in Sec. 5.11, are to be considered as maintained values. Thus they should be thought of as the minimum illuminance on the task at any time. Choosing the 70% lifepoint (the point at which group relamping is usually done) for LLD insures that the illuminance level will not fall below the recommended value at any time.

Luminaire Dirt Depreciation

Dirt accumulation on a luminaire will reduce the reflectance of luminaire surfaces and the transmittance of luminaire lenses and diffusers. Extensive studies have resulted in defining six luminaire maintenance categories (I through VI) and five degrees of dirtiness of the luminaire environment (very clean, clean, medium, dirty, very dirty). The luminaire maintenance category is based on such factors as whether or not the luminaire is enclosed or open, whether or not it has louvers or diffusers, and what is the percentage of upward and downward lumens. Manufacturers publish a maintenance category for each of their luminaires (see Table 7.4).

With the maintenance category established, the depreciation factor may be determined from

$$\mathrm{LDD} = e^{-At^B} \tag{7.11}$$

where t is the time in years between cleanings and A and B are constants, values of which are given in Table 7.6.

Room Surface Dirt Depreciation

Dirt on room surfaces reduces the reflected component of luminous flux and thus the illuminance on the work plane. The resulting RSDD factor is a

TABLE 7.6

Luminaire maintenance category	B	A: Very clean	Clean	Medium	Dirty	Very dirty
I	0.69	0.038	0.071	0.111	0.162	0.301
II	0.62	0.033	0.068	0.102	0.147	0.188
III	0.70	0.079	0.106	0.143	0.184	0.236
IV	0.72	0.070	0.131	0.216	0.314	0.452
V	0.53	0.078	0.128	0.190	0.249	0.321
VI	0.88	0.076	0.145	0.218	0.284	0.396

function of the dirtiness of the environment, the time between cleanings, the luminaire flux distribution, and the room proportions.

Studies have shown that room surface dirt depreciation is essentially the same function of room cleanliness and time as the luminaire dirt depreciation for maintenance category V. Thus the percent depreciation can be obtained by using the constants for category V from Table 7.6 in the following modified form of Eq. (7.11):

$$\% \text{ Dirt depreciation} = 100(1 - e^{-At^{0.53}}) \tag{7.12}$$

Next one must determine the type of luminaire flux distribution from Table 7.2 and the manufacturer's intensity distribution data. For example, luminaire 4 in Table 7.4 delivers 65% of the lamp lumens downward and 22.5% upward (12.5% are absorbed). Thus, of the total luminaire output, 74% is down and 26% is up. From Table 7.2, the luminaire is classified as semidirect.

With this information in hand, one can enter Table 7.7 and obtain the RSDD. Table 7.7 relates percent dirt depreciation to luminaire flux distribution and room proportions.

EXAMPLE 7.5

A lighting installation in a large office area 160 × 50 × 10 ft consists of 200 no. 6 two-lamp luminaires containing 40-W standard cool white rapid-start fluorescent lamps. Maintenance policy is to clean luminaires and room surfaces annually and to replace burned-out lamps when 12 failures have occurred. Find the light loss factor for this installation, due to recoverable factors.

Solution.

$$\text{LBO} = \frac{388}{400} = 0.97 \qquad \text{LLD} = \frac{2650}{3150} = 0.84$$

From Table 7.4, the luminaire maintenance category is V. We will assume the office area is clean. Then, from Table 7.6, $A = 0.128$

TABLE 7.7 LUMINAIRE DISTRIBUTION TYPE

Percent expected dirt depreciation	Direct				Semidirect				Direct-indirect				Semiindirect				Indirect			
	10	20	30	40	10	20	30	40	10	20	30	40	10	20	30	40	10	20	30	40
Room cavity ratio																				
1	98	96	94	92	97	92	89	84	94	87	80	76	94	87	80	73	90	80	70	60
2	98	96	94	92	96	92	88	83	94	87	80	75	94	87	79	72	90	80	69	59
3	98	95	93	90	96	91	87	82	94	86	79	74	94	86	78	71	90	79	68	58
5	97	94	91	89	94	90	84	79	93	86	78	72	93	86	77	69	89	78	66	55
7	97	94	90	87	93	88	82	77	93	84	77	70	93	84	76	68	89	76	65	53
10	96	92	87	83	93	86	79	72	93	84	75	67	92	83	75	67	88	75	62	50

and $B = 0.53$. These give

$$\text{LDD} = e^{-0.128(1)^{0.53}} = 0.88$$

Luminaire 6 delivers 84% of its output lumens downard, 16% upward. From Table 7.2, it is classified as semidirect. The percent expected dirt depreciation, from Eq. (7.12), is

$$\%\ \text{Dirt depreciation} = 100(1 - e^{-0.128(1)^{0.53}}) = 12\%$$

Assuming a 30-in work plane, the room cavity ratio is, from Eq. (7.7),

$$\text{RCR} = \frac{5(7.5)(210)}{160(50)} = 0.98$$

Then Table 7.7 yields RSDD = 0.96. Finally the light loss factor is

$$\text{LLF} = 0.97(0.84)(0.88)(0.96) = 0.69$$

As noted earlier, the light loss factor should include unrecoverable, as well as recoverable, losses. Some of the unrecoverable causes of lowered light output are:

Luminaire Ambient Temperature (LAT) Fluorescent lamp and luminaire ratings are established in still air at 25°C. Furthermore, fluorescent luminaire photometry is relative in that the luminaire output is adjusted for a 25°C lamp operating temperature. In an actual installation, the luminaire operating temperature is likely to differ from the test condition. Operation above or below the test value decreases lumen output by about 1½% per degree Celsius. For example, if the plenum temperature for a fluorescent luminaire is 30°C, the lamp output would decrease by about 7½%, giving a temperature correction factor of 0.925. There is no significant variation with ambient temperature for incandescent and HID lamps.

Voltage at the Luminaire (VL) Deviations in supply voltage from rated value can affect the light output of all types of lamps. A 1% change in voltage causes a 3% change in lumen output for incandescent lamps and for mercury lamps with reactor ballasts, and a 0.4% change for fluorescent lamps.

Luminaire Surface Depreciation (LSD) Effects such as yellowing of plastic with age and pitting and discoloration of enamel over time can cause a reduction in light reflected within the luminaire and transmitted through the diffuser. No numbers are available for this factor.

Ballast Factor (B) Ballast factor is the ratio of the lamp lumens when the lamp is operated on its ballast to the rated lamp lumens determined on a standard lamp and ballast testing circuit. Commercial ballasts do not generally perform as efficiently as test ballasts. For fluorescent lamps, a ballast factor of .95 should be used for standard lamps and .89 for energy-saving lamps, assuming CBM ballasts are used. For some non-CBM and small inductor fluorescent ballasts, lower factors of .50 to .75 can apply. Ballast factors are not available for HID systems.

7.10 SPACING OF LUMINAIRES

One additional piece of information is given in Table 7.4 which we have not yet considered. This is the spacing criterion (SC) for each luminaire. SC is a ratio of the maximum spacing between luminaires to the mounting height of the luminaires above the work plane which is likely to give reasonable uniformity of horizontal illuminance throughout the space. SC is simply a guide to the lighting designer and it should not be interpreted as a specification of the actual spacing-to-mounting-height ratio to be used in a given installation.

The SC numbers in Table 7.4 were determined using the luminaire layouts in Fig. 7.8. In Fig. 7.8*a*, two identical luminaires with symmetrical intensity distributions, separated by a distance S_1, are shown. It is assumed that the illuminances on the work plane at point P directly under one luminaire and at point Q midway between the two luminaires are produced by just the direct components of luminous flux of just those luminaires. Thus, the direct components of other luminaires in the space are neglected and the reflected components from room surfaces produced by all luminaires are also neglected. The distance S_1 is set so that the illuminances at P, due to source 1, and at Q, due to both sources, are equal. The experiment is then repeated for the configuration in Fig. 7.8*b*, with the same assumptions. The

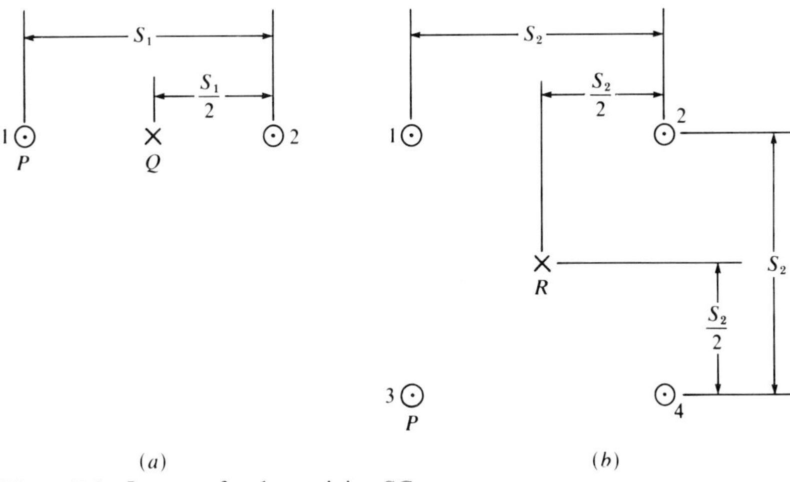

Figure 7.8 Layouts for determining SC.

distance S_2 is set so that the illuminances at P, due to source 1, and at R, due to all four sources, are equal. The spacing criterion is then obtained from

$$\mathrm{SC} = \frac{S}{\mathrm{MH}} \tag{7.13}$$

where MH is the mounting height above the work plane and S is the smallest of the two values, S_1 and S_2.

If the luminaire has an asymmetric intensity distribution, as is often the case with fluorescent units, only the test in Fig. 7.8*a* is performed, but it is done twice, once with the units perpendicular to the line between the two units and once with them parallel. Two values of SC are then listed in Table 7.4, the first denoting the maximum spacing between rows and the second giving the maximum spacing between units in a row.

Certain luminaires, such as those with "batwing" intensity distributions, may not fit the spacing criterion. These units may produce reasonably uniform illuminance only within a range of spacings, rather than for spacings less than some maximum value. In such cases, the manufacturer's particular spacing guideline should be used. Furthermore, the spacing criterion is not applicable for large spacings (SC $>$ 1.5). Here again, a specific manufacturer's spacing criterion is required.

7.11 AN ELEMENTARY LIGHTING DESIGN

We now have all of the necessary ingredients to carry out an elementary lighting design, whose purpose is to provide a prescribed average maintained work-plane illuminance in an interior space. Let us consider a 45 × 20 × 10 ft school room, with reflectances $\rho_c = 80\%$, $\rho_w = 50\%$, and $\rho_f = 30\%$. It is desired to light this area with luminaire 6 housing two 40-W standard cool-white rapid-start fluorescent lamps. Assume an average of 5 fc of daylighting is present. Our task is to determine the number of luminaires required and to propose a lighting layout.

The first step is to determine the required illuminance level, using the procedure in Sec. 5.15. The illuminance category from Table 5.3*a* is D or E, most likely the latter. The workers are young, the demand for speed and accuracy is, at most, important, and the background reflectance is presumed high. Thus the overall weighting factor from Table 5.3b is −2 and the appropriate illuminance is $(50 - 5) = 45$ fc.

Next the cavity ratios are calculated from Eq. (7.7), assuming a 30-in work plane.

$$\mathrm{RCR} = \frac{5(7.5)(65)}{45(20)} = 2.7$$

$$\mathrm{FCR} = 2.7\left(\frac{2.5}{7.5}\right) = 0.9$$

Knowing RCR, the CU for a 20% floor reflectance can be obtained from Table 7.4, in this case with a single interpolation. The result is CU = 0.59. The floor cavity reflectance is 27% from Table 7.3. From Table 7.5, the CU multiplier is 1.04 and thus the overall CU = 0.61.

Turning next to the light loss factor, we will allow 5% burnouts before replacement. Therefore LBO = 0.95. LLD is 0.84 and LDD is 0.88, the same as in Example 7.5. The percent dirt depreciation will also be the same as in that example (12%) and entering Table 7.7 yields RSDD = 0.95. We will assume a ballast factor of .95. Thus LLF = 0.63.

To determine the number of luminaires required, Eq. (7.6) is used.

$$\Phi_i = \frac{45(900)}{0.61(0.63)} = 105{,}000 \text{ lm}$$

From Table 6.7, each lamp delivers 3150 initial lumens. Therefore

$$N = \frac{105{,}000}{2(3150)} = 16 \text{ to } 17 \text{ luminaires}$$

The spacing criterion for luminaire 6 is 1.5MH = 11 ft between rows and 1.2MH = 9 ft between units in the same row, maximum. Three possible layouts are shown in Fig. 7.9. Many factors, which we will be discussing in later chapters, go into deciding which of these layouts, if any, is "best."

The design which we have just performed can be organized into a worksheet format, as shown below. The user first enters the room dimensions and reflectances and the lamp and luminaire information. Next he inserts maintenance information and the desired maintained illumination level.

The remainder of the worksheet involves calculations. Cavity ratios and reflectances are determined and, from these, the CU values are found. Next the light loss information is calculated and entered.

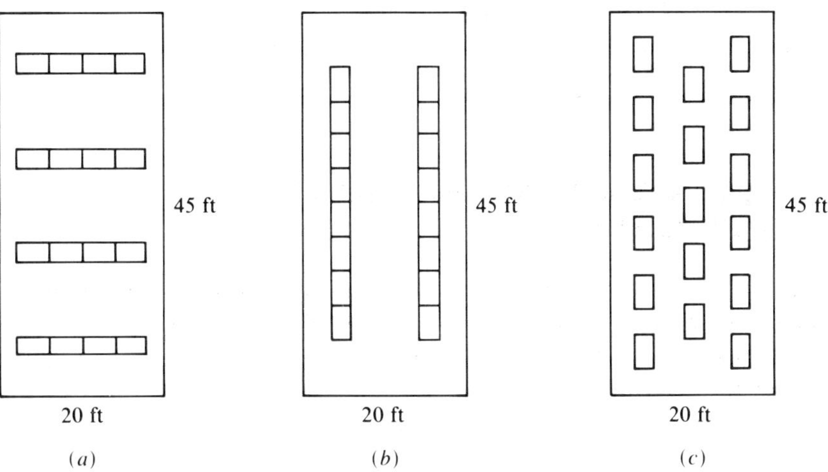

Figure 7.9 Possible school room lighting layouts.

Finally, the initial lumens and the required number of luminaires are determined, the spacing criteria are entered and the initial illumination level and watts per square foot of power consumption are calculated. This latter factor is a "benchmark" as to whether the designer has or has not created an energy-efficient lighting design. The value of 1.6 is reasonably good, although some contemporary fluorescent designs are being accomplished at under 1 W/ft². The use of energy-saving fluorescent lamps would reduce the 1.6 value obtained in the design presented here.

ZONAL CAVITY METHOD WORKSHEET

Room dimensions (ft):	l 45, w 20, h 10 WP height 2.5, WP area (A_{wp}) 900
Reflectances (%):	ρ_c 80, ρ_w 50, ρ_f 30
Lamp: F40CW	Initial lumens 3150 Lumens at 70% life 2650 Lamp watts (P_L) 40 Luminaire ballast watts (P_B) 12 (2 lamp ballast)
Luminaire: 6	Spacing/MH (SC) 1.5/1.2 Downward flux 84% Suspension distance (ft) 0 Number of lamps/luminaire (n) 2
Maintenance:	Room classification clean Luminaire maintenance category V Cleaning schedule 1/yr Interim lamp replacement 0%
Desired average maintained footcandles (E):	45 (50 total)
Calculations:	RCR 2.7, CCR 0, FCR 0.9 $\%\rho_{cc}$ 80%, $\%\rho_{fc}$ 27, CU_{20} 0.59, CU 0.61 LBO 0.95, LLD 0.84, B 0.95, VL 1.0 LDD 0.88, RSDD 0.95, LAT 1.00, LLF 0.63
Number of luminaires:	N 16–17
Maximum spacing (ft):	Between units 11, Between rows 9
Initial footcandles:	$E_i = \frac{\Phi_i(\text{CU})}{A_{wp}}$ 68 (16 units)
Maintained footcandles:	$E = E_i(\text{LLF})$ 43 (48 total)
Watts per square foot:	$W = \frac{(nP_L + P_B)N}{A_{wp}}$ 1.6 (16 units)

7.12 WALL AND CEILING LUMINOUS EXITANCE COEFFICIENTS

It is often desired to calculate the average luminous exitance of the wall and ceiling surfaces within a space. Equations similar in form to that used to find the average horizontal illuminance in a space [Eq. (7.6)] have been developed for this purpose. These are:

$$M_w = \frac{\Phi_i(\text{WEC})}{A} \tag{7.14}$$

$$M_c = \frac{\Phi_i(\text{CEC})}{A} \tag{7.15}$$

where M_w, M_c = average initial wall and ceiling cavity luminous exitances

ϕ_i = total initial lamp lumens

WEC, CEC = wall and ceiling cavity luminous exitance coefficients

A = work-plane area

If the luminaires are recessed or surface-mounted, M_c is the average luminous exitance of the ceiling plane; if they are suspended, of the luminaire plane. In either case, the luminous exitances of the luminaires are excluded from M_c.

If the maintained, rather than initial, average luminous exitance is desired, a light loss factor can be included in Eqs. (7.14) and (7.15), in the same fashion as it was for the illuminance calculation. Also, if the wall surfaces do not all have the same reflectance factor, the luminous exitance of a particular wall can be found from

$$M_i = M_w \frac{\rho_i}{\rho_w} \tag{7.16}$$

where the subscript i refers to the ith wall section. Finally, since all room surfaces are assumed to be perfectly diffusing, the luminance of a wall or ceiling surface can be found by dividing the luminous exitances in Eqs. (7.14) and (7.15) by π. If the luminous exitances are in lumens per square foot, the luminances will be in candelas per square foot.

Wall and ceiling luminous exitance coefficients for the luminaires in Table 7.4 are given in Table 7.8, based on $\rho_{fc} = 20\%$.

EXAMPLE 7.6

It is desired to find the initial average illuminances on the walls and ceiling of the room described in Sec. 7.11.

Solution. From Table 7.8, WEC = 0.167 and CEC = 0.201 for the given RCR and wall and ceiling reflectances. The floor correction factor from Table 7.5 is 1.04, which does not alter these values significantly. Then Eqs. (7.14) and (7.15) yield

$$M_w = \frac{32(3150)(0.167)}{20(45)} = 19 \text{ lm/ft}^2$$

$$M_c = \frac{32(3150)(0.201)}{20(45)} = 22 \text{ lm/ft}^2$$

The respective illuminances are

$$E_w = \frac{19}{0.5} = 38 \text{ fc} \qquad E_c = \frac{22}{0.8} = 28 \text{ fc}$$

TABLE 7.8 LUMINOUS EXITANCE COEFFICIENTS

Wall exitance coefficients for 20% effective floor cavity reflectance ($\rho_{fc} = 20$)

ρ_{cc}	80			70			50			30			10		
ρ_w	50	30	10	50	30	10	50	30	10	50	30	10	50	30	10
RCR															
1. 0															
1	.317	.181	.057	.301	.172	.055	.270	.155	.049	.241	.139	.045	.215	.125	.040
2	.269	.147	.045	.254	.140	.043	.225	.125	.039	.199	.112	.035	.175	.099	.031
3	.236	.126	.038	.223	.119	.036	.197	.107	.032	.173	.095	.029	.150	.083	.026
4	.210	.109	.032	.198	.103	.031	.174	.092	.028	.152	.082	.025	.131	.071	.022
5	.191	.097	.028	.179	.092	.027	.158	.082	.024	.137	.072	.022	.118	.063	.019
6	.175	.088	.025	.164	.083	.024	.144	.074	.022	.125	.065	.019	.108	.057	.017
7	.162	.080	.023	.152	.076	.022	.133	.067	.019	.116	.059	.017	.099	.052	.015
8	.150	.073	.021	.141	.069	.020	.123	.062	.018	.107	.054	.016	.092	.047	.014
9	.139	.067	.019	.131	.064	.018	.115	.057	.016	.100	.050	.014	.086	.044	.013
10	.130	.063	.017	.123	.059	.017	.108	.053	.015	.094	.046	.013	.080	.040	.012
2. 0															
1	.231	.131	.042	.219	.125	.040	.199	.114	.036	.179	.104	.033	.162	.094	.030
2	.204	.112	.034	.194	.107	.033	.175	.097	.030	.158	.089	.028	.142	.080	.025
3	.185	.099	.030	.177	.095	.028	.160	.087	.026	.144	.079	.024	.130	.072	.022
4	.170	.088	.026	.161	.085	.025	.146	.078	.023	.132	.071	.021	.119	.065	.020
5	.158	.080	.023	.150	.077	.022	.136	.071	.021	.124	.065	.019	.111	.060	.018
6	.147	.074	.021	.141	.071	.020	.128	.066	.019	.116	.060	.018	.105	.055	.016
7	.139	.069	.019	.133	.066	.019	.121	.061	.018	.110	.056	.016	.100	.052	.015
8	.131	.064	.018	.125	.062	.017	.114	.057	.016	.104	.053	.015	.095	.049	.014
9	.124	.060	.017	.119	.058	.016	.109	.054	.015	.099	.050	.014	.090	.046	.013
10	.118	.056	.016	.113	.055	.015	.104	.051	.014	.095	.047	.013	.086	.043	.012

Ceiling cavity exitance coefficients for 20% floor cavity reflectance ($\rho_{fc} = 20$)

ρ_{cc}	80			70			50			30			10		
ρ_w	50	30	10	50	30	10	50	30	10	50	30	10	50	30	10
RCR															
1. 0	.423	.423	.423	.361	.361	.361	.246	.246	.246	.142	.142	.142	.045	.045	.045
1	.421	.397	.374	.361	.340	.322	.247	.234	.223	.142	.136	.130	.045	.044	.042
2	.417	.379	.348	.357	.327	.301	.245	.226	.210	.141	.131	.123	.045	.043	.040
3	.411	.367	.332	.352	.317	.288	.242	.220	.202	.140	.128	.119	.045	.042	.039
4	.405	.358	.322	.347	.309	.279	.239	.215	.197	.138	.126	.116	.044	.041	.038
5	.399	.350	.314	.343	.303	.273	.236	.212	.193	.137	.124	.115	.044	.041	.038
6	.393	.344	.309	.338	.298	.269	.233	.209	.190	.135	.123	.113	.044	.040	.037
7	.388	.339	.305	.334	.294	.266	.231	.206	.188	.134	.122	.112	.043	.040	.037
8	.383	.335	.302	.330	.291	.264	.228	.204	.187	.133	.120	.111	.043	.039	.037
9	.378	.332	.300	.326	.288	.262	.226	.202	.186	.132	.119	.111	.043	.039	.037
10	.374	.328	.298	.322	.285	.260	.224	.201	.185	.130	.119	.110	.042	.039	.037
2. 0	.290	.290	.290	.248	.248	.248	.169	.169	.169	.097	.097	.097	.031	.031	.031
1	.282	.264	.248	.241	.227	.213	165	.156	.148	.095	.090	.086	.030	.029	.028
2	.274	.246	.222	.235	.212	.192	.161	.147	.134	.093	.085	.079	.030	.028	.026
3	.267	.233	.205	.229	.201	.178	.157	.139	.125	.091	.081	.074	.029	.026	.024
4	.260	.222	.193	.223	.192	.168	.154	.134	.118	089	.078	.070	.029	.026	.023
5	.255	.214	.185	.219	.185	.161	.151	.130	.113	.087	.076	.067	.028	.025	.022
6	.249	.208	.178	.214	.180	.155	.148	.126	.110	.086	.074	.065	.028	.024	.022
7	.245	.203	.173	.211	.176	.151	.145	.123	.107	.085	.073	.064	.027	.024	.021
8	.240	.198	.170	.207	.172	.148	.143	.121	.105	.083	.071	.063	.027	.023	.021
9	.236	.195	.167	.204	.169	.145	.141	.119	.103	.082	.070	.062	.027	.023	.020
10	.233	.192	.164	.200	.166	.143	.139	.117	.102	.081	.069	.061	.026	.023	.020

3.	0																.244	.244	.244	.209	.209	.209	.143	.143	.143	.082	.082	.082	.026	.026	.026
	1	.171	.098	.031	.162	.093	.029	.145	.083	.027	.129	.075	.024	.115	.067	.022	.232	.218	.206	.198	.187	.177	.135	.129	.123	.078	.075	.071	.025	.024	.023
	2	.160	.088	.027	.152	.084	.026	.137	.076	.024	.123	.069	.022	.111	.063	.020	.221	.199	.180	.189	.171	.156	.130	.118	.109	.075	.069	.064	.024	.022	.021
	3	.154	.082	.025	.147	.079	.024	.133	.072	.022	.121	.066	.020	.109	.061	.019	.213	.184	.161	.183	.159	.140	.125	.110	.098	.072	.065	.058	.023	.021	.019
	4	.146	.076	.022	.140	.073	.022	.128	.068	.020	.117	.063	.019	.106	.058	.018	.206	.173	.148	.177	.150	.129	.122	.104	.091	.070	.061	.054	.023	.020	.018
	5	.140	.072	.021	.135	.069	.020	.124	.064	.019	.113	.060	.018	.104	.056	.017	.200	.164	.138	.172	.142	.120	.119	.099	.085	.069	.058	.050	.022	.019	.017
	6	.135	.068	.019	.130	.065	.019	.120	.061	.018	.110	.057	.017	.102	.053	.016	.196	.158	.130	.168	.137	.114	.116	.096	.080	.067	.056	.048	.022	.018	.016
	7	.130	.064	.018	.126	.063	.018	.116	.059	.017	.108	.055	.016	.100	.052	.015	.192	.152	.125	.165	.132	.109	.114	.092	.077	.066	.054	.046	.021	.018	.015
	8	.126	.061	.017	.121	.060	.017	.113	.056	.016	.105	.053	.015	.097	.050	.015	.188	.148	.120	.162	.128	.105	.112	.090	.074	.065	.053	.044	.021	.017	.015
	9	.121	.059	.016	.117	.057	.016	.109	.054	.015	.102	.051	.015	.095	.048	.014	.185	.144	.117	.159	.125	.102	.110	.088	.072	.064	.052	.043	.021	.017	.014
	10	.117	.056	.016	.113	.055	.015	.106	.052	.015	.099	.049	.014	.093	.047	.013	.182	.141	.114	.157	.122	.099	.109	.086	.071	.063	.051	.042	.021	.017	.014
4.	0																.339	.339	.339	.290	.290	.290	.198	.198	.198	.114	.114	.114	.036	.036	.036
	1	.232	.132	.042	.219	.125	.040	.194	.112	.036	.172	.099	.032	.151	.088	.028	.329	.311	.294	.281	.267	.253	.192	.183	.175	.111	.106	.102	.035	.034	.033
	2	.217	.119	.037	.206	.113	.035	.184	.102	.032	.164	.092	.029	.145	.082	.026	.320	.290	.265	.274	.250	.229	.188	.173	.160	.108	.101	.094	.035	.033	.031
	3	.202	.108	.032	.192	.103	.031	.172	.093	.028	.154	.084	.026	.137	.076	.023	.313	.275	.245	.268	.238	.213	.184	.165	.149	.106	.096	.088	.034	.031	.029
	4	.190	.099	.029	.180	.094	.028	.162	.086	.026	.146	.078	.024	.130	.071	.022	.307	.264	.232	.263	.228	.201	.181	.159	.142	.105	.093	.084	.034	.030	.028
	5	.179	.091	.026	.171	.088	.026	.154	.080	.024	.139	.073	.022	.125	.067	.020	.301	.255	.222	.259	.221	.193	.178	.154	.136	.103	.091	.081	.033	.030	.027
	6	.168	.084	.024	.160	.081	.023	.145	.074	.022	.131	.068	.020	.118	.062	.018	.296	.248	.214	.254	.215	.187	.175	.151	.132	.102	.089	.079	.033	.029	.026
	7	.159	.079	.022	.152	.076	.022	.138	.070	.020	.125	.064	.019	.112	.058	.017	.291	.243	.209	.250	.210	.182	.173	.147	.129	.100	.087	.077	.032	.028	.025
	8	.151	.074	.021	.144	.071	.020	.131	.065	.019	.119	.060	.017	.108	.055	.016	.286	.238	.205	.246	.206	.179	.170	.145	.127	.099	.085	.075	.032	.028	.025
	9	.143	.069	.019	.137	.067	.019	.125	.062	.018	.114	.057	.016	.103	.052	.015	.282	.234	.202	.243	.203	.176	.168	.143	.125	.098	.084	.074	.032	.028	.025
	10	.136	.065	.018	.130	.063	.018	.119	.058	.016	.108	.054	.015	.098	.050	.014	.278	.231	.199	.239	.200	.174	.166	.141	.123	.097	.083	.074	.031	.027	.024
5.	0																.433	.433	.433	.370	.370	.370	.253	.253	.253	.145	.145	.145	.046	.046	.046
	1	.175	.100	.032	.158	.090	.029	.127	.073	.023	.098	.057	.018	.071	.041	.013	.425	.412	.399	.364	.353	.344	.249	.243	.238	.143	.141	.138	.046	.045	.045
	2	.164	.090	.028	.148	.082	.025	.120	.067	.021	.094	.053	.016	.069	.039	.012	.419	.396	.377	.359	.341	.326	.246	.236	.227	.142	.137	.133	.045	.044	.043
	3	.152	.081	.024	.139	.074	.022	.113	.061	.019	.089	.049	.015	.066	.037	.011	.413	.385	.362	.354	.332	.314	.243	.231	.220	.140	.135	.130	.045	.044	.043
	4	.143	.074	.022	.130	.068	.020	.106	.056	.017	.084	.045	.014	.063	.034	.010	.408	.376	.352	.350	.325	.306	.241	.227	.215	.139	.133	.127	.045	.043	.042
	5	.134	.068	.020	.123	.063	.018	.101	.052	.015	.080	.042	.013	.061	.033	.010	.404	.369	.344	.347	.320	.299	.239	.223	.211	.138	.131	.125	.045	.043	.041
	6	.126	.063	.018	.115	.058	.017	.095	.049	.014	.076	.039	.012	.058	.030	.009	.399	.364	.339	.343	.315	.295	.237	.221	.209	.137	.130	.124	.044	.042	.041
	7	.119	.059	.017	.109	.054	.015	.090	.045	.013	.072	.037	.011	.055	.029	.008	.396	.360	.335	.340	.312	.292	.235	.219	.207	.137	.129	.123	.044	.042	.041
	8	.112	.055	.015	.103	.051	.014	.085	.043	.012	.069	.035	.010	.053	.027	.008	.392	.356	.331	.337	.309	.289	.233	.217	.205	.136	.128	.122	.044	.042	.040
	9	.106	.051	.014	.098	.048	.013	.081	.040	.011	.066	.033	.009	.051	.026	.008	.389	.353	.329	.335	.306	.287	.232	.215	.204	.135	.127	.122	.044	.042	.040
	10	.101	.048	.013	.093	.045	.013	.077	.038	.011	.063	.031	.009	.049	.025	.007	.386	.350	.327	.332	.304	.285	.231	.214	.203	.134	.127	.121	.044	.042	.040

TABLE 7.8 *(continued)*

ρ_{cc}	80			70			50			30			10			80			70			50			30			10		
ρ_w	50	30	10	50	30	10	50	30	10	50	30	10	50	30	10	50	30	10	50	30	10	50	30	10	50	30	10	50	30	10
RCR	Wall exitance coefficients for 20% effective floor cavity reflectance ($\rho_{fc} = 20$)															Ceiling cavity exitance coefficients for 20% floor cavity reflectance ($\rho_{fc} = 20$)														
6. 0																.221	.221	.221	.189	.189	.189	.129	.129	.129	.074	.074	.074	.024	.024	.024
1	.192	.109	.035	.183	.105	.033	.168	.096	.031	.153	.088	.028	.140	.081	.026	.213	.198	.184	.182	.170	.158	.124	.117	.110	.072	.068	.064	.023	.022	.021
2	.176	.097	.030	.169	.093	.029	.155	.086	.027	.142	.080	.025	.130	.074	.023	.206	.181	.161	.176	.156	.139	.121	.108	.097	.070	.063	.057	.022	.020	.019
3	.163	.087	.026	.157	.084	.025	.144	.078	.024	.133	.073	.022	.122	.067	.021	.199	.169	.145	.171	.146	.126	.117	.101	.088	.068	.059	.052	.022	.019	.017
4	.153	.079	.023	.147	.077	.023	.135	.072	.021	.125	.067	.020	.115	.063	.019	.194	.160	.134	.167	.138	.116	.115	.096	.082	.066	.056	.048	.021	.018	.016
5	.144	.074	.021	.139	.071	.021	.129	.067	.020	.119	.063	.019	.110	.059	.018	.190	.153	.125	.163	.132	.109	.112	.092	.077	.065	.054	.046	.021	.018	.015
6	.136	.068	.019	.131	.066	.019	.121	.062	.018	.112	.058	.017	.104	.055	.016	.185	.147	.120	.159	.127	.104	.110	.089	.074	.064	.052	.044	.021	.017	.014
7	.128	.064	.018	.124	.062	.018	.115	.058	.017	.107	.055	.016	.100	.052	.015	.181	.142	.115	.156	.123	.100	.108	.086	.071	.063	.051	.042	.020	.017	.014
8	.122	.060	.017	.118	.058	.016	.110	.055	.016	.103	.052	.015	.096	.049	.014	.178	.138	.112	.153	.120	.097	.106	.084	.069	.062	.050	.041	.020	.016	.014
9	.117	.056	.016	.113	.055	.015	.105	.052	.015	.098	.049	.014	.092	.047	.013	.174	.135	.109	.150	.117	.095	.104	.083	.067	.061	.049	.040	.020	.016	.013
10	.111	.053	.015	.107	.052	.014	.100	.049	.014	.094	.047	.013	.088	.044	.013	.171	.133	.107	.147	.115	.093	.102	.081	.066	.060	.048	.040	.019	.016	.013
7. 0																.147	.147	.147	.125	.125	.125	.085	.085	.085	.049	.049	.049	.016	.016	.016
1	.152	.087	.027	.146	.084	.027	.136	.078	.025	.126	.073	.023	.117	.068	.022	.143	.131	.121	.123	.113	.104	.084	.078	.072	.048	.045	.042	.015	.014	.013
2	.138	.075	.023	.133	.073	.023	.123	.068	.021	.114	.064	.020	.106	.060	.019	.140	.121	.105	.120	.104	.091	.082	.072	.063	.047	.042	.037	.015	.014	.012
3	.124	.066	.020	.120	.064	.019	.111	.060	.018	.103	.057	.017	.096	.053	.016	.136	.113	.095	.117	.098	.082	.080	.068	.058	.046	.040	.034	.015	.013	.011
4	.114	.059	.017	.110	.057	.017	.102	.054	.016	.095	.051	.015	.088	.048	.015	.133	.107	.088	.114	.093	.076	.078	.065	.054	.045	.038	.032	.015	.012	.010
5	.106	.054	.016	.102	.052	.015	.095	.050	.015	.088	.047	.014	.082	.044	.013	.130	.103	.083	.112	.089	.072	.077	.062	.051	.045	.036	.030	.014	.012	.010
6	.098	.049	.014	.095	.048	.014	.088	.045	.013	.082	.043	.013	.077	.040	.012	.127	.099	.079	.109	.086	.069	.075	.060	.049	.044	.035	.029	.014	.012	.010
7	.091	.045	.013	.088	.044	.013	.082	.042	.012	.077	.039	.011	.072	.037	.011	.124	.096	.077	.107	.083	.067	.074	.058	.047	.043	.034	.038	.014	.011	.009
8	.086	.042	.012	.083	.041	.012	.077	.039	.011	.072	.037	.011	.067	.035	.010	.121	.094	.075	.104	.081	.065	.072	.057	.046	.042	.034	.028	.014	.011	.009
9	.081	.039	.011	.078	.038	.011	.073	.036	.010	.068	.034	.010	.064	.032	.009	.119	.092	.073	.102	.080	.064	.071	.056	.045	.041	.033	.027	.013	.011	.009
10	.076	.036	.010	.073	.035	.010	.069	.034	.009	.064	.032	.009	.060	.030	.009	.116	.090	.072	.100	.078	.063	.069	.055	.045	.041	.032	.027	.013	.011	.009

8.	0																.088	.088	.088	.075	.075	.075	.051	.051	.051	.029	.029	.029	.009	.009	.009
	1	.107	.061	.019	.104	.059	.019	.098	.056	.018	.093	.053	.017	.088	.051	.016	.081	.072	.065	.069	.062	.056	.047	.043	.039	.027	.025	.022	.009	.008	.007
	2	.103	.056	.017	.100	.055	.017	.095	.053	.016	.090	.051	.016	.086	.049	.015	.075	.061	.049	.064	.052	.042	.044	.036	.030	.025	.021	.017	.008	.007	.006
	3	.097	.051	.015	.094	.050	.015	.090	.049	.015	.086	.047	.014	.082	.045	.014	.070	.052	.038	.060	.045	.033	.041	.031	.023	.024	.018	.014	.008	.006	.004
	4	.091	.048	.014	.089	.047	.014	.085	.045	.014	.082	.044	.013	.079	.043	.013	.067	.046	.030	.057	.040	.026	.039	.028	.019	.023	.016	.011	.007	.005	.004
	5	.087	.044	.013	.085	.044	.013	.082	.043	.013	.079	.042	.012	.076	.040	.012	.063	.041	.025	.054	.035	.021	.037	.025	.015	.022	.015	.009	.007	.005	.003
	6	.082	.041	.012	.081	.041	.012	.078	.040	.012	.075	.039	.011	.072	.038	.011	.060	.037	.020	.052	.032	.018	.036	.022	.013	.021	.013	.007	.007	.004	.002
	7	.078	.039	.011	.077	.038	.011	.074	.037	.011	.071	.037	.011	.069	.036	.011	.057	.034	.017	.049	.029	.015	.034	.020	.011	.020	.012	.006	.006	.004	.002
	8	.075	.036	.010	.073	.036	.010	.071	.035	.010	.068	.035	.010	.066	.034	.010	.055	.031	.015	.047	.027	.013	.033	.019	.009	.019	.011	.005	.006	.004	.002
	9	.071	.035	.010	.070	.034	.010	.068	.034	.010	.066	.033	.009	.064	.032	.009	.053	.029	.013	.045	.025	.011	.031	.018	.008	.018	.010	.005	.006	.003	.002
	10	.068	.033	.009	.067	.032	.009	.065	.032	.009	.063	.031	.009	.061	.031	.009	.051	.027	.011	.044	.023	.010	.030	.016	.007	.018	.010	.004	.006	.003	.001
9.	0																.236	.236	.236	.201	.201	.201	.138	.138	.138	.079	.079	.079	.025	.025	.025
	1	.234	.133	.042	.225	.128	.041	.208	.119	.038	.192	.111	.036	.178	.103	.033	.229	.210	.194	.196	.181	.167	.134	.124	.115	.077	.072	.067	.025	.023	.022
	2	.213	.117	.036	.205	.113	.035	.190	.106	.033	.176	.099	.031	.163	.092	.029	.222	.193	.168	.190	.166	.145	.130	.115	.101	.075	.067	.059	.024	.022	.019
	3	.195	.104	.031	.188	.101	.030	.175	.095	.029	.162	.089	.027	.150	.083	.026	.216	.180	.151	.185	.155	.131	.127	.108	.092	.073	.063	.054	.024	.020	.018
	4	.181	.094	.028	.175	.091	.027	.162	.086	.026	.151	.081	.024	.140	.076	.023	.211	.170	.139	.181	.147	.121	.124	.102	.085	.072	.060	.050	.023	.020	.017
	5	.170	.087	.025	.164	.084	.025	.153	.080	.023	.142	.075	.022	.132	.071	.021	.206	.163	.131	.177	.141	.114	.122	.098	.080	.071	.058	.048	.023	.019	.016
	6	.159	.080	.023	.153	.078	.022	.143	.073	.021	.134	.069	.020	.125	.066	.019	.201	.157	.124	.173	.136	.108	.119	.095	.077	.069	.056	.046	.022	.018	.015
	7	.149	.074	.021	.144	.072	.020	.135	.068	.020	.126	.065	.019	.118	.061	.018	.197	.152	.120	.169	.131	.105	.117	.092	.074	.068	.054	.044	.022	.018	.015
	8	.141	.069	.019	.137	.067	.019	.128	.064	.018	.120	.061	.017	.112	.058	.017	.193	.148	.117	.166	.128	.102	.115	.090	.072	.067	.053	.043	.022	.017	.014
	9	.134	.065	.018	.129	.063	.018	.121	.060	.017	.114	.057	.016	.107	.054	.016	.189	.144	.114	.163	.125	.099	.113	.088	.071	.066	.052	.042	.021	.017	.014
	10	.126	.061	.017	.122	.059	.016	.115	.056	.016	.108	.054	.015	.101	.051	.015	.185	.141	.112	.159	.122	.098	.111	.086	.069	.064	.051	.041	.021	.017	.014
10.	0																.120	.120	.120	.103	.103	.103	.070	.070	.070	.040	.040	.040	.013	.013	.013
	1	.158	.090	.029	.154	.088	.028	.146	.084	.027	.138	.080	.026	.131	.076	.025	.111	.099	.087	.095	.085	.075	.065	.058	.052	.037	.034	.030	.012	.011	.010
	2	.150	.082	.025	.147	.081	.025	.139	.078	.024	.133	.075	.023	.127	.072	.023	.104	.083	.066	.089	.072	.057	.061	.050	.040	.035	.029	.023	.011	.009	.008
	3	.142	.076	.023	.139	.074	.022	.133	.072	.022	.127	.069	.021	.121	.067	.021	.098	.072	.051	.084	.062	.044	.058	.043	.031	.033	.025	.018	.011	.008	.006
	4	.135	.070	.021	.132	.069	.020	.126	.067	.020	.121	.065	.020	.117	.063	.019	.093	.063	.040	.080	.055	.035	.055	.038	.024	.032	.022	.014	.010	.007	.005
	5	.129	.066	.019	.126	.065	.019	.121	.063	.019	.117	.062	.018	.112	.060	.018	.089	.056	.032	.077	.049	.028	.053	.034	.020	.031	.020	.012	.010	.007	.004
	6	.122	.061	.017	.120	.060	.017	.115	.059	.017	.111	.058	.017	.107	.056	.017	.085	.051	.026	.073	.044	.023	.051	.031	.016	.029	.018	.010	.009	.006	.003
	7	.116	.057	.016	.114	.057	.016	.110	.056	.016	.106	.054	.016	.103	.053	.016	.082	.047	.022	.070	.041	.019	.049	.028	.014	.028	.017	.008	.009	.005	.003
	8	.111	.054	.015	.109	.054	.015	.105	.053	.015	.102	.052	.015	.099	.051	.015	.079	.043	.019	.068	.037	.016	.047	.026	.012	.027	.016	.007	.009	.005	.002
	9	.106	.051	.014	.104	.051	.014	.101	.050	.014	.098	.049	.014	.095	.048	.014	.076	.040	.016	.065	.035	.014	.045	.025	.010	.026	.015	.006	.009	.005	.002
	10	.101	.048	.013	.099	.048	.013	.096	.047	.013	.094	.047	.013	.091	.046	.013	.073	.038	.014	.063	.033	.013	.044	.023	.009	.025	.014	.005	.008	.004	.002

Source: J. E. Kaufman (ed.), *IES Lighting Handbook*, Reference volume, 1981 ed., IES, New York, 1981, fig. 9-12, with permission.

7.13 IRREGULARLY SHAPED ROOMS

We will consider first sloped ceilings and ceilings with beams present. The problem with these is to find the effective cavity reflectance at the luminaire plane from Eq. (7.8) or Table 7.3. Then the normal method for finding the coefficient of utilization can be applied. We will illustrate the procedure with two examples.

EXAMPLE 7.7

A sawtooth skylight, such as is often found in industrial and warehouse settings, is shown in Fig. 7.10. The dimensions are $a = 4$ ft, $b = 3$ ft, $w = 5$ ft, $l = 20$ ft. Surface a and the two ends have reflectances of 70%. Surface b is glass, with a reflectance of 10%. Find the effective cavity reflectance at the horizontal plane denoted by the dashed line.

Solution. It is necessary first to obtain the average reflectance of the cavity surfaces. Using the form of Eq. (7.9) gives

$$\rho_{\text{ave}} = \frac{80(0.7) + 12(0.7) + 60(0.1)}{80 + 12 + 60} = 0.463$$

Then Eq. (7.8) yields

$$\rho_{\text{cc}} = \frac{1}{1 + \frac{152}{100}\,[(1 - 0.463)/0.463]} = 0.36$$

EXAMPLE 7.8

A room with ceiling beams and suspended luminaires is shown in Fig. 7.11. The ceiling has a reflectance of 80%. The walls, and the sides and bottoms of the beams, have reflectances of 30%. Find ρ_{cc} at the luminaire plane.

Solution. The cavity ratio of the small cavities formed by the beams is

$$\text{CR} = \frac{5(2.5)(38)}{240} = 2.0$$

yielding $\rho_{\text{eff}} = 0.48$ from Table 7.3.

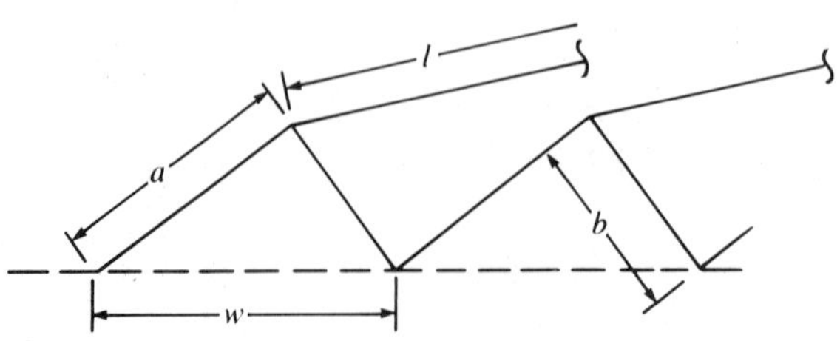

Figure 7.10

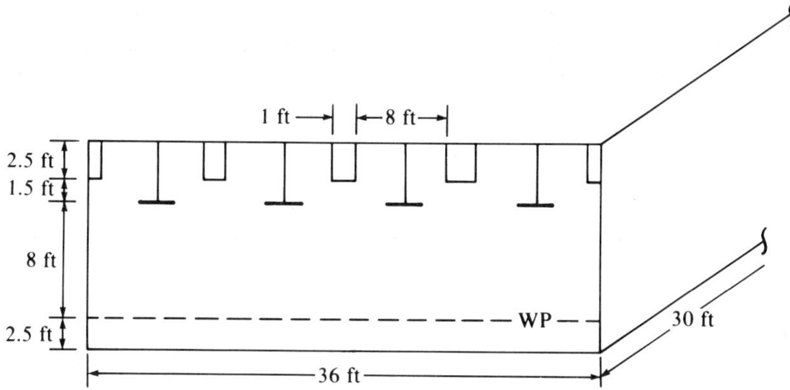

Figure 7.11

We consider next the shallow cavity between the bottom of the beams and the luminaires. The average reflectance of its "ceiling" is

$$\rho_{\text{ave}} = \frac{1(0.3) + 8(0.48)}{9} = 0.46$$

Its cavity ratio is

$$\text{CR} = \frac{5(1.5)(66)}{1080} = 0.46$$

From Table 7.3, after interpolation,

$$\rho_{cc} = 0.41$$

It is also possible to solve the problem using Eq. (7.8). The average reflectance of the small cavities is

$$\rho_{\text{ave}} = \frac{240(0.8) + 190(0.3)}{240 + 190} = 0.58$$

Then Eq. (7.8) yields

$$\rho_{\text{eff}} = \frac{1}{1 + \frac{430}{240}\left[(1 - 0.58)/0.58\right]} = 0.43$$

For the shallow cavity

$$\rho_{\text{ave}} = \frac{0.43(960) + 0.30(120) + 0.30(198)}{960 + 120 + 198} = 0.40$$

$$\rho_{cc} = \frac{1}{1 + \frac{1278}{1080}\left[(1 - 0.40)/0.40\right]} = 0.36$$

The answer using Eq. (7.8) is conservative, largely because the equation weights the 30% walls a bit too heavily.

We will next consider nonrectangular floor areas. In such cases, Eq. (7.7) is not directly applicable. But, if we recall that the basis for CR is the ratio of vertical to horizontal room areas, we can modify Eq. (7.7) to read

$$\mathrm{CR} = \frac{2.5A_1}{A_2} = \frac{2.5h\,P_1}{A_2} \tag{7.17}$$

where A_1 is the wall (vertical) area, A_2 is the work-plane (horizontal) area, which is also the floor area, P_1 is the work-plane perimeter, and h is cavity height.

EXAMPLE 7.9

The floor plan of a small auditorium is shown in Fig. 7.12. Assuming a room cavity height of 15 ft, find the room cavity ratio.

Solution. From Fig. 7.12, $A_1 = 2780\ \mathrm{ft}^2$ and $A_2 = 2187\ \mathrm{ft}^2$. Then from Eq. (7.17), RCR = 3.2.

As was mentioned earlier, the zonal cavity method was developed for rectangular parallelepipeds. Thus the developments in this section are approximations and should be treated as such. They do, however, yield reasonably accurate results for most situations.

7.14 FLUX TRANSFER AND CONFIGURATION AND FORM FACTORS

In this and the remaining sections of the chapter we will develop the analytical basis for the determination of coefficients of utilization and wall and ceiling luminous exitance coefficients.

The starting point for this discussion is the basic definition of luminance in Eq. (2.10), repeated here for ease of reference.

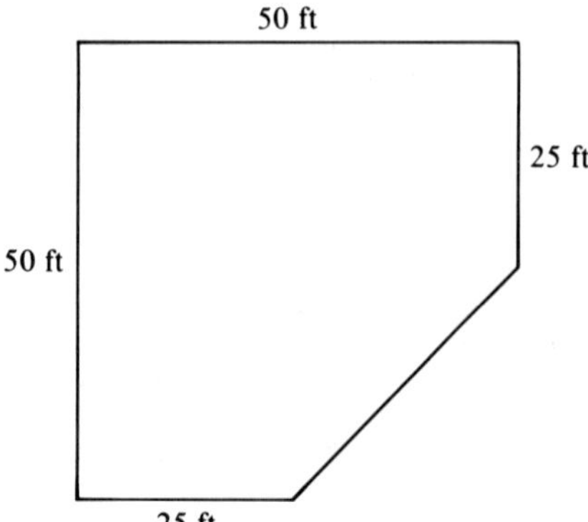

Figure 7.12

$$L = \frac{d^2\Phi}{d\omega\ dA \cos\theta} \tag{7.18}$$

This equation describes the luminance in a given direction as the luminous flux per unit of solid angle and per unit of projected area in that direction.

Consider luminous flux radiating from differential elements of two surfaces, as shown in Fig. 7.13. The cone defined by $d\omega_1$ interesects surface 2 in dA_2 and the cone defined by $d\omega_2$ intersects surface 1 in dA_1.

From Eq. (7.18) we can write

$$L_1 = \frac{d^2\ \Phi_{12}}{d\omega_1\ dA_1 \cos\theta_1} \tag{7.19}$$

$$L_2 = \frac{d^2\ \Phi_{21}}{d\omega_2\ dA_2 \cos\theta_2} \tag{7.20}$$

where Φ_{12} refers to the flux from dA_1 to dA_2 and Φ_{21} to the flux from dA_2 to dA_1. L_1 is the luminance of dA_1 in the direction of dA_2 and L_2 is the luminance of dA_2 in the direction of dA_1.

Each of these equations may be revised to yield

$$d^2\ \Phi_{12} = L_1 \cos\theta_1\ dA_1\left(\frac{dA_2 \cos\theta_2}{D^2}\right) \tag{7.21}$$

$$d^2\ \Phi_{21} = L_2 \cos\theta_2\ dA_2\left(\frac{dA_1 \cos\theta_1}{D^2}\right) \tag{7.22}$$

Consider now that surface 1 is a finite surface of area A_1, rather than a differential surface of area dA_1. From Eqs. (7.21) and (7.22)

$$d\Phi_{12} = L_1 \int_{A_1} \frac{\cos\theta_1 \cos\theta_2\ dA_2}{D^2}\ dA_1 \tag{7.23}$$

$$d\Phi_{21} = L_2 \int_{A_1} \frac{\cos\theta_1 \cos\theta_2\ dA_2}{D^2}\ dA_1 \tag{7.24}$$

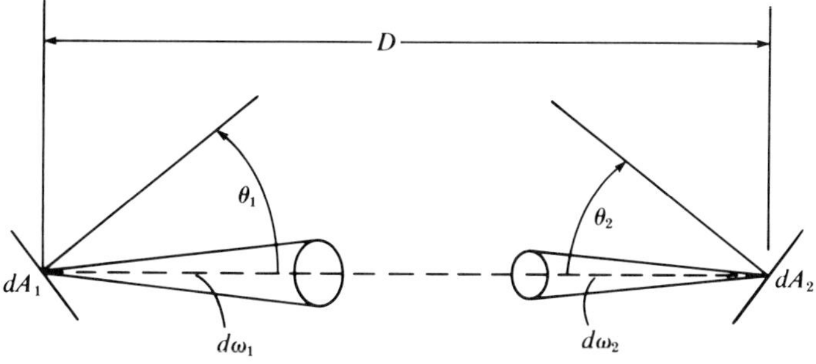

Figure 7.13 Flux transfer.

where we assume that the luminances L_1 and L_2 are constant (the surfaces A_1 and dA_2 are Lambertian). From Eqs. (7.23) and (7.24), we obtain

$$\frac{d\Phi_{12}}{d\Phi_{21}} = \frac{L_1}{L_2} = \frac{M_1}{M_2} = \frac{M_1}{d\Phi_2/dA_2} \tag{7.25}$$

where M_1 and M_2 are the luminous exitances of A_1 and dA_2, respectively, and $d\Phi_2$ is the total flux emitted by dA_2. Rearranging Eq. (7.25) yields

$$\frac{d\Phi_{21}}{d\Phi_2} = \frac{d\Phi_{12}}{M_1 dA_2} = \frac{E_2}{M_1} \equiv c_{21} \tag{7.26}$$

Equation (7.26) is the definition of configuration factor. The configuration factor c_{21} is the fraction of the flux emitted by dA_2 that is received directly by A_1. It is obtained as the ratio of the illuminance at dA_2 produced by the flux received directly from A_1 due to the luminous exitance of A_1. This is often called the *wrong-way law*, in the sense that the configuration factor from 2 to 1 is defined in terms of the illuminance at a point on 2 produced by the luminous exitance of an area at 1, but it is the direction of flux transfer that determines the subscripts.

Let us now repeat the derivation assuming both 1 and 2 are finite areas. We have

$$\Phi_{12} = L_1 \int_1 \int_2 \frac{\cos \theta_1 \cos \theta_2}{D^2} \, dA_1 \, dA_2 \tag{7.27}$$

$$\Phi_{21} = L_2 \int_1 \int_2 \frac{\cos \theta_1 \cos \theta_2}{D^2} \, dA_1 \, dA_2 \tag{7.28}$$

These yield

$$\frac{\Phi_{12}}{\Phi_{21}} = \frac{L_1}{L_2} = \frac{M_1}{M_2} = \frac{M_1}{\Phi_2/A_2} \tag{7.29}$$

from which we obtain

$$\frac{\Phi_{21}}{\Phi_2} = \frac{\Phi_{12}}{M_1 A_2} = \frac{\bar{E}_2}{M_1} \equiv f_{21} \tag{7.30}$$

Equation (7.30) defines form factor. The form factor f_{21} is the fraction of the flux emitted by area A_2 that is received directly by area A_1. It is obtained as the ratio of the average illuminance on A_2 produced by the flux received directly from A_1 due to the luminous exitance of A_1.

A reciprocity relationship can be developed for form factors by noting that the average illuminance at A_2 can also be expressed as

$$\bar{E}_2 = \frac{\Phi_{12}}{A_2} = \frac{\Phi_1 f_{12}}{A_2} = \frac{M_1 A_1 f_{12}}{A_2} \tag{7.31}$$

Then, from Eqs. (7.30) and (7.31),

$$A_2 f_{21} = A_1 f_{12} \tag{7.32}$$

Our thrust in the remainder of this chapter will be to deal with form factors, rather than configuration factors. This is because our interest is in the average illuminance on the work plane, not the illuminance at some point on the work plane. The latter topic will be explored in the next chapter and will require the use of configuration factors.

The derivation of form factors for most geometries is straightforward theoretically but often leads to multiple integrals with difficult boundary conditions. We will derive the form factor for the relatively simple case of parallel circular discs and give the form factors for other geometries without proof.

EXAMPLE 7.10

Two parallel circular discs of radius a and separated by distance d are shown in Fig. 7.14a. Find the form factor from one disc to the other, assuming the discs are perfectly diffusing.

Solution. In Fig. 7.14b, a perfectly diffusing sphere of radius r is circumscribed around the two discs. We observe that all flux impinging on one of the discs also impinges on the segment of the sphere intercepted by the disc. Thus the problem is altered to finding the flux transfer between two sphere segments, rather than two circular discs.

In Sec. 2.6 the integrating sphere was discussed and it was noted that the illuminances at all points within the sphere were the same and that the contributions to the illuminance at a given point from all other points were the same. Equation (2.37) expressed this result as

$$dE = \frac{L\ dA}{4r^2} \tag{7.33}$$

Each sphere segment has an area

$$A = 2\pi r^2(1 - \cos\theta) \tag{7.34}$$

Thus the illuminance at any point on the sphere due to one sphere segment (due to one disc) is

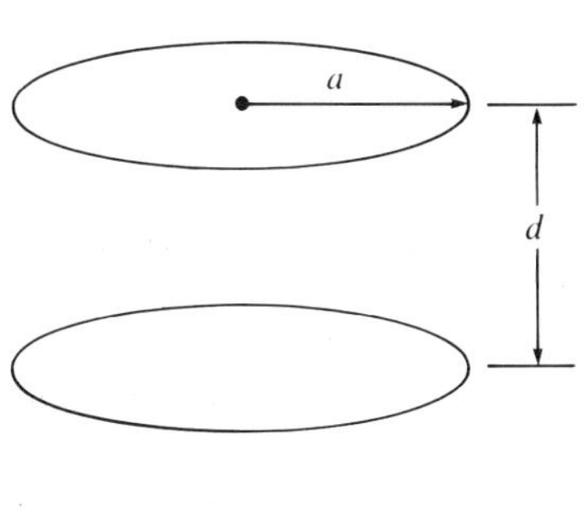

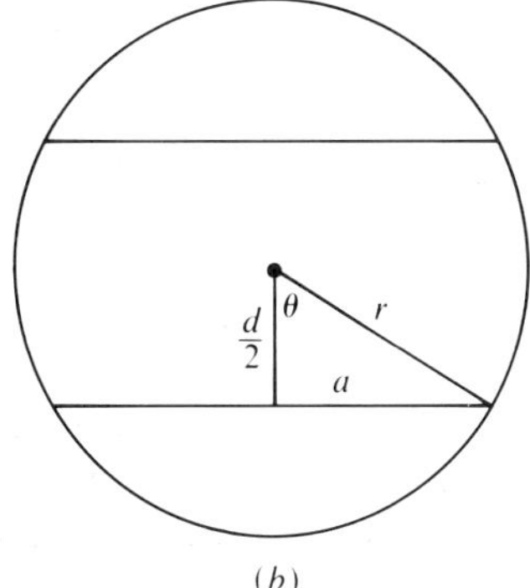

Figure 7.14 Sphere method for circular discs.

$$E = \frac{\pi L}{2}(1 - \cos\theta) \tag{7.35}$$

The flux from one sphere segment to the other is

$$\Phi = EA = \pi^2 L r^2 (1 - \cos\theta)^2 \tag{7.36}$$

The total flux from one sphere segment to the rest of the sphere is

$$\Phi_T = E(A_s - A) = \pi^2 L r^2 (1 - \cos\theta)(1 + \cos\theta) \tag{7.37}$$

where A_s is the total sphere surface area. The ratio of Φ to Φ_T is the desired form factor between the two discs.

$$f = \frac{\Phi}{\Phi_T} = \frac{1 - \cos\theta}{1 + \cos\theta} \tag{7.38}$$

It is usually desired to express this form factor in terms of $\alpha = d/a$. The angle θ is related to α through

$$\cos\theta = \frac{\alpha}{\sqrt{\alpha^2 + 4}} \tag{7.39}$$

which, when substituted into Eq. (7.38), yields

$$f = 1 + \frac{\alpha^2}{2} - \frac{\alpha}{2}\sqrt{\alpha^2 + 4} \tag{7.40}$$

7.15 FORM FACTORS FOR RECTANGULAR PARALLELEPIPEDS

Consider the two parallel rectangular surfaces shown in Fig. 7.15.

The form factor for this configuration is given without proof as

$$f = \frac{2}{\pi s t}\left[\frac{1}{2}\ln\frac{(1 + s^2)(1 + t^2)}{1 + s^2 + t^2} + t\sqrt{1 + s^2}\tan^{-1}\frac{t}{\sqrt{1 + s^2}} + s\sqrt{1 + t^2}\tan^{-1}\frac{s}{\sqrt{1 + t^2}} - t\tan^{-1} t - s\tan^{-1} s\right] \tag{7.41}$$

where

$$s = \frac{w}{h} \qquad t = \frac{l}{h} \tag{7.42}$$

Consider now the basic room cavity shown in Fig. 7.16. Its "ceiling" is the luminaire plane and its "floor" is the work plane.

Let the walls be numbered 1, the ceiling 2, and the floor 3. Then the form factor in Eq. (7.41) becomes $f_{23} = f_{32}$, because the areas of the floor and ceiling are the same and thus so are the form factors, from Eq. (7.32).

All other form factors for the space in Fig. 7.16 may be expressed in terms of f_{23}. Thus

$$f_{21} = f_{31} = 1 - f_{23} \tag{7.43}$$

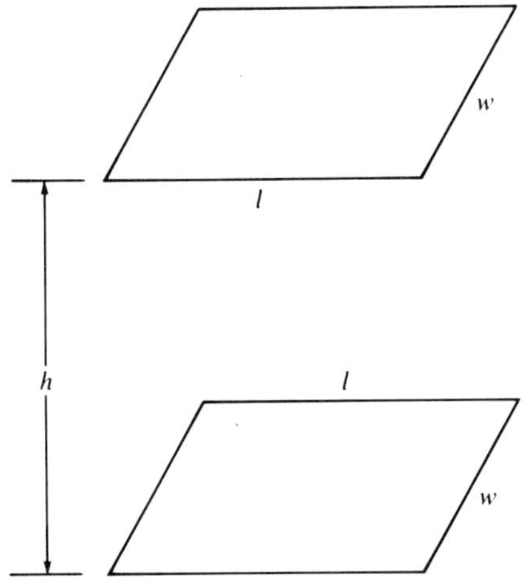

Figure 7.15 Parallel rectangles.

$$f_{12} = f_{13} = \frac{A_2}{A_1} f_{21} = \frac{A_2}{A_1} (1 - f_{23}) \tag{7.44}$$

$$f_{11} = 1 - 2f_{12} = 1 - 2 \frac{A_2}{A_1} (1 - f_{23}) \tag{7.45}$$

A word of justification for each of these equations is required. Equation (7.43) is true because the sum of the form factors for a given surface must be unity. Equation (7.44) arises because of the reciprocity relationships between the walls and the ceiling and floor. Equation (7.45) addresses the situation that the walls radiate to themselves, as well as to the floor and ceiling. Again, in that equation, we make use of the fact that the sum of the form factors for a given surface must be unity.

As shown by Eq. (7.42), the form factors for a rectangular space depend on the width:height and length:height ratios. Using Eq. (7.7), Eqs. (7.42) may be modified to

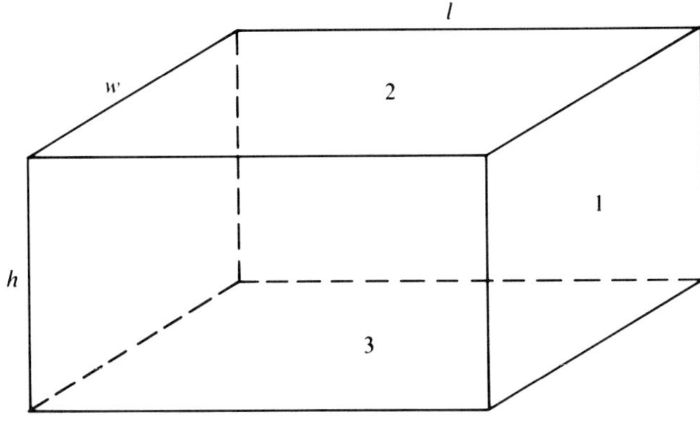

Figure 7.16 Basic room cavity.

$$s = \frac{5}{\mathrm{CR}}\left(1 + \frac{w}{l}\right) \qquad t = \frac{5}{\mathrm{CR}}\left(1 + \frac{l}{w}\right) \tag{7.46}$$

Thus f in Eq. (7.41), and all other form factors, may also be expressed in terms of cavity ratio and length:width ratio. The effect of l/w on f_{23} is small numerically. Thus the CU tables have been standardized for $l/w = 1.6$. The form factors for this length:width ratio are listed in Table 7.9. To generate the table, f_{23} is calculated from Eq. (7.41). Then Eqs. (7.43) to (7.45) are used to calculate the other form factors, after inserting A_2/A_1 from Eq. (7.17).

Equation (7.41) is complex analytically. The following approximation has been developed which is accurate within 0.4% for CRs of 20 or less:

$$f_{23} = 0.026 + 0.503\epsilon^{-0.270CR} + 0.470\epsilon^{-0.119CR} \tag{7.47}$$

7.16 CALCULATING CU, WEC, AND CEC

Consider a multisurfaced space with all surfaces perfectly diffusing. Assume we know the dimensions of the space and the reflectances of its surfaces. Consider two surfaces, m and n, within the space. Establish the following definitions:

M_{0n} = initial luminous exitance of surface n

M_n = final luminous exitance of surface n

ρ_n = reflectance of surface n

E_n = illuminance on surface n due to the other room surfaces

f_{mn} = fraction of flux emitted by surface m that falls on surface n

"Initial" means flux leaving the surface prior to reflections and "final" means the total flux ultimately leaving the surface, including reflected flux. Thus the final luminous exitance of surface n can be expressed as

TABLE 7.9

CR	f_{11}	f_{12}, f_{13}	f_{21}, f_{31}	f_{23}, f_{32}
0	0.000	0.500	0.000	1.000
1	0.133	0.434	0.173	0.827
2	0.224	0.388	0.311	0.689
3	0.298	0.351	0.421	0.579
4	0.361	0.320	0.511	0.489
5	0.415	0.292	0.585	0.415
6	0.463	0.269	0.645	0.355
7	0.504	0.248	0.694	0.306
8	0.540	0.230	0.735	0.265
9	0.573	0.214	0.769	0.231
10	0.601	0.199	0.798	0.202

$$M_n = M_{0n} + \rho_n E_n \tag{7.48}$$

Also

$$E_{n(m)} = \frac{M_m A_m f_{mn}}{A_n} = M_m f_{nm} \tag{7.49}$$

where $E_{n(m)}$ is the contribution of surface m to the illuminance E_n. Using Eqs. (7.48) and (7.49), we can write

$$M_n = M_{0n} + \rho_n \sum_m M_m F_{nm} \tag{7.50}$$

For the room cavity in Fig. 7.16, Eq. (7.50) yields the following three equations:

$$\begin{aligned} M_1 &= M_{01} + \rho_1(M_1 f_{11} + M_2 f_{12} + M_3 f_{13}) \\ M_2 &= M_{02} + \rho_2(M_1 f_{21} + M_3 f_{23}) \\ M_3 &= M_{03} + \rho_3(M_1 f_{31} + M_2 f_{32}) \end{aligned} \tag{7.51}$$

These may be rearranged and placed in matrix form to yield

$$\begin{bmatrix} 1-\rho_1 f_{11} & -\rho_1 f_{12} & -\rho_1 f_{13} \\ -\rho_2 f_{21} & 1 & -\rho_2 f_{23} \\ -\rho_3 f_{31} & -\rho_3 f_{32} & 1 \end{bmatrix} \begin{bmatrix} M_1 \\ M_2 \\ M_3 \end{bmatrix} = \begin{bmatrix} M_{01} \\ M_{02} \\ M_{03} \end{bmatrix} \tag{7.52}$$

All of the form factors in Eq. (7.52) may be expressed in terms of $f_{23} = f$ and CR, using Eqs. (7.17) and (7.43) through (7.45). This yields

$$\begin{bmatrix} 1-\rho_1\left[1-\frac{5}{\text{CR}}(1-f)\right] & -\rho_1\frac{2.5}{\text{CR}}(1-f) & -\rho_1\frac{2.5}{\text{CR}}(1-f) \\ -\rho_2(1-f) & 1 & -\rho_2 f \\ -\rho_3(1-f) & -\rho_3 f & 1 \end{bmatrix} \begin{bmatrix} M_1 \\ M_2 \\ M_3 \end{bmatrix} = \begin{bmatrix} M_{01} \\ M_{02} \\ M_{03} \end{bmatrix} \tag{7.53}$$

The inversion of the matrix in Eq. (7.53) is straightforward but algebraically complicated. The result is

$$\begin{bmatrix} M_{11} & M_{12} & M_{13} \\ M_{21} & M_{22} & M_{23} \\ M_{31} & M_{32} & M_{33} \end{bmatrix} \begin{bmatrix} M_{01} \\ M_{02} \\ M_{03} \end{bmatrix} = \begin{bmatrix} M_1 \\ M_2 \\ M_3 \end{bmatrix} \tag{7.54}$$

where

$$M_{11} = \frac{1}{1-\rho_1}\left\{1-\left(\frac{\rho_1}{1-\rho_1}\right)\left(\frac{2.5}{\text{CR}}\right)\left[\frac{C_1(C_2+C_3)}{C_0}\right]\right\}$$

$$M_{12} = \left(\frac{1}{1-\rho_2}\right)\left(\frac{\rho_1}{1-\rho_1}\right)\left(\frac{2.5}{\text{CR}}\right)\left(\frac{C_1 C_2}{C_0}\right)$$

$$M_{13} = \left(\frac{1}{1-\rho_3}\right)\left(\frac{\rho_1}{1-\rho_1}\right)\left(\frac{2.5}{\text{CR}}\right)\left(\frac{C_1 C_3}{C_0}\right)$$

$$M_{21} = \left(\frac{1}{1-\rho_1}\right)\left(\frac{\rho_2}{1-\rho_2}\right)\left(\frac{C_1 C_2}{C_0}\right)$$

$$M_{22} = \frac{1}{1-\rho_2}\left\{1 - \left(\frac{\rho_2}{1-\rho_2}\right)\left[\frac{C_2(C_1+C_3)}{C_0}\right]\right\}$$

$$M_{23} = \left(\frac{1}{1-\rho_3}\right)\left(\frac{\rho_2}{1-\rho_2}\right)\left(\frac{C_2C_3}{C_0}\right)$$

$$M_{31} = \left(\frac{1}{1-\rho_2}\right)\left(\frac{\rho_3}{1-\rho_3}\right)\left(\frac{C_2C_3}{C_0}\right)$$

$$M_{33} = \frac{1}{1-\rho_3}\left\{1 - \left(\frac{\rho_3}{1-\rho_3}\right)\left[\frac{C_3(C_1+C_2)}{C_0}\right]\right\} \tag{7.55}$$

and

$$C_1 = \frac{(1-\rho_1)(1-f^2)}{(2.5/\mathrm{CR})\,\rho_1\,(1-f^2) + f(1-\rho_1)}$$

$$C_2 = \frac{(1-\rho_2)(1+f)}{1+\rho_2 f}$$

$$C_3 = \frac{(1-\rho_3)(1+f)}{1+\rho_3 f}$$

$$C_0 = C_1 + C_2 + C_3 \tag{7.56}$$

Now let us place luminaires on the ceiling of the room cavity in Fig. 7.16. For each luminaire, we will define eighteen 10° flux zones, in the manner of Sec. 2.6. Let n be the zone index, where $1 \leq n \leq 18$, with zone 1 adjacent to the nadir and zone 18 adjacent to the zenith. The lumens in each zone are given by Eq. (2.32), and we can write

$$\Phi_D = \sum_{n=1}^{9} \Phi_n \tag{7.57}$$

$$\Phi_U = \sum_{n=10}^{18} \Phi_n \tag{7.58}$$

where Φ_n is the lumens in zone n, Φ_D is the total downward lumens, and Φ_U is the total upward lumens.

We desire to relate Φ_D and Φ_U to the total lamp lumens Φ_T. This is done by writing

$$\eta_D = \frac{\Phi_D}{\Phi_T} \tag{7.59}$$

$$\eta_U = \frac{\Phi_U}{\Phi_T} \tag{7.60}$$

Thus η_D is the fraction of the total lamp lumens which goes directly to the walls and floor in Fig. 7.16 and η_U is the fraction of the total lamp lumens that travels directly to the ceiling.

We need next to separate the downward lumens into two parts, those traveling directly to the walls and those going directly to the floor in Fig.

7.16. This requires determination of a parameter called the direct ratio (D_m), the fraction of the downward luminaire flux which reaches the work plane directly. We can write

$$D_m = \frac{1}{\Phi_D} \sum_{n=1}^{9} k_{mn} \Phi_n \tag{7.61}$$

where k_{mn} is called a zonal multiplier and is a function of both cavity ratio m and zone n. k_{mn} is the fraction of the downward flux in zone n which reaches the work plane directly for a given cavity ratio m.

In addition to being functions of cavity ratio and zone, the zonal multipliers also depend somewhat on the spacing:mounting height ratio (S:MH) of the luminaires. Until recently, coefficient of utilization tables (Table 7.4) were based on an S:MH of 0.4 for fluorescent luminaires and 0.7 for incandescent and HID luminaires. Zonal multipliers for these S:MH ratios were calculated and presented in tabular form.

Recent work has shown (Levin, 1981) that k_{mn} is not a significant function of S:MH. Rather it depends more on luminaire location. An approximating equation for k_{mn} has been developed.

$$k_{nm} = \epsilon^{-Am^B} \tag{7.62}$$

where A and B are constants for each flux zone and are given in Table 7.10.

With the direct ratio in hand, we can now calculate the coefficient of utilization. We return to Eq. (7.54) and write

$$M_3 = M_{31}M_{01} + M_{32}M_{02} + M_{33}M_{03} \tag{7.63}$$

M_{31}, M_{32}, and M_{33} are obtainable from Eqs. (7.55). It remains to determine M_{01}, M_{02}, and M_{03}. The fluxes reaching the cavity walls, ceiling, and floor are

$$\Phi_1 = (1 - D_m)\eta_D\Phi_T \tag{7.64a}$$

$$\Phi_2 = \eta_U\Phi_T \tag{7.64b}$$

TABLE 7.10
ZONE CONSTANTS

n	A	B
1	0	0
2	0.041	0.98
3	0.070	1.05
4	0.100	1.12
5	0.136	1.16
6	0.190	1.25
7	0.315	1.25
8	0.640	1.25
9	2.10	0.80

$$\Phi_3 = D_m \eta_D \Phi_T \tag{7.64c}$$

Thus the initial exitances are

$$M_{01} = \frac{\rho_1 (1 - D_m) \eta_D \Phi_T}{A_1}$$
$$M_{02} = \frac{\rho_2 \eta_U \Phi_T}{A_2}$$
$$M_{03} = \frac{\rho_3 D_m \eta_D \Phi_T}{A_3} \tag{7.65}$$

The coefficient of utilization is

$$\mathrm{CU} = \frac{M_3 A_3}{\rho_3 \Phi_T} \tag{7.66}$$

Using Eqs. (7.55), (7.63), and (7.65), Eq. (7.66) becomes

$$\mathrm{CU} = \frac{2.5 \rho_1 \eta_D (1 - D_m) C_1 C_3}{\mathrm{CR}(1 - \rho_1)(1 - \rho_2) C_0} + \frac{\rho_2 \eta_U C_2 C_3}{(1 - \rho_2)(1 - \rho_3) C_0} + \frac{D_m \eta_D}{1 - \rho_3} \left[1 - \frac{\rho_3 C_3 (C_1 + C_2)}{(1 - \rho_3) C_0} \right] \tag{7.67}$$

We consider next the wall and ceiling exitance coefficients. Equations (7.54) yield

$$M_1 = M_{11} M_{01} + M_{12} M_{02} + M_{13} M_{03} \tag{7.68}$$
$$M_2 = M_{21} M_{01} + M_{22} M_{02} + M_{23} M_{03} \tag{7.69}$$

From Eqs. (7.14) and (7.15),

$$\mathrm{WEC} = \frac{M_1 A_3}{\Phi_T} \tag{7.70}$$

$$\mathrm{CEC} = \frac{M_2 A_3}{\Phi_T} \tag{7.71}$$

Then, using Eqs. (7.55), (7.65), (7.68), and (7.69), we obtain

$$\mathrm{WEC} = \frac{2.5}{\mathrm{CR}} \left\{ \frac{\rho_1 (1 - D_m) \eta_D}{1 - \rho_1} \left[1 - \frac{2.5 \rho_1 C_1 (C_2 + C_3)}{\mathrm{CR}(1 - \rho_1) C_0} \right] + \frac{\rho_1 \rho_2 \eta_U C_1 C_2}{(1 - \rho_1)(1 - \rho_2) C_0} + \frac{\rho_1 \rho_3 D_m \eta_D C_1 C_3}{(1 - \rho_1)(1 - \rho_3) C_0} \right\} \tag{7.72}$$

$$\mathrm{CEC} = \frac{2.5 \rho_1 \rho_2 (1 - D_m) \eta_D C_1 C_2}{\mathrm{CR}(1 - \rho_1)(1 - \rho_2) C_0} + \frac{\rho_2 \eta_U}{1 - \rho_2} \left[1 - \frac{\rho_2 C_2 (C_1 + C_3)}{(1 - r_2) C_0} \right] + \frac{\rho_2 \rho_3 D_m \eta_D C_2 C_3}{(1 - \rho_2)(1 - \rho_3) C_0} \tag{7.73}$$

7.17 RADIATION COEFFICIENTS

In addition to the quantities we have already calculated, there are two others that are of use in interior lighting computations. These are the wall direct radiation coefficient (WDRC) and the wall reflected radiation coefficient (WRRC).

The wall luminous exitance M_1 in Eq. (7.68) is the result of two wall illuminances, one resulting from flux going directly from the luminaires to the walls and the second due to flux reflected from the room surfaces to the walls. These two illuminances may be expressed as

$$E_{1D} = \frac{\Phi_T \text{WDRC}}{A_3} \tag{7.74}$$

$$E_{1R} = \frac{\Phi_T \text{WRRC}}{A_3} \tag{7.75}$$

where the D and R denote direct and reflected, respectively. We also note that

$$E_1 = \frac{M_1}{\rho_1} = E_{1D} + E_{1R} \tag{7.76}$$

Now, Eqs. (7.70), (7.74), and (7.75) may be inserted into Eq. (7.76) to yield

$$\frac{\text{WEC}}{\rho_1} = \text{WDRC} + \text{WRRC} \tag{7.77}$$

If we can determine WDRC, then Eq. (7.77) may be used to find WRRC. The flux going directly to the walls is given by Eq. (7.64*a*). Thus

$$E_{1D} = \frac{(1 - D_m)\eta_D \Phi_T}{A_1} \tag{7.78}$$

Equating the right sides of Eqs. (7.74) and (7.78) and employing Eq. (7.17) gives

$$\text{WDRC} = (1 - D_m)\eta_D \frac{A_3}{A_1} = \frac{2.5}{\text{CR}} \eta_D (1 - D_m) \tag{7.79}$$

7.18 EFFECTIVE CAVITY REFLECTANCE

In Sec. 7.7, it was pointed out that the analytical expression from which Table 7.3 was developed was complex. With the concepts of flux transfer and form factor established, we are now in a position to derive that expression.

Figure 7.17 depicts a cavity. Surfaces 1 represent the walls of the cavity, surface 2 the base, and the dashed surface 3 the cavity opening. These surfaces have areas A_1, A_2, and A_3 and reflectances ρ_1, ρ_2, and 0, respec-

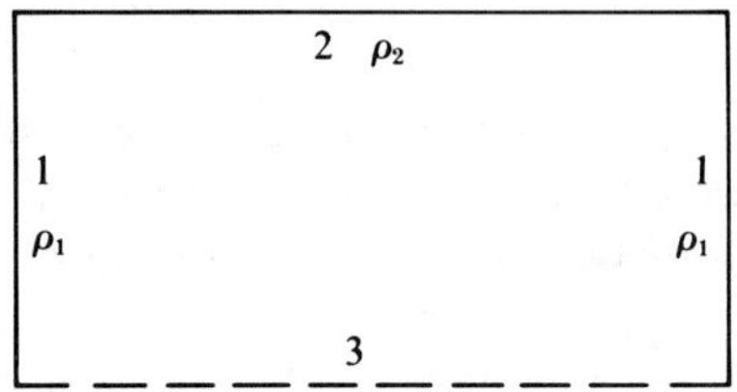

Figure 7.17 Cavity.

tively. Also, the initial exitances M_{01} and M_{02} are zero, as there are no sources of light within the cavity.

The flux into the cavity is $M_{03}A_3$; that out of the cavity is $(M_1 f_{31} + M_2 f_{32})A_3$ where M_1 and M_2 are the final exitances, after interreflections, of the walls and the base of the cavity. Thus the effective cavity reflectance is

$$\rho_{\text{eff}} = \frac{M_1 f_{31} + M_2 f_{32}}{M_{03}} \tag{7.80}$$

We now must find M_1 and M_2, under the given constraints. From Eq. (7.52),

$$\begin{bmatrix} 1-\rho_1 f_{11} & -\rho_1 f_{12} & -\rho_1 f_{13} \\ -\rho_2 f_{21} & 1 & -\rho_1 f_{23} \\ 0 & 0 & 1 \end{bmatrix} \begin{bmatrix} M_1 \\ M_2 \\ M_3 \end{bmatrix} = \begin{bmatrix} 0 \\ 0 \\ M_{03} \end{bmatrix} \tag{7.81}$$

This yields

$$\begin{aligned} (1-\rho_1 f_{11})M_1 - \rho_1 f_{12} M_2 &= \rho_1 f_{13} M_{03} \\ -\rho_2 f_{21} M_1 \quad + \quad M_2 &= \rho_2 f_{23} M_{03} \end{aligned} \tag{7.82}$$

which gives

$$\begin{aligned} M_1 &= \frac{(\rho_1 f_{13} + \rho_1\rho_2 f_{12} f_{23})M_{03}}{1-\rho_1 f_{11} - \rho_1\rho_2 f_{12} f_{21}} \\ M_2 &= \frac{[(1-\rho_1 f_{11})\rho_2 f_{23} + \rho_1\rho_2 f_{21} f_{13}]M_{03}}{1-\rho_1 f_{11} - \rho_1\rho_2 f_{12} f_{21}} \end{aligned} \tag{7.83}$$

Inserting these equations into Eq. (7.80) yields

$$\rho_{\text{eff}} = \frac{f_{31}(\rho_1 f_{13} + \rho_1\rho_2 f_{12} f_{13}) + f_{32}(1-\rho_1 f_{11})\rho_2 f_{23} + \rho_1\rho_2 f_{21} f_{13}}{1-\rho_1 f_{11} - \rho_1\rho_2 f_{12} f_{21}} \tag{7.84}$$

The result in Eq. (7.84) can be simplified by using Eqs. (7.43) to (7.45) to express all of the form factors in terms of $f = f_{23}$. Letting $A_2/A_1 = A$, we have

$$\rho_{\text{eff}} = \frac{\rho_1\rho_2 f[2A(1-f)-f] + \rho_1 A(1-f)^2 + \rho_2 f^2}{1-\rho_1[1-2A(1-f)] - \rho_1\rho_2 A(1-f)^2}$$

REFERENCES

Jones, J. R., and B. F. Jones: "Using the Zonal-Cavity System in Lighting Calculations," *Illuminating Engineering*, Part 1, May 1964; Part 2, June 1964; Part 3, July 1964; Part 4, August 1964.

Kaufman, J. E.: *IES Lighting Handbook*, 1981 ed., Illuminating Engineering Society, New York, 1981.

Levere, R. C., R. E. Levin, and W. C. Primrose: "Spacing Criteria for Interior Luminaires—the Practices and Pitfalls," *IES Journal*, October 1973.

Levin, R. E.: "Revision of the S/MH Concept," *Lighting Design and Application*, August 1977.

———: "Cavities, Coefficients and Direct Ratios," IES Journal, April, 1982.

———: "The Photometric Connection," Parts 1–4, *Lighting Design and Application*, September-December 1982.

Moon, P.: *The Scientific Basis for Illuminating Engineering*, McGraw-Hill, New York, 1936.

O'Brien, P. F.: "Lighting Calculations for Thirty-Five Thousand Rooms," *Illuminating Engineering*, April 1960.

———: "Numerical Analysis for Lighting Design," *Illuminating Engineering*, April 1965.

———: "Effective Reflectance of Room Cavities with Specular and Diffuse Surfaces," *Illuminating Engineering*, April 1966.

Chapter 8

Interior Lighting Design: Additional Metrics

8.1 THE SCOPE OF THE CHAPTER

In Chap. 7, attention was focused on obtaining the average horizontal illuminance (E_{ave}) on the work plane in an interior space. No effort was made to ascertain the illuminance at a particular point, such as on a worker's desk. Nor was any consideration given to the effects of luminances of tasks and backgrounds viewed by the worker. Last, no assessment was made of whether or not the worker found the space visually acceptable and comfortable.

In this chapter each of the above issues will be addressed. First, we will derive the configuration factors for several common source geometries. These will be used in the development of the configuration factor method for computing the direct component of the horizontal illuminance (E_P) at a given point P within a space from a perfectly diffusing ceiling element. Then we will present the infinite plane exitance method for obtaining the reflected component of the horizontal illuminance at a point. This latter technique will also involve the use of configuration factors.

Next, two additional procedures for obtaining the direct component of E_P that do not require perfectly diffusing sources will be presented. These are called the inverse-square-law approximation method and the angular coordinate method. The former involves subdividing a luminaire into elements small enough so that each satisfies the constraints of the ISL delineated in Chap. 2. The latter, applicable mainly to point and linear luminaires, presents the illuminance at a point as a function of two angular coordinates and the mounting height of the luminaire above the work plane.

Following this, we will develop a procedure for predetermining the contrast rendition factor (CRF) and visibility level (VL) in a proposed lighting installation. The inverse-square-law approximation method will be used to calculate expected task and background luminances, from which CRF and VL will be determined. This will permit us to compare proposed lighting systems on a VL basis before they are installed.

Finally we will develop a procedure for determining whether or not an observer is likely to be comfortable, in a visual sense, within a given interior space. This will lead us to a parameter called visual comfort probability (VCP), which is the percentage of observers at a specified location in a room that will find that room above the borderline between visual comfort and discomfort.

The reader should not infer that if a worker has adequate E_{ave}, E_P, VL, and VCP, as we will come to define the term "adequate," the lighting design for the space is necessarily optimal, or even good. There are many other variables, particularly those involving lighting aesthetics, which must be considered by the lighting designer. Among these are:

1. *The visual composition of the space.* How well will the luminance patterns merge with the structural patterns?
2. *The focal centers of the space.* Are there "attention centers" to which the viewer should be directed?
3. *The visual zones of the space.* These are generally taken as the overhead zone, the perimeter zone, and the occupied zone. What are the lighting requirements of each zone and how should these be blended?
4. *The color and atmosphere of the space.* What is the sun exposure? Do we wish to create a cool or a warm atmosphere? What degree of "whiteness" is desired?
5. *The desired appearance of objects within the space.* Is color rendition of objects important? Are certain objects to sparkle; others to appear diffuse? Are shade and shadow desirable?

These are just a few of the myriad of factors which the good lighting designer will take into account in creating a lighting solution for a space. Most of these cannot be quantified, in the sense that we calculate E_{ave}, E_P, VL, and VCP. Rather they are learned through experience and observation, hopefully under the guidance of a qualified professional in the field.

8.2 MORE ON CONFIGURATION FACTORS

The configuration factor c_{21} was defined in Sec. 7.14 as the ratio of the illuminance at point 2 produced by the flux received directly from surface 1 to the luminous exitance of surface 1. Thus the illuminance at point 2 is

$$E_2 = c_{21} M_1 \tag{8.1}$$

In Secs. 2.8 and 2.9, we derived formulas for the illuminance at a point produced by disc, strip, tube, and rectangular sources. We can now see that all we need to do to obtain configuration factors for these geometries is to divide the illuminance in each of these equations by the luminous exitance of the source.

Infinite Horizontal Plane

For the perfectly diffusing infinite horizontal plane shown in Fig. 8.1, the configuration factor to the point P beneath it is obtained by letting $R \to \infty$ in Eq. (2.46). The result is

$$E = \pi L = M \tag{8.2}$$

and thus

$$c = 1 \tag{8.3}$$

EXAMPLE 8.1

Assume the wall and ceiling surfaces of the rectangular box in Fig. 8.2*a* are perfectly diffusing and have the same luminance. Find the configuration factor at a point P anywhere on the floor of the box.

Solution. The configuration factor for this geometry is the same as for the infinite plane, because, with the walls perfectly diffusing, it is impossible for the observer at P to distinguish the vertical walls from their projection on an infinite horizontal plane, as shown in Fig. 8.2*b*. It is as though the walls were rotated 90° about their intersection lines with the ceiling and then extended to infinity. Thus the configuration factor for the equal-luminance, perfectly diffusing rectangular box is $c = 1$, irrespective of the location of P on the floor of the box.

Rectangle

The illuminance produced by a perfectly diffusing rectangular source at a point P on a line perpendicular to one corner of the source is given by Eq. (2.62) if P is in a plane parallel to the source and by Eq. (2.64) if P is in a

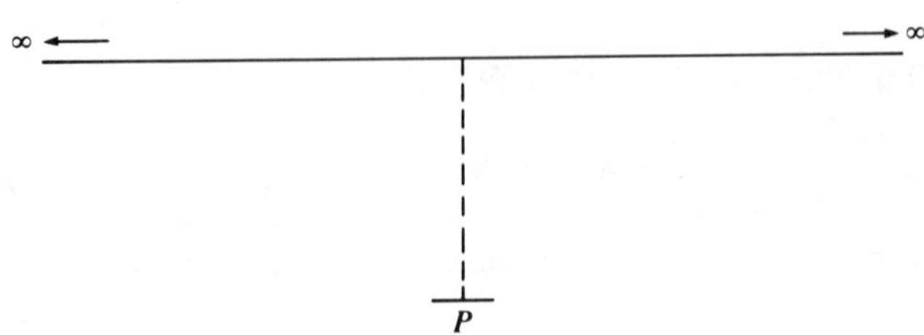

Figure 8.1 Infinite horizontal plane.

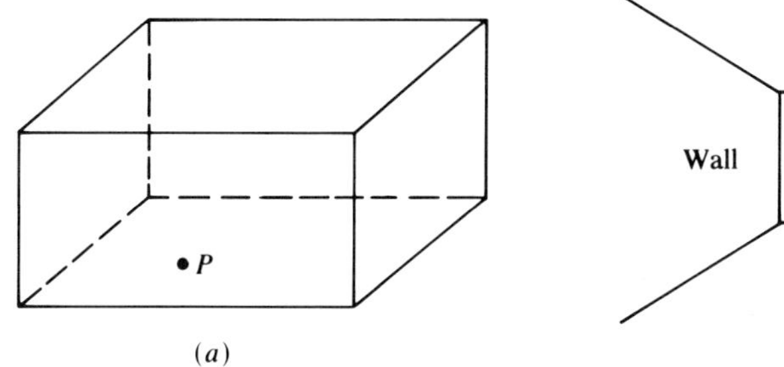

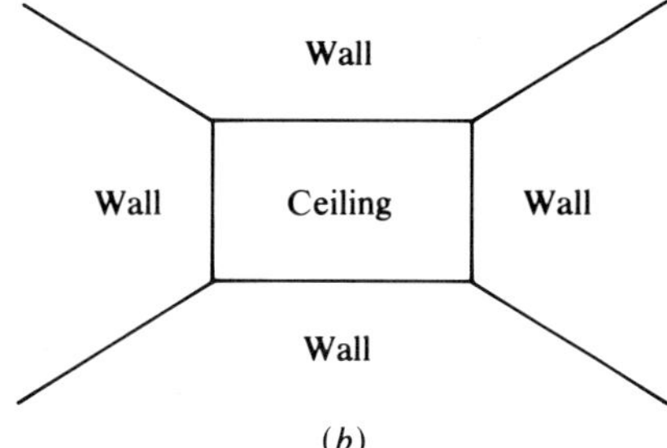

Figure 8.2

plane perpendicular to the source. The configuration factor in each case is obtained by dividing the illuminance by $M = \pi L$.

These two configuration factors may be placed in an alternative useful form by making the following identifications in Fig. 8.3:

$$\sin \gamma = \frac{h}{\sqrt{h^2 + q^2}}$$

$$\sin \beta = \frac{w}{\sqrt{w^2 + q^2}}$$

$$\sin \gamma_1 = \frac{h}{\sqrt{w^2 + h^2 + q^2}}$$

$$\sin \beta_1 = \frac{w}{\sqrt{w^2 + h^2 + q^2}} \tag{8.4}$$

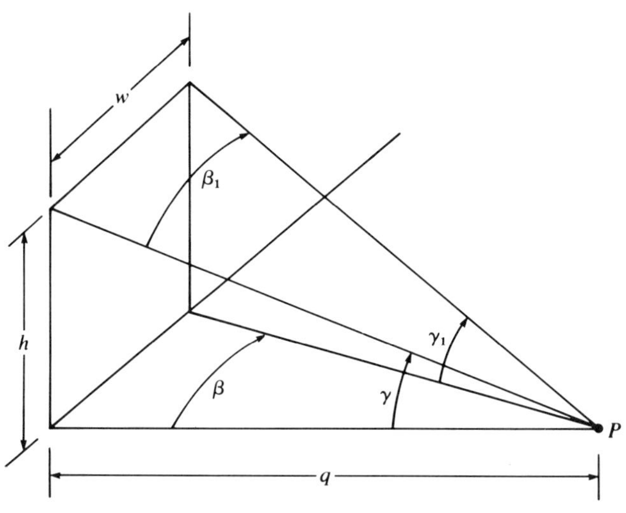

Figure 8.3

Inserting these into Eqs. (2.62) and (2.64) yields

$$c_{\parallel} = \frac{1}{2\pi} (\beta_1 \sin \gamma + \gamma_1 \sin \beta) \tag{8.5}$$

$$c_{\perp} = \frac{1}{2\pi} (\beta - \beta_1 \cos \gamma) \tag{8.6}$$

If the configuration factor is desired for a point on a perpendicular line that does not originate at a corner of the source, we can use the technique of superposition, as was done in Example. 2.13.

EXAMPLE 8.2

Assume the box in Fig. 8.2 is a cube and that point P is at the center of the floor. The luminous exitance of the walls is M_w, that of the ceiling $M_c > M_w$. Find the component of the horizontal illuminance at point P produced by these exitances.

Solution. The illuminance may be obtained as the superposition of the illuminances produced by an infinite plane of luminance exitance M_w and a square of luminous exitance $M_c - M_w$. The square is composed of four of the areas in Fig. 8.3, all equal. From Eqs. (8.4),

$$\sin \gamma = \sin \beta = \frac{0.5}{\sqrt{1.25}} = 0.447$$

$$\sin \gamma_1 = \sin \beta_1 = \frac{0.5}{\sqrt{1.50}} = 0.408$$

$$\gamma_1 = \beta_1 = 0.421 \text{ rad}$$

Then, from Eq. (8.5), multiplied by 4, the configuration factor for the square ceiling is

$$c_{\parallel} = \frac{2}{\pi} [0.421(0.447) + 0.421(0.447)] = 0.240$$

The configuration factor for the infinite plane is 1.0. Thus

$$\begin{aligned} E_P &= M_w(1) + 0.240(M_c - M_w) \\ &= 0.240M_c + 0.760M_w \end{aligned}$$

Circular Disc

The illuminance produced at P in Fig. 8.4 by a perfectly diffusing disc source is given in Eq. (2.46). Thus the configuration factor is

$$c = \frac{R^2}{R^2 + D^2} = \frac{1 - \cos \theta}{2} \tag{8.7}$$

If the point is displaced a distance S horizontally from the normal to the disc's center to a point P_1, c becomes

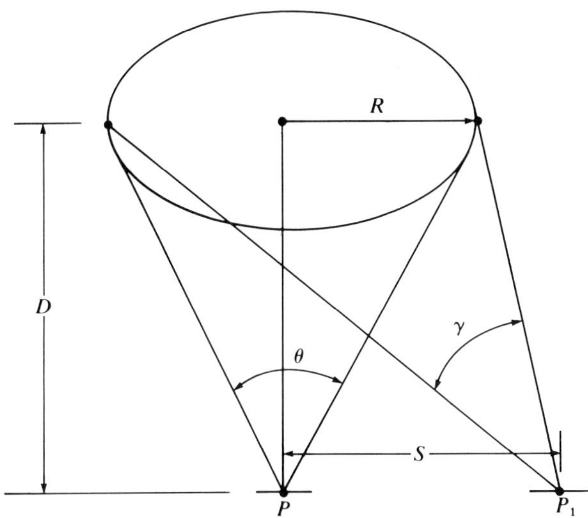

Figure 8.4

$$c = \frac{1 - \cos\gamma}{2} = \frac{1}{2}\left[1 - \frac{D^2 + S^2 - R^2}{\sqrt{(D^2 + S^2 + R^2)^2 - 4R^2S^2}}\right] \tag{8.8}$$

Equation (8.8) is given without proof, but may be derived using the sphere method outlined in Example 7.10.

Strip and Tube Sources

The illuminances produced in various planes at P in Fig. 8.5 by a perfectly diffusing strip source are given by Eqs. (2.54), (2.57), and (2.59). We note from Fig. 8.5 that

$$\tan\beta = \frac{h}{\sqrt{q^2 + r^2}}$$

$$\sin^2\beta = \frac{h^2}{q^2 + r^2 + h^2}$$

$$\sin^2\gamma = \frac{q^2}{q^2 + r^2} \tag{8.9}$$

Inserting Eqs. (8.9) into the three strip-source equations yields

$$c_H = \frac{W}{2\pi h}\sin^2\gamma\,(\sin^2\beta + \beta\tan\beta) \tag{8.10}$$

$$c_{V_\parallel} = \frac{r}{q}\,c_H \tag{8.11}$$

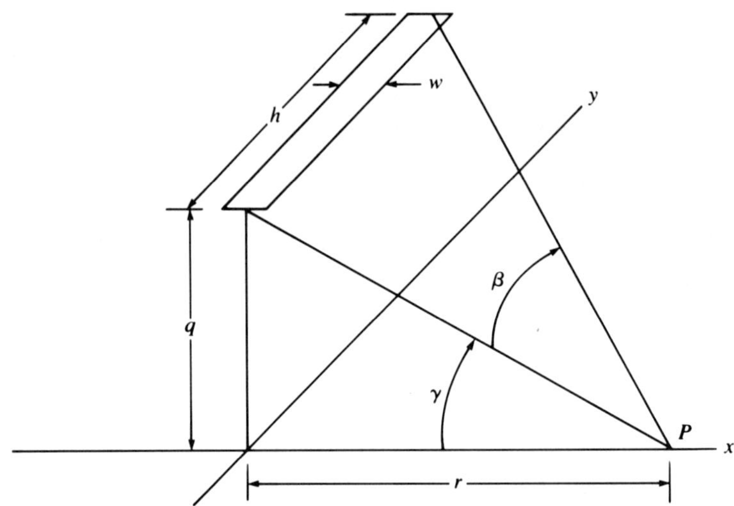

Figure 8.5

$$c_{V\perp} = \frac{W}{2\pi q} \sin^2 \gamma \sin^2 \beta \tag{8.12}$$

For tube sources, the only change is that noted in Eq. (2.60). Thus

$$c_{\text{tube}} = \frac{1}{\sin \gamma} c_{\text{strip}} \tag{8.13}$$

Sloped Planes

A plane of dimensions m by f and sloped at angle ϕ is shown in Fig. 8.6. The configuration factor from the plane to point P in a horizontal plane a distance a from a corner of the intersection of the two planes can be shown to be

$$c = \frac{1}{2\pi}\left[\tan^{-1}\frac{m}{a} + \frac{f\cos\phi - a}{A}\tan^{-1}\frac{m}{A} + \frac{m\cos\phi}{B}\left(\tan^{-1}\frac{f - a\cos\phi}{B} + \tan^{-1}\frac{a\cos\phi}{B}\right)\right] \tag{8.14}$$

where

$$A = \sqrt{f^2 + a^2 - 2af\cos\phi} \qquad B = \sqrt{m^2 + a^2 \sin^2\phi} \tag{8.15}$$

Letting $m \to \infty$ in Eq. (8.14) and doubling the answer gives the configuration factor for a sloped plane of infinite extent in and out of the page. The result is

$$c = \frac{1}{2}\left(1 + \frac{f\cos\phi - a}{\sqrt{f^2 + a^2 - 2af\cos\phi}}\right) \tag{8.16}$$

Now letting $f \to \infty$ in Eq. (8.16) gives the configuration factor for a sloped plane which is infinite in both directions. For this case we have

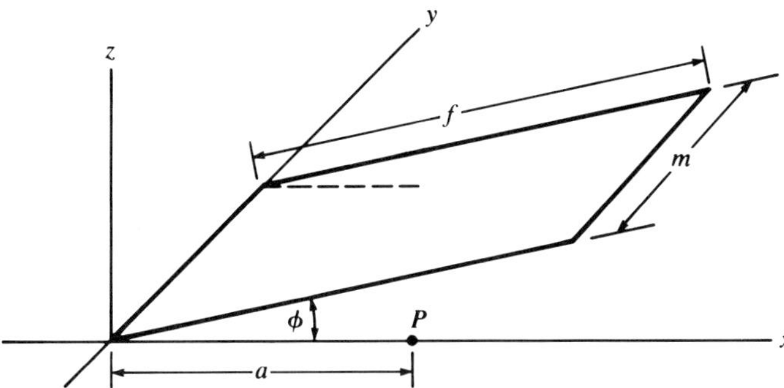

Figure 8.6

$$c = \tfrac{1}{2}(1 + \cos \phi) \tag{8.17}$$

EXAMPLE 8.3

The perfectly diffusing skylight in Fig. 8.7 is 5 ft in width, of infinite extent in and out of the page, and has a luminance of 200 cd/ft². Find the horizontal illuminance at point P.

Solution. Using superposition and Eq. (8.16), the configuration factor is

$$c = \frac{1}{2}\left[\frac{25(0.866) - 20}{\sqrt{625 + 400 - 1000(0.866)}} - \frac{20(0.866) - 20}{\sqrt{400 + 400 - 800(0.866)}}\right]$$
$$= 0.194$$

Then $E_p = 200\pi(0.194) = 122$ fc.

The procedure up to now in this section has been to derive configuration factors in rectangular coordinates and then transform them to angular coordinates. It is reasonable to inquire if some of the configuration factors could be derived directly in angular coordinates. Such a development is possible (Levin, 1971) and will now be presented.

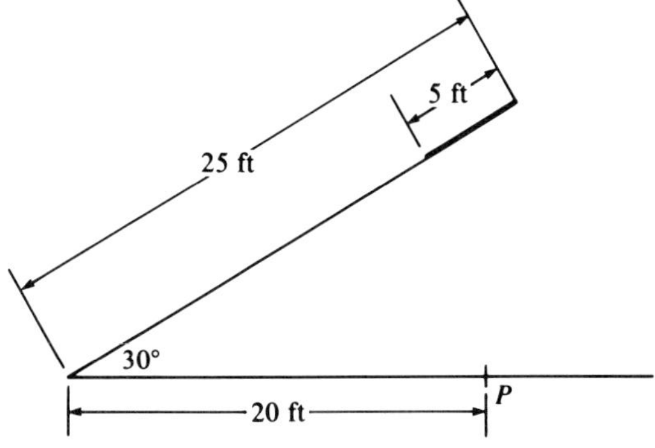

Figure 8.7

A perfectly diffusing area source S of arbitrary shape but prescribed angular boundaries (α_1, α_2, β_1, β_2) is shown in Fig. 8.8. It lights a point P. The differential solid angle from a small element dA of the source to the point P is

$$d\omega = \cos\beta \, d\alpha \, d\beta \tag{8.18}$$

The line from the source element to P makes an angle γ with the normal to the horizontal plane given by

$$\cos\gamma = \cos\alpha \cos\beta \tag{8.19}$$

Then, from Eq. (2.13), the horizontal illuminance at P is

$$E_P = \int_\beta \int_\alpha L \cos\alpha \cos^2\beta \, d\alpha \, d\beta \tag{8.19}$$

Performing the double integration gives

$$E_p = L(\sin\alpha_2 - \sin\alpha_1)\,[f(\beta_2) - f(\beta_1)] \tag{8.20}$$

where

$$f(\beta) = \tfrac{1}{2}\beta + \tfrac{1}{4}\sin 2\beta \tag{8.21}$$

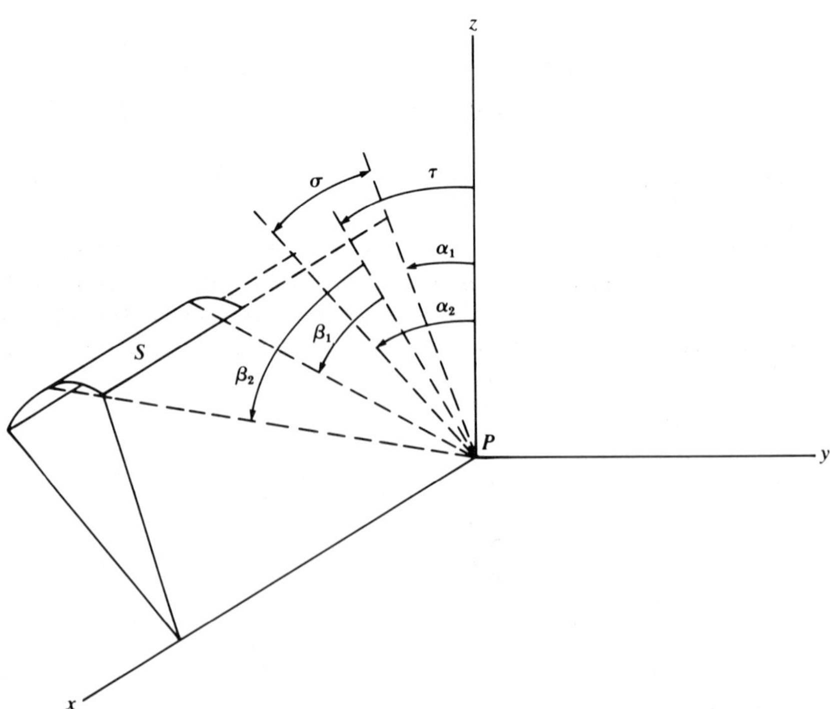

Figure 8.8 (*Source*: R. E. Levin, "Illumination Due to Area Sources Expressed in Angular Coordinates." Reprinted from the October 1971 issue of the *Journal of IES* with permission of the Illuminating Engineering Society of North America.)

Because the source is perfectly diffusing, the result in Eq. (8.20) is independent of source shape within the angular boundaries. However it does assume that β_1 and β_2 are independent of α. This means that the ends of S must be circular arcs in planes normal to the x axis. Deviations from this requirement are not serious when the source is reasonably narrow.

It is often useful to express the result in Eq. (8.20) in terms of the angles σ and τ in Fig. 8.8, rather than α_1 and α_2. Noting that

$$\alpha_1 = \tau - \frac{\sigma}{2} \qquad \alpha_2 = \tau + \frac{\sigma}{2} \tag{8.22}$$

Eq. (8.20) can be rewritten as

$$E_P = 2L \cos \tau \sin \frac{\sigma}{2} \left[f(\beta_2) - f(\beta_1) \right] \tag{8.23}$$

Let us use the result in Eq. (8.23) to obtain the horizontal illuminance at point P in Fig. 8.5 produced by the given strip source. The result should verify the configuration factor in Eq. (8.10). From Fig. 8.5,

$$\beta_1 = 0 \qquad \beta_2 = \beta \tag{8.24a}$$

$$\cos \tau = \sin \gamma \tag{8.24b}$$

$$\sin \frac{\sigma}{2} \simeq \frac{\sigma}{2} = \frac{W \sin \gamma}{2\sqrt{q^2 + r^2}} \tag{8.24c}$$

Equation (8.24c) results from first observing that, with W small compared with the other distances present, the angle σ is small and thus the angle and its sine are approximately equal. Then σ is found by dividing projected arc length by radius.

Inserting Eqs. (8.24) into Eq. (8.23) yields

$$\begin{aligned} E_P &= 2L \sin \gamma \frac{W \sin \gamma}{2\sqrt{q^2 + r^2}} \left(\tfrac{1}{2}\beta + \tfrac{1}{4} \sin 2\beta \right) \\ &= \frac{LW \sin^2 \gamma}{2h} \left(\frac{h}{\sqrt{q^2 + r^2}} \beta + \frac{h}{2\sqrt{q^2 + r^2}} \sin 2\beta \right) \\ &= \frac{LW \sin^2 \gamma}{2h} (\beta \tan \beta + \sin^2 \beta) \end{aligned} \tag{8.25}$$

Dividing by πL [Eq. (8.1)] yields c_H.

8.3 CONFIGURATION FACTOR METHOD: DIRECT COMPONENT

The configuration factor method permits the calculation of the direct component of the horizontal illuminance at a point on the work plane from a perfectly diffusing source located on the ceiling.

Consider Fig. 8.9. It shows a row of five 4×4 ft fluorescent luminaires, whose intensity is assumed to be a cosine distribution, mounted 8 ft above the work plane. The horizontal illuminance on the work plane at point P is desired.

The first step in the procedure is to create a room cavity, indicated by the dashed lines in Fig. 8.9, whose width w is the length of the luminaire row and whose length l is twice that amount. The room cavity is always drawn so that its left wall contains the center line of the luminaires. The height of the cavity is the mounting height of the luminaires above the work plane. For the situation in Fig. 8.9,

$$\text{RCR} = \frac{5(8)(60)}{800} = 3.0$$

Next, we create the grid shown in Fig. 8.10 and locate the coordinates of the point P. The grid is 10×20 squares, so P is located

Along l: $\frac{12}{40}(20) = 6$ squares in from the left
Along w: $\frac{8}{20}(10) = 4$ squares down from the top

Since "6" corresponds to "G" in Fig. 8.10, P lies at coordinates (G, 4). Note that the vertical numbering of squares need only be from 0 to 5 because of symmetry.

A set of normalized configuration factors (c_n) for a strip source is given in Table 8.1. These were calculated from Eq. (8.10) for rooms whose length-to-width ratio is $l/w = 2$ and for strips whose width to length ratio is $W/w = 0.1$. For the given problem, we find from Table 8.1 that $c_n = 0.0144$. The luminaire width-to-length ratio is $\frac{4}{20} = 0.2$; thus $c = 2(0.0144) = 0.0288$.

Now assume that each fluorescent luminaire in Fig. 8.9 has an intensity $I = I_0 \cos\theta$ candelas, where θ is the angle with the normal to the luminaire. Using the method of Sec. 2.6, we find that each luminaire emits πI_0 lumens

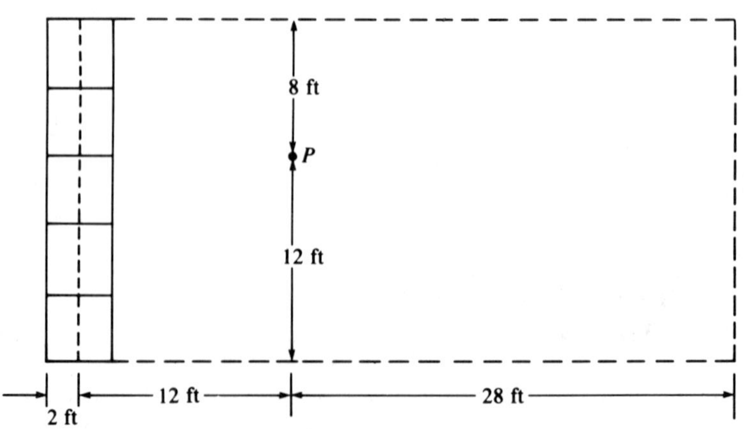

Figure 8.9

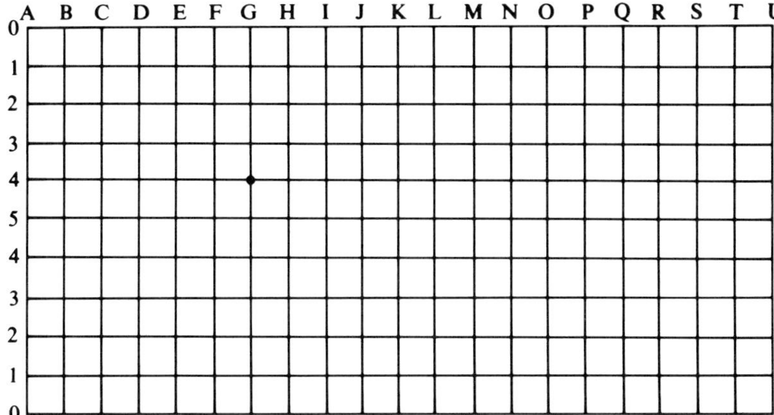

Figure 8.10

and thus has a luminous exitance of $\pi I_0/16$ lumens per square foot (a constant luminance of $I_0/16$ candelas per square foot). Assume that $I_0 = 2400$ cd is given. Then the horizontal illuminance at point P is

$$E_P = \frac{\pi}{16}(2400)(0.0288) = 13.6 \text{ fc}$$

EXAMPLE 8.4

A row of five 2 × 4 ft fluorescent luminaires, each of whose intensity is $I = 2000 \cos \theta$ candelas, is mounted 7 ft above the work plane, as shown in Fig. 8.11. Find the horizontal illuminance at point P.

Solution. The luminaire row length is 20 ft. Thus the room cavity ratio is

$$\text{RCR} = \frac{5(7)(60)}{20(40)} = 2.625$$

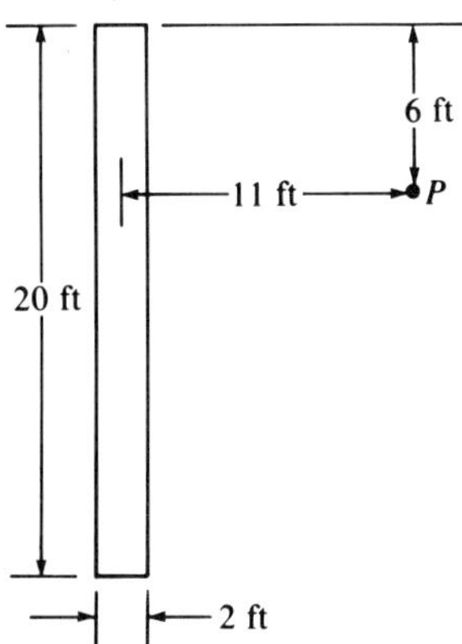

Figure 8.11

TABLE 8.1 NORMALIZED CONFIGURATION FACTORS
Strip Source, $l/w = 2$, $W/w = 0.1$

RCR	A0	B0	C0	D0	E0	F0	G0	I0	K0	P0	U0
1	0.1873	0.0958	0.0318	0.0124	0.0058	0.0031	0.0018	0.0007	0.0004	0.0001	0
2	0.0931	0.0763	0.0473	0.0269	0.0154	0.0092	0.0058	0.0026	0.0013	0.0003	0.0001
3	0.0611	0.0557	0.0434	0.0307	0.0209	0.0141	0.0097	0.0048	0.0025	0.0007	0.0003
4	0.0446	0.0423	0.0363	0.0290	0.0221	0.0164	0.0121	0.0067	0.0038	0.0011	0.0004
5	0.0345	0.0333	0.0300	0.0256	0.0210	0.0167	0.0131	0.0079	0.0048	0.0016	0.0006
6	0.0275	0.0268	0.0249	0.0222	0.0190	0.0159	0.0131	0.0086	0.0056	0.0020	0.0008
7	0.0225	0.0221	0.0209	0.0191	0.0169	0.0147	0.0125	0.0088	0.0060	0.0024	0.0010
8	0.0187	0.0184	0.0176	0.0164	0.0149	0.0133	0.0116	0.0086	0.0062	0.0027	0.0012
9	0.0157	0.0155	0.0150	0.0142	0.0131	0.0119	0.0106	0.0082	0.0062	0.0029	0.0014
10	0.0134	0.0133	0.0129	0.0123	0.0115	0.0106	0.0097	0.0077	0.0060	0.0030	0.0015

RCR	A1	B1	C1	D1	E1	F1	G1	I1	K1	P1	U1
1	0.3124	0.1557	0.0470	0.0169	0.0075	0.0038	0.0021	0.0008	0.0004	0.0001	0
2	0.1338	0.1079	0.0644	0.0350	0.0194	0.0112	0.0069	0.0029	0.0014	0.0004	0.0001
3	0.0797	0.0722	0.0553	0.0383	0.0254	0.0168	0.0113	0.0054	0.0028	0.0007	0.0003
4	0.0549	0.0519	0.0442	0.0348	0.0261	0.0191	0.0139	0.0074	0.0042	0.0012	0.0004
5	0.0408	0.0393	0.0352	0.0298	0.0242	0.0191	0.0148	0.0088	0.0053	0.0017	0.0006
6	0.0316	0.0308	0.0285	0.0252	0.0215	0.0179	0.0146	0.0094	0.0060	0.0021	0.0009
7	0.0252	0.0247	0.0233	0.0212	0.0188	0.0162	0.0137	0.0095	0.0064	0.0025	0.0011
8	0.0206	0.0203	0.0194	0.0180	0.0163	0.0144	0.0126	0.0092	0.0066	0.0028	0.0013
9	0.0171	0.0169	0.0163	0.0153	0.0141	0.0128	0.0114	0.0088	0.0065	0.0030	0.0014
10	0.0144	0.0143	0.0138	0.0132	0.0123	0.0113	0.0103	0.0082	0.0063	0.0031	0.0016

RCR	A2	B2	C2	D2	E2	F2	G2	I2	K2	P2	U2
1	0.3595	0.1792	0.0558	0.0202	0.0088	0.0044	0.0025	0.0009	0.0004	0.0001	0
2	0.1595	0.1287	0.0768	0.0414	0.0226	0.0129	0.0078	0.0032	0.0016	0.0004	0.0001
3	0.0943	0.0852	0.0649	0.0446	0.0293	0.0191	0.0126	0.0059	0.0030	0.0008	0.0003
4	0.0636	0.0600	0.0508	0.0397	0.0296	0.0214	0.0154	0.0081	0.0045	0.0013	0.0005
5	0.0462	0.0444	0.0398	0.0335	0.0270	0.0211	0.0162	0.0095	0.0056	0.0017	0.0007
6	0.0351	0.0342	0.0316	0.0278	0.0236	0.0195	0.0158	0.0101	0.0064	0.0022	0.0009
7	0.0276	0.0270	0.0255	0.0231	0.0204	0.0175	0.0147	0.0101	0.0068	0.0026	0.0011
8	0.0222	0.0219	0.0209	0.0193	0.0175	0.0154	0.0134	0.0098	0.0069	0.0029	0.0013
9	0.0183	0.0180	0.0174	0.0163	0.0150	0.0136	0.0121	0.0092	0.0068	0.0031	0.0015
10	0.0152	0.0151	0.0146	0.0139	0.0130	0.0119	0.0108	0.0086	0.0066	0.0032	0.0016

RCR	A3	B3	C3	D3	E3	F3	G3	I3	K3	P3	U3
1	0.3689	0.1864	0.0597	0.0220	0.0097	0.0048	0.0027	0.0010	0.0005	0.0001	0
2	0.1719	0.1393	0.0839	0.0456	0.0249	0.0142	0.0085	0.0035	0.0016	0.0004	0.0001
3	0.1037	0.0938	0.0714	0.0490	0.0321	0.0209	0.0137	0.0063	0.0032	0.0008	0.0003
4	0.0699	0.0659	0.0557	0.0434	0.0322	0.0232	0.0166	0.0086	0.0047	0.0013	0.0005
5	0.0503	0.0484	0.0432	0.0363	0.0291	0.0227	0.0174	0.0100	0.0059	0.0018	0.0007
6	0.0378	0.0368	0.0339	0.0298	0.0253	0.0209	0.0168	0.0106	0.0067	0.0023	0.0009
7	0.0294	0.0288	0.0271	0.0246	0.0216	0.0185	0.0155	0.0106	0.0071	0.0026	0.0011
8	0.0235	0.0231	0.0220	0.0204	0.0184	0.0162	0.0140	0.0102	0.0072	0.0030	0.0013
9	0.0191	0.0189	0.0182	0.0171	0.0157	0.0142	0.0126	0.0095	0.0070	0.0032	0.0015
10	0.0159	0.0157	0.0152	0.0145	0.0135	0.0124	0.0112	0.0088	0.0068	0.0033	0.0016

RCR	A4	B4	C4	D4	E4	F4	G4	I4	K4	P4	U4
1	0.3716	0.1888	0.0612	0.0229	0.0101	0.0051	0.0028	0.0010	0.0005	0.0001	0
2	0.1771	0.1440	0.0873	0.0477	0.0262	0.0149	0.0089	0.0036	0.0017	0.0004	0.0001
3	0.1086	0.0983	0.0751	0.0515	0.0338	0.0219	0.0144	0.0066	0.0033	0.0008	0.0003
4	0.0735	0.0693	0.0586	0.0457	0.0338	0.0243	0.0174	0.0089	0.0048	0.0013	0.0005
5	0.0528	0.0508	0.0453	0.0380	0.0304	0.0237	0.0181	0.0104	0.0061	0.0018	0.0007
6	0.0395	0.0384	0.0354	0.0311	0.0263	0.0216	0.0174	0.0109	0.0068	0.0023	0.0009
7	0.0306	0.0299	0.0281	0.0255	0.0224	0.0191	0.0160	0.0109	0.0072	0.0027	0.0011
8	0.0243	0.0239	0.0228	0.0210	0.0190	0.0167	0.0144	0.0104	0.0073	0.0030	0.0013
9	0.0197	0.0194	0.0187	0.0176	0.0161	0.0145	0.0129	0.0098	0.0072	0.0032	0.0015
10	0.0163	0.0161	0.0156	0.0148	0.0138	0.0126	0.0114	0.0090	0.0069	0.0033	0.0016

RCR	A5	B5	C5	D5	E5	F5	G5	I5	K5	P5	U5
1	0.3722	0.1893	0.0617	0.0232	0.0103	0.0052	0.0029	0.0011	0.0005	0.0001	0
2	0.1786	0.1453	0.0883	0.0484	0.0266	0.0152	0.0091	0.0037	0.0017	0.0004	0.0001
3	0.1101	0.0997	0.0762	0.0524	0.0343	0.0223	0.0146	0.0067	0.0033	0.0008	0.0003
4	0.0747	0.0704	0.0596	0.0464	0.0343	0.0247	0.0176	0.0091	0.0049	0.0013	0.0005
5	0.0536	0.0516	0.0460	0.0386	0.0309	0.0240	0.0183	0.0105	0.0061	0.0018	0.0007
6	0.0401	0.0390	0.0359	0.0315	0.0266	0.0219	0.0176	0.0111	0.0069	0.0023	0.0009
7	0.0310	0.0303	0.0285	0.0258	0.0226	0.0193	0.0162	0.0110	0.0073	0.0027	0.0011
8	0.0245	0.0241	0.0230	0.0213	0.0191	0.0169	0.0146	0.0105	0.0074	0.0030	0.0013
9	0.0199	0.0196	0.0189	0.0177	0.0163	0.0147	0.0130	0.0098	0.0072	0.0032	0.0015
10	0.0164	0.0162	0.0157	0.0149	0.0139	0.0127	0.0115	0.0091	0.0069	0.0034	0.0016

Point P is located

Along l: $\frac{11}{40}(20) = 5.5$ squares in
Along w: $\frac{6}{20}(10) = 3$ squares down

Thus P lies between F and G on the grid at F–G,3.

Interpolation is required to obtain c_n from Table 8.1:

RCR	F3	F–G,3	G3
2	0.0142		0.0085
2.625	0.0184	0.0151	0.0118
3	0.0209		0.0137

$$c_n = 0.0151$$

The luminaire ratio is $W/w = \frac{2}{20} = 0.1$ giving $c = 1(0.0151) = 0.0151$.

Each luminaire emits 2000π lumens and thus has a luminous exitance of $2000\pi/8 = 250\pi$ lumens per square foot. Thus

$$E_P = 250\pi(0.0151) = 11.9 \text{ fc}$$

In addition to its basic limitation to perfectly diffusing sources, the configuration factor method using strip sources is limited by the accuracy of Eq. (8.10). This equation is valid only as long as the width W is small compared with the distance from source to receiver. For applications involving rows of fluorescent luminaires on ceilings, the method is generally quite accurate for luminaires up to 4 ft in width. For example, the exact illuminance at P in Fig. 8.9 is 14.0 fc, which represents less than a 3% error. For luminaires of greater width, such as luminous ceiling panels, another configuration factor should be used, namely that in Eq. (8.5) for a point on a plane to a parallel rectangular source. We will now develop that procedure.

A rectangular luminous ceiling is shown in Fig. 8.12. It is assumed to be perfectly diffusing with a luminous exitance $M = 300$ lumens per square foot. The horizontal illuminance at point P, on a work plane 10 ft below the ceiling, is desired.

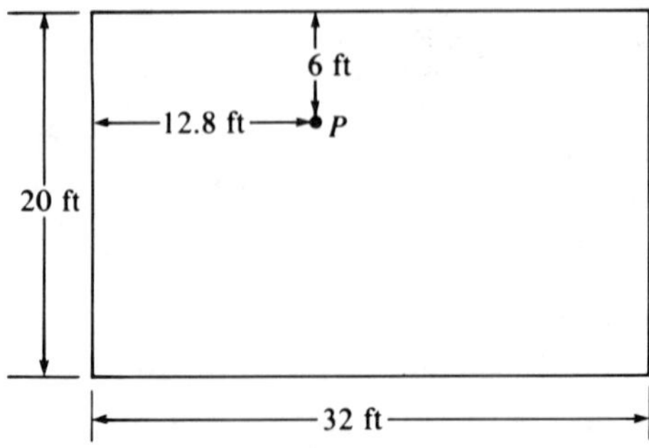

Figure 8.12

As in the previous procedure, the first step is to obtain the room cavity ratio.

$$\text{RCR} = \frac{5(10)(52)}{20(32)} = 4.0625$$

Next a grid is created as shown in Fig. 8.13, where each block represents 10% of the room length and width. The lettering and numbering of the coordinates indicate the symmetry that exists with a rectangular element.

The location of point P is shown on the grid in Fig. 8.13. Its coordinates are determined from

Along l: $$\tfrac{12.8}{32}(10) = 4 \quad (\text{E})$$

Along w: $$\tfrac{6}{20}(10) = 3$$

A set of configuration factors for a square ceiling element to a point on a horizontal work plane below is given in Table 8.2. For the given coordinates and room cavity ratio, we obtain

RCR	E3
4	0.6096
4.0625	0.6034
5	0.5102

$$c = 0.6034$$

Thus $$E_P = 300(0.6034) = 180 \text{ fc}$$

In this and the following two examples, it is justifiable to use four significant figures in the computations but not in the ultimate illuminance because the configuration factors in Table 8.2 are for square rooms whereas the rooms in the examples are rectangular.

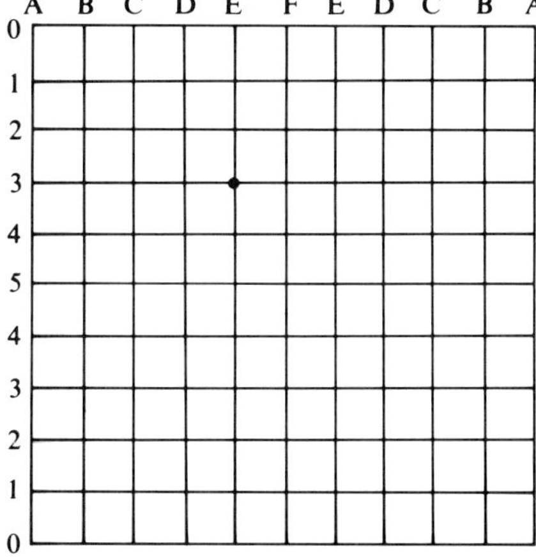

Figure 8.13

TABLE 8.2 CONFIGURATION FACTORS
Square Room ($l/w = 1$). Ceiling to Point on the Floor

RCR	A0	B0,A1	C0,A2	D0,A3	E0,A4	F0,A5	B1
1	0.2480	0.4244	0.4708	0.4837	0.4882	0.4893	0.7393
2	0.2421	0.3526	0.4159	0.4448	0.4571	0.4605	0.5227
3	0.2329	0.3092	0.3654	0.3989	0.4158	0.4208	0.4159
4	0.2212	0.2774	0.3231	0.3543	0.3718	0.3773	0.3513
5	0.2078	0.2507	0.2872	0.3138	0.3296	0.3348	0.3048
6	0.1935	0.2270	0.2559	0.2777	0.2911	0.2955	0.2678
7	0.1790	0.2054	0.2284	0.2459	0.2569	0.2606	0.2369
8	0.1648	0.1859	0.2041	0.2181	0.2269	0.2299	0.2105
9	0.1512	0.1681	0.1827	0.1939	0.2009	0.2033	0.1875
10	0.1385	0.1522	0.1639	0.1728	0.1785	0.1804	0.1676
RCR	C1,B2	D1,B3	E1,B4	F1,B5	C2	D2,C3	E2,C4
1	0.8144	0.8333	0.8394	0.8410	0.9010	0.9235	0.9308
2	0.6160	0.6563	0.6728	0.6773	0.7269	0.7752	0.7949
3	0.4928	0.5375	0.5595	0.5660	0.5848	0.6381	0.6641
4	0.4108	0.4808	0.4729	0.4799	0.4813	0.5286	0.5546
5	0.3505	0.3836	0.4031	0.4095	0.4039	0.4426	0.4653
6	0.3031	0.3295	0.3457	0.3511	0.3437	0.3742	0.3928
7	0.2642	0.2851	0.2981	0.3024	0.2953	0.3191	0.3338
8	0.2318	0.2481	0.2584	0.2619	0.2557	0.2742	0.2857
9	0.2043	0.2172	0.2252	0.2280	0.2229	0.2372	0.2462
10	0.1809	0.1910	0.1974	0.1996	0.1954	0.2066	0.2137
RCR	F2,C5	D3	E3,D4	F3,D5	E4	F4,E5	F5
1	0.9327	0.9478	0.9558	0.9587	0.9641	0.9662	0.9683
2	0.8002	0.8279	0.8495	0.8554	0.8721	0.8783	0.8847
3	0.6719	0.6967	0.7255	0.7340	0.7557	0.7647	0.7739
4	0.5628	0.5808	0.6096	0.6186	0.6399	0.6494	0.6591
5	0.4727	0.4852	0.5102	0.5184	0.5367	0.5453	0.5541
6	0.3990	0.4077	0.4281	0.4349	0.4496	0.4568	0.4642
7	0.3388	0.3450	0.3611	0.3666	0.3781	0.3838	0.3896
8	0.2896	0.2942	0.3067	0.3110	0.3199	0.3243	0.3289
9	0.2493	0.2527	0.2624	0.2657	0.2726	0.2760	0.2795
10	0.2161	0.2187	0.2262	0.2288	0.2341	0.2367	0.2395

EXAMPLE 8.5

Assume the ceiling indicated by the dashed lines in Fig. 8.9 is uniformly luminous with a luminous exitance of 200 lm/ft². Find the horizontal illuminance at point P.

Solution. The room cavity ratio is 3.0, as before. P is located

Along l: $\frac{12}{40}(10) = 3$ (D)
Along w: $\frac{8}{20}(10) = 4$

From Table 8.2, $c = 0.7255$. Then $E_P = 200(0.7255) = 150$ fc.

EXAMPLE 8.6

Use the configuration factor method for rectangular sources to find an approximate value for the illuminance at P in Example 8.4.

Solution. To solve this problem using rectangular sources, create two nearly square room cavities, as shown in Fig. 8.14, whose widths are $w = 20$ ft and whose lengths are $l = w \pm W/2 = 19$ ft and 21 ft. The room cavity ratios are

$$\mathrm{RCR}_{19} = \frac{5(7)(39)}{20(19)} = 3.592 \qquad \mathrm{RCR}_{21} = \frac{5(7)(41)}{20(21)} = 3.417$$

The locations of P are obtained from

19-ft room:	$10(\frac{10}{19}) = 5.263$	$10(\frac{6}{20}) = 3$
21-ft room:	$10(\frac{12}{21}) = 5.714$	$10(\frac{6}{20}) = 3$

giving coordinates in the range F–E,3 for both rooms.

Then Table 8.2 yields

19-FT ROOM

	F3		E3
3	0.7340		0.7255
3.592	0.6657	0.6634	0.6569
4	0.6186		0.6096

21-FT ROOM

	F3		E3
3	0.7340		0.7255
3.417	0.6859	0.6797	0.6772
4	0.6186		0.6096

Thus the net configuration factor is $c = 0.6797 - 0.6634 = 0.0163$, giving an illuminance of

$$E_P = 250\pi(0.0163) = 13 \text{ fc}$$

This answer is slightly in error because Table 8.2 was derived for square elements, whereas those in Fig. 8.14 are slightly rectangular.

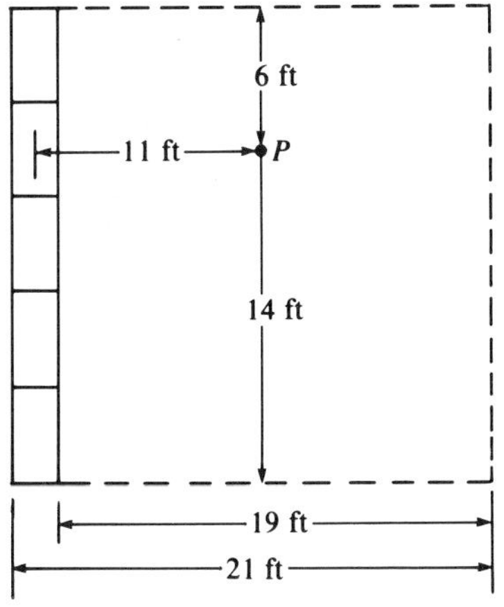

Figure 8.14

Slight errors in the configuration factors for the 19-ft and 21-ft rooms can cause a larger error in their difference. Thus, as mentioned earlier, we should use strip source configuration factors for luminaires 4 ft or less in width.

It may seem surprising that configuration factors for square ceiling elements are used in rectangular room problems. To explore this, replace the room in Fig. 8.12 by a square room 20 × 20 ft. For the same RCR, its height would be, from Eq. (7.7),

$$h = \frac{lw(\text{RCR})}{5(l+w)} = \frac{w(\text{RCR})}{10} = \frac{20(4.0625)}{10} = 8.125 \text{ ft}$$

The relative location of the point P′ in the square room would be 8 ft in from one edge and 6 ft in from the other. What we are saying is that the configuration factor for the point P in Fig. 8.12 is very nearly the same as for the point P' in Fig. 8.15. For this to be true requires that the length of the room be not excessively greater than its width (less than 3:1).

8.4 INFINITE PLANE METHOD: REFLECTED COMPONENT

Determination of the horizontal illuminance at a point on the work plane from reflected flux also involves the use of configuration factors and, in addition, the concept of an infinite plane, discussed earlier in Example 8.2. The method assumes that the walls, because they are perfectly diffusing, can be replaced by their infinite projections on the ceiling plane. Then, assuming the walls all have the same luminous exitance, the total reflected illuminance at the point on the work plane is the illuminance produced by the infinite plane plus the illuminance produced by a finite ceiling plane whose luminous exitance is the difference between its actual luminous exitance and that of the walls. In equation form,

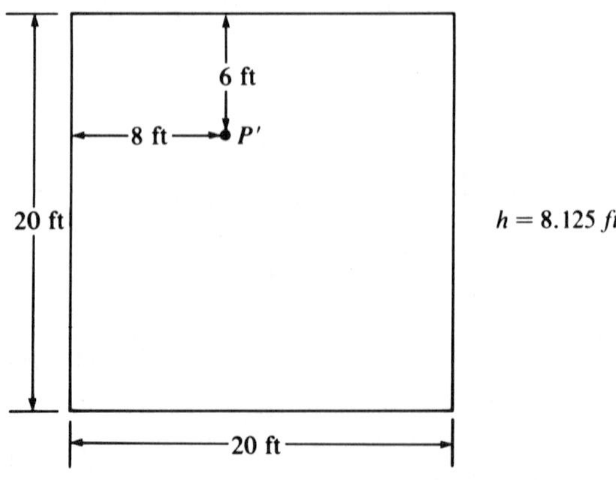

Figure 8.15

$$E_P = 1.0M_w + c(M_c - M_w) \tag{8.26}$$

where c is the configuration factor from the point to the ceiling plane and M_w and M_c are given by Eqs. (7.14) and (7.15), respectively. Inserting these equations into Eq. (8.26) yields

$$E_p = \frac{\Phi_i}{A}[\text{WEC} + c(\text{CEC} - \text{WEC})] \tag{8.27}$$

To illustrate the procedure, consider a room 20 × 30 ft with luminaires mounted 8 ft above the work plane. Assume the room reflectances are 70% for the ceiling, 50% for the walls, and 20% for the floor. Assume further that luminaire 6 in Table 7.8 is used, that there are 18 of them, and that each houses two 40-W cool white, rapid-start fluorescent lamps. The horizontal illuminance due to reflected flux on a desk located 2 ft from the long wall at its midpoint is desired.

The calculations follow:

$$\phi_i = 18(2)(3150) = 113{,}400 \text{ lm}$$

$$A = 20(30) = 600 \text{ ft}^2$$

Desk location (Fig. 8.13): Fl

$$\text{RCR} = \frac{5(8)(50)}{600} = 3.333$$

$$c = 0.537 \quad \text{(Table 8.2)}$$

$$\text{WEC} = 0.154 \quad \text{(Table 7.8)}$$

$$\text{CEC} = 0.172 \quad \text{(Table 7.8)}$$

$$M_w = \frac{113{,}400(0.154)}{600} = 29.1 \text{ lm/ft}^2 \quad \text{[Eq. (7.14)]}$$

$$M_c = \frac{113{,}400(0.172)}{600} = 32.5 \text{ lm/ft}^2 \quad \text{[Eq. (7.15)]}$$

$$E_P = 29.1 + 0.537(32.5 - 29.1) = 30.9 \text{ fc} \quad \text{[Eq. (8.26)]}$$

The procedure just developed does not require that the luminaire be perfectly diffusing but it does assume that the walls are perfectly diffusing and have the same luminous exitance. There is a procedure for perfectly diffusing luminous walls which do not have the same luminous exitance (see Balogh, 1966; O'Brien and Balogh, 1967). This procedure uses Eq. (8.6) to develop a table of configuration factors, similar to Table 8.2, for the case of a point lying in a plane perpendicular to the luminous element. Then a method very similar to that in Sec. 8.3 for rectangular ceiling elements is used to find the illuminance at the point on the work plane.

8.5 INVERSE-SQUARE-LAW APPROXIMATIONS

In recent years, computer procedures have been developed (DiLaura, 1975) which calculate the illuminance on a visual task from an array of luminaires. In these procedures, each luminaire is divided into elements, such that the maximum dimension of each element is small in comparison with the distance from the center of the element to the task. This constraint permits use of the inverse-square law of illumination (ISL) to calculate the illuminance at the task as the summation of the illuminances from the luminaire elements.

The horizontal illuminance E on a visual task produced by a point source of light can be calculated by the ISL as

$$E = \frac{I(\theta, \phi) \cos \theta}{d^2} \tag{8.28}$$

where $I(\theta, \phi)$ is the intensity of the source in the direction of the task, θ and ϕ are the polar and azimuth angles describing the location of the source with respect to the task, and d is the distance from source to task. In interior spaces the sources are customarily all at the same mounting height h above the task and it is more useful to have Eq. (8.28) in the form

$$E = \frac{I(\theta, \phi) \cos^3 \theta}{h^2} \tag{8.29}$$

The geometry to be used in this section is shown in Fig. 8.16. The task is located at the origin of an xy coordinate system such that $+y$ is always the viewing direction. A typical luminaire element is shown, with its center at x_e, y_e, h. The location of the center may also be described in terms of h, the polar angle θ, and the azimuth angle ϕ, as shown. θ can vary from 0 to 90°, whereas ϕ can vary from 0 to ±180°. A luminaire angle ψ is also shown. ψ is 0° when the luminaire is viewed lengthwise and is 90° for crosswise viewing.

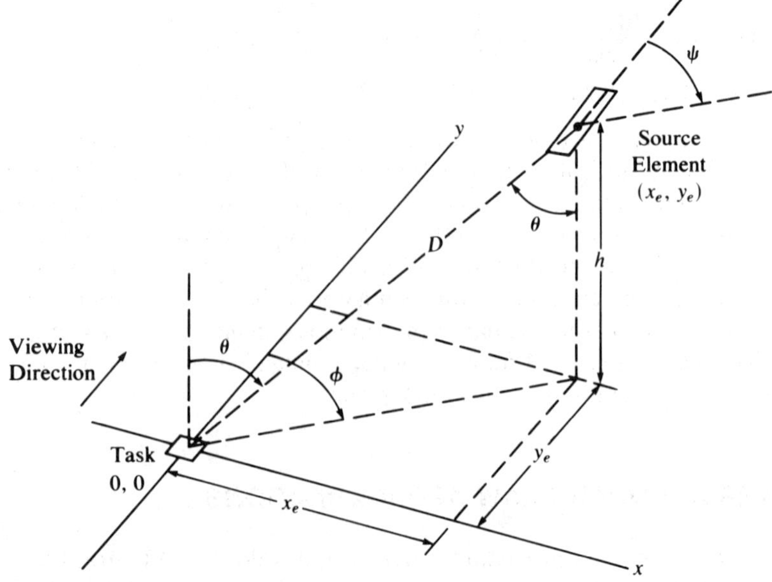

Figure 8.16

ϕ and ψ are not necessarily the same angles. ϕ is the azimuth angle at the task with respect to the vewing direction whereas ψ is the azimuth angle at the luminaire with respect to the luminaire axis.

Consider that the element shown in Fig. 8.16 has an area A_e and is part of a larger luminaire of area A_l. The approximate illuminance on the task from this element is, from Eq. (8.29),

$$E_{\text{ISL}} = \frac{A_e}{A_l}\left(\frac{I_{\text{mid}}(\theta, \phi)\cos^3\theta_{\text{mid}}}{h^2}\right) \tag{8.30}$$

where A_e/A_l is the fraction of the luminaire intensity attributable to the element and "mid" denotes that I, θ, and ϕ are with reference to the element's center. The subscript "ISL" denotes that E is an approximate illuminance obtained from the inverse-square law. The total illuminance on the task is a summation of the illuminances from all the luminaire elements.

In using Eq. (8.30), two problems arise. The first involves the determination of I_{mid} for each source element. The second has to do with deciding how many source elements are needed.

Luminaire manufacturers generally present intensity distribution data in 5 or 10° increments in polar angle θ for three values of luminaire azimuth angle ψ, namely parallel (0°), perpendicular (90°), and at 45° to the luminaire axis. It is thus necessary to develop interpolation procedures for obtaining intensity values at other θ and ψ angles.

First-degree polynomial interpolations, while easy to perform, are often quite inaccurate. Second- and third-degree polynomial interpolations, on the other hand, have been widely used with good results if appropriate cautions are observed. We will use a second-degree polynomial for ψ interpolations (since only three values of intensity are generally available for each θ) and a third-degree polynomial for those in θ. For the ψ interpolation,

$$I(\psi) = a_0 + a_1\psi + a_2\psi^2 \tag{8.31}$$

where

$$a_0 = I_0$$
$$a_1 = \frac{-3I_0 + 4I_{45} - I_{90}}{90}$$
$$a_2 = \frac{I_0 - 2I_{45} + I_{90}}{4050} \tag{8.32}$$

For each value of θ, this interpolation passes a parabola through the intensity values at $\psi = 0$, 45, and 90°, resulting in three coefficients which are then used to calculate $I(\psi)$ for the given value of ψ. This interpolation is done prior to the θ interpolation, and is repeated four times for each intensity determination, yielding intensity values labeled I_1, I_2, I_3, and I_4 at four consecutive θ angles, each $\Delta\theta$ apart, with two on each side of the given θ value. The procedure is illustrated in Table 8.3, where it is desired to obtain the intensity at $\psi = 56.9°$ and $\theta = 43.9°$. Here $\Delta\theta = 5°$ and it is necessary to perform the ψ interpolation at $\theta = 35$, 40, 45, and 50°. The resulting intensity values are $I_1 = 947$ cd, $I_2 = 857$ cd, $I_3 = 699$ cd, and $I_4 = 505$ cd.

TABLE 8.3 ILLUSTRATING DOUBLE INTERPOLATION TO OBTAIN $I(\psi, \theta)$

θ	I_0	I_{45}	$I_{56.9}$	I_{90}
35°	290	860	(I_1)947	1050
40°	260	780	(I_2)857	940
43.9°			(I) 738	
45°	220	650	(I_3)669	690
50°	180	490	(I_4)505	400

First interpolation: $\psi = 56.9°$

θ	a_0	a_1	a_2	
35°	290	16.89	−0.09383	$I_1 = 947$
40°	260	15.56	−0.08889	$I_2 = 857$
45°	220	13.89	−0.09630	$I_3 = 699$
50°	180	11.33	−0.09877	$I_4 = 505$

Second interpolation: $\theta = 43.9°$, $\theta' = 8.9°$

θ	A_0	A_1	A_2	A_3	I
43.9°	947	−9.067	−2.000	0.0427	738

For the θ interpolation there are two cases, namely $\Delta\theta = 5°$ and $\Delta\theta = 10°$. The pertinent equations are

$\Delta\theta = 5°$:
$$I(\theta') = A_0 + A_1\theta' + A_2(\theta')^2 + A_3(\theta')^3 \tag{8.33}$$

where
$$A_0 = I_1$$
$$A_1 = \frac{-11I_1 + 18I_2 - 9I_3 + 2I_4}{30}$$
$$A_2 = \frac{2I_1 - 5I_2 + 4I_3 - I_4}{50}$$
$$A_3 = \frac{-I_1 + 3I_2 - 3I_3 + I_4}{750} \tag{8.34}$$

$\Delta\theta = 10°$:
$$I(\theta') = A_0' + A_1'\theta' + A_2'(\theta')^2 + A_3'(\theta')^3 \tag{8.35}$$

where
$$A_0' = A_0 \qquad A_1' = \tfrac{1}{2}A_1$$
$$A_2' = \tfrac{1}{4}A_2 \qquad A_3' = \tfrac{1}{8}A_3 \tag{8.36}$$

The use of θ' requires explanation. To avoid large numbers in the calculation of $I(\theta')$, it is desirable to shift the $\theta = 0°$ axis to the value of θ at I_1. Thus in the example, $\theta' = 0$ corresponds to $\theta = 35°$. This yields values of θ' between 5 and 10° for $\Delta\theta = 5°$ and between 10 and 20° for $\Delta\theta = 10°$ in Eqs. (8.33) and (8.35), respectively. Referring again to Table 8.3, for $\theta = 43.9°$, $\theta' = 8.9°$ $(43.9 - 35)$.

It was mentioned earlier that certain cautions must be used with polynomial interpolations. To examine these constraints for a parabolic interpolation, recast Eq. (8.31) into a standard form for a parabola and obtain

$$I(\psi) - \left(a_0 - \frac{a_1^2}{4a_2}\right) = a_2\left(\psi + \frac{a_1}{2a_2}\right)^2 \tag{8.37}$$

For the values of θ and intensity in Table 8.3, Eq. (8.37) becomes

$$\begin{aligned}
\theta = 35°&: \quad I(\psi) - 1050 = -0.09383(\psi - 90°)^2 \\
\theta = 40°&: \quad I(\psi) - 941 = -0.08889(\psi - 87.5°)^2 \\
\theta = 45°&: \quad I(\psi) - 721 = -0.09630(\psi - 72.1°)^2 \\
\theta = 50°&: \quad I(\psi) - 505 = -0.09877(\psi - 57.4°)^2
\end{aligned} \tag{8.38}$$

Equations (8.38) are plotted in Fig. 8.17 and the twelve intensity values used in their calculation are marked, as are the four calculated intensities. Note that each curve has one point of zero slope [at the values in parentheses in Eqs. (8.38)] and that the curves do not have zero slope at $\psi = 0°$ or at $\psi = 90°$ (except by coincidence for the $\theta = 35°$ curve at $\psi = 90°$). This poses a problem because most luminaires have a symmetry which requires that the slopes of the intensity distribution curves be zero at $\psi = 0°$ and $\psi = 90°$. The parabolic interpolation violates this constraint. Also the parabolic interpolation permits an intensity value at a given ψ which is greater than the three values of intensity used to determine it. This may or may not be correct. The parabolic interpolation is quite good for luminaires possessing near-rotational symmetry but is likely to be in error for batwing luminaires, corridor luminaires, and luminaires designed to provide relatively uniform illuminance for rectangular areas.

An alternative to the parabolic interpolation on ψ has been suggested (Levin, 1984). The desire is to have an interpolation which (*a*) passes through three points, (*b*) has zero slope at its end points, and (*c*) possesses

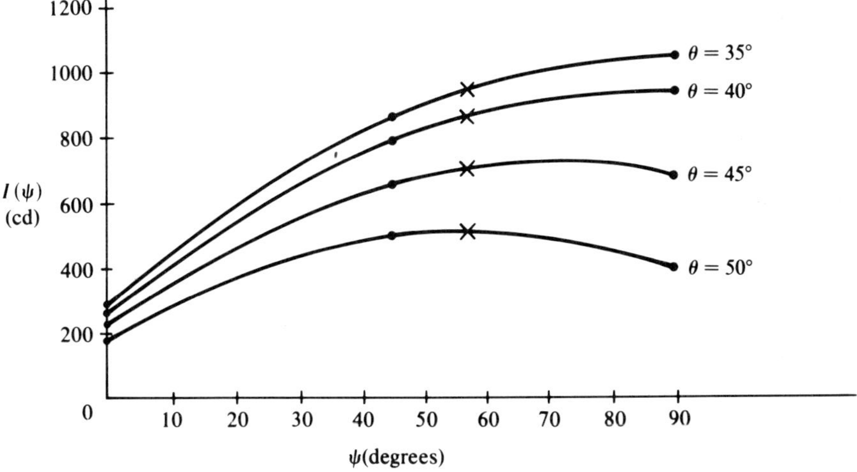

Figure 8.17

a derivative with no more than one zero in the interval. Such a function is

$$y = ax^4 + bx^3 + cx^2 \tag{8.39}$$

To utilize this equation for the ψ interpolation, we normalize so that the range $0 < \psi < 90$ becomes $0 < x < 1$ and $y(0) = I(0°)$, $y(\frac{1}{2}) = I(45°) - I(0°) \equiv A$ and $y(1) = I(90°) - I(0°) \equiv B$. Then we can write

$$A = \tfrac{1}{16}a + \tfrac{1}{8}b + \tfrac{1}{4}c \tag{8.40a}$$

$$B = a + b + c \tag{8.40b}$$

$$0 = 4a + 3b + 2c \tag{8.40c}$$

where Eq. (8.40c) arises from requiring the derivative to be zero at $x = 1$. Solution of Eqs. (8.40) yields

$$\begin{aligned} a &= 16A - 8B \\ b &= -32A + 14B \\ c &= 16A - 5B \end{aligned} \tag{8.41}$$

and inserting these results into Eq. (8.39) gives

$$y = (16A - 8B)\,x^4 - (32A - 14B)\,x^3 + (16A - 5B)\,x^2 \tag{8.42}$$

For the data in Table 8.3 at $\theta = 35°$,

$$A = 860 - 290 = 570 \qquad B = 1050 - 290 = 760$$

$$x = \frac{56.9}{90} = 0.632$$

$$a = 3040 \qquad b = -7600 \qquad c = 5320$$

$$y = 692$$

$$I_1 = 982 \text{ cd}$$

For the other values of θ,

$$\begin{aligned} I_2 &= 882 \text{ cd} \qquad \text{for } \theta = 40° \\ I_3 &= 715 \text{ cd} \qquad \text{for } \theta = 45° \\ I_4 &= 506 \text{ cd} \qquad \text{for } \theta = 50° \end{aligned}$$

Then, performing the θ interpolation indicated by Eq. (8.33) and (8.34) gives $I = 765$ *cd*. This compares with $I = 738$ cd obtained earlier using the parabolic interpolation. The difference in this case is slight (less than 3%). It is greater for small ψ.

There are other interpolation methods, most of them more complicated, that can be used with intensity distribution data. For further information, the reader is directed to "Interpolation of Intensity Distributions" by R. E. Levin (1984).

We turn next to the number of source elements required, which of course depends on the accuracy we wish to achieve. From Eq. (8.30), this accuracy depends on how I and θ vary from one portion of the element to another. Much has been written in the past about requiring that the distance from source to task be at least five times the maximum source dimension for the inverse-square law to yield accurate answers. This is true, if it is desired to have an error in E_{ISL} less than ±1%. But if a ±5% error is tolerable, then this distance can often be equal to the maximum source dimension if the intensity is reasonably constant over the source element and the source is reasonably Lambertian (Murdoch, 1981).

Table 8.4 has been prepared to display the variables which must be considered in selecting element lengths for rows of fluorescent luminaires.

TABLE 8.4 ELEMENT LENGTHS FOR ROWS PARALLEL (PERPENDICULAR) TO VIEWING DIRECTION

x_e, ft* (y_e, ft)	y_1, ft* (x_1, ft)	y_2, ft* (x_2, ft)	$\Delta\theta°$	$\Delta\phi°$
0	0	1.2	10	0
	1.2	2.5	10	0
	2.5	4	10	0
	4	6	11	0
	6	8	8	0
	8	12	11	0
	12	20	11	0
2	0	1.2	3	31
	1.2	2.5	6	20
	2.5	4	8	12
	4	6	9	8
	6	8	8	4
	8	12	10	4
	12	20	11	4
4	0	1.7	2	23
	1.7	4	7	22
	4	6	7	11
	6	8	6	7
	8	12	9	9
	12	20	10	7
6	0	1.7	2	16
	1.7	4	4	18
	4	8	9	19
	8	12	7	10
	12	20	9	10
8	0	4	2	27
	4	12	8	30
	12	20	8	12
10	0	4	2	22
	4	12	9	28
	12	20	7	13
12	0	8	4	34
	8	20	9	25
14	0	20	11	55

*All x and y values can be either + or −.

The table is based on $h = 7$ ft (Fig. 8.16) and can be used for rows parallel or perpendicular to the viewing direction. Dimensions for the latter are shown in parentheses.

In selecting element lengths, four factors have been considered.

1. *Luminaire length.* Element lengths were chosen to fit the standard 4-ft luminaire, or integral multiples thereof. For example, in Table 8.4 there is no element with y_1 $(x_1) = 2$ ft and y_2 $(x_2) = 6$ ft, where y_1 (x_1) and y_2 (x_2) are the coordinates of the ends of the element (see Fig. 8.16).
2. *Variation in polar angle θ.* A row directly over the task was considered first. For such a row, x_e $(y_e) = 0$ and there is no variation with azimuth angle ϕ, except in switching from $\phi = 0°$ directly in front of the task to $\phi = 180°$ directly behind it for $x_e = 0$. Element lengths in this row were chosen such that the change in θ would be limited to approximately 10° in traversing each element length. Without knowing how the intensity of a given luminaire will vary with θ, it was felt that $\Delta\theta \simeq 10°$ was the maximum change in θ that should be allowed. Also this choice is consistent with contemporary luminaire intensity distribution data, which is provided by the manufacturer in no more than 10° increments in θ. Values of $\Delta\theta$ for the element lengths chosen are given in Table 8.4.
3. *Variation in azimuth angle ϕ.* Intensity distribution data is normally provided in no less than three azimuth planes, each separated by 45° in ϕ. Again because the actual intensity distribution is not known, it was decided to limit variations in ϕ across an element to approximately half this value, or 22.5°, although in a very few cases, where variations in $\Delta\theta$ were small, a $\Delta\phi$ greater than 22.5° was allowed. Values of $\Delta\phi$ for the element lengths chosen are also given in Table 8.4
4. *% Error.* As a general guideline, element lengths were chosen to keep % error below 1% for elements within 8 ft horizontally of the task. For $h = 7$ ft, this means less than 1% error in E_{ISL} for all θ less than about 50°. Beyond 50°, it was felt that, in general, luminaire intensity would be lower and a greater % error in E_{ISL} (up to as much as 8%) could be tolerated.

EXAMPLE 8.7

To illustrate the use of Table 8.4, suppose it is desired to select element lengths for a row of fluorescent luminaires parallel to the viewing direction and displaced 5 ft horizontally from the task ($x_e = 5$ ft). The row runs from 8 ft behind the task to 16 ft in front of the task ($y = -8$ ft to $+16$ ft).

Solution. With $x_e = 5$ ft, the $x_e = 4$ column will be used, rather than the $x_e = 6$ column, to be conservative. The first element lies between

−8 ft and −6 ft, the last element between +12 ft and +16 ft (it would lie between +12 and +20 ft if the row went that far). Thus 10 elements are needed to represent the row.

Table 8.4 may be used also in the selection of element sizes for arrays of square luminaires. Consider the case of a 3 × 3 ft square luminaire. Reference to Table 8.4 shows that for x less than 6 ft and y less than 6 ft, suggested element lengths are 1.2 ft, 1.3 ft, 1.5 ft, 1.7 ft, 2 ft, and 2.3 ft. Roughly, the average element size is 1.5 ft. Thus, within a 6-ft-radius horizontal circle centered at the task, a 3 × 3 ft square luminaire will be divided into four 1.5 × 1.5 ft elements. Outside of this circle, it is permissible to use the entire 3 × 3 ft unit as a single element.

For a 4 × 4 ft square luminaire, similar arguments show that it should be separated into nine 1.33 × 1.33 ft elements if it lies inside a 3-ft-radius circle, four 2 × 2 ft elements if it lies outside a 3-ft-radius circle but within a 6-ft-radius circle, and considered as a single element if it lies outside a 6-ft-radius circle.

The inverse-square-law approximation procedures developed in this section may be used to find the horizontal illuminance at a point provided by the luminaires in Table 7.4. Intensity values for these luminaires are given in candelas per 1000 lamp lumens in Table 8.5. These are average intensities at each polar angle, obtained by dividing the sum of $I_{0°}$, $2I_{45°}$ and $I_{90°}$ by 4.

TABLE 8.5 INTENSITY IN CANDELAS (PER 1000 LAMP LUMENS) FOR LUMINAIRES IN TABLE 7.4

	Luminaire Number									
$\theta°$	1	2	3	4	5	6	7	8	9	10
5	73	238	274	263	189	210	107	312	132	253
15	73	264	302	258	176	211	104	268	144	249
25	73	248	349	236	147	212	99	213	181	236
35	73	191	321	210	110	204	90	148	202	214
45	73	122	209	163	64	164	80	87	173	172
55	73	63	46	98	35	79	67	51	113	96
65	72	46	8	56	21	37	52	30	63	45
75	71	38	3	30	10	26	36	16	43	19
85	70	32	3	11	3	18	22	4	28	7
95	67	28	4	8	4	16	15	0	23	0
105	63	28	8	15	19	22	15	0	31	0
115	58	41	16	22	41	27	14	0	30	0
125	55	43	23	31	67	23	13	0	20	0
135	51	33	29	47	93	19	12	0	10	0
145	48	23	34	60	117	12	10	0	8	0
155	47	9	42	83	136	8	9	0	5	0
165	45	3	28	105	151	5	7	0	2	0
175	44	1	3	111	155	4	6	0	0	0

To illustrate the calculation of illuminance from these data, reconsider the situation in Fig. 8.11, but assume now that luminaire No. 10 in Table 7.4 is used instead of a luminaire with a cosine intensity distribution. Assume further that each luminaire contains four 40-W cool white rapid start fluorescent lamps.

The luminaire row is 20 ft long and the point P is displaced 11 ft horizontally. From Fig. 8.6 and Table 8.4, with x_e chosen to be 10 ft and letting positive y be toward the top of the page, the luminaire elements will lie between −14 ft and −12 ft, −12 ft and −4 ft, −4 ft and 0, 0 and +4 ft, and +4 and +6 ft. Utilizing the notation in Fig. 8.6, the calculations can be presented in tabular form as follows:

H,ft	y_1,ft	y_2,ft	x_e,ft	y_e,ft	$\theta°$	I,cd	E,fc	E',fc
7	−14	−12	11	−13	67.7°	38	.0426	.27
7	−12	−4	11	−8	62.8°	56	.1095	2.76
7	−4	0	11	−2	57.9°	80	.2440	3.07
7	0	4	11	2	57.9°	80	.2440	3.07
7	4	6	11	5	59.9°	71	.1825	1.15
								10.32

First, values of θ are calculated from the given dimensions. Next, intensities are found from Table 8.5 in candelas per 1000 lamp lumens for a 4-ft-long luminaire. Linear interpolations have been used in this case. Then the illuminances (E) are calculated using Eq. (8.29). These illuminances must be scaled by two factors to obtain the actual illuminances (E'). The first factor is the number of thousands of lumens per 4-ft luminaire, given by $4(3150)/1000 = 12.6$. The second factor is the length of the luminaire element divided by 4 ft, giving $\frac{1}{2}$ for the first and fifth elements, 2 for the second element, and 1 for the third and fourth elements. The resulting illuminance at point P is approximately 10 fc.

8.6 ANGULAR COORDINATE DIRECT ILLUMINANCE COMPONENT (DIC) METHOD

The DIC method (IES, 1974; Jones et al., 1969; Kaufman, 1981) is an additional procedure for determining the direct component of illuminance at a point that does not require that the sources be Lambertian. It is most applicable to continuous rows of fluorescent luminaires, although it may also be used with individual luminaires that are separated.

Inherent in the method is locating the point P with respect to the luminaire in Fig. 8.18a. This location is specified in terms of the mounting height H, the longitudinal angle α, and the lateral angle β, where

$$\alpha = \tan^{-1}\frac{L}{H} \qquad \beta = \tan^{-1}\frac{R}{H} \tag{8.43}$$

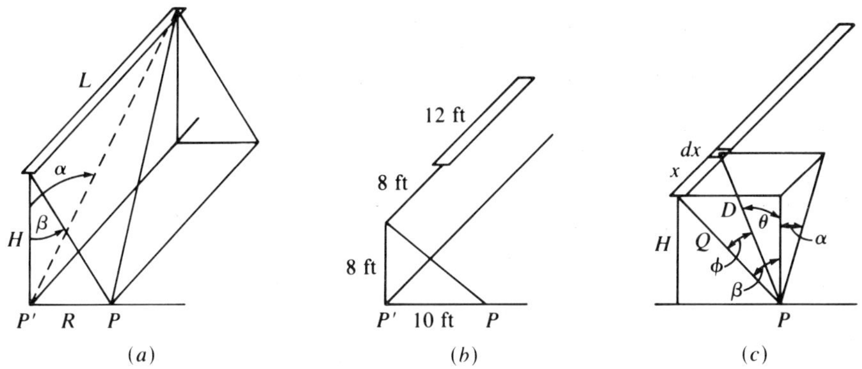

Figure 8.18 Angular coordinates for DIC method.

Note that the plane containing β lies at the end of the luminaire row and is perpendicular to it.

Data for use in calculating the direct component of horizontal illuminance are prepared by the manufacturer. An example for a prismatic wrap-around luminaire (Jones et al., 1969) is shown in Table 8.6(a). The numbers are horizontal footcandles on the work plane at point P for selected displacements R and luminaire row lengths L, as expressed by β and α, with H set at 6 ft. For other mounting heights, the values listed must be multiplied by $6/H$. Since the point P does not normally lie directly out from the end of a luminaire row, it is often necessary to use superposition to find the net illuminance at P.

EXAMPLE 8.8

For the situation in Fig. 8.18b, determine the horizontal illuminance at point P produced by the luminaire in Table 8.6(a) using the DIC method.

Solution. The problem must be solved by finding the illuminance produced by a 20-ft luminaire and subtracting that produced by an 8-ft luminaire. From Eqs. (8.43), the necessary angles are

$L = 20$ ft	$L = 8$ ft
$H = 8$ ft	$H = 8$ ft
$R = 10$ ft	$R = 10$ ft
$\alpha = 68.2°$	$\alpha = 45.0°$
$\beta = 51.3°$	$\beta = 51.3°$

The required interpolations in Table 8.6(a) are

	45	51.3	55
60	13.0		4.9
68.2		8.18	
70	13.4		5.2

	45	51.3	55
40	11.0		3.6
45.0		6.83	
50	12.3		4.4

TABLE 8.6(a) DIRECT HORIZONTAL ILLUMINANCE COMPONENTS FOR PRISMATIC WRAP-AROUND FLUORESCENT LUMINAIRE (IN FOOTCANDLES)

α \ β	5	15	25	35	45	55	65
0–10	9.4	8.9	7.8	5.8	3.1	0.9	0.2
0–20	18.4	17.3	15.2	11.4	6.1	1.8	0.4
0–30	25.9	24.6	22.0	16.5	8.8	2.7	0.6
0–40	31.7	30.1	27.1	20.8	11.0	3.6	0.8
0–50	35.0	33.2	29.7	22.8	12.3	4.4	1.0
0–60	36.1	34.3	30.7	23.7	13.0	4.9	1.2
0–70	36.5	34.7	31.1	24.1	13.4	5.2	1.4
0–80	36.7	34.9	31.3	24.3	13.6	5.4	1.5
0–90	36.7	34.9	31.3	24.3	13.6	5.4	1.5

Source: J. R. Jones, R. C. Levere, N. Ivanicki, and P. Chesebrough, "Angular Coordinate System and Computing Illumination at a Point," *Illuminating Engineering*, April 1969.

Then

$$E_P = \tfrac{6}{8}(8.18 - 6.83) = 1.0 \text{ fc}$$

In order to develop data such as that in Table 8.6(a) for a given luminaire, it is first necessary to calculate the luminaire intensities at the given α and β angles from published intensity values in the parallel, perpendicular, and 45° planes. This is done using an intensity interpolation technique, such as discussed in Sec. 8.5. In this particular instance, a piecewise cosine intensity interpolation was used (Jones et al., 1969), with the α angles taken at the midpoints of the zones in Table 8.6(a) (at $\alpha = 5°, 15°, \ldots, 85°$). The resulting intensity data for the luminaire in Table 8.6(a) are given in Table 8.6(b).

The next step is to obtain a set of zone multipliers, each of which gives the illuminance at point P per 1000 cd provided by an element of a luminaire at a given β and included within a given α range. These multipliers for horizontal illuminance are given in Table 8.7.

The zone multipliers may be derived with the aid of Fig. 8.18*c*. The differential horizontal illuminance at P produced by the luminaire element dx is

$$dE = \frac{dI \cos^3 \theta}{H^2} \tag{8.44}$$

The differential intensity is

$$dI = LW \cos \phi \, dx \tag{8.45}$$

where L is the luminance of the element in the direction of P and W is the projected width of the element toward P. Combining gives

$$dE = \frac{LW \cos \phi \cos^3 \theta \, dx}{H^2} \tag{8.46}$$

TABLE 8.6(b) INTENSITY DATA FOR LUMINAIRE IN TABLE 8.6(a)

Polar angle	Parallel	Perpendicular	45°
5	1300	1310	1310
15	1250	1370	1310
25	1150	1445	1330
35	980	1455	1310
45	640	1220	1150
55	275	660	450
65	135	290	240
75	85	200	180
85	40	145	125

α \ β	5	15	25	35	45	55	65	75	85
5	1310	1365	1440	1453	1213	658	290	200	145
15	1260	1320	1400	1390	1150	640	286	199	145
25	1155	1210	1320	1282	1022	600	280	196	145
35	982	1000	1045	1105	810	530	272	194	144
45	638	625	595	546	475	420	260	191	144
55	270	268	262	260	254	250	227	188	143
65	135	137	140	146	152	177	200	178	142
75	85	86	87	88	90	100	112	150	140
85	40	80	40	41	41	43	48	51	114

Source: J. R. Jones, R. C. Levere, N. Ivanicki, and P. Chesebrough, "Angular Coordinate System and Computing Illumination at a Point," *Illuminating Engineering*, April 1969.

We note that

$$\tan\phi = \frac{x}{Q} \qquad \cos\beta = \frac{H}{Q} \tag{8.47}$$

giving

$$x = \frac{H\tan\phi}{\cos\beta} \qquad dx = \frac{H\sec^2\phi\, d\phi}{\cos\beta} \tag{8.48}$$

TABLE 8.7 ZONE MULTIPLIERS

α \ β	5	15	25	35	45	55	65	75	85
0–10	7.144	6.519	5.395	3.993	2.576	1.379	0.533	0.127	0.0049
10–20	6.934	6.360	5.315	3.988	2.613	1.421	0.578	0.135	0.0052
20–30	6.515	6.033	5.141	3.961	2.663	1.507	0.632	0.151	0.0059
30–40	5.904	5.545	4.847	3.880	2.753	1.633	0.723	0.180	0.0072
40–50	5.113	4.875	4.400	3.691	2.790	1.787	0.858	0.230	0.0096
50–60	4.163	4.033	3.760	3.324	2.708	1.918	1.044	0.318	0.0145
60–70	3.093	2.038	2.917	2.712	2.391	1.910	1.234	0.471	0.0264
70–80	1.942	1.925	1.893	1.833	1.722	1.549	1.221	0.675	0.0657
80–90	0.842	0.842	0.837	0.830	0.823	0.790	0.749	0.578	0.1596

Source: J. R. Jones, R. C. Levere, N. Ivanicki, and P. Chesebrough, "Angular Coordinate System and Computing Illumination at a Point," *Illuminating Engineering*, April 1969.

We further note that

$$\cos\theta = \frac{H}{D} \qquad \cos\phi = \frac{Q}{D} \tag{8.49}$$

and thus

$$\cos\theta = \cos\phi \cos\beta \tag{8.50}$$

Inserting Eqs. (8.48) and (8.50) into Eq. (8.46) gives

$$dE = \frac{LW \cos^2\beta}{H} \cos^2\phi \, d\phi \tag{8.51}$$

and integrating yields

$$E = \frac{LW \cos^2\beta}{2H} (\phi + \sin\phi \cos\phi) \Big|_{\phi_1}^{\phi_2} \tag{8.52}$$

The intensity of a 4-ft element located at

$$\phi_m = \frac{\phi_1 + \phi_2}{2} \tag{8.53}$$

is

$$I = 4LW \cos\phi_m \tag{8.54}$$

Inserting Eq. (8.54) into Eq. (8.52) gives

$$E = \frac{I \cos^2\beta}{8H \cos\phi_m} (\phi + \sin\phi \cos\phi) \Big|_{\phi_1}^{\phi_2} \tag{8.55}$$

The zone multipliers are calculated for $I = 1000$ cd and $H = 6$ ft. Inserting these numbers yields

$$\text{ZM} = \frac{1000 \cos^2\beta}{48 \cos\phi_m} (\phi + \sin\phi \cos\phi) \Big|_{\phi_1}^{\phi_2} \tag{8.56}$$

EXAMPLE 8.9

Calculate the zone multiplier for $\beta = 25°$ and $\alpha = 0$ to $10°$ and use it to find the horizontal component of illuminance produced by the luminaire in Table 8.6(a). Repeat for $\alpha = 0$ to $20°$.

Solution. The first step is to find the ϕ angles for each α. From Fig. 8.18*c* and Eqs. (8.47),

$$\tan\alpha = \frac{x}{H} = \frac{\tan\phi}{\cos\beta} \tag{8.57}$$

$\alpha = 0°$:

$$\tan\phi = 0 \qquad \phi = 0°$$
$$\sin\phi = 0 \qquad \cos\phi = 1$$

$\alpha = 10°$:

$$\tan\phi = 0.1763(0.9063) = 0.1598$$
$$\phi = 9.078° = 0.1584 \text{ rad}$$
$$\sin\phi = 0.1578 \qquad \cos\phi = 0.9875$$
$$\phi_m = \frac{9.0780 + 0}{2} = 4.5390° \qquad \cos\phi_m = 0.9969$$

Inserting these values into Eq. (8.56) gives

$$ZM = \frac{1000(0.8214)}{48(0.9969)}[0.1584 + 0.1578(0.9875)] = 5.395$$

This is the zone multiplier appearing in Table 8.7 for $\beta = 25°$ and $\alpha = 0$ to $10°$. Multiplying by the intensity value in thousands for $\beta = 25°$ and $\alpha = 5°$ from Table 8.6(b) gives

$$E = 5.395(1.440) = 7.8 \text{ fc}$$

which appears in Table 8.6(a).

For $\alpha = 0$ to $20°$, we must first calculate the zone multiplier for $\alpha = 10$ to $20°$. The necessary quantities are

$\alpha = 20°$:

$$\phi = 18.256° = 0.3186 \text{ rad}$$

$$\sin\phi = 0.3133 \qquad \cos\phi = 0.9497$$

$\alpha = 10°$:

$$\phi = 9.078° = 0.1584 \text{ rad}$$

$$\sin\phi = 0.1578 \qquad \cos\phi = 0.9875$$

$$\phi_m = \frac{9.078 + 18.256}{2} = 13.667°$$

$$ZM = 5.315$$

Then

$$E = 7.8 + 5.315(1.400) = 7.8 + 7.4 = 15.2 \text{ fc}$$

8.7 PREDETERMINATION OF CRF AND VL

The concepts of contrast rendition factor (CRF) and visibility level (VL) were introduced in Chap. 5 and a method was presented for calculating these two quantities for an existing lighting installation, when task and background (adaptation) luminances are known. In this section we will develop a method for predetermining these quantities for a proposed lighting system.

The procedure for obtaining CRF and VL was presented in Sec. 5.8 and illustrated in Example 5.4. In that example, $\tilde{C}$ was obtained from the known task and background luminances in the real environment; $\tilde{C}_0$ was obtained from measurement in the perfectly diffusing sphere and the ratio $\tilde{C}/\tilde{C}_0$ gave CRF. Then, from Table 5.1, $\bar{C}$ for VL1 was found, and the ratio $\tilde{C}/\bar{C}$ gave VL.

For a proposed, but nonexisting, lighting installation, we cannot measure the task and background luminances. Rather we must calculate them, as a function of source and viewer locations with respect to the given tasks. Then the procedure of Sec. 5.8 can be applied to obtain CRF and VL.

At this point we must distinguish between reflectance factor ρ and luminance factor β. The former is the ratio of the total flux reflected from a surface to the total flux incident on the surface. The directivity of the reflected flux is of no concern.

Luminance factor is the ratio of the luminance of a surface under specified conditions of incidence and observation to the luminance of a completely reflecting, perfectly diffusing surface under the same conditions. For the normal viewing angle of 25° with the vertical, we have

$$\beta(25, \theta, \phi) = \frac{L(25, \theta, \phi)}{L_{rd}(\theta, \phi)} = \frac{L(25, \theta, \phi)}{E(\theta, \phi)} \tag{8.58}$$

where the quantities in parentheses show the dependence of β on the polar, azimuth, and viewing angles and the label "rd" refers to the perfectly reflecting and diffusing surface. The second form of the equation arises because the luminance in footlamberts of a perfectly reflecting and diffusing surface is identical to its illuminace in footcandles.

Referring back to Fig. 8.16, if we wish to obtain measured values of luminance factors for the given viewing angle as functions of source and viewer locations (with respect to a task location), it is necessary to place a collimated source of light at location θ, ϕ with respect to the task and a receiver (photometer) at a chosen task polar viewing angle (usually 25°) and at $\phi = 180°$. Then Eq. (8.58) can be rewritten in the forms:

$$\beta_{ti}(\theta, \phi) = \frac{L_{ti}(\theta, \phi)}{E_i} \tag{8.59}$$

$$\beta_{bi}(\theta, \phi) = \frac{L_{bi}(\theta, \phi)}{E_i} \tag{8.60}$$

where the i refers to the ith source element, t to task, and b to background.

Complete luminance factor data have been published for a concentric ring no. 2 pencil task. Approximating equations for the two β's for this task for 25° viewing have been developed (Murdoch, 1978) in the form

$$\beta = k + (a_0 + a_1\sqrt{\phi}) \ln\theta + (b_0 + b_1\sqrt{\phi})(\ln\theta)^2 + (c_0 + b_1\sqrt{\phi})(\ln\theta)^3 \tag{8.61}$$

where the angles are in degrees and the coefficients are

	Task	Background
k	0.736	0.852
a_0	−0.214	−0.125
a_1	0.0150	0.00597
b_0	0.188	0.105
b_1	−0.0135	−0.00530
c_0	−0.0309	−0.0149
c_1	0.00204	0.000524

Equation (8.61) and the associated coefficients do not include polarization effects, which are present to some extent in the pencil task.

With β_{ti} and β_{bi} known, we can use Eqs. (8.59) and (8.60) to calculate L_{ti} and L_{bi} for each source element present, after having calculated E_i using the method of Sec. 8.5. Then

$$\begin{aligned} E &= \Sigma E_i \\ L_t &= \Sigma L_{ti} \\ L_b &= \Sigma L_{bi} \end{aligned} \tag{8.62}$$

after which $\tilde{C}$ is obtained from Eq. (5.3).

In the above procedure we have found only the direct components of

E, L_t, and L_b. A similar procedure could be developed to take into account the contributions of reflections from room surface to these quantities. Such a procedure would replace the source elements with small sections of room surfaces and would be complicated by the fact that interreflections between room surfaces should be included, necessitating the use of form factors and matrix solutions. Large computer programs have been developed (DiLaura, 1975) to calculate both the direct and reflected components of E, L_t, and L_b.

As noted in Sec. 5.8, small changes in contrast can produce relatively large changes in visibility. Thus any general computer modeling procedure for obtaining task and background luminances should state clearly which of the following parameters (Kaufman, 1981) are included (no procedure currently extant takes into account the full range of all of these variables).

1. Room size and shape
2. Room surface reflectances
3. Room partitions and furniture
4. Luminaire characteristics
5. Number and location of luminaires
6. Nature and luminance of surfaces present in the environment
7. Observer location, line of sight, and viewing angle
8. Nature of the visual task
9. Effects of viewer shadowing of the task
10. Luminaire and task polarization effects

On the other hand, it should not be inferred from the above that simple models are not welcome. There is a tendency, exacerbated by the ubiquitous computer, to overquantify. Very often less complicated models of spaces can be useful in giving the designer "ball park" estimates of lighting parameters, particularly if the results are displayed graphically. Then designer judgment based on experience can supplant more complex calculations.

EXAMPLE 8.10

The standard no. 2 pencil task ($\tilde{C}_0 = 0.1675$) is placed on a desk at point T in a room and lighted by four identical luminaires whose intensity distribution is given in Fig. 8.19. The luminaires are 6.2 ft above the task and the viewer is facing in the direction of the arrow and viewing at the customary 25°. Calculate the CRF and VL present, ignoring any reflected components.

Solution. The solution is presented in tabular form below:

Source	$\theta°$	$\phi°$	I(cd)	E_i(fc)	β_{bi}	β_{ti}	L_{bi} (fL)	L_{ti} (fL)
1	48.75	45	2362	17.6	0.903	0.755	15.9	13.3
2	48.75	45	2362	17.6	0.903	0.755	15.9	13.3
3	48.75	135	2362	17.6	0.775	0.629	13.6	11.1
4	48.75	135	2362	17.6	0.775	0.629	13.6	11.1
				70.4			59.0	48.8

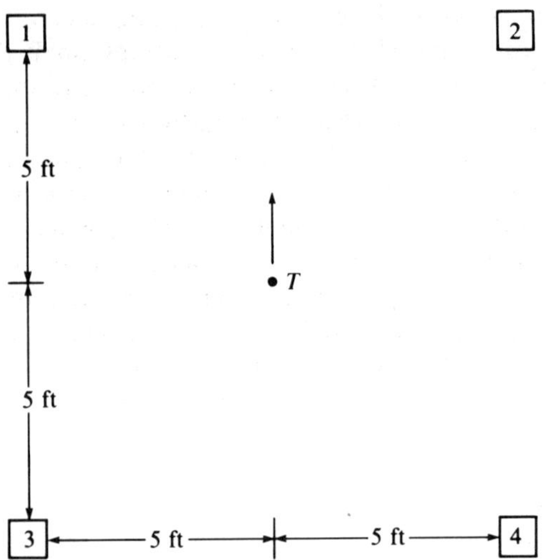

Figure 8.19

In this table, I has been calculated from Eqs. (8.33) and (8.34), E_i from Eq. (8.29), β_{bi} and β_{ti} from Eq. (8.61), and L_{bi} and L_{ti} from Eqs. (8.59) and (8.60).

Now, from Eq. (5.3)

$$\tilde{C} = \frac{59.0 - 48.8}{59.0} = 0.1729$$

Then, from Eq. (5.10),

$$\text{CRF} = \frac{0.1729}{0.1675} = 1.032$$

Entering Table 5.1 at $L_b = 59.0$ fL yields $\bar{C} = 0.0839$. This gives, from Eq. (5.7),

$$\text{VL} = \frac{0.1729}{0.0839} = 2.1$$

Note that the CRF provided is greater than one. This is because two of the luminaires are located behind the viewer and the other two are located well away from the viewer's line of sight, giving a viewing condition slightly better than in the reference sphere.

8.8 VISUAL COMFORT

Investigations into the matter of discomfort glare have been going on for the past 30 years. Throughout this time the objective has been to develop a universally applicable, easy-to-use procedure for evaluating quantitatively the discomfort produced by glare sources in the field of view. The problem is a difficult one. What is desired is a reasonably simple empirical formula that yields numerical values of glare sensation which can be related to estimates of glare effect by a variety of observers under a variety of seeing conditions. Such a formula has been derived and a procedure has been developed to permit computation of the percentage of observers who will be visually comfortable in a given seeing environment. It is the purpose of this section to describe this formula and procedure, and its limitations.

Discomfort glare is caused by direct glare from luminaires which are too bright, inadequately shielded, or too great in area. It is also caused by reflected glare from specular surfaces lighted by sources having concentrated beams. The procedure described herein considers the former but not the latter. There are five major factors influencing the amount of discomfort glare and resulting visual discomfort. These are field luminance (F) in candelas per square meter; position index (P) of each glare source; visual size (ω) in steradians of each glare source; luminance (L) in caldelas per square meter of each glare source; and the number of glare sources (n).

Visual comfort can be evaluated for any location within a space and for any direction of viewing from that location. It has been found that the worst viewing condition is typically near the center of a wall with the viewer looking horizontally across the room. In order to have a common basis for the comparison of lighting systems, it has been decided to base visual comfort evaluations on a standard viewing position 4 ft out from the center of a wall and 4 ft above the floor with the viewer looking forward horizontally. This evaluation is normally made both lengthwise and crosswise in the room, to include effects of room shape and luminaire orientation.

Field Luminance

This is the background luminance against which the glare sources are viewed. It should be the solid angle weighted average of all luminances in the field of view, given by

$$F = \frac{L_w\omega_w + L_f\omega_f + L_c\omega_c + \Sigma L_s\omega_s}{5} \tag{8.63}$$

where w, f, c, and s denote walls, floor, ceiling, and source, respectively, and the 5 in the denominator is based on the assumption that the total field of view is 5 steradians (a cone angle of about 78°). From the standard viewing position, the average luminance of the opposite wall is sometimes used in place of F in Eq. (8.63) as an approximation of field luminance, particularly in small rooms with high ceilings.

Logically, a higher field luminance should mean a lesser glare effect, because the background (adaptation) luminance is higher and the luminances of the glare sources are closer in value to the luminance of their background.

Position Index

This is a measure of the location of the glare source with reference to the horizontal line of sight of the observer. We graph position index versus location in Fig. 8.20, noting from the graph that P increases as H or Y increases but decreases as X increases. This means that glare sources making a large angle at the eye with the horizontal line of sight have a larger position index than those making smaller angles. Such sources should produce less glare effect and thus we anticipate that the glare effect is inversely proportional to position index.

The data in Fig. 8.20 may be represented by the equation:

$$\ln P = \left(\frac{317}{9} - \frac{287}{900}\tau - \frac{11}{9}\epsilon^{-2\tau/9}\right) 10^{-3}\sigma + \left(21 + \frac{4}{15}\tau - \frac{2}{675}\tau^2\right) 10^{-5}\sigma^2 \tag{8.64}$$

where

$$\tau = \tan^{-1}\frac{Y}{H} \qquad \sigma = \tan^{-1}\frac{\sqrt{H^2 + Y^2}}{X} \tag{8.65}$$

The angles τ and σ in Eqs. (8.65) are to be entered in degrees.

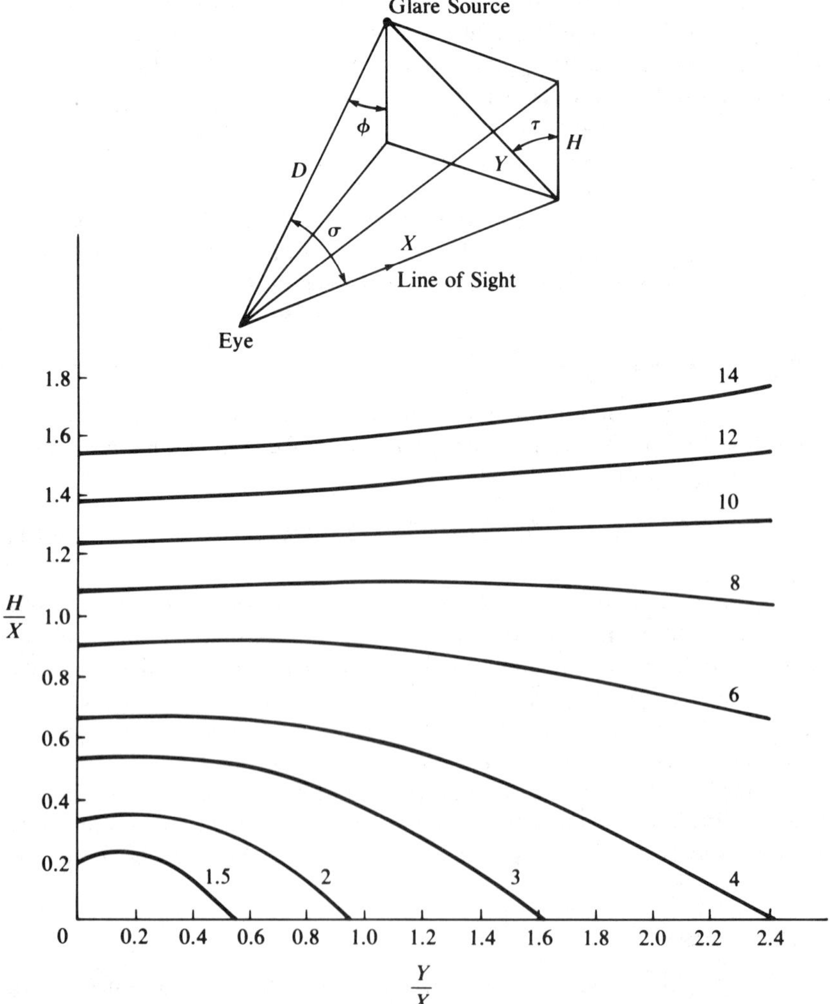

Figure 8.20 Position index P for discomfort glare evaluation.

Visual Size

The visual size of a glare source is the solid angle in steradians that it subtends at the eye. From the diagram in Fig. 8.20, this angle is given by

$$\omega = \frac{A_p}{H^2 + X^2 + Y^2} \tag{8.66}$$

where A_p is the projected area of the glare source in the direction of the observer, given by

$$A_p = A \cos \phi = A \frac{H}{\sqrt{H^2 + X^2 + Y^2}} \tag{8.67}$$

with A being the actual glare source area and ϕ the angle between the vertical and the line between glare source and observer.

Data taken from many observers indicate that the glare effect is related to ω in a complicated way. An empirical formula has been developed to express this relationship. The result is that the glare effect is directly proportional to a quantity Q, given by

$$Q = 20.4\omega + 1.52\omega^{0.2} - 0.075 \tag{8.68}$$

A plot of Q versus ω is shown in Fig. 8.21.

Table 8.8 is included to relate a solid angle to its plane angle at the tip of a cone. For example, a solid angle of 0.1 steradian means an angular spread at the observer's eye of 10.2°.

Source Luminance

This is the average luminance L of the glare source in the direction of the observer and is usually computed from the candlepower distribution data of the

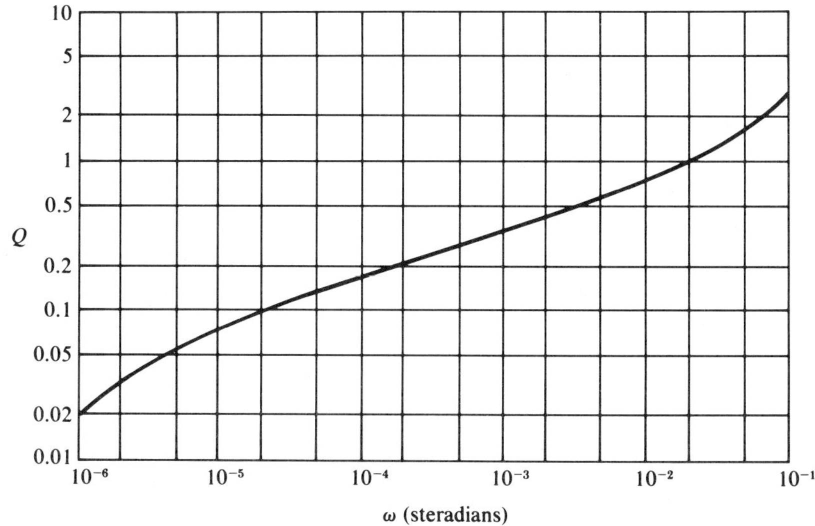

Figure 8.21

TABLE 8.8

Solid angle (steradians)	Cone angle (degrees)
5.0	78.2
3.0	58.5
2.0	47.0
1.0	32.7
0.5	23.0
0.2	14.5
0.1	10.2
0.05	7.2
0.02	4.6
0.01	3.2
0.005	2.3
0.002	1.4
0.001	1.0
0.0001	0.3

luminaire. If I is the luminaire intensity in candelas in the direction of the observer and A_p is the projected area of the luminaire in that direction, the luminance is given by I/A_p. It is reasonable to assume that the glare effect will be directly proportional to L.

Number of Sources

Certainly the glare effect will increase as the number of glare sources increases, but because glare sources differ in size, luminance, and position, the relationship should not be expected to be a linear one.

The fundamental glare formula was developed by S. K. Guth (1963) and gives the glare sensation M of a single source as

$$M = \frac{0.5LQ}{PF^{0.44}} \tag{8.69}$$

where L and F are in candelas per square meter. We note in this formula that the glare sensation varies directly with source luminance (L) and source size (Q) and inversely with position index (P) and field luminance (F), the latter to the 0.44 power.

In determining the glare effect of a source, it has been determined that the most meaningful measure is that value of source luminance which produces a sensation at the borderline between comfort and discomfort for the average observer. We call such a luminance a BCD luminance. The relationship between BCD luminance and source solid angular size for sources on the line of sight and with constant field luminance has been determined experimentally and is shown in Fig. 8.22. This rather unusual looking curve can be described as follows: For small sources (less than 10^{-2} steradian), the curve is reasonably linear on a log-log plot. For larger sources, source area becomes increasingly more important and the curve slopes downward. In other words, as source area becomes quite large, an observer can tolerate considerably less luminance. The curve then reaches a minimum and turns upward, showing that the source is now becoming a significant part of the visual field and is less identified as a spot of glare by an observer. Eventually the curve levels when the entire field is filled by the source.

The basic curve can be expanded into families of parallel curves by including field luminance and position index as additional variables. Either increasing field luminance or displacing the source from the line of sight gives higher values of BCD luminance for sources of 10^{-1} steradian or less. For example, increasing field luminance by 5:1 or moving the source 20° above the line of sight doubles the BCD luminance. For sources larger than 10^{-1} steradian, the curves are no longer parallel and actually come together when the source fills the field. The region up to 10^{-1} steradian (10°) includes most common luminaires and it is this region for which the glare sensation formula and the equation for Q apply. Large area glare sources remain a subject for further research inquiry.

We turn next to multiple sources. In developing a procedure for these, two approaches are possible:

1. Derive a new formula for multiple sources.
2. Develop a summation procedure using the single source formula.

It is the latter course that has been pursued.

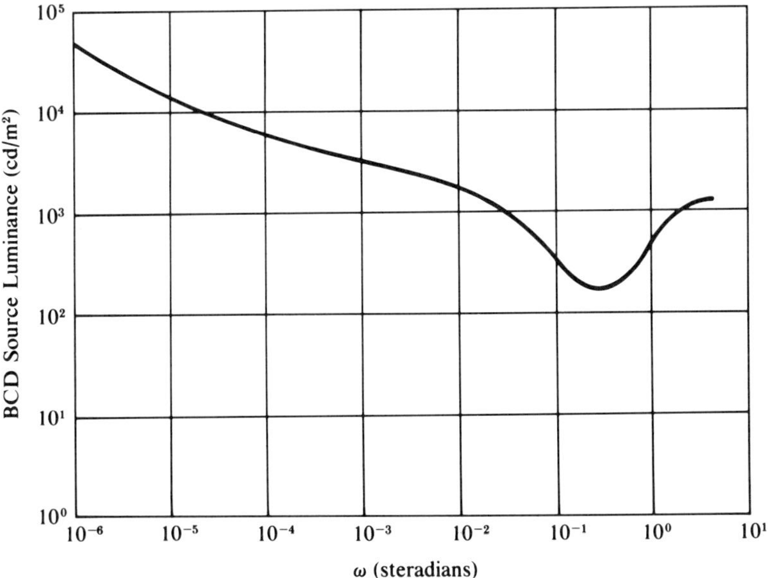

Figure 8.22 BCD luminance versus source solid angle.

It is tempting to simply sum the glare sensations of the several glare sources in a given installation to get the overall glare sensation. But this procedure does not yield correct results, largely because of the nonlinear relationship between glare sensation and source size ω. Rather it is necessary to develop experimental data in which observers are asked to establish their comfort or discomfort for groups of luminaires of varying sizes, luminances, and positions, much as was done originally for a single glare source. This has been done by Guth, using 210 observers and 48 lighting conditions. Furthermore, Guth has developed a variable exponential method of computing the overall glare sensation which gives good agreement with experimental results. First the individual glare sensations are summed to obtain

$$M_t = M_1 + M_2 + \cdots + M_n \tag{8.70}$$

Then an overall discomfort glare rating is defined by

$$\mathrm{DGR} = (M_t)^a \tag{8.71}$$

in which the exponent a is a function of the number of glare sources present in the area being evaluated. This relationship has been determined experimentally to be

$$a = n^{-0.0914} \tag{8.72}$$

A few values of a as a function of n are given in Table 8.9. Note that a decreases rapidly at first as n increases but then levels off.

Glare sensations for a single source and discomfort glare ratings for many sources are numbers to which it is hard to give physical interpretation. Higher values of M and DGR indicate greater discomfort, but how much greater is difficult to say.

Guth has devised a method of expressing discomfort glare in terms of the percentage of a group of observers which would likely find a given M or DGR at, or more comfortable than, the borderline between comfort and dis-

TABLE 8.9

n	a
1	1
2	0.939
3	0.904
4	0.881
6	0.849
8	0.827
10	0.810
20	0.760
50	0.699
100	0.656
200	0.616

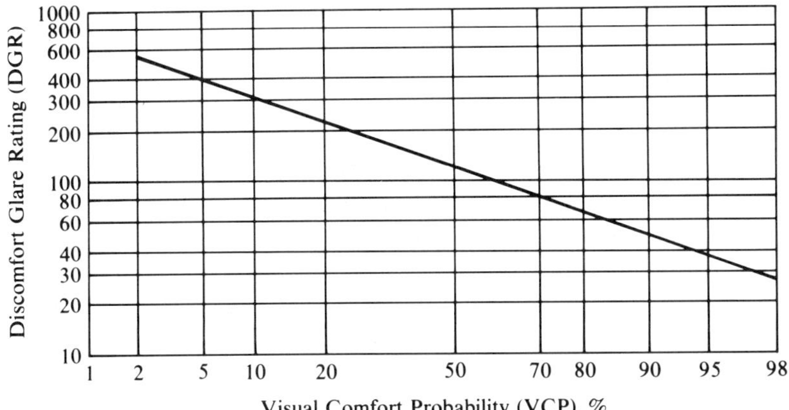

Figure 8.23 Discomfort glare evaluation.

comfort. The experimental relationship between discomfort glare rating (DGR) and visual comfort probability (VCP) is shown in Fig. 8.23, which is based on 107 observers in 35 different experimental conditions. It may be expressed analytically by (Kaufman, 1981)

$$\text{VCP} = \frac{100}{\sqrt{2\pi}} \int_{-\infty}^{K} \epsilon^{-k^2/2}\, dk \tag{8.73}$$

where k is a dummy variable of integration and

$$K = 6.374 - 1.3227 \ln \text{DGR} \tag{8.74}$$

The graph in Fig. 8.23 shows that 50% of the observers are likely to find a given installation at or more comfortable than BCD when the DGR is 120. For 90% of the observers to be at or more comfortable than BCD, the DGR must be 45, about 37% of the value for 50% VCP. A summary of glare sensation descriptions and DGR and VCP values is given in Table 8.10.

The Illuminating Engineering Society has established an alternative glare criterion. Their statement is that direct glare will not pose a visual com-

TABLE 8.10

Glare description	DGR	VCP
Unnoticeable	35	95
Perceptible	50	87
Acceptable	65	75
Distracting	90	64
BCD	120	50
Barely uncomfortable	160	34
Perceptibily uncomfortable	220	20
Uncomfortable	300	11
Intolerable	400	5

fort problem in a lighting installation if the following three conditions are met:

1. The VCP is 70 or greater.
2. The ratio of maximum to average luminaire luminance does not exceed 5:1 at angles of 45°, 55°, 65°, 75°, and 85° from the vertical for both lengthwise and crosswise viewing.
3. Maximum luminaire luminances lengthwise and crosswise do not exceed the following values:

Angle from vertical	Maximum (cd/m^2)
45°	7710
55°	5500
65°	3860
75°	2570
85°	1695

EXAMPLE 8.11

Assume a glare source is located 30° above the line of sight and has a luminance of 2000 cd/m^2 in the direction of an observer in the rear of the room. Further assume that projected source size is 10^{-2} steradian and that the field luminance is 100 cd/m^2. Compute the glare sensation M and the visual comfort probability.

Solution. First, from Eq. (8.64) with $\tau = 0$ and $\sigma = 30°$, we obtain a position index of 3.35. Next, from Eq. (8.68), we find $Q = 0.734$ for $\omega = 0.01$ steradian. Now using Eq. (8.69) we have

$$M = \frac{0.5LQ}{PF^{0.44}} = \frac{0.5(2000)(0.734)}{3.35(100^{0.44})} = 29$$

and entering Fig. 8.23 tells us that about 97% of the occupants of the area will be comfortable.

To see the relative effect of the various factors, let us first halve the field luminance to 50 cd/m^2. We find that M becomes 39 and VCP 94. Thus a 50% decrease in F produces only a 34% increase in M and a 3% decrease in VCP. The mathematical reason for this relatively small influence of F is that it is raised to the 0.44 power in the glare sensation formula.

If the glare source is located only 10° above the line of sight instead of 30°, P becomes 1.40. Then M becomes 69 and VCP about 75. Thus a 3:1 decrease in angle of elevation results in a 2.4:1 increase in M and 23% decrease in VCP.

Next, if the projected source size increases to 0.05 steradian, Q becomes 1.78, M becomes 70, and VCP 75. Lastly, if source luminance is increased 2.4 times, M is 69 and VCP is again 75.

Thus a 3:1 decrease in elevation angle, a 5:1 increase in source size, and a 2.4:1 increase in source luminance all produce a 2:1 increase in M and a 16% decrease in VCP. For field luminance to produce this same effect, F would have to decrease from 100 to about 20 cd/m^2, a 5:1 decrease.

EXAMPLE 8.12

Assume four identical sources are located on the ceiling of a room in the following positions:

Source	X(ft)	Y(ft)	H(ft)
1	8	4	6
2	8	−4	6
3	14	4	6
4	14	−4	6

Furthermore assume each source is a square panel 1.5 ft on a side with a luminance in the direction of the observer of 1700 cd/m^2 and that the field luminance is 170 cd/m^2. Compute the glare sensation (M), disability glare rating (DGR), and the visual comfort probability (VCP).

Solution. We will first obtain the position indexes. The computations may be tabulated as follows:

Source	$\tau(°)$	$\sigma(°)$	P
1	33.7	42.0	4.5
2	33.7	42.0	4.5
3	33.7	27.3	2.4
4	33.7	27.3	2.4

Next we need values for Q. Again, in tabulated form

Source	D(ft)	A_p(ft^2)	ω	Q
1	10.8	1.25	0.011	0,77
2	10.8	1.25	0.011	0.77
3	15.8	0.85	0.0034	0.48
4	15.8	0.85	0.0034	0.48

Now the glare sensation can be calculated from Eq. (8.69) as

$$M = \frac{0.5(1700Q)}{P(170^{0.44})} = \frac{88.7Q}{P}$$

This yields $M = 15.2$ for the first two sources and $M = 17.7$ for the last two. Summing gives $M_t = 65.8$.

Noting that $4^{-0.0914} = 0.881$, the discomfort glare rating is $\text{DGR} = 65.8^{0.881} = 40$ and, from Fig. 8.23, the visual comfort probability is about 93%.

In order to systematize VCP calculations and permit comparisons of luminaires under similar conditions, the IESNA has developed a set of standard conditions for VCP determination (RQQ, 1973). These are

1. An initial average horizontal illuminance of 100 fc
2. Reflectance values of $\rho_{cc} = 80\%$, $\rho_w = 50\%$, and $\rho_{fc} = 20\%$
3. Mounting heights of luminaires of 8.5, 10, 13, and 16 ft above the floor
4. Standardized room lengths and widths
5. A standard layout of uniformly distributed luminaires
6. A limit to the field of view at the standard observer location of 53° upward from the forward viewing direction

Even with these standardizations, the VCP system should be used with some caution. We have already mentioned limiting its use to glare sources of less than 0.1 steradian. In addition, it should be noted that most of the research leading to VCP involved luminances such as one obtains from fluorescent lamps and verification largely involved fluorescent luminaires. Thus use of the method with high luminance, small area sources may not yield useful results. Finally, two lighting systems should differ by more than 5 in VCP if they are to be judged different in a visual comfort sense.

REFERENCES

Balogh, E.: "Infinite Plane Luminance Difference Technique for Computing Illumination," *Illuminating Engineering,* April 1966.

DiLaura, D. L.: "On the Computation of Equivalent Sphere Illumination," *Journal of the IES,* January 1975.

Guth, S. K.: "A Method for the Evaluation of Discomfort Glare," *Illuminating Engineering,* May 1963.

———: "Computing Visual Comfort Ratings for a Specific Interior Lighting Installation," *Illuminating Engineering,* October 1966.

IES Design Practice Committee: "The Determination of Illumination at a Point in Interior Spaces," *Journal of the IES,* January 1974.

Jones, J. R., R. C. Levere, N. Ivanicki, and P. Chesebrough: "Angular Coordinate System and Computing Illumination at a Point," *Illuminating Engineering,* April 1969.

Kaufman, J. E.: *IES Lighting Handbook,* 1981 ed., Illuminating Engineering Society, New York, 1981.

Levin, R. E.: "Illumination Due to Area Sources Expressed in Angular Coordinates," *Journal of the IES*, October 1971.

———: "Position Index in VCP Calculations," *Journal of the IES,* January 1975.

———: "Interpolation of Intensity Distributions," Annual Conference of the Illuminating Engineering Society, St. Louis, August 1984.

McGowan, T. K., and S. K. Guth: "Extending and Applying the IES Visual Comfort Rating Procedure," *Illuminating Engineering,* April 1969.

Moon, P.: *The Scientific Basis of Illuminating Engineering,* McGraw-Hill, New York, 1936.

Murdoch, J. B.: "A Procedure for Calculating the Direct Components of CRF and ESI with a Programmable Hand Calculator," *Journal of the IES,* October 1978.

———: "Inverse Square Law Approximation of Illuminance," *Journal of the IES,* January 1981.

O'Brien, P. F. and E. Balogh: "Configuration Factors for Computing Illumination Within Interiors," *Illuminating Engineering,* April 1967.

RQQ Committee: "Report No. 2" and "Appendix to Report No. 2," *Journal of the IES,* April and July 1973.

Chapter 9

Daylighting

9.1 BACKGROUND MATERIAL

Prior to the 1973 Arab oil embargo, interest in daylighting in the United States was at a low level. With energy inexpensive, the electric lighting industry in this country was quick to point out the disadvantages of daylighting—glare, variability, difficulty of control, excessive illuminances. Although there was a modest amount of analytically based work going on in daylighting, the major use of daylight was by architects, who allowed it to spill into interior spaces to create visual effects, rather than to provide quality lighting for doing visual work.

All of that has of course changed in the intervening years. There has been a resurgence of interest in admitting daylight into interior spaces, not just to enhance those spaces psychologically and aesthetically, but to provide quality lighting at the appropriate quantitative level for excellence of seeing. In fact we have come to realize that properly designed daylighting can do a better job than electric lighting in minimizing veiling reflections, if the light is supplied from the side rather than from overhead. However, the major reason for the renaissance in daylighting is energy conservation and cost reduction arising from being able to turn the electric lights off in an interior space for a significant portion of the day.

The qualitative, architectural aspects of daylighting are well-covered in the literature (Evans, 1981; Hopkinson et al., 1966; IES, 1978; Kaufman, 1981; Libbey-Owens-Ford, 1976). We will review briefly the major daylighting configurations in the next section, pointing out that each is an application of toplighting, sidelighting, or a combination thereof. Then in the remainder of the chapter, we will concentrate on the quantitative aspects of daylighting, specifically toward providing the designer with sufficient information so that the average work-plane illuminance and the illuminances at several key points on the work plane in a space may be estimated.

The calculation procedures to be presented have four major components. First, information on the luminances and illuminances to be expected from the sun and sky is assembled. Conventionally, two sky conditions have been considered, namely overcast and clear, and these will be addressed here. Recently, interest has arisen in providing data for a third sky condition, the partly cloudy sky.

With the sun and sky data in hand, the designer can predict the external illuminance, vertical, horizontal, or sloped, on the fenestration (window, skylight, etc.). This may require inclusion of not only direct radiation from the sun and sky but also reflected radiation from the ground and adjacent structures.

The third major component of the daylighting design process is to ascertain the visual transmission characteristics of the fenestration material. Generally there are two types of transmittances to consider, namely direct transmittance of sunlight and diffuse transmittance of clear or overcast skylight. The transmittances of the fenestration for these two types of input may indeed be different and the former may depend on solar altitude.

The final step is to process the luminous flux which enters the interior space. Three procedures will be presented, the lumen method for skylighting, the lumen method for sidelighting, and the daylight factor method. The first of these permits calculation of the average horizontal illuminance on the work plane and is very similar to the zonal cavity method of electric lighting design presented in Chap. 7. The second procedure yields the horizontal illuminance at three selected points on the work plane: in the middle of the space, 5 ft from the window wall, and 5 ft from the wall opposite the window wall. The third procedure permits calculation of the illuminance at any work-plane point within the space as a fraction of the horizontal illuminance outdoors at the same instant of time from an unobstructed, generally overcast sky.

9.2 DAYLIGHTING DESIGN CONSIDERATIONS

Daylighting may be provided through a variety of architectural forms, a few of which are shown in Fig. 9.1. In sidelighting, the amount of light admitted depends on the area of the opening, the glazing material used, and whether or not any control elements such as blinds or shades are employed. The distribution of the light depends on the location of the opening in the side wall. Generally, useful light levels can be achieved at distances into the room of up to $2\frac{1}{2}$ times the height of the window opening. This is illustrated by the unilateral design in Fig. 9.1*a*. The uniformity is improved by the use of a bilateral design (Fig. 9.1*b*), but it is somewhat rare that a room will have two outside walls opposite each other.

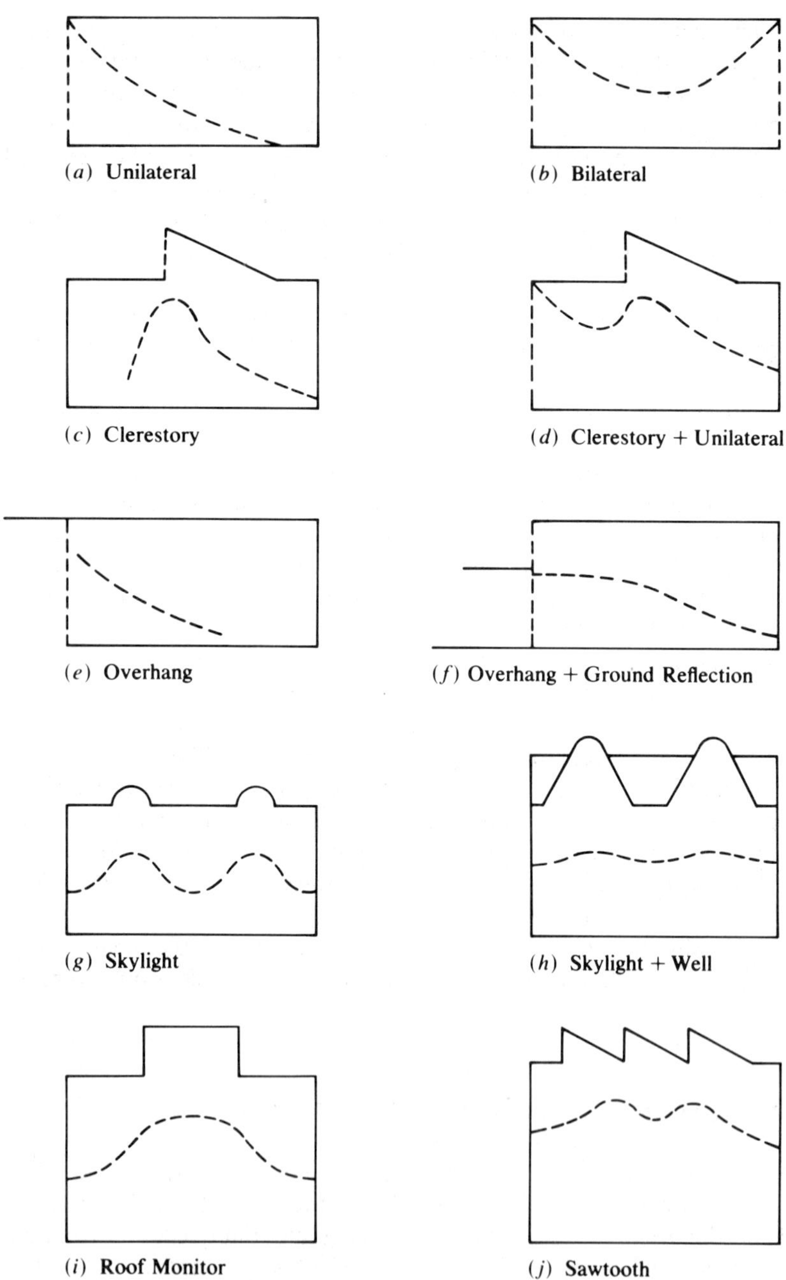

Figure 9.1 (Dashed lines represent illuminance distributions.)

Use of a clerestory element, either alone or in combination with a unilateral element (Fig. 9.1*c* and *d*) can provide deeper penetration of daylight while still maintaining uniformity. The clerestory provides light to the portion of the room most remote from the window wall whereas the unilateral element lights the portion nearest the window wall.

Direct sunlight on windows is generally to be avoided, because of the glare it produces. Sometimes windows may be oriented so as to avoid this problem and also translucent or tinted glazing materials may be used. The latter, however, severely reduce the amount of skylight admitted also.

Another solution is to use overhangs (Fig. 9.1*e*) to shield the room from direct sunlight. These also reduce the illumination from the sky and the penetration of the daylight into the space. A slightly more complicated procedure (Fig. 9.1*f*) is to use the overhang as a reflective surface to direct light deeper into the room. Also, reflections from ground surfaces can be used to replace the light near the window which is lost due to the overhang.

In toplighting, the double-domed skylight (Fig. 9.1*g*) is the most popular unit. When combined with a light well (Fig. 9.1*h*), it can provide quite uniform horizontal illuminance. The interior of the well should be painted a matte-finish white and the well angle should be between 45 and 60°.

A roof monitor (Fig. 9.1*i*) is another method of toplighting but does not generally achieve the uniformity of a skylighting installation. It can be used in high-ceiling industrial settings, as can the sawtooth arrangement (Fig. 9.1*j*). The latter is quite popular in warehouse and storage areas.

9.3 SOLAR ILLUMINANCE

The average illuminance on a surface perpendicular to the sun's rays and lying just outside the earth's atmosphere can be calculated from

$$\bar{E}_0 = k \int_{0.38}^{0.77} G_{s\lambda} \, V(\lambda) \, d\lambda \qquad (9.1)$$

where $G_{s\lambda}$ is the solar spectral irradiance in watts per unit of surface area per micron, $V(\lambda)$ is the spectral luminous efficiency of the eye, and k is the maximum luminous efficacy (683 lm/W). Note the similarity of Eq. (9.1) to Eq. (3.21).

The $G_{s\lambda}$ function has been standardized (see Gillette and Pierpoint, 1982), and when it and $V(\lambda)$ are inserted into Eq. (9.1) and the integration is performed the result is

$$\bar{E}_0 = 127.5 \text{ Klx} = 11{,}850 \text{ fc}$$

These numbers can be thought of as the lighting equivalents of the solar constant (1377 W/m²), obtained by performing $\int_0^\infty G_{s\lambda} \, d\lambda$.

The quantity $\bar{E}_0$ is an average illuminance. It must be adjusted for a given date to account for the elliptical shape of the earth's orbit around the sun. The actual illuminance on day n of the year is given by

$$E_0 = \bar{E}_0 \left(1 + 0.033 \cos \frac{360n}{365}\right) \qquad (9.2)$$

E_0 is the solar illuminance outside the earth's atmosphere on day n on a surface perpendicular to the sun's rays. To obtain the solar illuminance E_p at sea level on the same day on a similarly oriented surface requires that we account for attenuation through the earth's atmosphere. Atmospheric attenuation is a function of the composition of the atmosphere and of the length of the path traversed by the sun's rays. Thus E_p can be expressed as

$$E_p = E_0 \, e^{-cm} \tag{9.3}$$

where c is the optical atmospheric extinction coefficient and m is the relative optical air mass. This latter quantity is the ratio of the mass of atmosphere in the actual earth-sun path to the mass when the sun is directly overhead at sea level, expressed by

$$m = \frac{1}{\sin h} \tag{9.4}$$

where h is the angle of elevation of the sun above the horizon.

The atmospheric extinction coefficient has been measured at several locations throughout the world with a wide variety of results. Recent measurements in the United States give a value of $c = 0.210$ for the clear sky case. We will use this as a typical value for c.

Combining Eqs. (9.2), (9.3), and (9.4) yields the final expression for solar illuminance at sea level on a surface perpendicular to the sun rays:

$$E_p = \bar{E}_0 \left(1 + 0.033 \cos \frac{360n}{365}\right) e^{-0.210/\sin h} \tag{9.5}$$

We are seldom interested in the solar illuminance on this perpendicular surface. Rather we wish to know the illuminance provided by sunlight on horizontal and vertical surfaces (skylights and windows). To obtain these latter illuminances requires that we position the sun accurately in the sky at a given time, date, and location on the earth in terms of its altitude (vertical) angle h and its azimuth (horizontal) angle ϕ. These angles can be expressed in terms of the local latitude L, the solar declination δ, and the local solar time, given in terms of the hour angle H.

The five angles are shown in Fig. 9.2, where we have placed a celestial sphere around and centered at an observer's point on the earth's surface. The celestial equator is in the plane of the earth's equator and the celestial poles are at the ends of the earth's axis of rotation. The local latitude L is shown as the angle between the plane of the equator and the observer's zenith. It is also the angle between the earth's axis of rotation and the observer's north horizon line. Solar declination δ is the angular distance of the sun north of the celestial equator, the complement of the angle between the earth-sun line and the earth's axis of rotation. This angle arises because the earth's axis is tilted at an angle of 23.45° with respect to the normal to its orbital plane. Each year in northern latitudes, δ is +23.45° around June 21, −23.45° around December 21, and 0° around March 21 and September 21. On any date, δ may be obtained approximately from

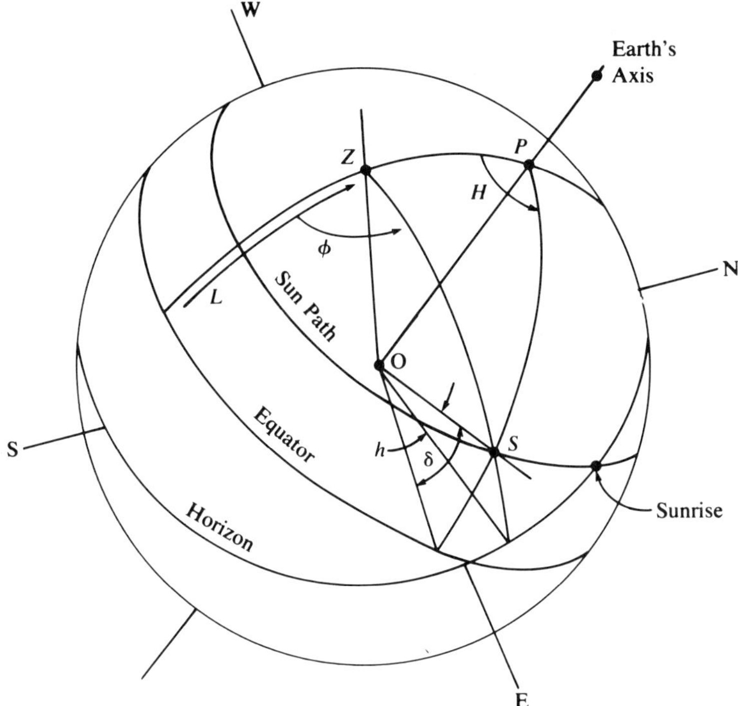

Figure 9.2

$$\delta = 23.45 \sin \beta \qquad \beta = \frac{360(n - 81)}{365} \quad \text{degrees} \tag{9.6}$$

For example, on March 21, n is 80 and Eq. (9.6) gives a declination of $-0.4°$.

Because an observer's location is generally not at the center of a time zone and because the earth's orbit around the sun is slightly elliptical and thus its orbital velocity varies slightly throughout the year, solar time, as would be measured by a sun dial, varies slightly from local clock time. There are two corrections to make in going from the observer's clock time to solar time. The first is a longitude time correction for the observer's location and is given by

$$\Delta T = 4(\lambda_{TZ} - \lambda_0) \tag{9.7}$$

where λ_{TZ} is the longitude of the center of the time zone, λ_0 is the longitude of the observer, and ΔT is the time difference in minutes between the center of the time zone and the observer's location. The 4 arises because each time zone is 15° wide and it takes 1 h for the sun to traverse a time zone. Thus the sun travels 1° in 4 min.

The second correction, which addresses the variation due to the elliptical orbit, is given by the equation of time, which can be approximated as

$$\text{ET} = 9.87 \sin 2\beta - 7.53 \cos \beta - 1.5 \sin \beta \tag{9.8}$$

The equation of time yields a maximum variation of about ±15 min during the course of a year, with the sun running behind local time by about 15 min in February and ahead of local time by about 15 min in October and November.

With the two corrections defined, we can write

$$\mathrm{ST} = \mathrm{TZ} + \mathrm{ET} + \Delta T \tag{9.9}$$

where ST is solar time and TZ is time zone time, both in minutes. To equate ST to the hour angle H in degrees shown in Fig. 9.2, we first convert ST to the number of minutes before or after noon, with before noon positive and after noon negative, and then multiply the result by 0.25, since 60 min of time corresponds to 15 degrees of angle.

EXAMPLE 9.1

Find the solar declination δ and the hour angle H at 8.30 A.M. Central Standard time on October 21 at 32 degrees north latitude and 95 degrees west longitude.

Solution. Tables 9.1 and 9.2 will be of assistance in solving this and similar problems. From Table 9.1, we see that $n = 294$ for October 21. Thus, from Eq. (9.6)

$$\delta = 23.45 \sin \frac{360(213)}{365} = 23.45 \sin 210 = -11.8°$$

From Table 9.2, the middle of the Central time zone is at 90 degrees west longitude. From Eq. (9.7)

$$\Delta T = 4(90 - 95) = -20 \text{ min}$$

From Eq. (9.8)

$$\begin{aligned}\mathrm{ET} &= 9.87 \sin 420 - 7.53 \cos 210 - 1.5 \sin 210 \\ &= 8.55 + 6.52 + 0.75 = 16 \text{ min}\end{aligned}$$

TABLE 9.1

Month	Days in month	n
January	31	i
February	28	$31 + i$
March	31	$59 + i$
April	30	$90 + i$
May	31	$120 + i$
June	30	$151 + i$
July	31	$181 + i$
August	31	$212 + i$
September	30	$243 + i$
October	31	$273 + i$
November	30	$304 + i$
December	31	$334 + i$

TABLE 9.2

Zone	Name	Standard meridian
+4	Atlantic	60° *W*
+5	Eastern	75° *W*
+6	Central	90° *W*
+7	Mountain	105° *W*
+8	Pacific	120° *W*

Then, from Eq. (9.9)

$$\text{ST} = 8{:}30 + 0{:}16 - 0{:}20 = 8{:}26 \text{ A.M.}$$

This is 214 min before noon, and thus

$$H = 0.25(214) = 53.5^\circ$$

With L, δ, and H known, we are now in a position to return to Fig. 9.2 and obtain expressions for solar altitude h and azimuth ϕ. The desired relationships can be obtained by applying the cosine and sine laws of spherical trigonometry to the spherical triangle *ZPS* in Fig. 9.2 The resulting equations are

$$\sin h = \cos L \cos \delta \cos H + \sin L \sin \delta \tag{9.10a}$$

$$\sin \phi = \frac{\cos \delta \sin H}{\cos h} \tag{9.10b}$$

There is a possible ambiguity in Eq. 9.10b. For example, if $\sin \phi$ in Eq. 9.10b is .99, $|\phi|$ is either 81.8° or 98.2°. From March 21 to September 21, the magnitude of the azimuth angle is greater than 90° in the early morning and late afternoon. For other dates during the year it is not.

A single test can be made to determine which ϕ angle is correct. In Eq. 9.10b, let $\phi = 90^\circ$ ($\sin \phi = 1$). Combine this equation with Eq. 9.10a, eliminate h, and solve for H in terms of the remaining variables, δ and L. The result is

$$\cos H = \frac{\tan \delta}{\tan L} \tag{9.11}$$

Then, for the given δ and L, find H from Eq. 9.11 and divide by .25 to obtain the solar time ST in minutes before or after noon. Compare this ST with the actual ST to determine which of the two ϕs given by Eq. 9.10b is correct.

EXAMPLE 9.2

Find the solar altitude and azimuth for the date, time, and location in Example 9.1.

Solution. Applying Eqs. (9.10) directly gives

$$\begin{aligned}\sin h &= \cos 32 \cos(-11.8) \cos 53.5 + \sin 32 \sin(-11.8)\\ &= 0.848(0.979)(0.595) + 0.530(-0.204) = 0.386\\ h &= 22.7^\circ\end{aligned}$$

$$\sin \phi = \frac{\cos(-11.8) \sin 53.5}{\cos 22.7} = \frac{0.979(0.804)}{0.923} = 0.853$$

$$\phi = 58.5^\circ$$

With a means now available for calculating solar altitude and azimuth, we return to the original problem of determining the illuminances on horizontal and vertical surfaces produced by the sun when we know the solar illuminance on a surface perpendicular to the sun's rays. The situation is depicted in Fig. 9.3. Knowing E_p, we obtain

$$E_H = E_p \cos(90 - h) = E_p \sin h \tag{9.12}$$

and

$$E_V = E_p \cos \psi = E_p \cos h \cos \gamma \tag{9.13}$$

EXAMPLE 9.3

For the date, time, and location of Example 9.1, find the solar illuminance on a horizontal skylight and on a vertical window facing east.

Solution. From Eq. (9.5), the perpendicular solar illuminance is

$$\begin{aligned}E_p &= 11{,}850\left[1 + 0.033 \cos \frac{360(294)}{365}\right] e^{-0.210/\sin 22.7}\\ &= 11{,}850(1.011)(0.580) = 6950 \text{ fc}\end{aligned}$$

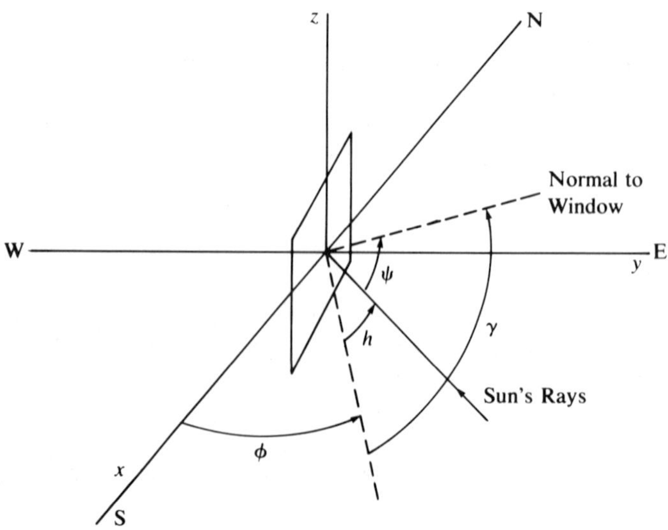

Figure 9.3

The illuminance on the skylight is, from Eq. (9.12),

$$E_H = 6950 \sin 22.7 = 2680 \text{ fc}$$

For the vertical window, γ is $90 - 58.5 = 31.5°$, and Eq. (9.13) yields

$$E_V = 6950 \cos 22.7 \cos 31.5 = 5460 \text{ fc}$$

It is instructive to consider which of the parameters are most significant in determining the values of perpendicular and horizontal solar illuminance. In Eq. (9.5), the quantity in parentheses has very little effect on E_p. Thus, as a close approximation, E_p is a function of solar altitude only. With this approximation, E_H in Eq. (9.12) is also a function of solar altitude only. E_p, with the parenthetical term omitted, is plotted in Fig. 9.4. Also plotted there are E_H, from Eq. (9.12), and E_V, from Eq. (9.13), the latter for various values of γ, the horizontal angle between the sun's location and the normal to the vertical surface.

Values of solar altitude for selected dates, times, and latitudes are given in Table 9.3 to facilitate use of Fig. 9.4.

9.4 OVERCAST AND CLEAR SKY LUMINANCES

Until quite recently, the major source of data on sky luminance, both overcast and clear, in the United State was the work done by H. H. Kimball and I. F. Hand at the U.S. Weather Bureau from 1919 to 1922 (1923). The results of these measurements are presented in Tables 9.4 and 9.5 in the form of equivalent sky luminances for overcast and clear sky conditions. As Kimball and Hand point out, these are average values. They state that for

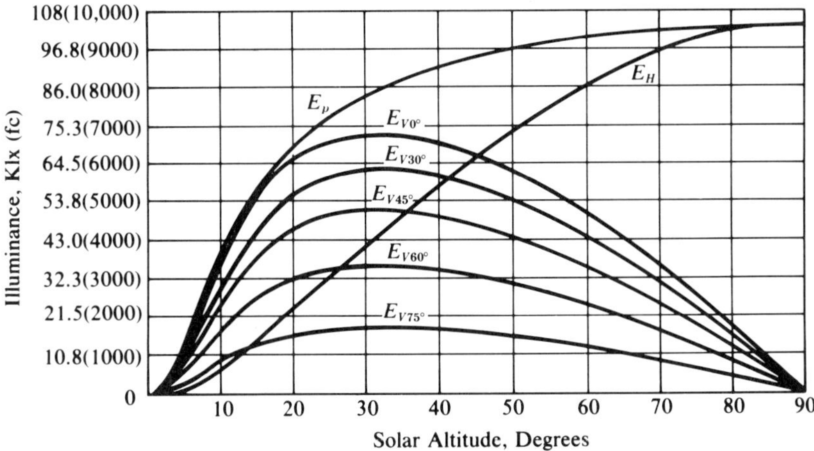

Figure 9.4 Solar illuminance. (*Source:* IES Daylighting Committee, *Recommended Practice of Daylighting,* IES, New York, 1979, fig. 37-A, with permission.)

TABLE 9.3 DEGREES OF SOLAR ALTITUDE VERSUS LATITUDE, DATE, AND TIME

Latitude	Date	A.M. / P.M.	6 / 6	7 / 5	8 / 4	9 / 3	10 / 2	11 / 1	Noon / Noon
30°N	June 21		12	24	37	50	63	75	83
	Mar.–Sept. 21			13	26	38	49	57	60
	Dec. 21				12	21	29	35	37
34°N	June 21		13	25	37	50	62	74	79
	Mar.–Sept. 21			12	25	36	46	53	56
	Dec. 21				9	18	26	31	33
38°N	June 21		14	26	37	49	61	71	75
	Mar.–Sept. 21			12	23	34	43	50	52
	Dec. 21				7	16	23	27	28
42°N	June 21		16	26	38	49	60	68	71
	Mar.–Sept. 21			11	22	32	40	46	48
	Dec. 21				4	13	19	23	25
46°N	June 21		17	27	37	48	57	65	67
	Mar.–Sept. 21			10	20	30	37	42	44
	Dec. 21				2	10	15	20	21

clear skies, their measurements varied between 60 and 150% of the average; for cloudy skies between 30 and 200%. They also comment that the illuminance under a rainy sky is about one-half that of a cloudy sky and that the illuminance from a partly cloudy sky may be three to four times as great as from a clear sky.

An overcast sky has been defined rather vaguely by the IESNA as one that has 100% cloud cover with no sunlight visible. This definition does not address the uniformity of the cloud cover, its ceiling, its density, nor its thickness. An overcast sky normally has a luminance distribution which varies with polar angle θ (angle from the zenith) but is relatively constant with changes in the azimuth angle ϕ. The luminance at the zenith is often 2.5 to 3 times that at the horizon. Nevertheless, it has become customary in the United States to use a single value of overcast sky luminance for a given latitude, date, and time, which is called equivalent sky luminance and is chosen so as to produce the same horizontal illuminance as would be produced by the actual nonuniform overcast sky. It is values of equivalent sky luminance that appear in Table 9.4.

A word of explanation about the units in Tables 9.4 and 9.5 is in order. As noted in Chap. 2, the footlambert (fL) is not a unit of luminance. However it has been common practice in daylighting circles to use it as such. When it is used in this way, the number of footlamberts is simply π times the number of candelas per square foot. Thus, to convert from footlamberts to candelas per square meter in Tables 9.4 and 9.5, we multiply the number of footlamberts by $10.76/\pi = 3.425$.

In Sec. 2.5, we derived an expression for the horizontal illuminance E_H produced by a sky of constant luminance. The result was

$$E_H = \pi L \tag{9.14}$$

If L is in candelas per square meter, E_H is in lux; if L is in candelas per square foot, E_H is in footcandles; if L is in footlamberts, we omit the π from Eq. (9.14) and E_H in footcandles is numerically equal to L in footlamberts.

The dominant force in determining the equivalent sky luminance of an overcast sky is the solar altitude h. If the data in Table 9.4 are replotted versus h, a single curve results which is shown as the solid curve in Fig. 9.5. It has been found (Dietz, et al., 1981) that this curve may be approximated within 1% by the equations (h entered in degrees)

$$\begin{aligned} L_0 &= 120h \quad \text{cd/m}^2 = 35h \quad \text{fL} \qquad 0° < h < 60° \\ L_0 &= 9.11h^{1.63} \quad \text{cd/m}^2 = 2.66h^{1.63} \quad \text{fL} \qquad 60° < h < 90° \end{aligned} \tag{9.15}$$

Use of Fig. 9.5 or Eqs. (9.15), along with Eq. (9.10a), permits a determination of equivalent overcast sky luminance for any latitude, date, and time, in contrast to Table 9.4 which applies only to the equinoxes and solstices. Also, from Eq. (9.14), the values in footlamberts obtained from Fig. 9.5 or from Eqs. (9.15) are numerically equal to the horizontal illuminances in footcandles.

Thus far we have presented overcast sky luminance in terms of the Kimball and Hand data only. While this data has been widely used in the United States, it has not been accepted worldwide and recent work in this country seems to indicate that it is in error at the larger solar altitudes.

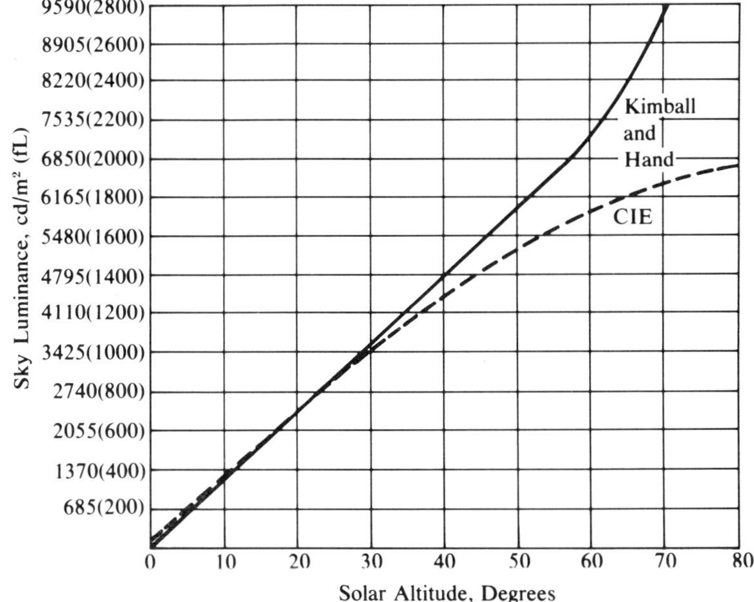

Figure 9.5 Equivalent sky luminance—overcast sky.

TABLE 9.4 EQUIVALENT SKY LUMINANCE—AVERAGE OVERCAST DAY
In Candelas per Square Meter (cd/m^2) and Footlamberts (fL)

	8 A.M. 4 P.M.		9 A.M. 3 P.M.		10 A.M. 2 P.M.		11 A.M. 1 P.M.		Noon	
Latitude	cd/m^2	fL	cd/m^2	fL	cd/m^2	fL	cd/m^2	fL	cd/m^2	fL
December 21										
30° N	1440	420	2540	740	3490	1020	4150	1210	4350	1270
32	1200	350	2400	700	3290	960	3940	1150	4110	1200
34	1100	320	2230	650	3120	910	3770	1100	3910	1140
36	890	260	2060	600	2880	840	3490	1020	3670	1070
38	790	230	1880	550	2710	790	3220	940	3430	1000
40	650	190	1710	500	2540	740	3080	900	3190	930
42	510	150	1540	450	2260	660	2810	820	2950	860
44	340	100	1300	380	2060	600	2600	760	2710	790
46	210	60	1160	340	1880	550	2330	680	2500	730
48	140	40	990	290	1610	470	2160	630	2230	650
50	0	0	820	240	1440	420	1920	560	1990	580
March 21 or September 21										
30° N	3120	910	4520	1320	5860	1710	6890	2010	7330	2140
32	3010	880	4420	1290	5650	1650	6650	1940	7096	2070
34	2950	860	4280	1250	5480	1600	6410	1870	6780	1980
36	2880	840	4180	1220	5340	1560	6170	1800	6510	1900
38	2740	800	4110	1200	5140	1500	5960	1740	6300	1840
40	2710	790	3910	1140	5000	1460	5720	1670	6030	1760
42	2600	760	3840	1120	4830	1410	5480	1600	5790	1690
44	2540	740	3700	1080	4590	1340	5280	1540	5550	1620
46	2430	710	3530	1030	4420	1230	5040	1470	5310	1550
48	2360	690	3390	990	4250	1240	4830	1410	5070	1480
50	2230	650	3220	940	4040	1180	4560	1330	4800	1400

June 21										
30° N	4350	1270	5930	1730	7710	2250				
32	4390	1280	5930	1730	7670	2240				
34	4420	1290	5930	1730	7610	2220				
36	4420	1290	5930	1730	7540	2200	10140	2960		
38	4420	1290	5890	1720	7400	2160	9730	2840		
40	4420	1290	5820	1700	7260	2120	9080	2650	10480	3060
42	4450	1300	5790	1690	7130	2080	8700	2540	9800	2860
44	4420	1290	5720	1670	7020	2050	8330	2430	9110	2660
46	4420	1290	5620	1640	6890	2010	7980	2330	8630	2520
48	4420	1290	5550	1620	6710	1960	7710	2250	8220	2400
50	4320	1260	5450	1590	6510	1900	7400	2160	7810	2280

Source: J. E. Kaufman (ed.), *IES Lighting Handbook*, Reference volume, 1981 ed., IES, New York, 1981, fig. 7-9, with permission.

TABLE 9.5 EQUIVALENT SKY LUMINANCE—CLEAR DAYS†
In Candelas per Square Meter (Footlamberts)

	December 21					March and September 21					June 21				
Latitude	8 A.M.	10 A.M.	Noon	2 P.M.	4 P.M.	8 A.M.	10 A.M.	Noon	2 P.M.	4 P.M.	8 A.M.	10 A.M.	Noon	2 P.M.	4 P.M.
North															
30° N	1540	2060	2060	2060	1540	2400	3430	3600	3430	2400	5310	4800	3430	4800	5310
	(450)	(600)	(600)	(600)	(450)	(700)	(1000)	(1050)	(1000)	(700)	(1550)	(1400)	(1000)	(1400)	(1550)
34° N	1200	1880	1880	1880	1200	2740	2740	3080	2740	2740	4630	4800	3250	4800	4630
	(350)	(550)	(550)	(550)	(350)	(800)	(800)	(900)	(800)	(800)	(1350)	(1400)	(950)	(1400)	(1350)
38° N	1030	1880	1880	1880	1030	2570	2740	3080	2740	2570	4630	4450	3250	4450	4630
	(300)	(550)	(550)	(550)	(300)	(750)	(800)	(900)	(800)	(750)	(1350)	(1300)	(950)	(1300)	(1350)
42° N	860	1710	1710	1710	860	2400	2570	2740	2570	2400	4450	4450	3250	4450	4450
	(250)	(500)	(500)	(500)	(250)	(700)	(750)	(800)	(750)	(700)	(1300)	(1300)	(950)	(1300)	(1300)
46° N	510	1540	1710	1540	510	2400	2570	2570	2570	2400	4450	4280	3250	4280	4450
	(150)	(450)	(500)	(450)	(150)	(700)	(750)	(750)	(750)	(700)	(1300)	(1250)	(950)	(1250)	(1300)
South															
30° N	3770	6680	7710	6680	3770	5820	7880	9590	7880	5820	4110	5480	8220	5480	4110
	(1100)	(1950)	(2250)	(1950)	(1100)	(1700)	(2300)	(2800)	(2300)	(1700)	(1200)	(1600)	(2400)	(1600)	(1200)
34° N	3770	6510	7540	6510	3770	5820	9080	9940	9080	5820	4630	5650	7880	5650	4630
	(1100)	(1900)	(2200)	(1900)	(1100)	(1700)	(2650)	(2900)	(2650)	(1700)	(1350)	(1650)	(2300)	(1650)	(1350)
38° N	3080	7880	7540	7880	3080	5820	9250	10100	9250	5820	4630	5650	7880	5650	4630
	(900)	(2300)	(2200)	(2300)	(900)	(1700)	(2700)	(2950)	(2700)	(1700)	(1350)	(1650)	(2300)	(1650)	(1350)
42° N	2060	7190	7370	7190	2060	5820	9250	8390	9250	5820	4630	6850	8570	6850	4630
	(600)	(2100)	(2150)	(2100)	(600)	(1700)	(2700)	(2450)	(2700)	(1700)	(1350)	(2000)	(2500)	(2000)	(1350)
46° N	1370	6510	7190	6510	1370	5820	9250	9940	9280	5820	4630	7190	9250	7190	4630
	(400)	(1900)	(2100)	(1900)	(400)	(1700)	(2700)	(2900)	(2710)	(1700)	(1350)	(2100)	(2700)	(2100)	(1350)

East															
30° N	5310	5140	3430	2400	1370	6850	8570	5140	3080	2400	9590	9080	4800	3430	2400
	(1550)	(1500)	(1000)	(700)	(400)	(2000)	(2500)	(1500)	(900)	(700)	(2800)	(2650)	(1400)	(1000)	(700)
34° N	4630	4800	3250	2400	1370	8220	8910	5480	3250	2230	9590	9250	4970	3430	2400
	(1350)	(1400)	(950)	(700)	(400)	(2400)	(2600)	(1600)	(950)	(650)	(2800)	(2700)	(1450)	(1000)	(700)
38° N	4110	4450	3080	2230	1200	8570	8910	5140	3080	2060	9590	9250	4800	3600	2400
	(1200)	(1300)	(900)	(650)	(350)	(2500)	(2600)	(1500)	(900)	(600)	(2800)	(2700)	(1400)	(1050)	(700)
42° N	2570	4110	2910	2060	860	8220	8220	4970	2740	2060	9940	8910	4800	3430	2400
	(750)	(1200)	(850)	(600)	(250)	(2400)	(2400)	(1450)	(800)	(600)	(2900)	(2600)	(1400)	(1000)	(700)
46° N	1710	3770	2740	1710	510	7880	7190	4800	2400	2060	9760	8910	4800	3430	2400
	(500)	(1100)	(800)	(500)	(150)	(2300)	(2100)	(1400)	(700)	(600)	(2850)	(2600)	(1400)	(1000)	(700)
West															
30° N	1370	2400	3430	5140	5310	2400	3080	5140	8570	6850	2400	3430	4930	9080	9590
	(400)	(700)	(1000)	(1500)	(1550)	(700)	(900)	(1500)	(2500)	(2000)	(700)	(1000)	(1440)	(2650)	(2800)
34° N	1370	2400	3250	4800	4630	2230	3080	5480	8910	8220	2400	3430	4800	9250	9590
	(400)	(700)	(950)	(1400)	(1350)	(650)	(900)	(1600)	(2600)	(2400)	(700)	(1000)	(1400)	(2700)	(2800)
38° N	1200	2230	3080	4450	4110	2060	3080	5140	8910	8570	2400	3600	4800	9250	9590
	(350)	(650)	(900)	(1300)	(1200)	(600)	(900)	(1500)	(2600)	(2500)	(700)	(1050)	(1400)	(2700)	(2800)
42° N	860	2060	2910	4110	2570	2060	2740	4970	8220	8220	2400	3430	4800	8910	9940
	(250)	(600)	(850)	(1200)	(750)	(600)	(800)	(1450)	(2400)	(2400)	(700)	(1000)	(1400)	(2600)	(2900)
46° N	510	1710	2740	3770	1710	2060	2400	4800	7190	7880	2400	3430	4800	8910	9760
	(150)	(500)	(500)	(1100)	(500)	(600)	(700)	(1400)	(2100)	(2300)	(700)	(1000)	(1400)	(2600)	(2850)

†Average values, direct sunlight excluded

Source: J. E. Kaufman (ed.), *IES Lighting Handbook*, Reference volume, 1981 ed., IES, New York, 1981, fig. 7-10, with permission.

The International Commission on Illumination, based on the work of P. Moon and D. Spencer, has adopted the following equations for overcast sky luminance (Gillette and Kusuda, 1982):

$$L_0 = \frac{L_{z0}}{3}\,(1 + 2\cos\theta) \tag{9.16}$$

and

$$L_{z0} = 123 + 8600\sin h \tag{9.17}$$

In these equations L_0 and L_{z0}, the zenith luminance, are in candelas per square meter.

The horizontal illuminance produced by the CIE overcast sky may be found by the technique presented in Sec. 2.5. Inserting Eq. (9.16) into Eq. (2.27) yields

$$dE_{H0} = \frac{2\pi L_{z0}}{3}\,(1 + 2\cos\theta)\cos\theta\sin\theta\,d\theta \tag{9.18}$$

which, when integrated between the limits of 0 and $\pi/2$, gives

$$E_{H0} = \frac{7\pi L_{z0}}{9} = 0.30 + 21.0\sin h \qquad \text{Klx}$$

$$= 28 + 1950 \sin \text{h} \qquad \text{fc} \tag{9.19}$$

where use has been made of Eq. (9.17).

The equivalent sky luminance in footlamberts that would produce the last form of Eq. (9.19) is equal to E_{H0} in footcandles. This permits us to plot the last Eq. (9.19) as the dashed curve in Fig. 9.5. Note that these CIE values agree quite well with the Kimball and Hand values for small solar altitudes. Recent work in the United States (Gillette and Pierpoint, 1982) suggests that the 8600 in Eq. (9.17) should be replaced by 10,600. This would lift the dashed curve in Fig. 9.5 and bring it more closely in line with the solid curve.

We turn now to the clear sky, defined by the IESNA as a sky with less than 30% cloud cover. Here again the IESNA definition leaves something to be desired in that it does not address the issue of where the partial cloud cover is located. If it blankets the sun, the situation is obviously quite different than if it is opposite the sun in the sky.

Clear sky luminance varies with both polar and azimuth angles. The luminance near the sun and the horizon may exceed the zenith luminance by as much as 12 to 1 and be greater than the luminance near the opposite horizon by as much as 6 to 1. Nonetheless, as in the overcast sky case, it has become the custom in the United States to use the concept of equivalent sky luminance to describe a clear sky, and values of this quantity as a function of compass direction, latitude, date, and time are presented in Table 9.5.

If the four luminances for a given latitude, date, and time in Table 9.5 are averaged, a single equivalent luminance results. Values of equivalent luminance may then be plotted versus solar altitude. The plotted points are considerably scattered but an average curve may be passed through them

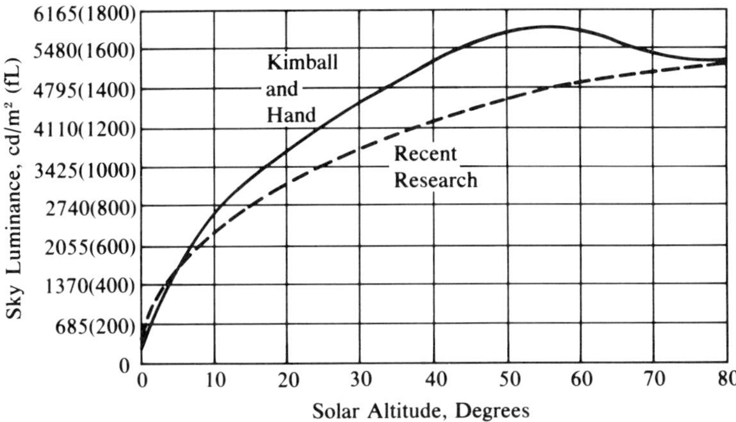

Figure 9.6 Equivalent sky luminance—clear sky (no direct sun).

and this is shown as the solid curve in Fig. 9.6. The footlambert values, as in Fig. 9.5, are numerically equal to the horizontal illuminances in footcandles.

Very often, especially in energy analyses, it is desired to know the horizontal illuminance provided by the clear sky and the direct sun in combination. Illuminance values from Figs. 9.4 and 9.6 may be added to obtain this quantity. The result is shown in Fig. 9.7.

As in the overcast sky case, the international community has not embraced the Kimball and Hand data. The CIE has adopted an equation de-

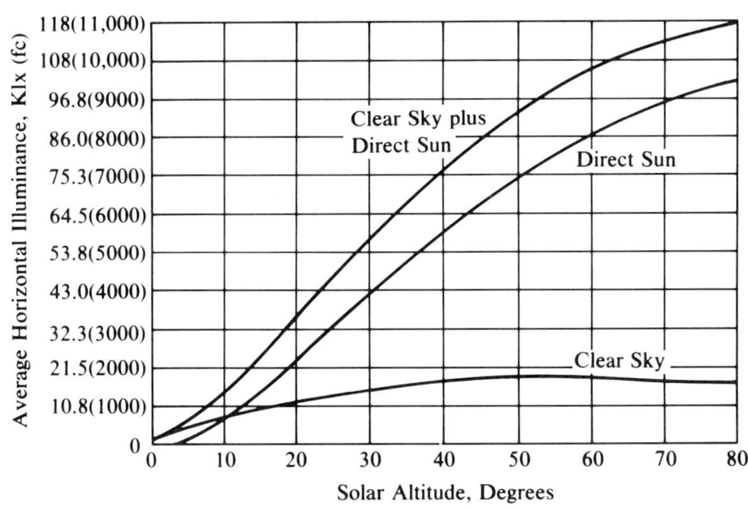

Figure 9.7 Horizontal luminance—clear sky plus direct sun.

veloped by R. Kittler for the luminance of a clear sky

$$L_c = L_{zc} \frac{(0.91 + 10e^{-3\gamma} + 0.45 \cos^2 \gamma)(1 - e^{-0.32 \sec \epsilon})}{0.274[0.91 + 10e^{-3[(\pi/2)-h]} + 0.45 \sin^2 h]} \tag{9.20}$$

where the pertinent angles are shown in Fig. 9.8.

Zenith luminance is a function of solar altitude. From empirical data, it can be written in the general form

$$L_{zc} = a_0 + a_1 h^2 \qquad \text{cd/m}^2 \tag{9.21}$$

The coefficients a_0 and a_1 vary with atmospheric turbidity and water content. There seems to be no general agreement worldwide on their values. One set recently advanced (Gillette and Kusuda, 1982) is $a_0 = 514$ and $a_1 = 3611$, with h entered in radians. This is for a semirural region with relatively humid air. As humidity decreases, a_0 and a_1 increase. Also for urban and industrial areas a_0 and a_1 increase (Gillette, 1983).

It is not a simple matter to find the horizontal illuminance through integration of Eq. (9.20), as we did in the overcast sky case, from which the equivalent clear sky luminance could be obtained. Instead, recent research (Gillette and Pierpoint, 1982) has proceeded to gather horizontal illuminance data from measurements taken around the world, plot it as a function of solar altitude, and fit an equation through the data points. The most acceptable equation currently is

$$\begin{aligned} E_{HC} &= 0.8 + 15.5 \sin^{1/2} h \qquad \text{Klx} \\ &= 74 + 1440 \sin^{1/2} h \qquad \text{fc} \end{aligned} \tag{9.22}$$

where E_{HC} is the horizontal illuminance produced by the clear sky. Then using Eq. (9.14), the equivalent luminance for the clear sky case is calculated as

$$L_{c,\text{ave}} = 255 + 4930 \sin^{1/2} h \qquad \text{cd/m}^2 \tag{9.23}$$

Equation (9.23) is plotted as the dashed curve in Fig. 9.6. Note that its values fall somewhat below the Kimball and Hand data, but not excessively so.

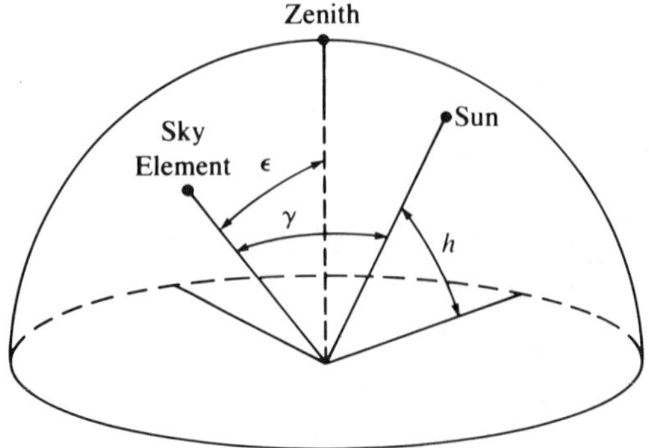

Figure 9.8 Angles for Eq. (9.20).

EXAMPLE 9.4

For the situation in Example 9.1, find the horizontal illuminance due to an overcast sky and a clear sky plus direct sun.

Solution. From Example 9.2, the solar altitude and azimuth are 22.7° and 58.5°, respectively. From Example 9.3, the horizontal illuminance due to the sun is 2680 fc.

From the Kimball and Hand curve in Fig. 9.5, $E_{H0} = 800$ fc. The second Eq. (9.19) yields

$$E_{H0} = 28 + 1950 \sin 22.7 = 780 \text{ fc}$$

which may also be obtained from the CIE curve in Fig. 9.5.

For the clear sky, the Kimball and Hand curve in Fig. 9.6 yields $E_{HC} = 1140$ fc. The second Eq. (9.22) gives

$$E_{HC} = 74 + 1440 \sin^{1/2} 22.7 = 970 \text{ fc}$$

which is also the value obtained from the dashed curve in Fig. 9.6. Thus the horizontal illuminance for the clear sky plus direct sun is 3820 or 3650 fc, depending on which data base is chosen.

In recent years, sky luminance and illuminance measurements have been conducted at several locations throughout the world. It is too early to tell whether or not these measurements will supplant the Kimball and Hand and/or CIE data. More likely, modifications in the current data bases will be made to accommodate variations in the atmosphere better and to include the partly cloudy sky.

9.5 LUMEN METHOD OF SKYLIGHTING

This section is the first of several devoted to presenting calculation procedures for obtaining horizontal illuminance on the work plane in daylit interiors. In it we develop a method for calculating the average work-plane illuminance provided by diffusing plastic skylights mounted at or slightly above roof level. The procedure is known as the lumen method of skylighting (Kaufman, 1981) and is very similar to the zonal cavity method of interior lighting design presented in Chap. 7, with the ceiling-mounted luminaires being replaced by skylights.

The basic formula, which should be compared with Eq. (7.6) for the zonal cavity procedure, is

$$E_t = E_h\left(\frac{A_t}{A_w}\right)(K_u)(K_m) \tag{9.24}$$

where E_t = average illuminance on the work plane from skylights
E_h = horizontal illuminance on the exterior of the skylighting elements

A_t = gross area of the skylighting elements
A_w = area of the work plane
K_u = utilization coefficient
K_m = light loss factor

The calculation procedure may be divided into four steps:

1. Determination of the horizontal illuminance on the exterior of the skylighting elements
2. Determination of the net transmittances, direct and diffuse, of the skylighting elements
3. Determination of the utilization coefficient and light loss factor
4. Determination of
   a. The average horizontal illuminance on the work plane (if the number and size of the skylighting elements are specified) or
   b. The number and size of the skylighting elements (if the average horizontal illuminance is specified)

Step 1 Incident Horizontal Illuminance

The procedures outlined in Secs. 9.3 and 9.4 are used to find the horizontal illuminance on the exterior of the skylights. Generally the analysis is done for both the overcast sky and clear sky plus direct sun conditions and then a design judgment is made as to how many skylights to use, how large each skylight should be, and whether or not controls, such as louvers or lenses, should be installed at the bottoms of the skylight well openings.

Step 2 Net Skylight Transmittances

There are two transmittances to consider when dealing with skylights. One is the direct transmittance (T_D), which is used whenever direct sunlight impinges on the skylight and which varies with the angle of incidence of the incoming solar radiation, decreasing rapidly for flat sheets at high values of this angle. The second is the diffuse transmittance (T_d), which is used with the uniform clear and overcast sky luminances and which does not vary with angle of incidence. Occasionally manufacturers provide transmittance data for flat sheets of their plastic in the form of a single value of T_d and a curve showing the variation of T_D with angle. But more often they provide a single number, assuming that the differences between direct and diffuse transmission are relatively small and that the direct transmittance is relatively independent of solar altitude. Generally these assumptions do not introduce serious errors. Manufacturers' single-number flat sheet transmittance data for typical plastics used in skylights are presented in Table 9.6.

Most skylights are domed, and this affects transmittance in three ways. First, the process of doming significantly decreases sheet thickness at the center of the dome. Second, the angle of incidence of the direct sunlight

TABLE 9.6

Type	Thickness (in)	Transmittance (%)
Transparent	$\frac{1}{8}$–$\frac{3}{16}$	92
Dense translucent	$\frac{1}{8}$	32
Dense translucent	$\frac{3}{16}$	24
Medium translucent	$\frac{1}{8}$	56
Medium translucent	$\frac{3}{16}$	52
Light translucent	$\frac{1}{8}$	72
Light translucent	$\frac{3}{16}$	68

varies over the dome's surface. Third, the dome, because it extends above the roof line, has greater light-gathering surface area than a flat sheet.

The first of these factors can be included by modifying the flat sheet transmittance to (Architectural Aluminum Manufacturers Association, 1977)

$$T_{\mathrm{DM}} = 1.25 T_{\mathrm{FS}}(1.18 - 0.416 T_{\mathrm{FS}}) \tag{9.25}$$

where T_{DM} is dome transmittance and T_{FS} is flat sheet transmittance. This equation does not change the transmittance of a transparent sheet with 92% transmittance but increases the transmittance of a translucent sheet with 44% transmittance by about 25%, in conformity with what actually happens in practice. The second and third factors may be considered together by noting that the effect of doming is to cause T_D to become constant within 10% for all angles of incidence less than 70° (sun altitudes greater than 20°). Thus for most dome applications, we can use a single number for T_D equal to its value at 0° angle of incidence, justifying the second assumption made in connection with Table 9.6.

Because of energy considerations, most contemporary skylights are double domed, usually a transparent dome over a translucent dome. The overall transmittance of such a unit may be obtained from (Pierson, 1962)

$$T = \frac{T_1 T_2}{1 - R_1 R_2} \tag{9.26}$$

where T_1 and T_2 are the transmittances of the individual domes, R_1 is the reflectance from the bottom side of the upper dome, and R_2 is the reflectance from the top side of the lower dome. This formula takes into account the interreflections between the two domes. Generally, it is assumed that $R_1 = 1 - T_1$ and $R_2 = 1 - T_2$.

It remains to include the effect of any light well present between the dome and the ceiling plane of the room. The reflections within such a well will reduce the overall transmittance of the skylighting system. First a well index is calculated from

$$\mathrm{WI} = \frac{h(w + l)}{2wl} \tag{9.27}$$

where h, w, and l are well height, width, and length, respectively. Second, well efficiency (N_w) is defined as the fraction of the luminous flux emerging from the skylight dome which enters the room from the well opening. Well efficiency as a function of well index and well-wall reflectance is displayed in Fig. 9.9 (Fig. 9-75 in Kaufman, 1981). Third, because a skylight has a non-light-transmitting frame to hold the dome in place, it is necessary to include the ratio of net to gross skylight area (R_a). Last, if any diffusers, lenses, louvers or other controls are present, their transmittances (T_c) must be included. Then the net transmittance of the skylight-well system may be found from

$$T_n = \mathrm{T}(N_w)(R_a)(T_c) \tag{9.28}$$

where T is the transmittance of the dome unit.

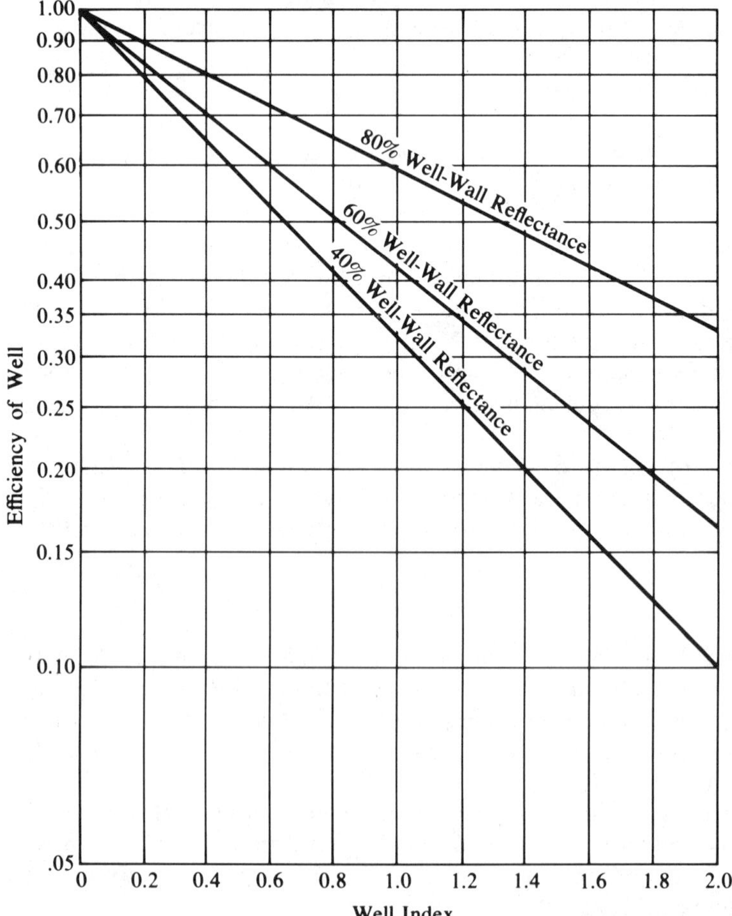

Figure 9.9 Efficiency of well versus well index. [*Source:* J. E. Kaufman (ed.), *IES Lighting Handbook*, Reference volume, 1981 ed., IES, New York, 1981, fig. 9-75, with permission.]

Step 3 Utilization Coefficient and Light Loss Factor

In this step we seek the fraction (K_u) of the luminous flux incident on the skylight which reaches the work plane. This is made up of two parts, the net transmittance determined in step 2 and the room coefficient of utilization (RCU), which is the fraction of the luminous flux entering the room which reaches the work plane. RCU is determined in much the same manner as was CU in the zonal cavity method. First the room cavity ratio is found from Eq. (7.7), repeated here for convenience.

$$\text{RCR} = \frac{5h_c(l + w)}{l(w)} \tag{9.29}$$

where h_c is the ceiling height above the work plane.

With the room cavity ratio determined, RCU for a 20% floor reflectance may be obtained from Table 9.7. For other floor reflectances, Table 7.5 may then be used to obtain the appropriate multiplying factor, as in the zonal cavity method.

Equations have been developed (Dietz et al., 1981) for obtaining RCU values for typical office and warehouse interiors. These are based on ceiling, wall, and floor reflectances of 75, 50, and 30% for the former and 50, 30, and 20% for the latter.

Office: $$\text{RCU} = \frac{1}{1 + 0.0288\ \text{RCR}^{1.560}} \qquad \text{RCR} < 8 \tag{9.30}$$

Warehouse: $$\text{RCU} = \frac{1}{1 + 0.0995\ \text{RCR}^{1.087}} \qquad \text{RCR} < 8 \tag{9.31}$$

TABLE 9.7 ROOM COEFFICIENTS OF UTILIZATION FOR SKYLIGHTING Based on 20% Floor Reflectance

	Ceiling reflectance			
	75%		50%	
	Wall reflectance			
Room cavity ratio	50%	30%	50%	30%
1.0	0.95	0.93	0.93	0.92
2.0	0.87	0.84	0.85	0.82
3.0	0.79	0.77	0.78	0.75
4.0	0.74	0.72	0.73	0.69
5.0	0.68	0.66	0.68	0.63
6.0	0.63	0.61	0.63	0.59
7.0	0.58	0.56	0.58	0.54
8.0	0.53	0.51	0.53	0.50
9.0	0.49	0.47	0.49	0.47
10.0	0.45	0.43	0.45	0.43

These equations give RCUs within ±3% of actual values for a wide range of room cavity ratios.

With RCU determined, the utilization coefficient is obtained from

$$K_u = \text{RCU}(T_n) \tag{9.32}$$

This gives the fraction of the external flux incident on the skylight which reaches the work plane initially. It remains to take into account lumen depreciation over time. The light loss factor K_m is the product of room surface dirt depreciation (RSDD) and skylight dirt depreciation (SDD). RSDD may be obtained using the procedure presented in Sec. 7.9. Little data is available on SDD factors. As a rough guide, we will use 0.75 for office areas and 0.65 for industrial areas, based on cleaning at least once annually.

Step 4 Illuminance or Number and Size of Skylights

Let the number of skylights be N and the gross area of each skylight be A. Then the gross skylight area is $A_t = N(A)$.

On an overcast day, the average horizontal work-plane illuminance for a particular date, time, and latitude is

$$E_{to} = E_{ho}\left[\frac{N(A)}{A_w}\right](K_{ud})(K_m) \tag{9.33}$$

where the o denotes overcast and the d denotes that the net diffuse transmittance is used in computing the utilization coefficient.

For a day with clear sky and direct sun, the illuminance equation becomes

$$E_{tcs} = E_{hc}\left[\frac{N(A)}{A_w}\right](K_{ud})(K_m) + E_{hs}\left[\frac{N(A)}{A_w}\right](K_{uD})(K_m) \tag{9.34}$$

where cs denotes clear sky plus direct sun, c denotes clear sky, s denotes direct sun, and d and D denote that the net diffuse transmittance was used in computing the utilization coefficient for the clear sky contribution to illuminance and the net direct transmittance was used in computing the utilization coefficient for the contribution of the direct sunlight. If only one transmittance is given by the manufacturer, Eq. (9.34) becomes

$$E_{tcs} = (E_{hc} + E_{hs})\left[\frac{N(A)}{A_w}\right](K_u)(K_m) \tag{9.35}$$

EXAMPLE 9.5

Consider an office area 40 × 30 ft with a 10-ft ceiling. Assume reflectances of 75% ceiling, 50% walls, and 20% floor.

The room is to be lighted by six 3 × 3 ft double-domed transparent over translucent skylights mounted on 9-in-high light wells whose wall reflectance is 60%. Each skylight has a 1.5-in-wide retaining frame to hold the dome in place.

The transmittances of the flat sheets of plastic from which the dome is made are given as

	T_d	T_D
Transparent	0.79	0.92
Translucent	0.46	0.45

The depreciation factors are RSDD = 0.95 and SDD = 0.75.

Find the average horizontal illuminance on the work plane at 10 A.M. on March and September 21 at 42° north latitude for both overcast sky and clear sky plus direct sun conditions.

Solution.

Step 1 The horizontal illuminances on the exterior of the skylights may be found from Table 9.3 and Figs. 9.4, 9.5, and 9.6. Using the Kimball and Hand data, the results are

Solar altitude	E_{ho}(fc)	E_{hc}(fc)	E_{hs}(fc)
40°	1400	1540	5500

Step 2 When the flat sheets are domed, their transmittances [from Eq. (9.25)] become

	T_d	T_D
Transparent	0.84	0.92
Translucent	0.57	0.56

Combining them into a double dome yields [from Eq. (9.26)] $T_d = 0.51$, $T_D = 0.53$. Each well is 3 × 3 × 0.75 ft, giving a well index of 0.25 and, from Fig. 9.9, a well efficiency of 0.80. With a 1.5-in frame, the ratio of net to gross skylight area is 0.84. No louvers or lenses are present so the net transmittances are, from Eq. (9.28),

$$T_{dn} = 0.51(0.80)(0.84) = 0.34 \qquad T_{Dn} = 0.53(0.80)(0.84) = 0.36$$

Step 3 From Eq. (9.29), the room cavity ratio is 2.19. Entering Table 9.7 yields an RCU of 0.85. Then

$$K_{ud} = 0.85(0.34) = 0.29 \qquad K_{uD} = 0.85(0.36) = 0.31$$

The light loss factor is

$$K_m = 0.95(0.75) = 0.71$$

Step 4 The average horizontal illuminances on the work plane are

$$E_{to} = 1400\left[\frac{6(9)}{1200}\right](0.29)(0.71) = 13 \text{ fc}$$

$$E_{tcs} = 1540\left[\frac{6(9)}{1200}\right](0.29)(0.71) + 5500\left[\frac{6(9)}{1200}\right](0.31)(0.71)$$

$$= 14 + 54 = 68 \text{ fc}$$

EXAMPLE 9.6

For the office area in Example 9.5, it is desired to have a minimum of 30 fc on the work plane on June 21 at 10 A.M. at 42° north latitude. How many skylights of the type described in Example 9.5 are required?

Solution. For the minimum illuminance we assume an overcast sky. From Table 9.3, the solar altitude is 60°. Then from Fig. 9.5, the horizontal illuminance on the skylights is 2100 fc. Only the diffuse transmittance and utilization coefficient are needed and these are the same as in Example 9.5. The light loss factor is unchanged. Thus, from Eq. (9.33),

$$30 = 2100\left[\frac{N(9)}{1200}\right](0.29)(0.71)$$

$$N = 9 \text{ skylights}$$

8′ 12′

5′

10′

These could be located as shown. A general guideline for spacing skylights to provide uniform work-plane illuminance is not to exceed 1.5 times the ceiling height above the work plane. The suggested layout does not quite meet this criterion. The portion of the roof area covered by skylights is $\frac{81}{1200} = 0.07$, which is within a second general guideline of 4 to 8%.

9.6 LUMEN METHOD OF SIDELIGHTING

The lumen method of sidelighting (IES, 1978; Kaufman, 1981; Libbey-Owens-Ford, 1976) is a procedure for obtaining the horizontal illuminance at three prescribed points on the work plane produced by daylight entering the space through vertical windows. It takes into account not only the direct sunlight and skylight impinging on the window but also the sunlight and skylight reflected onto the window from nearby ground surfaces.

Whereas in skylighting we needed to know only the horizontal illuminance provided externally by the sun and sky, in sidelighting we require both the vertical and horizontal external illuminances, the former to deal with direct sunlight and skylight and the latter to account for reflections from ground surfaces. The horizontal illuminances are obtained, as before, from Fig. 9.4 for the sun, Fig. 9.5 for the overcast sky, and Fig. 9.6 for the clear sky, or from Fig. 9.7 for the clear sky plus direct sun. The vertical illuminances are obtained from Figs. 9.4 (IES Daylighting Comm., 1979) and 9.10 (Libbey-Owens-Ford, 1976), where the degree notation in each figure denotes the horizontal angle between the normal to the vertical surface and the sun's azimuth. The overcast sky does not undergo seasonal changes and thus only one curve is given, in Fig. 9.10*a*. For a given solar altitude, clear sky vertical illuminance does vary with season and thus three representative curves are given, in Fig. 9.10*b*, *c*, and *d*.

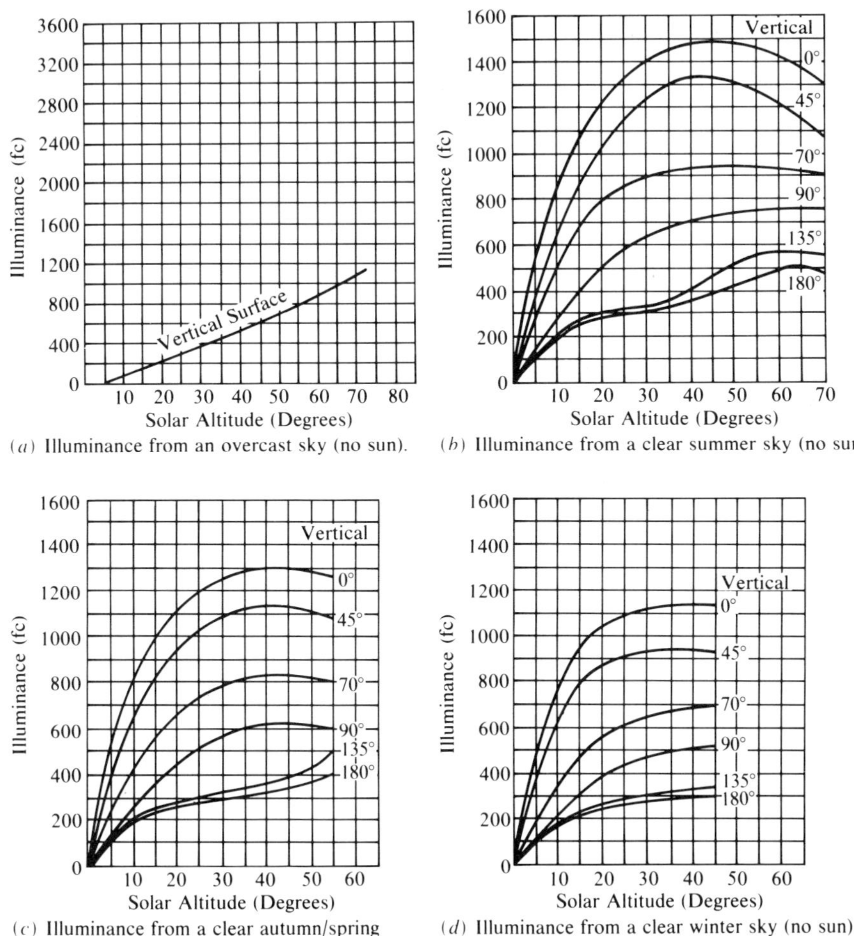

(*a*) Illuminance from an overcast sky (no sun).

(*b*) Illuminance from a clear summer sky (no sun).

(*c*) Illuminance from a clear autumn/spring sky (no sun).

(*d*) Illuminance from a clear winter sky (no sun).

Figure 9.10 Vertical illuminance. (*Source:* IES Daylighting Committee, *Recommended Practice of Daylighting*, IES, New York, 1979, figs. 36A–D, with permission.)

We will proceed through the lumen method of sidelighting in a step-by-step fashion.

Step 1 Direct Vertical Illuminances

For the given date, time, and latitude, obtain solar altitude and azimuth from Eqs. (9.10 a and b) and (9.11). Determine the vertical illuminance provided directly by the overcast sky from Fig. 9.10*a* and, for the given window direction, determine the vertical illuminances provided directly by the sun and clear sky from Figs. 9.4 and 9.10*b*, *c*, or *d*, depending on what season is of interest. The direct illuminance on the window caused by the sky is customarily labeled E_{kw}, that from the sun E_{uw}, and the total E_{sw}.

Step 2 Reflected Vertical Illuminances

Determine the horizontal illuminances on ground surfaces provided by the sun (E_{ug}) from Fig. 9.4 and by the overcast and clear skies (E_{kg}) from Figs. 9.5 and 9.6. The total is denoted by E_{sg}. A fraction of the luminous flux represented by E_{sg} wiil be reflected to the window. Consider Fig. 9.11, in which we show a ground surface G extending to infinity in and out of the page. We will assume that the surface is perfectly diffusing and that its reflectance is known. Also known is the horizontal illuminance E_{sg}. We desire the vertical illuminance on the window at point P, produced by G.

The problem may be solved using Eq. (2.64), repeated below, for the illuminance at a point on a perpendicular plane directly out from a corner of a rectangular source of constant luminance.

$$E = \frac{L}{2}\left(\tan^{-1}\frac{w}{q} - \frac{q}{\sqrt{h^2+q^2}}\tan^{-1}\frac{w}{\sqrt{h^2+q^2}}\right) \tag{9.36}$$

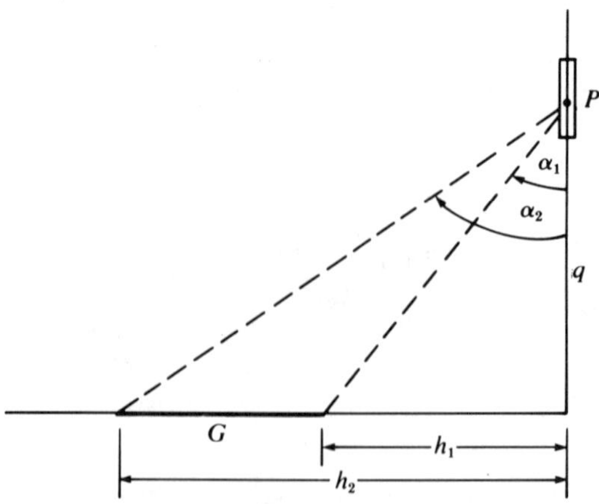

Figure 9.11 Ground plane reflections.

In this case, w is infinite and we must subtract the illuminance produced by a ground surface of width h_1 from that produced by one of width h_2. The result is

$$E_{gw} = L\left[\frac{\pi}{2} - \frac{q}{\sqrt{h_2^2 + q^2}}\left(\frac{\pi}{2}\right) - \frac{\pi}{2} + \frac{q}{\sqrt{h_1^2 + q^2}}\left(\frac{\pi}{2}\right)\right]$$

$$= \frac{\pi L}{2}(\cos\alpha_1 - \cos\alpha_2) \tag{9.37}$$

where gw means on the window from the ground. It remains to express L in terms of the horizontal illuminance on the ground and the ground surface reflectance. The basic relationship is

$$M = \pi L = \rho E_{sg} \tag{9.38}$$

where M is the luminous exitance of the surface and ρ is its reflectance. Substituting Eq. (9.38) into Eq. (9.37) yields

$$E_{gw} = \frac{\rho E_{sg}}{2}(\cos\alpha_1 - \cos\alpha_2) \tag{9.39}$$

If the ground surface is the same (for example, all grass) from the window outward, Eq. (9.39) reduces to the simple form

$$E_{gw} = \frac{\rho E_{sg}}{2} \tag{9.40}$$

Several typical ground surfaces reflectances are given in Table 9.8.

EXAMPLE 9.7

In Fig. 9.11 assume there are three ground surfaces. The ones nearest and farthest from the wall are grass with a reflectance of 20%. The middle surface is light concrete with a reflectance of 40%. The angles are $\alpha_1 = 45°$ and $\alpha_2 = 74°$. If the horizontal illuminance is 4000 fc, find the vertical illuminance on the window at P.

TABLE 9.8

Material	ρ(%)
Cement	27
Concrete	20–40
Asphalt	7–14
Earth	10
Grass	6–20
Vegetation	25
Snow	70
Red brick	30
Gravel	15
White paint	55–75

Solution. From Eq. (9.39),

$$E_{gw} = 2000[0.2(\cos 0 - \cos 45) + 0.4(\cos 45 - \cos 74) + 0.2(\cos 74 - \cos 90)]$$

$$= 2000(0.059 + 0.173 + 0.055) = 574 \text{ fc}$$

Step 3 Luminous Flux Transmittance

With the external vertical illuminance on the window, the window transmittance and the window area known, the luminous fluxes entering the space from the sky-sun and from the ground may be calculated as

$$\phi_{sw} = E_{sw}(\tau)(A_w) \qquad \phi_{gw} = E_{gw}(\tau)(A_w) \tag{9.41}$$

where τ is the window transmittance and A_w is the area of the glazed portion of the window. Values of transmittance for typical window glasses are shown in Table 9.9.

Step 4 Light Loss Factor

There are two factors that contribute to the loss of light due to dirt depreciation. The first is room surface dirt depreciation (RSDD) and is obtained in the same manner as it was in Sec. 7.9 for the zonal cavity method. In using Table 7.7, one can assume that the daylighting has a direct-indirect intensity distribution.

The second factor is window dirt depreciation (WDD). Values of this parameter for vertical windows appear in Table 9.10. The overall light loss factor is

$$K_m = \text{RSDD(WDD)} \tag{9.42}$$

TABLE 9.9

Glass	Thickness (in)	τ(%)
Clear	$\frac{1}{8}$	89
Clear	$\frac{3}{16}$	88
Clear	$\frac{1}{4}$	87
Clear	$\frac{5}{16}$	86
Grey	$\frac{1}{8}$	61
Grey	$\frac{3}{16}$	51
Grey	$\frac{1}{4}$	44
Grey	$\frac{5}{16}$	35
Bronze	$\frac{1}{8}$	68
Bronze	$\frac{3}{16}$	59
Bronze	$\frac{1}{4}$	52
Bronze	$\frac{5}{16}$	44
Thermopane	$\frac{1}{8}$	80
Thermopane	$\frac{3}{16}$	79
Thermopane	$\frac{1}{4}$	77

TABLE 9.10

	WDD (%)	
	Office	Factory
Average value over 6 months	83	71
Value after 3 months	82	69
Value after 6 months	73	55

Step 5 Work-Plane Illuminances

The lumen method of sidelighting is designed to provide the horizontal work-plane illuminance at three points, as shown in Fig. 9.12. The illuminance at point A is known as E_{max}; that at point B, E_{mid}; that at point C, E_{min}. The work plane is assumed to be 30 in above the floor and the three points are on the center line of the room and of the window(s). The windows are assumed to extend from 36 in above the floor to the ceiling.

The method hinges on obtaining several coefficients of utilization. For each source of daylight (overcast sky, clear sky plus direct sun, and ground surface), two coefficients, a C factor and a K factor, are obtained. The C factor has room length, wall reflectance, and room width as parameters. The K factor includes ceiling height (from the floor), wall reflectance, and room width. Their product is a measure of the fraction of the input luminous flux that reaches the work plane. In equation form:

$$E_{sp} = \phi_{sw}(C_s)(K_s)(K_m) \qquad E_{gp} = \phi_{gw}(C_g)(K_g)(K_m) \tag{9.43}$$

where sp and gp mean on the work plane from the sky component and ground component, respectively.

Coefficients of utilization for an overcast sky, a clear sky (with or without direct sun), a uniform ground and diffuse window shades are shown

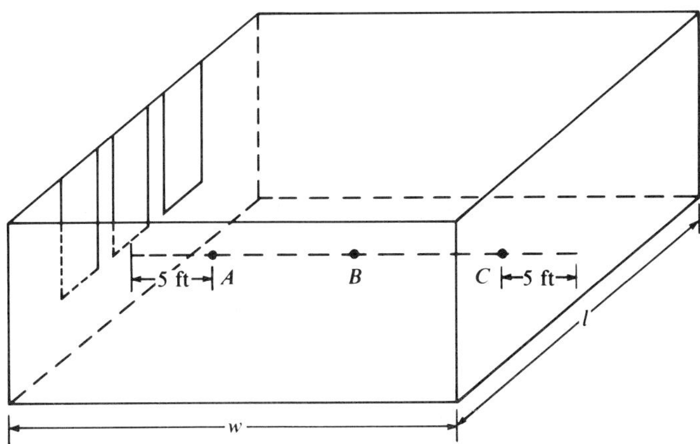

Figure 9.12

in Fig. 9.13 (Libbey-Owens-Ford, 1976). For the diffuse shade case, the direction from which the incident light comes is unimportant and the vertical illuminances from ground and sky at the window are added, with one-half the total assigned to the sky and the other half to the ground.

Modifications of the lumen method of sidelighting have been developed and additional tables of coefficients of utilization have been generated to handle the cases of overhangs and venetian blinds. The reader is referred to the literature (IES, 1978; Kaufman, 1981; Libbey-Owens-Ford, 1976) for these extensions.

We will illustrate the basic calculation procedure through two examples.

EXAMPLE 9.8

Carry out the lumen method of sidelighting for the following data:

40° N latitude, September 21, noon

Room: $40 \times 30 \times 12$ ft

$\rho_c = 80\%,\ \rho_w = 70\%,\ \rho_f = 30\%$

Overcast sky

Light concrete ground surface

Window: 3/16-in grey glass
Sill height: 3 ft
Length: 40 ft
Frame: 20% of opening

Solution.

Step 1 From Table 9.3, the solar altitude is 50°. Then from Fig. 9.10*a*, $E_{sw} = E_{kw} = 700$ fc.

Step 2 From Fig. 9.5, $E_{sg} = E_{kg} = 1740$ fc. The ground surface is assumed to extend to infinity in all three directions. From Table 9.8, its reflectance is 40%. Thus from Eq. (9.40),

$$E_{gw} = \frac{0.40(1740)}{2} = 348 \text{ fc}$$

Step 3 From Table 9.9, the glass transmittance is $\tau = 51\%$. The window glazing area is $A_w = 0.8(40)(9) = 288$ ft². Then from Eqs. (9.41)

$$\phi_{sw} = 700(0.51)(288) = 102{,}800 \text{ lm}$$
$$\phi_{gw} = 348(0.51)(288) = 51{,}100 \text{ lm}$$

Step 4 We will assume a medium office area and use the average value of WDD = 0.83. From Eq. (7.12), assuming cleaning every 6 months,

$$\%\text{DD} = 100[1 - e^{-0.19(0.5^{0.53})}] = 12\%$$

Room Length		20'		30'		40'	
Wall Reflectance		70%	30%	70%	30%	70%	30%
	Room Width						
MAX	20'	.0276	.0251	.0191	.0173	.0143	.0137
	30'	.0272	.0248	.0188	.0172	.0137	.0131
	40'	.0269	.0246	.0182	.0171	.0133	.0130
MID	20'	.0159	.0117	.0101	.0087	.0081	.0071
	30'	.0058	.0050	.0054	.0040	.0034	.0033
	40'	.0039	.0027	.0030	.0023	.0022	.0019
MIN	20'	.0087	.0053	.0063	.0043	.0050	.0037
	30'	.0032	.0019	.0029	.0017	.0020	.0014
	40'	.0019	.0009	.0016	.0009	.0012	.0008

C Coefficients of utilization for room length and width, illuminated by overcast sky.

Ceiling Ht.		8'		10'		12'		14'	
Wall Reflectance		70%	30%	70%	30%	70%	30%	70%	30%
	Room Width								
MAX	20'	.125	.129	.121	.123	.111	.111	.0991	.0973
	30'	.122	.131	.122	.121	.111	.111	.0945	.0973
	40'	.145	.133	.131	.126	.111	.111	.0973	.0982
MID	20'	.0908	.0982	.107	.115	.111	.111	.105	.122
	30'	.156	.102	.0939	.113	.111	.111	.121	.134
	40'	.106	.0948	.123	.107	.111	.111	.135	.127
MIN	20'	.0908	.102	.0951	.114	.111	.111	.118	.134
	30'	.0924	.119	.101	.114	.111	.111	.125	.126
	40'	.111	.0926	.125	.109	.111	.111	.133	.130

K Coefficients of utilization for ceiling height and room width, illuminated by overcast sky.

[A] Coefficients of utilization: "C" and "K" factors. Illumination from an *overcast sky, without window controls.* Ceiling reflectance—80 percent; floor reflectance—30 percent.

Room Length		20'		30'		40'	
Wall Reflectance		70%	30%	70%	30%	70%	30%
	Room Width						
MAX	20'	.0206	.0173	.0143	.0123	.0110	.0098
	30'	.0203	.0173	.0137	.0120	.0098	.0092
	40'	.0200	.0168	.0131	.0119	.0096	.0091
MID	20'	.0153	.0104	.0100	.0079	.0083	.0067
	30'	.0082	.0054	.0062	.0043	.0046	.0037
	40'	.0052	.0032	.0040	.0028	.0029	.0023
M	20'	.0106	.0060	.0079	.0049	.0067	.0043
	30'	.0054	.0028	.0047	.0023	.0032	.0021
	40'	.0031	.0014	.0027	.0013	.0021	.0012

C Coefficients of utilization for room length and width, illuminated by clear sky (with or without direct sun).

Ceiling Ht.		8'		10'		12'		14'	
Wall Reflectance		70%	30%	70%	30%	70%	30%	70%	30%
	Room Width								
MAX	20'	.145	.155	.129	.132	.111	.111	.101	.0982
	30'	.141	.149	.125	.130	.111	.111	.0954	.101
	40'	.157	.157	.135	.134	.111	.111	.0964	.0991
MID	20'	.110	.128	.116	.126	.111	.111	.103	.108
	30'	.106	.125	.110	.129	.111	.111	.112	.120
	40'	.117	.118	.122	.118	.111	.111	.123	.122
MIN	20'	.105	.129	.112	.130	.111	.111	.111	.116
	30'	.0994	.144	.107	.126	.111	.111	.107	.124
	40'	.119	.116	.130	.118	.111	.111	.120	.118

K Coefficients of utilization for ceiling height and room width, illuminated by clear sky (with or without direct sun).

[B] Coefficients of utilization: "C" and "K" factors. Illumination from a *clear sky, without window control.* Ceiling reflectance—80 percent; floor reflectance—30 percent.

Room Length		20'		30'		40'	
Wall Reflectance		70%	30%	70%	30%	70%	30%
	Room Width						
MAX	20'	.0147	.0112	.0102	.0088	.0081	.0071
	30'	.0141	.0112	.0098	.0088	.0077	.0070
	40'	.0137	.0112	.0093	.0086	.0072	.0069
MID	20'	.0128	.0090	.0094	.0071	.0073	.0060
	30'	.0083	.0057	.0062	.0048	.0050	.0041
	40'	.0055	.0037	.0044	.0033	.0042	.0026
MIN	20'	.0106	.0071	.0082	.0054	.0067	.0044
	30'	.0051	.0026	.0041	.0023	.0033	.0021
	40'	.0029	.0018	.0026	.0012	.0022	.0011

C Coefficients of utilization for room length and width, illuminated by uniform ground light.

Ceiling Ht.		8'		10'		12'		14'	
Wall Reflectance		70%	30%	70%	30%	70%	30%	70%	30%
	Room Width								
MAX	20'	.124	.206	.140	.135	.111	.111	.0909	.0859
	30'	.182	.188	.140	.143	.111	.111	.0918	.0878
	40'	.124	.182	.140	.142	.111	.111	.0936	.0879
MID	20'	.123	.145	.122	.129	.111	.111	.100	.0945
	30'	.0966	.104	.107	.112	.111	.111	.110	.105
	40'	.0790	.0786	.0999	.106	.111	.111	.118	.118
MIN	20'	.0994	.108	.110	.114	.111	.111	.107	.104
	30'	.0816	.0822	.0984	.105	.111	.111	.121	.116
	40'	.0700	.0656	.0946	.0986	.111	.111	.125	.132

K Coefficients of utilization for ceiling height and room width, illuminated by uniform ground light.

[C] Coefficients of utilization: "C" and "K" factors. Illumination from a *uniform ground, without window controls.* Ceiling reflectance—80 percent; floor reflectance—30 percent.

Room Length		20'		30'		40'	
Wall Reflectance		70%	30%	70%	30%	70%	30%
	Room Width						
MAX	20'	.0247	.0217	.0174	.0152	.0128	.0120
	30'	.0241	.0214	.0166	.0151	.0120	.0116
	40'	.0237	.0212	.0161	.0150	.0118	.0113
MID	20'	.0169	.0122	.0110	.0092	.0089	.0077
	30'	.0078	.0060	.0067	.0048	.0044	.0041
	40'	.0053	.0033	.0039	.0028	.0029	.0024
MIN	20'	.0108	.0066	.0080	.0052	.0063	.0047
	30'	.0047	.0026	.0042	.0023	.0029	.0020
	40'	.0027	.0013	.0022	.0012	.0018	.0011

C Coefficients of utilization for room length and width, illuminated by uniform sky light through diffuse window shade.

Ceiling Ht.		8'		10'		12'		14'	
Wall Reflectance		70%	30%	70%	30%	70%	30%	70%	30%
	Room Width								
MAX	20'	.145	.154	.123	.128	.111	.111	.0991	.0964
	30'	.141	.151	.126	.128	.111	.111	.0945	.0964
	40'	.159	.157	.137	.127	.111	.111	.0973	.0964
MID	20'	.101	.116	.115	.125	.111	.111	.101	.110
	30'	.0952	.113	.105	.122	.111	.111	.110	.122
	40'	.111	.105	.124	.107	.111	.111	.130	.124
MIN	20'	.0974	.111	.107	.121	.111	.111	.112	.119
	30'	.0956	.125	.103	.117	.111	.111	.115	.125
	40'	.111	.105	.125	.111	.111	.111	.133	.124

K Coefficients of utilization for ceiling height and room width, illuminated by uniform sky light through diffuse window shade.

[D] Coefficients of utilization: "C" and "K" factors. Illumination from the *"uniform sky," with diffuse window shades.* Ceiling reflectance—80 percent; floor reflectance—30 percent.

Figure 9.13 Coefficients of utilization. (*Source:* IES Daylighting Committee, *Recommended Practice of Daylighting*, IES, New York, 1979, figs. 40A–C, 41A, with permission.)

The room cavity ratio from Eq. (7.7) is 2.8. Thus, from Table 7.7, RSDD = 0.93. Finally, from Eq. (9.42),

$$K_m = 0.83(0.93) = 0.77$$

Step 5 From Fig. 9.13*a* for the overcast sky,

$$C_{s,\max} = 0.0137 \qquad K_{s,\max} = 0.111$$
$$C_{s,\text{mid}} = 0.0034 \qquad K_{s,\text{mid}} = 0.111$$
$$C_{s,\min} = 0.0020 \qquad K_{s,\min} = 0.111$$

and from Fig. 9.13*c* for the ground surface,

$$C_{g,\max} = 0.0077 \qquad K_{g,\max} = 0.111$$
$$C_{g,\text{mid}} = 0.0050 \qquad K_{g,\text{mid}} = 0.111$$
$$C_{g,\min} = 0.0033 \qquad K_{g,\min} = 0.111$$

Then, from Eqs. (9.43),

$$E_{sp,\max} = 102{,}800(0.0137)(0.111)(0.77) = 120 \text{ fc}$$
$$E_{sp,\text{mid}} = 102{,}800(0.0034)(0.111)(0.77) = 30 \text{ fc}$$
$$E_{sp,\min} = 102{,}800(0.0020)(0.111)(0.77) = 18 \text{ fc}$$
$$E_{gp,\max} = 51{,}100(0.0077)(0.111)(0.77) = 34 \text{ fc}$$
$$E_{gp,\text{mid}} = 51{,}100(0.0050)(0.111)(0.77) = 22 \text{ fc}$$
$$E_{gp,\min} = 51{,}100(0.0033)(0.111)(0.77) = 14 \text{ fc}$$

Therefore

$$E_{\max} = 154 \text{ fc}$$
$$E_{\text{mid}} = 52 \text{ fc}$$
$$E_{\min} = 32 \text{ fc}$$

EXAMPLE 9.9

Repeat Example 9.8 for a clear sky, with the window facing east and with a dark grass ground surface.

Solution.

Step 1 The solar altitude is still 50° and the solar azimuth is 0°. From Fig. 9.10*c*, $E_{kw} = 610$ fc. Because of the 0° azimuth, there is no direct sun on the window. Thus $E_{sw} = 610$ fc.

Step 2 From Fig. 9.4, $E_{ug} = 6900$ fc and from Fig. 9.6, $E_{kg} = 1690$ fc. From Table 9.8, $\rho = 6\%$ and from Eq. (9.40),

$$E_{gw} = \frac{0.06(6900 + 1690)}{2} = 260 \text{ fc}$$

Step 3 From Eqs. (9.41)

$$\phi_{sw} = 610(0.51)(288) = 89{,}600 \text{ lm}$$
$$\phi_{gw} = 260(0.51)(288) = 38{,}200 \text{ lm}$$

Step 4 From Example 9.8, $K_m = 0.77$.

Step 5 From Fig. 9.13*b* for the clear sky

$$\begin{aligned} C_{s,\max} &= 0.0098 & K_{s,\max} &= 0.111 \\ C_{s,\text{mid}} &= 0.0046 & K_{s,\text{mid}} &= 0.111 \\ C_{s,\min} &= 0.0032 & K_{s,\min} &= 0.111 \end{aligned}$$

and from Fig. 9.13*c* for the ground surface (values unchanged from Example 9.8)

$$\begin{aligned} C_{g,\max} &= 0.0077 & K_{g,\max} &= 0.111 \\ C_{g,\text{mid}} &= 0.0050 & K_{s,\text{mid}} &= 0.111 \\ C_{g,\min} &= 0.0033 & K_{g,\min} &= 0.111 \end{aligned}$$

Then, from Eqs. (9.43)

$$\begin{aligned} E_{sp,\max} &= 89{,}600(0.0098)(0.111)(0.77) = 75 \text{ fc} \\ E_{sp,\text{mid}} &= 89{,}600(0.0046)(0.111)(0.77) = 35 \text{ fc} \\ E_{sp,\min} &= 89{,}600(0.0032)(0.111)(0.77) = 25 \text{ fc} \\ E_{gp,\max} &= 38{,}200(0.0077)(0.111)(0.77) = 25 \text{ fc} \\ E_{gp,\text{mid}} &= 38{,}200(0.0050)(0.111)(0.77) = 16 \text{ fc} \\ E_{gp,\min} &= 38{,}200(0.0033)(0.111)(0.77) = 11 \text{ fc} \end{aligned}$$

and

$$\begin{aligned} E_{\max} &= 100 \text{ fc} \\ E_{\text{mid}} &= 51 \text{ fc} \\ E_{\min} &= 36 \text{ fc} \end{aligned}$$

9.7 DAYLIGHT FACTOR METHOD

The procedures described in the previous two sections are popular in the United States for obtaining horizontal work-plane illuminance in daylit interiors. In Europe, where cloudy skies predominate, another procedure, called the daylight factor method, has been widely used.

Daylight factor (DF) is defined as the ratio of the illuminance at a point on a plane produced by the luminous flux received directly or indirectly at that point from a sky of a given luminance distribution to the illuminance on a horizontal plane produced by an unobstructed hemisphere of this same sky.

In the above definition, direct sunlight is excluded. Thus the basic daylight factor method, as it has been applied abroad, has been limited to the overcast sky case, in particular to the uniform overcast sky and to the CIE overcast sky. Recent work in this country has sought to extend the method to include clear skies as well, but the complications are severe. In this section, we will limit our treatment to overcast skies.

The three ways in which daylight may reach a point on a horizontal plane within a room are shown in Fig. 9.14. The sky component (SC) is that portion of the daylight factor due to daylight received directly at the point

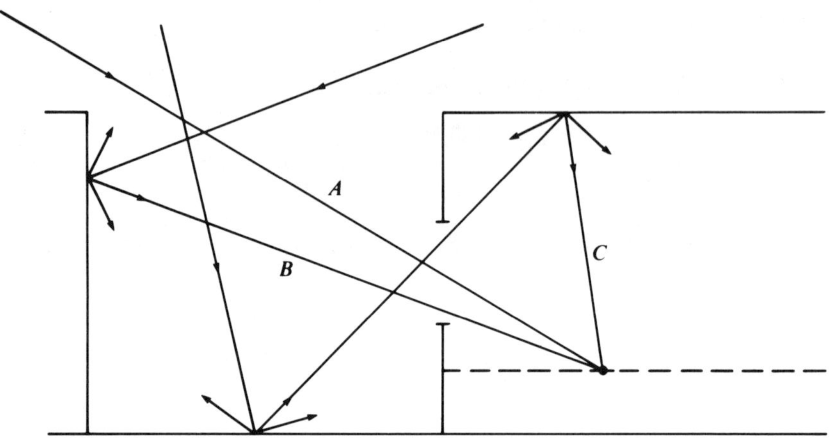

Figure 9.14

from the sky. It is indicated by ray A in Fig. 9.14. The externally reflected component (ERC) is that portion of the daylight factor due to daylight received directly at the point from external reflecting surfaces. Ray B in Fig. 9.14 depicts this situation. The internally reflected component (IRC) is that portion of the daylight factor due to daylight that reaches the point from internal reflecting surfaces. A typical situation of this type is shown by ray C in Fig. 9.14. Note that the IRC includes light that enters the space directly and then is reflected to the point as well as light that is reflected into the space from external surfaces and then is further reflected before reaching the point.

The daylight factor is the sum of its three components, given by

$$\mathrm{DF} = \mathrm{SC} + \mathrm{ERC} + \mathrm{IRC} \tag{9.44}$$

We will now examine each of these components in greater detail and develop procedures for calculating them.

Sky Component

From its definition, the sky component is basically a configuration factor. In Sec. 8.2, the configuration factor for a perfectly diffusing rectangular source to a point P on a line perpendicular to one corner of the source and lying in a plane perpendicular to the source was shown to be

$$C_{\perp} = \frac{1}{2\pi} (\beta - \beta_1 \cos \gamma) \tag{9.45}$$

A diagram of the situation was shown in Fig. 8.4. $C_{\perp}$ is the ratio of the illuminance at P to the luminance exitance of the rectangle. To convert $C_{\perp}$ into the sky component for the uniform overcast sky case, we note that the horizontal illuminance produced by such a sky is given by Eq. (9.14) as $E_H = \pi L$. If there were no glazing material, the luminance of the window would be that of the sky. With glazing, it is τL, where τ is the transmittance of the glazing material. Thus the luminous exitance of the window is $M_w = \pi\tau L$ and the sky component for the uniform overcast sky case is

$$SC_{UNI} = \frac{E_p}{E_H} = \frac{C_\perp M_w}{E_H} = \frac{C_\perp \pi\tau L}{\pi L} = C_\perp \tau \tag{9.46}$$

$$SC_{UNI} = \frac{\tau}{2\pi} (\beta - \beta_1 \cos \gamma) \tag{9.47}$$

For the CIE overcast sky case, for which the sky luminance is given by Eqs. (9.16) and (9.17), a derivation similar to that leading to Eq. (9.45) yields

$$SC_{CIE} = \frac{\tau}{7\pi} [\tfrac{3}{2} (\beta - \beta_1 \cos \gamma) + 2 \sin^{-1}(\sin \beta \sin \gamma) - \sin 2\gamma \sin \beta_1] \tag{9.48}$$

The Building Research Station at Watford in England has prepared tables of SC values (in %) for both the uniform and CIE overcast skies, as functions of h/q and w/q (Hopkinson et al., 1966). These are reproduced as Tables 9.11 and 9.12. The tables include an allowance for the transmittance of a single sheet of clear glass (about 85%). They are based on the diagram in Fig. 8.4, with the point P on a line perpendicular to the lower left corner of the window. For other locations of point P, the method of superposition illustrated in Example 2.13 should be used to obtain the net sky component,

TABLE 9.11 SKY COMPONENTS (%)—UNIFORM OVERCAST SKY

h/q \ w/q	0.1	0.2	0.3	0.4	0.5	0.6	0.8	1.0	1.5	2.0	3.0	6.0	$\alpha°$
∞	1.3	2.6	3.9	5.0	6.1	7.0	8.8	10.1	12.3	13.6	14.7	15.3	90
5.0	1.3	2.6	3.8	5.0	6.0	7.0	8.7	10.0	12.0	13.1	14.0	14.6	79
3.0	1.2	2.5	3.6	4.8	5.7	6.7	8.2	9.4	11.4	12.4	13.1	13.5	72
2.0	1.1	2.3	3.2	4.2	5.1	5.9	7.2	8.3	9.9	10.7	11.3	11.5	63
1.8	1.0	2.2	3.0	4.0	4.8	5.6	6.8	7.8	9.4	10.1	10.6	10.8	61
1.6	0.97	2.0	2.9	3.8	4.5	5.2	6.4	7.3	8.6	9.3	9.8	10.0	58
1.4	0.91	1.9	2.7	3.5	4.2	4.8	5.9	6.7	7.8	8.5	8.8	9.0	54
1.2	0.82	1.7	2.4	3.1	3.8	4.3	5.2	5.9	6.9	7.4	7.7	7.8	50
1.0	0.72	1.5	2.1	2.7	3.2	3.7	4.5	5.0	5.8	6.2	6.3	6.5	45
0.9	0.65	1.4	1.9	2.4	2.9	3.4	4.0	4.5	5.1	5.5	5.6	5.7	42
0.8	0.57	1.2	1.6	2.1	2.6	3.0	3.6	4.0	4.5	4.7	4.8	4.9	39
0.7	0.50	1.0	1.4	1.8	2.2	2.5	3.0	3.3	3.7	3.9	4.0	4.0	35
0.6	0.40	0.83	1.1	1.5	1.8	2.1	2.4	2.7	2.9	3.1	3.2	3.2	31
0.5	0.30	0.63	0.86	1.2	1.4	1.6	1.8	2.0	2.1	2.3	2.3	2.4	27
0.4	0.21	0.43	0.59	0.80	0.94	1.1	1.2	1.4	1.5	1.6	1.6	1.7	22
0.3	0.13	0.25	0.33	0.46	0.54	0.64	0.73	0.82	0.87	0.92	0.94	0.96	17
0.2	0.05	0.11	0.16	0.22	0.27	0.31	0.36	0.40	0.44	0.46	0.47	0.48	11
0.1	0.01	0.02	0.04	0.07	0.09	0.11	0.13	0.14	0.15	0.16	0.17	0.18	6

Source: R. G. Hopkinson, P. Petherbridge, and J. Longmore, "Simplified Daylight Tables," National Building Studies Report #26, HMSO, London, 1958; reprinted in R. G. Hopkinson, P. Petherbridge, and J. Longmore, *Daylighting*, Heineman, London, 1966.

TABLE 9.12 SKY COMPONENTS (%)—CIE OVERCAST SKY

h/q \ w/q	0.1	0.2	0.3	0.4	0.5	0.6	0.8	1.0	1.5	2.0	3.0	6.0	$\alpha°$
∞	1.3	2.5	3.7	4.9	5.9	6.9	8.4	9.6	11.9	13.0	14.2	14.9	90
5.0	1.2	2.4	3.7	4.8	5.9	6.8	8.3	9.4	11.4	12.7	13.7	14.1	79
3.0	1.2	2.3	3.5	4.5	5.5	6.4	7.8	8.7	10.4	11.7	12.4	12.6	72
2.0	1.0	2.0	3.1	4.0	4.8	5.6	6.7	7.5	8.9	9.7	10.0	10.2	63
1.8	0.97	1.9	2.9	3.8	4.6	5.3	6.3	7.1	8.3	9.0	9.3	9.5	61
1.6	0.90	1.8	2.7	3.5	4.2	4.9	5.8	6.5	7.6	8.2	8.5	8.6	58
1.4	0.82	1.6	2.4	3.2	3.8	4.4	5.2	5.9	6.8	7.3	7.5	7.6	54
1.2	0.71	1.4	2.1	2.7	3.3	3.8	4.5	5.0	5.8	6.1	6.2	6.3	50
1.0	0.57	1.1	1.7	2.2	2.6	3.0	3.6	4.0	4.5	4.7	4.8	5.0	45
0.9	0.50	0.99	1.5	1.9	2.2	2.6	3.1	3.4	3.8	4.0	4.1	4.2	42
0.8	0.42	0.83	1.2	1.6	1.9	2.2	2.6	2.9	3.2	3.3	3.4	3.4	39
0.7	0.33	0.68	0.97	1.3	1.5	1.7	2.1	2.3	2.5	2.6	2.7	2.8	35
0.6	0.24	0.53	0.74	0.98	1.2	1.3	1.6	1.8	1.9	2.0	2.1	2.1	31
0.5	0.16	0.39	0.52	0.70	0.82	0.97	1.1	1.3	1.4	1.5	1.5	1.5	27
0.4	0.10	0.25	0.34	0.45	0.54	0.62	0.75	0.89	0.95	0.96	0.97	0.98	22
0.3	0.06	0.14	0.18	0.26	0.30	0.34	0.42	0.47	0.50	0.51	0.52	0.53	17
0.2	0.03	0.06	0.09	0.11	0.12	0.14	0.20	0.21	0.22	0.23	0.23	0.24	11
0.1	0.01	0.02	0.02	0.03	0.03	0.04	0.05	0.05	0.06	0.07	0.07	0.08	6

Source: R. G. Hopkinson, P. Petherbridge, and J. Longmore, "Simplified Daylight Tables," National Building Studies Report #26, HMSO, London, 1958; reprinted in R. G. Hopkinson, P. Petherbridge, and J. Longmore, *Daylighting*, Heinemann, London, 1966.

except that, if a portion of the window lies below the work plane, that portion should be neglected in sky component calculations.

It remains to consider the effect of external obstructions (buildings, etc.) on the sky component. Consider the situation depicted in Fig. 9.15. A building of height H located a distance D from the window partially obscures the sky from the reference point P. The angle of obstruction is defined by

$$\alpha = \tan^{-1} \frac{H - 2.5}{D + q} \tag{9.49}$$

Obstruction angles are included in Tables 9.11 and 9.12. For a given w/q, one enters either table at the appropriate value of α and obtains the sky component to be subtracted because of the obstruction. Alternatively, one may

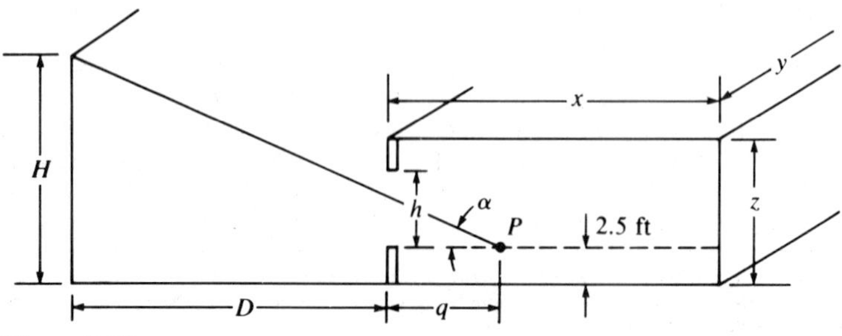

Figure 9.15

compute the $(H - 2.5)/(D + q)$ ratio and enter the tables directly to obtain the amount to be subtracted.

EXAMPLE 9.10

In Fig. 9.15, let $H = 27.5$ ft, $D = 40$ ft, $h = 7$ ft and $q = 10$ ft. If the window is 16 ft wide and point P is on a line perpendicular to the bottom center of the window, find the net sky component if the glazing is a single clear glass sheet and the sky is CIE overcast.

Solution.

$$\frac{H - 2.5}{D + q} = 0.5$$

$$\frac{h}{q} = 0.7 \qquad \frac{w}{q} = 0.8$$

From Table 9.12, the sky component for the unobstructed window is $2(2.1) = 4.2\%$. For the obstruction, $\alpha = 27°$ and the sky component is $2(1.1) = 2.2\%$. Thus the net sky component is

$$\text{SC} = 4.2 - 2.2 = 2.0\%$$

Externally Reflected Component

The externally reflected component is also a configuration factor, but a very difficult one to calculate because of its dependence on the geometry of the external obstructions, given in terms of the angles these surfaces subtend at reference point P. Also, the luminances of these surfaces can vary widely in practice, making calculation of the resulting illuminance at P difficult.

As long as we have overcast skies and surfaces of medium to low reflectances, the contribution of the externally reflected light to the total illuminance at P is small and approximations may be used in the calculation of the externally reflected component. For a uniform overcast sky, it is customary to take the luminance of the obstruction as 0.10 of the (uniform) luminance of the overcast sky. For the CIE overcast sky, the average luminance is, from Eq. (9.19), $0.778L_{z0}$. Very roughly, the near-horizon average sky luminance is one-half of this value. Thus, for the CIE overcast sky, it is customary to take the luminance of obstructions as 0.20 of the near-horizon average sky luminance if the obstruction angle is less than 20°. Otherwise a value of 0.10 is used.

With these luminance modifications, we can obtain the externally reflected component by using Tables 9.11 and 9.12 and the procedure explained previously for obtaining the sky component due to an obstruction. For the uniform sky, we multiply the sky component for the obstruction by 0.10; for the CIE sky by 0.20.

EXAMPLE 9.11

Find the externally reflected components for the situation in Example 9.10.

Solution. The sky is CIE overcast. From Example 9.10, the sky component for the obstruction was 2.2%. The obstruction angle is greater than 20°, so the externally reflected component is

$$\text{ERC} = 0.1(2.2) = 0.22\%$$

Note that the ERC is indeed relatively small, 11% of the net sky component.

Internally Reflected Component

The procedure for obtaining the internally reflected component is based on the theory of the integrating sphere, presented in Sec. 2.6, and the split-flux principle (Hopkinson et al., 1966). As shown in Fig. 9.16, the split-flux principle divides the flux entering the room through the window into two components by inserting a horizontal plane at the middle of the window. One component of flux is that entering the room directly from the sky or by reflection from obstructions above the dividing plane. The other component is produced entirely by reflections, from ground surfaces and from obstructions below the dividing plane.

In Sec. 2.6, the indirect illuminance at any point on the inside of an integrating sphere was shown to be [Eq. (2.40)]

$$E = \frac{\rho\phi}{4\pi r^2(1-\rho)} \tag{9.50}$$

where ρ is the sphere reflectance, r is its radius, and ϕ is the flux entering the space. It follows that $4\pi r^2$ is the surface area of the enclosure and $\rho\phi$ is the flux resulting from the first reflection from that surface area.

The split-flux procedure assumes that the result in Eq. (9.50), which is for a perfectly diffusing sphere with a single value of reflectance, may be applied to a rectangular room with a variety of reflectances. Thus, in the denominator of Eq. (9.50), $4\pi r^2$ is replaced by A, the sum of the areas of the ceiling, floor, and walls, including the window, and ρ is replaced by ρ_{av}, the weighted-average reflectance of the room surfaces, again including the window, with an assumed reflectance of 15%. The numerator of Eq. (9.50) is altered to express the first reflected flux ϕ_r as the sum of two fluxes:

$$\phi_r = \tau E_{w1} A_w \rho_{fw} + \tau E_{w2} A_w \rho_{cw} \tag{9.51}$$

where E_{w1} and E_{w2} are the illuminances on the outside of the window due to

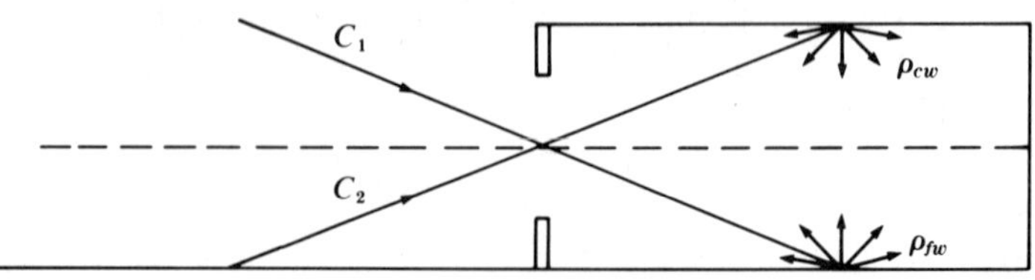

Figure 9.16

the fluxes above and below, respectively, the dividing plane in Fig. 9.16 and ρ_{fw} and ρ_{cw} are the weighted-average reflectances of the lower and upper portions, respectively, of the room, each excluding the window wall.

The internally reflected component is the ratio of the illuminance in the modified form of Eq. (9.50) to the illuminance E_H from the unobstructed sky hemisphere. Thus we can write

$$\begin{aligned} \mathrm{IRC} &= \frac{\tau E_{w1} A_w \rho_{fw} + \tau E_{w2} A_w \rho_{cw}}{E_H A\ (1-\rho_{\mathrm{av}})} \\ &= \frac{\tau A_w}{A(1-\rho_{\mathrm{av}})} \left(\frac{\rho_{fw} E_{w1}}{E_H} + \frac{\rho_{cw} E_{w2}}{E_H} \right) \\ &= \frac{\tau A_w}{A(1-\rho_{\mathrm{av}})} (C \rho_{fw} + 0.05 \rho_{cw}) \end{aligned} \tag{9.52}$$

The final form of Eq. (9.52) requires explanation. In the first term inside the parentheses, E_{w1}/E_H has been replaced by C. Values of this constant have been calculated for various obstruction angles and are tabulated in Table 9.13 for the CIE overcast sky, with the luminance of the obstruction assumed to be 0.10 of the average sky luminance and with the obstruction assumed to extend to infinity laterally.

In the second term in parentheses, the ratio E_{w2}/E_H is replaced by 0.05. The assumption is made that E_{w2} is produced by a perfectly diffusing ground surface whose reflectance is 10%. From Eq. (9.39), with $\alpha_1 = 0°$, $\alpha_2 = 90°$ and $\rho = 0.1$, the value of E_{w2} is seen to be $0.05E_H$.

EXAMPLE 9.12

For the situation in Example 9.10, assume the room has dimensions $x = 20$ ft, $y = 30$ ft, and $z = 11$ ft. Further assume that $\rho_c = 80\%$, $\rho_w = 50\%$, and $\rho_f = 30\%$. Find the internally reflected component and the overall daylight factor.

Solution.

$$A_w = 7(16) = 112 \text{ ft}^2$$
$$A = 2(30)(20) + 2(30)(11) + 2(20)(11) = 2300 \text{ ft}^2$$

TABLE 9.13

$\alpha°$	C(%)
0	39
10	35
20	31
30	25
40	20
50	14
60	10
70	7
80	5

The dividing plane is located 6 ft above the floor. Denoting the areas above and below this plane by A_a and A_b, we have

$$A_a = 30(20) + 30(5) + 2(20)(5) = 950 \text{ ft}^2$$
$$A_b = 30(20) + 30(6) + 2(20)(6) = 1020 \text{ ft}^2$$

The weighted-average reflectances are

$$\rho_{\text{av}} = \frac{0.8(600) + 0.3(600) + 0.5(988) + 0.15(112)}{2300} = 0.51$$

$$\rho_{cw} = \frac{0.8(600) + 0.5(350)}{950} = 0.69$$

$$\rho_{fw} = \frac{0.3(600) + 0.5(420)}{1020} = 0.38$$

where $\tau = 85\%$ and thus window reflectance is 15%.

From Table 9.13 for the given obstruction angle of 27°, interpolation yields $C = 27\%$. Now Eq. (9.52) gives

$$\text{IRC} = \frac{0.85(112)}{2300(1 - 0.51)} [0.27(0.38) + 0.05(0.69)] = 0.012$$

Thus the internally reflected component is 1.2% and, from Eq. (9.44), the daylight factor is

$$\text{DF} = 2.0 + 0.2 + 1.2 = 3.4\%$$

We have presented two methods of computing the horizontal illuminance from a window, the lumen method of sidelighting (LMS) and the daylight factor method (DFM). Each has its strengths and weaknesses. The LMS handles both clear and cloudy skies and includes direct sunlight. The DFM is basically limited to cloudy skies. The LMS is limited to certain types of window controls and geometries (those for which coefficients have been developed) whereas the DFM is not. The LMS provides the horizontal illuminance only at three specified locations. With the DFM, the horizontal illuminance at any location within the space can be calculated. Last, the LMS is largely empirical; the DFM is largely analytical.

9.8 ENERGY SAVING WITH DAYLIGHTING

In this section a procedure will be presented for calculating the lighting energy that can potentially be saved by turning off some or all of the electric lighting in a room that has skylights. It is not a procedure for determining the total energy balance within a space, although it could be extended to do so by including heating and air conditioning considerations and the thermal losses of skylights. Rather it is a procedure for determining the lighting energy tradeoffs between electric lighting and skylighting based on the ability of each to provide average horizontal footcandles on a work plane.

To illustrate how the procedure works, consider a certain skylighting installation that provides 30 fc on the work plane at a specified latitude, date, time, and sky condition. Further assume that the electric lighting provides 50 fc. Then, if the electric lighting can be switched on and off in parts, it is possible to turn off 60 percent of the electric lighting under the assumed conditions. If the entire electric lighting installation, including ballasting, consume P watts, then $0.6P$ watts can be saved. Since each watt represents 3.413 Btu/h, the saving is $2.048P$ Btu/h.

If, on another day, we find that the skylighting produces 60 fc, then all of the electric lighting can be turned off and the saving is $3.413P$ Btu/h. In this latter case, we have more illumination in the room space than required, but we can only take credit for the first 50 fc.

It is relatively straightforward, though computationally laborious, to find the number of Btu's saved per year. In the procedure to be presented here, the simplest situation will be considered, namely turning none of the electric lighting off until the work-plane illuminance provided by skylighting exceeds the design illuminance, and then turning it all off. Thus we will either save $3.413P$ Btu/h or we will save nothing. The procedure will be presented in a step-by-step manner and an example worked to illustrate the necessary calculations.

Step 1 Use the zonal cavity method described in Chap. 7 to find the number of watts (including ballast watts) of electric lighting required to produce the desired average work-plane illuminance level within the given room.

Step 2 Use the lumen method of skylighting described in Sec. 9.5 to find the external horizontal illuminance on the skylights required to produce the desired average work-plane illuminance. This is done for both the overcast sky and the clear sky plus direct sun cases.

Step 3 From the graphs in Figs. 9.5 and 9.7, determine the solar altitude at which the illuminance determined in step 2 is achieved, again for both the overcast sky and clear sky plus direct sun cases.

Step 4 For the given latitude, determine the number of hours on a given day that the solar altitude exceeds the value obtained in step 2 for both sky cases. Generally this is done on the 21st day of each month using data such as that presented in Table 9.14 (for 40° north latitude). Such data may be calculated from Eq. (9.10a).

Step 5 For the location of interest, determine the percentages of clear and overcast days each month. This data, by city, is available from the U.S. Weather Bureau. In Table 9.15 we show the percentages for New York City (40°N latitude).

From this data and the data of step 4, determine the total number of hours each month, for both sky cases, that the work-plane illuminance provided by skylighting exceeds the prescribed value. Sum these figures to

TABLE 9.14 SOLAR ALTITUDE (IN DEGREES) VERSUS DATE AND TIME FOR 40°N LATITUDE

	Solar time (A.M.)							
Date	5	6	7	8	9	10	11	12
Jan. 21				8.1	16.8	23.8	28.4	30.0
Feb. 21			4.3	14.8	24.3	32.1	37.3	39.2
Mar. 21			11.4	22.5	32.8	41.6	47.7	50.0
Apr. 21		7.4	18.9	30.3	41.3	51.2	58.7	61.6
May 21	1.9	12.7	24.0	35.4	46.8	57.5	66.2	70.0
June 21	4.2	14.8	26.0	37.4	48.8	59.8	69.2	73.5
July 21	2.3	13.1	24.3	35.8	47.2	57.9	66.7	70.6
Aug. 21		7.9	19.3	30.7	41.8	51.7	59.3	62.3
Sept. 21			11.4	22.5	32.8	41.6	47.7	50.0
Oct. 21			4.5	15.0	24.5	32.4	37.6	39.5
Nov. 21				8.2	17.0	24.0	28.6	30.2
Dec. 21				5.5	14.0	20.7	25.0	26.6

determine the number of hours per year that this is the case. Multiply the number of hours by $3.413P$ to determine the number of Btu's per year saved by turning off the electric lighting.

EXAMPLE 9.13

Assume the installation in Example 9.5 is changed from 6 to 12 skylights, all other factors remaining the same, and is located in New York City. Further assume that an average horizontal illuminance of 30 fc is desired. The electric lighting is to be provided by luminaire 6 in Table 7.4, with F40CW lamps and LLF = 0.74. Find the number of Btu's per year which may be saved if the electric lighting is turned off whenever the illuminance level from skylighting exceeds 30 fc.

TABLE 9.15 PERCENTAGE OF CLEAR AND OVERCAST DAYS New York City (40°N Latitude)

Period	Days	% Clear	% Overcast
Dec. 22–Jan. 21	31	52	48
Jan. 22–Feb. 21	31	57	43
Feb. 22–Mar. 21	28	59	41
Mar. 22–Apr. 21	31	60	40
Apr. 22–May 21	30	61	39
May 22–June 21	31	62	38
June 22–July 21	30	64	36
July 22–Aug. 21	31	63	37
Aug. 22–Sept. 21	31	63	37
Sept. 22–Oct. 21	30	62	38
Oct. 22–Nov. 21	31	57	43
Nov. 22–Dec. 21	30	53	47

Solution.

Step 1 Proceeding through the steps of the zonal cavity procedure in Secs. 7.6 to 7.9 shows that 13 two-lamp fluorescent luminaires are needed, each consuming 92 W. Thus the total wattage consumption is 1196 W.

Step 2 With little error, the value of T_{Dn} in Example 9.5 may be used for both the direct sun and clear sky transmittances. With this simplification, the horizontal illuminances on the work plane for 12 skylights from step 4 in Example 9.5 are

$$E_{to} = 0.0185E_{ho}$$
$$E_{tcs} = 0.0198(E_{hc} + E_{hs})$$

Thus

$$E_{ho} = 54.0E_{to} = 1620 \text{ fc}$$
$$E_{hc} + E_{hs} = 50.5E_{tcs} = 1515 \text{ fc}$$

Step 3 From Figs. 9.5 and 9.7, the solar altitudes at which the illuminance levels in step 2 are reached are

$h = 46°$ (overcast sky)
$h = 11°$ (clear sky plus direct sun)

Step 4 It is now necessary to use Table 9.14 to determine the average number of hours during a given month that the solar altitude exceeds the values in step 3. This information is tabulated below, where H_s is the number of hours per day at the start of the period, H_e is the number of hours per day at the end of the period, and H_{av} is the average number of hours per day during the period.

	Overcast (46°)			Clear (11°)		
Period	H_s	H_e	H_{av}	H_s	H_e	H_{av}
Dec. 22–Jan. 21	0	0	0	6.6	7.3	6.9
Jan. 22–Feb. 21	0	0	0	7.3	8.7	8.0
Feb. 22–Mar. 21	0	2.8	1.4	8.7	10.1	8.9
Mar. 22–Apr. 21	2.8	4.5	3.6	10.1	11.4	10.7
Apr. 22–May 21	4.5	6.2	5.4	11.4	12.3	11.8
May 22–June 21	6.2	6.5	6.3	12.3	12.6	12.4
June 22–July 21	6.5	6.3	6.4	12.6	12.4	12.5
July 22–Aug. 21	6.3	5.2	5.8	12.4	11.5	11.9
Aug. 22–Sept. 21	5.2	2.8	4.0	11.5	10.1	10.8
Sept. 22–Oct. 21	2.8	0	1.4	10.1	8.8	9.4
Oct. 22–Nov. 21	0	0	0	8.8	7.4	8.1
Nov. 22–Dec. 21	0	0	0	7.4	6.6	7.0

Step 5 The numbers of clear and overcast days are determined from Table 9.15. From these and the average hours per day that the solar

altitude exceeds the required amount, the total number of hours per year that the electric lighting can be extinguished may be found. The tabulation appears below.

	Overcast (46°)			Clear (11°)		
Period	H_{av}	Days	Hours	H_{av}	Days	Hours
Dec. 22–Jan. 21	0	14.9	0	6.9	16.1	111.1
Jan. 22–Feb. 21	0	13.3	0	8.0	17.7	141.6
Feb. 22–Mar. 21	1.4	11.5	16.1	8.9	16.5	146.9
Mar. 22–Apr. 21	3.6	12.4	44.6	10.7	18.6	199.0
Apr. 22–May 21	5.4	11.7	63.2	11.8	18.3	215.9
May 22–June 21	6.3	11.8	74.3	12.4	19.2	238.1
June 22–July 21	6.4	10.8	69.1	12.5	19.2	240.0
July 22–Aug. 21	5.8	11.5	66.7	11.9	19.5	232.1
Aug. 22–Sept. 21	4.0	11.5	46.0	10.8	19.5	210.6
Sept. 22–Oct. 21	1.4	11.4	16.0	9.4	18.6	174.8
Oct. 22–Nov. 21	0	13.3	0	8.1	17.7	143.4
Nov. 22–Dec. 21	0	14.1	0	7.0	15.9	111.3
			396			2165

The total hours are 2561. Thus the potential Btu per year savings are $2561(1196)(3.413) = 10.45 \times 10^6$ Btu.

REFERENCES

Architectural Aluminum Manufacturers Association: "Voluntary Standard Procedure for Calculating Skylight Annual Energy Balance," AAMA Publication 1602. 1. 1977, Chicago, Ill., 1977.

Dickinson, W. C., and P. N. Cheremisinoff: *Solar Energy Technology Handbook*, Part A, Dekker, New York, 1980.

Dietz, P. S., J. B. Murdoch, J. L. Pokoski,and J. R. Boyle: "A Skylight Energy Balance Analysis Procedure," *Journal of the Illuminating Engineering Society*, October 1981.

Duffie, J. A., and W. A. Beckman: *Solar Energy Thermal Processes*, Wiley, New York, 1974.

Evans, B. H.: *Daylight in Architecture*, McGraw-Hill, New York, 1981.

Gillette, G.: *A Daylight Model for Building Energy Simulation,* NBS Building Science Series 152, Washington, 1983.

Gillette, G., and T. Kusuda: "A Daylighting Computation Procedure for Use in DOE-2," *IES Journal*, October 1982.

Gillette, G., and W. Pierpoint: "A General Illuminance Model for Daylight Availability," IES Technical Conference, Atlanta, 1982.

Hopkinson, R. G., P. Petherbridge, and J. Longmore: *Daylighting*, Heinemann, London, 1966.

IES Daylighting Committee: *Recommended Practice of Daylighting*, Illuminating Engineering Society of North America, New York, 1978.

International Commission on Illumination: "Daylight: International Recommendations for the Calculation of Natural Light," CIE Report #16, Paris, 1970.

International Commission on Illumination: "Standardization of Luminance Distribution of Clear Skies," CIE Report #22, Paris, 1973.

Kaufman, J. E.: *IES Lighting Handbook*, Reference volume, 1981 ed., Illuminating Engineering Society of North America, New York, 1981.

Kimball, H. H., and I. F. Hand: "Daylight Illumination on Horizontal, Vertical and Sloping Surfaces," *Transactions of the Illuminating Engineering Society*, vol. XVIII, May 1923.

Kittler, R.: "A Universal Calculations Method for Simple Predetermination of Natural Radiation on Building Surfaces and Solar Collectors," *Building and Environment*, vol. 16, no. 3, Great Britain, 1981.

Libbey-Owens-Ford Co.: "How to Predict Interior Daylight Illumination," Toledo, Ohio, 1976.

Linforth, E.: "Efficiency of Domed Acrylic Skylights," *Illuminating Engineering*, October 1958.

Murdoch, J. B.: "A Procedure for Calculating the Potential Savings in Lighting Energy from the Use of Skylights," *Journal of the Illuminating Engineering Society*, July 1977.

Pierson, O. L.: *Acrylics for the Architectural Control of Solar Energy*, Rohm and Haas, Philadelphia, 1962.

Chapter 10

Optics and Control of Light

10.1 INTRODUCTION

In this chapter we will review the fundamentals of geometric and physical optics and discuss their application to the optical control of luminous flux distributions in interior and exterior spaces.

Light may be controlled optically in a number of ways, the major of which are by reflection, diffusion, transmission, absorption, refraction, interference, diffraction, and polarization. The first five of these are based on geometric optics; the last three are phenomena from physical optics. We will discuss each of these control mechanisms in some detail, including both their overall effect and their selective effect on certain wavelengths.

Much of geometric optics is based on Huygens' principle* which states that as an electromagnetic wavefront advances each point on it may be treated as a source of secondary wavefronts, which then advance in all directions in the same fashion as did the original wavefront. The new overall wavefront is then obtained by constructing the envelope of the secondary wavefronts. The process is illustrated in Fig. 10.1 to demonstrate a plane light wave reflecting from a plane polished surface. The effect of the reflecting surface is to reverse the direction of travel of the secondary wavefronts as they impinge upon it.

10.2 REFLECTION

Reflection is a generic term which describes the process by which a portion of the electromagnetic radiation (luminous flux) incident on a surface leaves

*Christian Huygens (1629–1695), Dutch astronomer, mathematician, and physicist.

Figure 10.1 Plane wave reflection.

that surface without passing through it. The different types of reflection are illustrated in Fig. 10.2. Specular reflection is illustrated in Fig. 10.2*a*, and is, by definition, the situation where the law of reflection for plane waves is obeyed. This law states that the incident and reflected rays and the normal to the surface lie in the same plane and that the angle of reflection is equal to the angle of incidence (θ). The law may be derived from the construction in Fig. 10.1.

Examples of specular materials are silvered mirrors, polished metals, aluminum foil, and stainless steel. Reflectances of specular surfaces range from 90 to 95% for vaporized silver to 50% for stainless steel and 5% for clear glass.

At the other extreme from specular reflection is diffuse reflection, which is the characteristic of matte-finished surfaces and surfaces composed of small crystal particles. It is illustrated in Fig. 10.2*b*. A perfectly diffusing surface, as was noted in Chap. 2, is defined as one for which surface luminance is constant with angle of observation. Flat paints and white plaster closely approximate this condition, except at large angles of incidence.

Between perfectly diffuse and specular lie all other types of reflection, one of which is spread reflection, shown in Fig. 10.2*c*. Here there is no mirror image of the source, as in the specular case, but the angle of maximum reflected intensity does equal the angle of incidence. There is some diffusion, but insufficient to create constant luminance. Spread reflection occurs when polished aluminum or similar material is brushed, etched, pebbled, or corrugated.

It is also possible to have combinations of the three types. Thus diffuse-specular, diffuse-spread, and specular-spread surfaces exist. These may provide almost any pattern of reflected light and the pattern may be a function of angle of incidence. Porcelain enamel, semigloss and gloss paint,

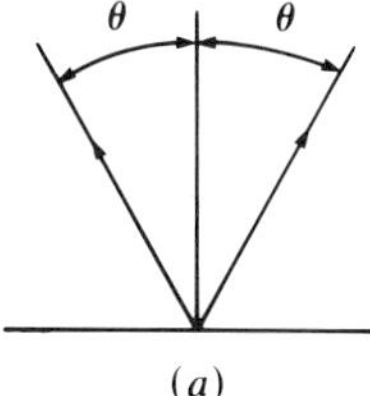

(*a*)

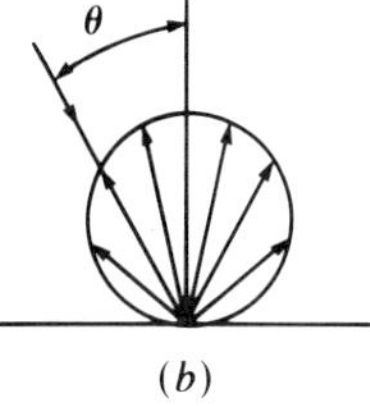

(*b*)

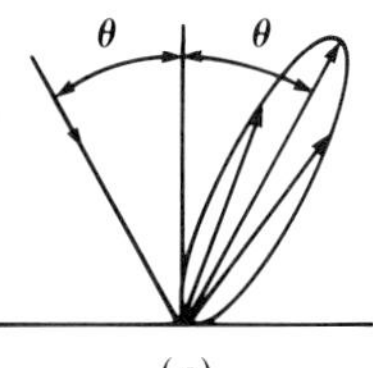

(*c*)

Figure 10.2 Types of reflection.

and glossy paper are examples of materials which can produce these compound reflections.

10.3 MIRRORS AND RAY TRACING

Mirrors may be plane or curved, with the curvature of the latter in either one or two dimensions. The curved reflecting surfaces may be circular, parabolic, elliptical, hyperbolic, or some other contour to suit a particular application.

Inherent in any reflector design is the ability to do ray tracing. The concave spherical mirror in Fig. 10.3 will be used as a vehicle for illustrating how this is done. An object AX is located a distance p from the vertex O of the mirror, whose radius is r. Three rays are shown emanating from point A. The first is directed through the mirror's center at C, encounters the mirror at D, and reflects back along itself through C. The second is directed parallel to the axis of the mirror toward E, where it obeys the law of reflection and departs through point F. The third, directed toward the mirror's vertex at O, also obeys the law of reflection and proceeds along the line OB.

For small θ, the three reflected rays cross at point B and thus the image of A will lie at B. Similarly, the image of X will lie at Y and thus the object AX is imaged as BY, an inverted image smaller than the original object.

We need to investigate the point F more fully. To obtain its distance f from O, we observe from Fig. 10.3 that angle $OCE = \theta$ and angle $OFE = 2\theta$.

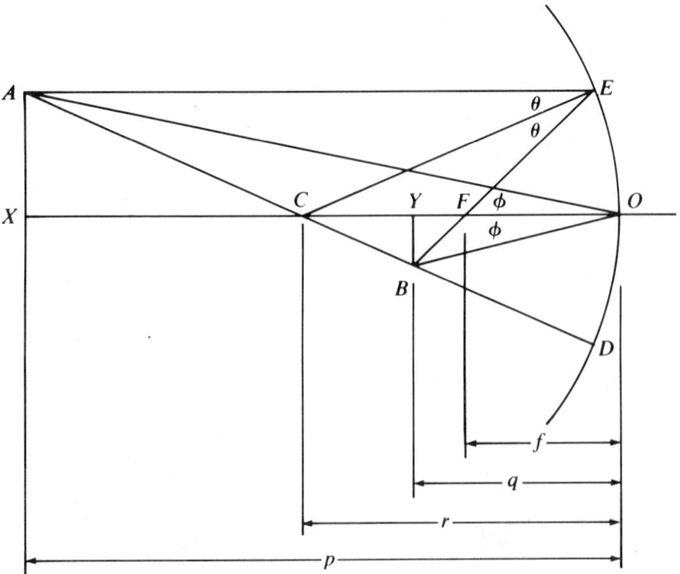

Figure 10.3 A concave spherical mirror.

Thus

$$EF = \frac{r \sin \theta}{\sin 2\theta} = \frac{r}{2 \cos \theta} = CF \tag{10.1}$$

and

$$f = CO - CF = r\left(1 - \frac{1}{2 \cos \theta}\right) \tag{10.2}$$

For small θ, $\cos \theta \simeq 1$ and

$$f \simeq r(1 - \tfrac{1}{2}) = \frac{r}{2} \tag{10.3}$$

Thus for a small-aperture mirror, all parallel rays, such as AE, reflect through point F, midway between C and O. By definition, F is the focal point of the mirror and f is its focal length.

As the ray AE moves farther from the axis XO, the focal length from Eq. (10.2) decreases. This property of a spherical mirror is called spherical aberration and can cause such a mirror to form a blurred image of an object.

For the small-mirror case, it is useful to relate p, q, and r. From Fig. 10.3

$$\frac{AX}{BY} = \frac{XC}{CY} = \frac{p - r}{r - q}$$

$$\frac{AX}{BY} = \frac{XO}{YO} = \frac{p}{q} \tag{10.4}$$

Combining yields

$$\frac{1}{p} + \frac{1}{q} = \frac{2}{r} = \frac{1}{f} \tag{10.5}$$

Thus if p and r are known, q and f can be found. If $r = \infty$, $q = -p$, which is the equation for a plane mirror and states that the image is located behind the mirror at a distance equal to the object distance in front of the mirror.

In doing ray tracing, it is desirable to establish certain conventions. These are generally stated as follows:

1. Draw all diagrams with incident rays from left to right.
2. Assume object and image distances are positive if the object and image are to the left of the mirror.
3. Assume radius and focal length are positive if the center of curvature and focal point lie to the left of the mirror.
4. Assume images have positive size if they are above the horizontal axis.

This last stipulation and the geometry in Fig. 10.3 allow us to express the magnification M of the mirror as

$$M = \frac{BY}{AX} = -\frac{q}{p} \tag{10.6}$$

It is instructive to plot magnification from Eq. (10.6) versus object distance for a concave spherical mirror. The result is shown in Fig. 10.4, where object distance is expressed in focal lengths. If the object is an incandescent lamp filament, then the graph shows the relative size of the image of the filament as a function of filament distance from the vertex of the spherical reflecting surface.

EXAMPLE 10.1

In Fig. 10.3, let $p = 30$ cm and $r = 20$ cm. Find the image distance and magnification. Repeat for a convex mirror with the same radius and object distance. Find the focal length for each mirror.

Solution. For the concave mirror, from Eq. (10.5),

$$\frac{1}{30} + \frac{1}{q} = \frac{2}{20}$$

$$q = 15 \text{ cm}$$

A positive value for q indicates that the image in Fig. 10.3 is to the left of the vertex. From Eq. (10.6) the magnification is -0.5 and from Eq. (10.3), the focal length is 10 cm.

For the convex mirror case, the center of curvature is to the right of the vertex. Thus, from Eq. (10.5)

$$\frac{1}{30} + \frac{1}{q} = \frac{2}{-20}$$

$$q = -7.5 \text{ cm}$$

$$f = \frac{-20}{2} = -10 \text{ cm}$$

$$M = \frac{-7.5}{30} = 0.25$$

In this case, the image is to the right of the mirror and, with a positive magnification, is erect and above the horizontal axis. The construction is shown in Fig. 10.5.

The image in Fig. 10.3 is called *real*, while that in Fig. 10.5 is called *virtual*. When the rays originating at a point converge after reflection, as in Fig. 10.3, a real image is produced. When they diverge, a virtual image results. A virtual image does not exist physically; it would not be formed on a screen placed at the image location. It is simply the apparent source of light rays leaving the mirror.

10.4 CONIC SECTIONS

Reflector contours can be placed into two major categories, conic and general. Conic sections, often called basic contours, can be described easily mathematically and include the circle, ellipse, parabola, hyperbola, and combina-

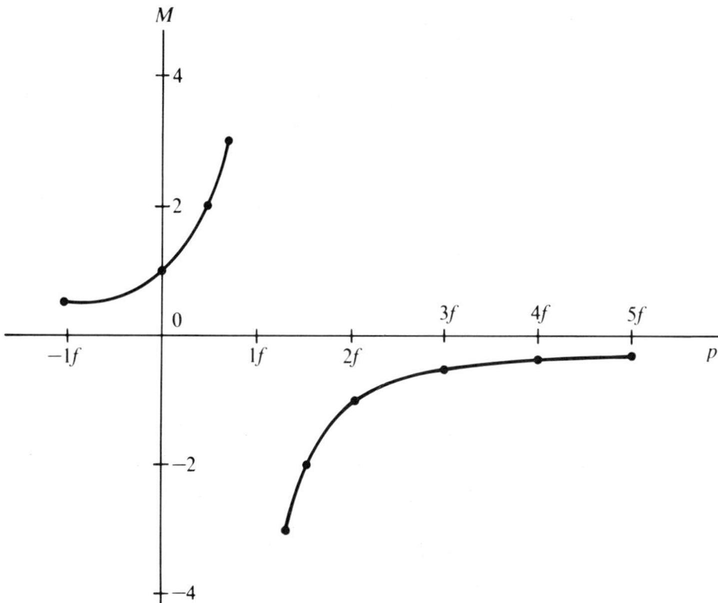

Figure 10.4 Magnification versus object distance for a concave spherical mirror.

tions thereof. General contours are those designed through general mathematical analysis or through computerized ray-tracing techniques to provide specific reflected intensity distributions.

The basic contours are often used in reflector design, at least as a starting point, because they are relatively easy to understand and analyze and because they often yield ray patterns close to those which are ultimately desired.

All of the basic contours are formed as the intersection of a plane and a right circular cone, as shown in Fig. 10.6. In Fig. 10.6*a*, a plane parallel to

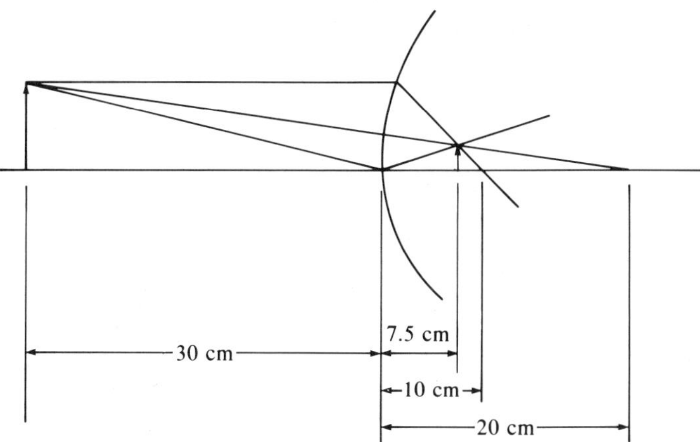

Figure 10.5 A convex spherical mirror.

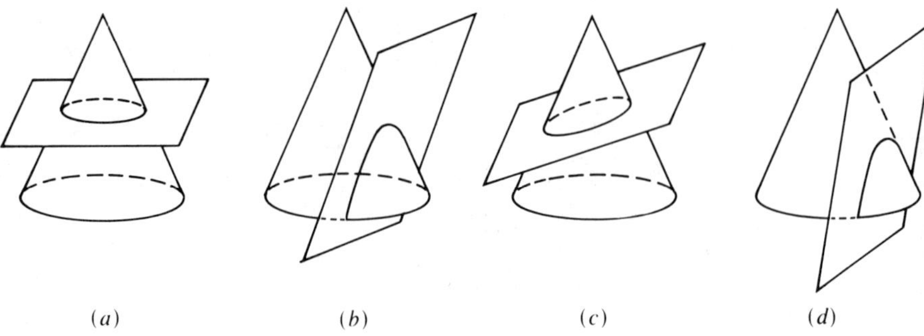

Figure 10.6 Conic sections.

the base of the cone intersects the cone in a circle. In Fig. 10.6*b*, a plane parallel to the slant face of the cone intersects the cone in a parabola. Any plane of intermediate orientation intersects the cone in an ellipse, as shown in Fig. 10.6*c*. If the plane in Fig. 10.6*b* becomes more upright than for a parabola, a hyperbola results, as shown in Fig. 10.6*d*. We will now consider each of these basic contours in further detail.

A circle is the locus of a point at a constant distance from a fixed point. If the fixed point (the center of the circle) is placed at point F in an xy coordinate system, as shown in Fig. 10.7*a*, the locus equation is

$$(x - a)^2 + y^2 = a^2 \tag{10.7}$$

The point F is the focal point of the circle and a is the focal length.

Assume that a point source of light is placed at the center of a half-circular reflector. The rays incident on the reflector will be directed back through the source, as shown in Fig. 10.7*b*. If there were no losses, the intensity of the source in all directions to the right of the reflector would be doubled.

Spherical reflectors, which are in reality three-dimensional circular reflectors, are widely used in projection systems to increase the intensity of the light collected by the lens system. Cylindrical reflectors are used in tubular incandescent showcase lamps to redirect light from the linear filament to the items on display.

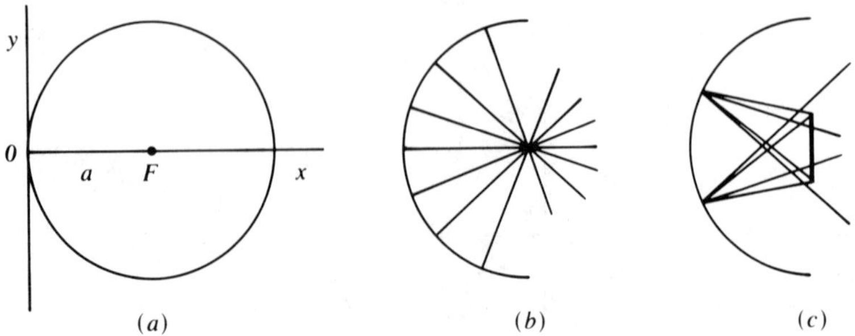

Figure 10.7 Ray patterns of a circular reflector.

A lamp filament is not a point source. Thus the ray pattern in Fig. 10.7*b* is idealized. Very often, as in projection lamp systems, the filament is planar, and lies in the position indicated in Fig. 10.7*c*. The resulting ray pattern is distorted. As shown in Fig. 10.7*c*, rays from portions of the filament above the horizontal axis emerge from the reflector below the horizontal axis. The reverse is true for portions of the filament below the horizontal axis. In a projection lamp system, the rays emerging from the reflector are gathered and focused by a lens system. If the lamp filament becomes too large, rays from its ends will not be intercepted by the lens system. Thus there is a maximum size filament a given lens system can accommodate and increases in filament size beyond this point will simply result in wasted light.

The optical gain of spherical reflector systems depends on the opacity of the source and how well it is centered. For open filament forms, the use of a spherical reflector to enhance intensity is quite effective; in compact forms it is relatively ineffective. Also the reflected rays can increase the temperature of the filament, increasing light output but decreasing life.

We turn next to the parabola, which is the locus of a point equally distant from a fixed point and a fixed straight line. The situation is shown in Fig. 10.8. The fixed point F is called the focus, located a distance f, the focal length, from the origin. The fixed straight line NM is also located a distance f from the origin. This line is called the directrix.

From the definition of a parabola, the criterion in Fig. 10.8 is that $FP = NP$. Thus the locus equation is

$$(x - f)^2 + y^2 = (x + f)^2 \tag{10.8}$$

which yields

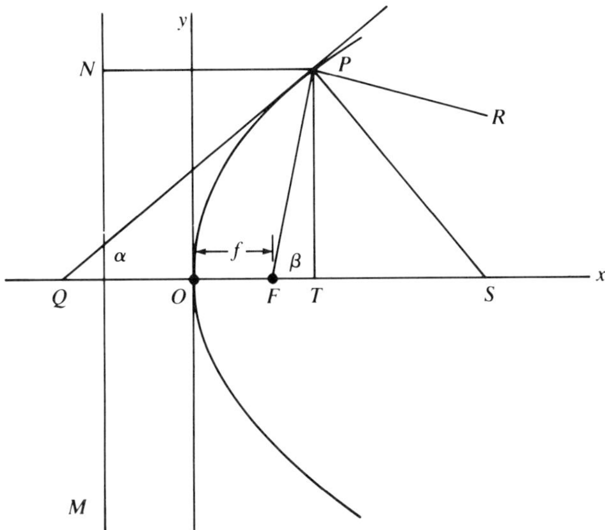

Figure 10.8 Parabolic contour.

$$y^2 = 4fx \tag{10.9}$$

Assume a point source of light is located at the focus of the parabola in Fig. 10.8. We wish to show that any ray of light from the source will be reflected horizontally from the reflector. Consider the ray FPR. Draw a tangent PQ and a normal PS to the parabola and a perpendicular PT to the x axis, all at point P. For ray PR to be horizontal, we must show that angle $TPR = 90°$. Define angle PQO as α, angle TPF as β. From Eq. (10.9),

$$\tan \alpha = \frac{dy}{dx} = \sqrt{\frac{f}{x}} \tag{10.10}$$

From Fig. 10.8

$$\tan \beta = \frac{y}{x-f} = \frac{2\sqrt{fx}}{x-f} \tag{10.11}$$

From the law of tangents

$$\tan 2\alpha = \frac{2 \tan \alpha}{1 - \tan^2 \alpha} = \frac{2\sqrt{fx}}{x-f} \tag{10.12}$$

Therefore

$$\beta = 2\alpha \tag{10.13}$$

and the tangent line QP bisects the angle NPF. From similar triangles, angle $TPS = \alpha$ and, from the law of reflection, angle $SPR =$ angle $FPS = 90 - \beta + \alpha = 90 - \alpha$. The sum of angles TPS and SPR is 90°, completing the proof.

It is relatively easy to design a parabola to fit into a given rectangular space. Consider Fig. 10.9a in which we show a rectangle of depth d and half-width w. The parabola to be inserted is to pass through points A and B. At point B, Eq. (10.9) yields

$$w^2 = 4fd \tag{10.14}$$

Thus the focal length of the desired parabola is

$$f = \frac{w^2}{4d} \tag{10.15}$$

and Eq. (10.9) can be rewritten as

$$y^2 = \frac{w^2}{d}\, x \tag{10.16}$$

For any other point P on the parabola for which $y = kw$ $(k < 1)$, we have, from Eq. (10.16),

$$k^2w^2 = \frac{w^2}{d}\, x$$

$$x = k^2 d \tag{10.17}$$

Thus at P

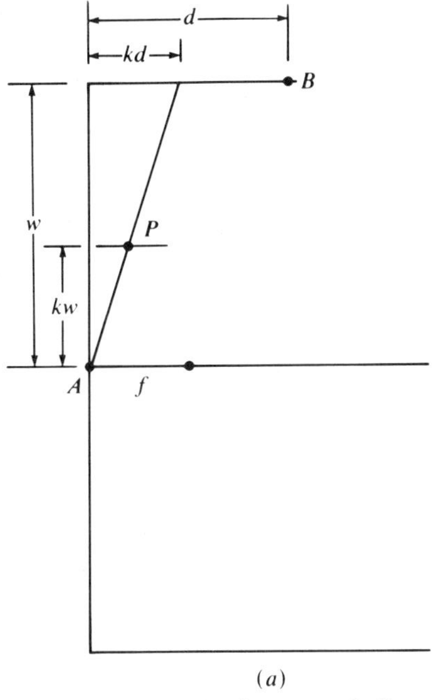

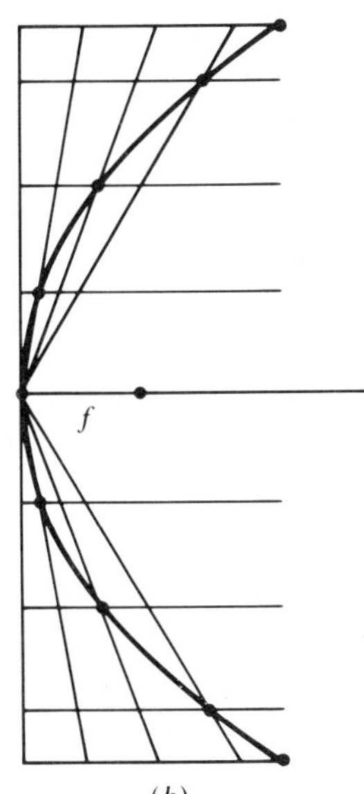

Figure 10.9 Designing a parabolic reflector.

$$\frac{y}{x} = \frac{kw}{k^2d} = \frac{w}{kd} \tag{10.18}$$

The result in Eq. (10.18) suggests a way of constructing the parabola through similar triangles. At $y = kw$ we draw a horizontal line. Next we mark off a distance kd on d and draw a line to A. The intersection of the two lines is at P. In Fig. 10.9*b*, the entire parabolic construction is shown.

A parabola with a point source of light at its focal point will produce parallel reflected rays. This is the ideal case and can be closely approximated if a compact incandescent lamp filament is the light source. If the source, or part of it, is not at the focal point, the reflected rays are somewhat non-parallel. This is illustrated in Fig. 10.10*a* where the source is in front of the focal point and in Fig. 10.10*b* where it is behind it.

In Fig. 10.10*c*, the effects shown in Fig. 10.10*a* and *b* are combined for a spherical source of finite radius whose center is at the focal point but whose extremes are not. The divergence of the reflected rays is seen to decrease as the distance from the source to the point of reflection increases.

Parabolic contours are used widely in incandescent lighting (in PAR and R lamps, search lights, automobile headlamps, photoflood lamps, and infrared lamps). They are also used as the reflector in fluorescent luminaires, in the parabolic wedge louver, and in fluorescent cove lighting systems. With HID lamps, they are used to create high-intensity floodlights and spotlights.

Two interesting combinations of circular and parabolic reflectors are

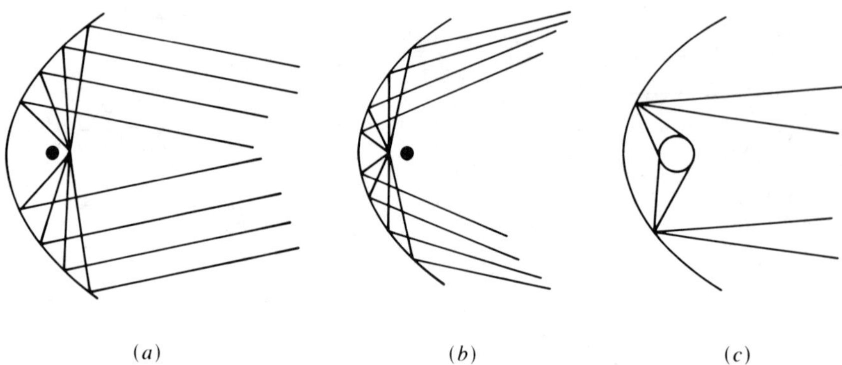

Figure 10.10 Parabola: Source in front of or behind focus.

shown in Fig. 10.11. In each combination, the focal point of the circle and the parabola is the same. The circular reflector intensifies the rays from the source prior to their impinging on the parabolic reflector. The unit in Fig. 10.11*b* is particularly useful in lighting vertical surfaces. If the unit, in the form of a fluorescent cove, is mounted above eye level, the source will be shielded from viewers in the space. In the unit in Fig. 10.11*a*, the intensity at the periphery of the reflector will be greater than that in the center of the beam.

We turn next to elliptical reflector contours. An ellipse is the locus of a point, the sum of whose distances from two fixed points is constant. In Fig. 10.12, the two fixed points are labeled F and F' and are termed the foci of the ellipse, with x coordinates f and $f + 2c$. If P is any point on the ellipse and $2a$ is the constant sum of its distances from the foci, then

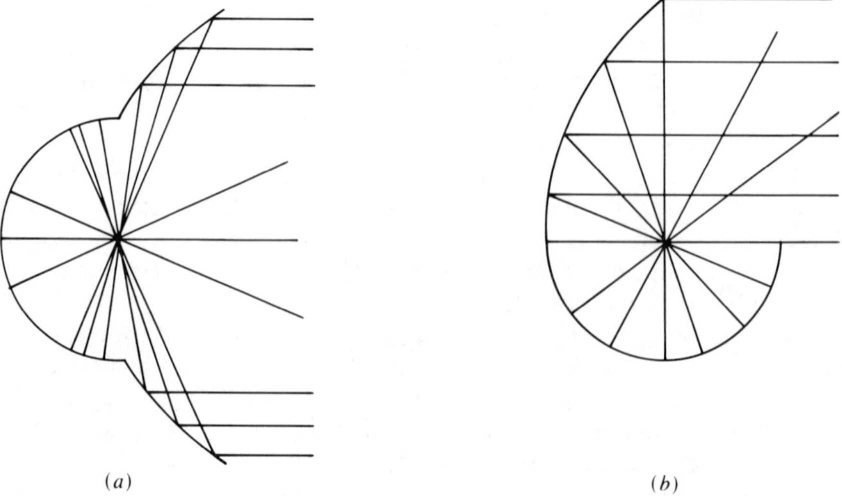

Figure 10.11 Circular-parabolic reflectors.

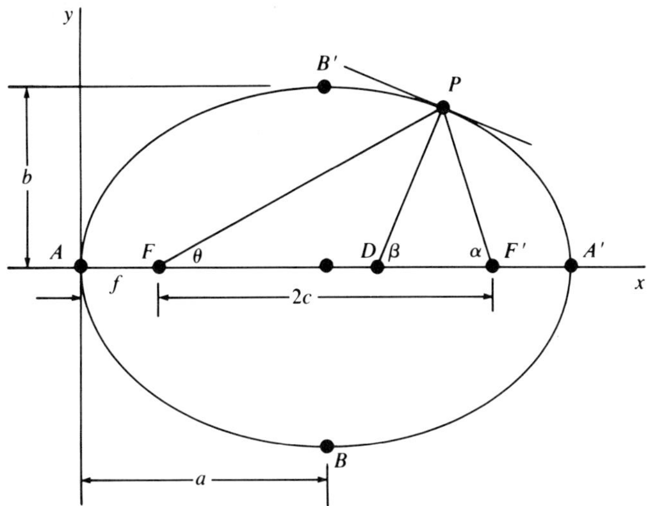

Figure 10.12 Elliptical contour.

$$FP + F'P = 2a \tag{10.19a}$$

$$\sqrt{(x-f)^2 + y^2} + \sqrt{(x-f-2c)^2 + y^2} = 2a \tag{10.19b}$$

The quantity $2a$ can be given physical meaning. Let $y = 0$ in Eq. (10.19b). The result is

$$x = 0,\ 2a \tag{10.20}$$

Thus the quantity $2a$ is the distance AA' in Fig. 10.12 and

$$a = c + f \tag{10.21}$$

The distance AA' is called the major axis of the ellipse.

The minor axis of the ellipse is BB'. To find its length, let $x = f + c$ in Eq. (10.19b). This yields

$$y = \pm\sqrt{a^2 - c^2} = \pm b \tag{10.22}$$

giving

$$BB' = 2\sqrt{a^2 - c^2} = 2b \tag{10.23}$$

Finally, using Eq. (10.21) and after considerable algebraic simplification, Eq. (10.19b) can be placed in the more useful form

$$\frac{(x-a)^2}{a^2} + \frac{y^2}{a^2 - c^2} = 1 \tag{10.24}$$

An elliptical reflector has the unique property that, if a source of light is placed at one focal point, all reflected rays pass through the second focal point. To prove this property, it is necessary to show that the three angles indicated in Fig. 10.12 are related through the law of reflection by

$$\beta - \theta = 180 - \beta - \alpha \tag{10.25}$$

The angle β is defined from the line DP, which is drawn perpendicular to the ellipse at P. The slope of this line, which is tan β, is the negative reciprocal of the slope of the tangent to the ellipse at P. To obtain the latter, Eq. (10.24) is differentiated to obtain

$$\frac{2(x-a)\,dx}{a^2} + \frac{1}{a^2-c^2}\,(2y)\,dy = 0 \tag{10.26}$$

$$\frac{dy}{dx} = \frac{(x-a)(c^2-a^2)}{a^2 y} \tag{10.27}$$

Thus β is given by

$$\tan\beta = \frac{a^2 y}{(x-a)(a^2-c^2)} \tag{10.28}$$

The other two angles in Fig. 10.12 are expressed by

$$\tan\theta = \frac{y}{x-a+c} \tag{10.29}$$

$$\tan\alpha = \frac{y}{a+c-x} \tag{10.30}$$

Then, noting that

$$\tan(180-\beta-\alpha) = -\tan(\beta+\alpha) \tag{10.31}$$

Eq. (10.25) can be rewritten as

$$\tan(\beta-\theta) = -\tan(\beta+\alpha) \tag{10.32}$$

From the law of tangents, each side of Eq. (10.32) can be expressed as

$$\tan(\beta-\theta) = \frac{\dfrac{a^2 y}{(x-a)(a^2-c^2)} - \dfrac{y}{x-a+c}}{1 + \dfrac{a^2 y^2}{(x-a)(x-a+c)(a^2-c^2)}} \tag{10.33}$$

$$-\tan(\beta+\alpha) = \frac{\dfrac{a^2 y}{(x-a)(a^2-c^2)} - \dfrac{y}{x-a-c}}{1 + \dfrac{a^2 y^2}{(x-a)(x-a-c)(a^2-c^2)}} \tag{10.34}$$

After algebraic simplifications, Eqs. (10.33) and (10.34) can be shown to be equal, completing the proof.

In Fig. 10.13*a* we desire to place an elliptical reflector inside a rectangular enclosure of half-width w and depth d such that one focal point is in the plane of the enclosure opening. First we must determine the location of the other focal point and the length of the major axis. We note that

$$\begin{aligned} d &= 2c + f \\ 2a &= d + f \\ w &= \sqrt{a^2 - c^2} \end{aligned} \tag{10.35}$$

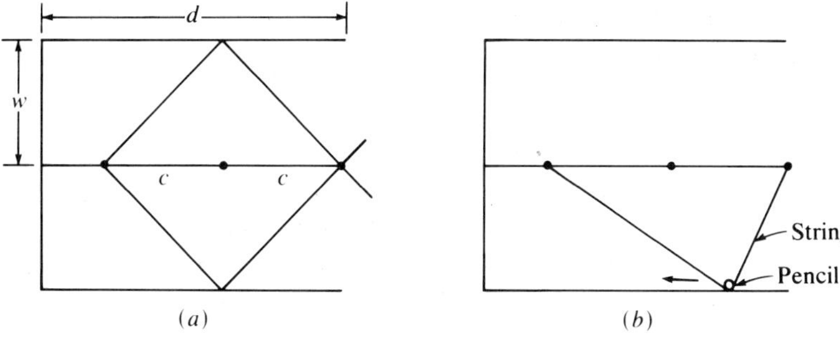

Figure 10.13 Designing an elliptical reflector.

where the last of these relations comes from Eq. (10.22). Solving Eqs. (10.35) for f and $2a$ yields

$$f = \frac{w^2}{d} \tag{10.36}$$

$$2a = d + \frac{w^2}{d} \tag{10.37}$$

To draw the elliptical contour, we could calculate points from Eq. (10.24). A simpler procedure is illustrated in Fig. 10.13b. Two pins are placed at the focal points and a string of length $2a$ is anchored between them. The ellipse is obtained by placing a pencil against the string to keep it taut and then tracing the locus.

Suppose it is desired that the emerging divergent beam extend only 60° above and below the horizontal. What must be the distance x to the ends of the elliptical contour? The angle of interest is α in Fig. 10.12. We solve for y in terms of x in Eq. (10.30) and insert the result in Eq. (10.24).

$$y = (a + c - x) \tan \alpha = (d - x) \tan \alpha \tag{10.38}$$

$$(x - a)^2 + \frac{a^2}{w^2} (d - x)^2 \tan^2 a = a^2$$

$$\left(1 + \frac{a^2 \tan 2a}{w^2}\right)x^2 - 2a\left(1 + \frac{ad \tan 2a}{w^2}\right)x + \frac{a^2 d^2 \tan 2a}{w^2} = 0 \tag{10.39}$$

All quantities in Eq. (10.39) are known except x.

EXAMPLE 10.2

Design an elliptical reflector to fit into a rectangular enclosure of depth 24 cm and half-width 12 cm, as in Fig. 10.13. It is desired that the beam angle α be 60°.

Solution. From Eqs. (10.36) and (10.37),

$$f = \frac{144}{24} = 6 \text{ cm} \qquad 2a = 24 + 6 = 30 \text{ cm}$$

Thus in Fig. 10.13, the two focal points are at 6 cm and 24 cm and the string length for the locus tracing is 30 cm. The ends of the ellipse are obtained from Eq. (10.39).

$$5.69x^2 - 255x + 2700 = 0$$
$$x = 27.7 \text{ cm},\ 17.1 \text{ cm}$$

The first answer is rejected. It comes about when x is greater than $a + c$ in Eq. (10.30), giving a negative value for $\tan \alpha$. However it is $\tan^2 \alpha$ that appears in Eq. (10.39). Thus there are two values of x satisfying Eq. (10.39) but only one of these satisfies Eq. (10.38).

The design in Example 10.2 illustrates the two major attributes of an elliptical reflector, namely its ability to converge all reflected rays to a small region and its ability to provide a divergent beam of controllable angular size.

As with the other reflector contours, the ray patterns of an ellipse are often not the ideal shown in Fig. 10.13*a* because the source of light is usually not a point source. The resulting patterns when the source is ahead of or behind a focal point are shown in Fig. 10.14*a* and *b*, respectively. In Fig. 10.14*c*, the effect of finite source size is demonstrated. Here a small spherical source is centered at the left focal point. The rays from each point on the reflector diverge somewhat upon reflection and the rays from all points on the reflector converge in the vicinity of the focal point, but not at it.

The elliptical reflector is widely used in light control. A few examples are shown in Fig. 10.15. Figure 10.15*a* shows the very common pinhole downlighting application, used extensively in architectural lighting. The light source is at one focal point and a small hole in the ceiling at the other. The source and reflector unit, which may be quite large, is nearly completely concealed from the viewer.

In Fig. 10.15*b*, a combination elliptical and circular reflector unit is shown. The lamp is at the focal point of the circle and one focal point of the ellipse. Rays hitting the circular reflector are reflected back through the source (assuming a noncompact filament), as in a projection lamp system. All reflected rays ultimately pass through the vicinity of the second focal point of the ellipse.

In Fig. 10.15*c*, a special surgical spotlight is shown, a combination elliptical and parabolic reflector unit. The source is at one focal point of the

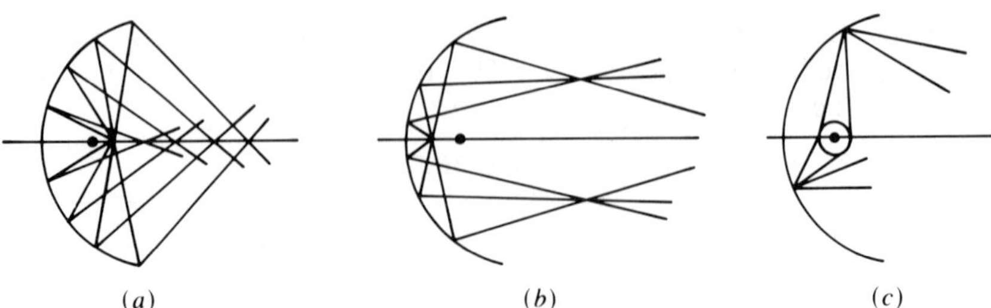

Figure 10.14 Ellipse: Source in front of or behind focus.

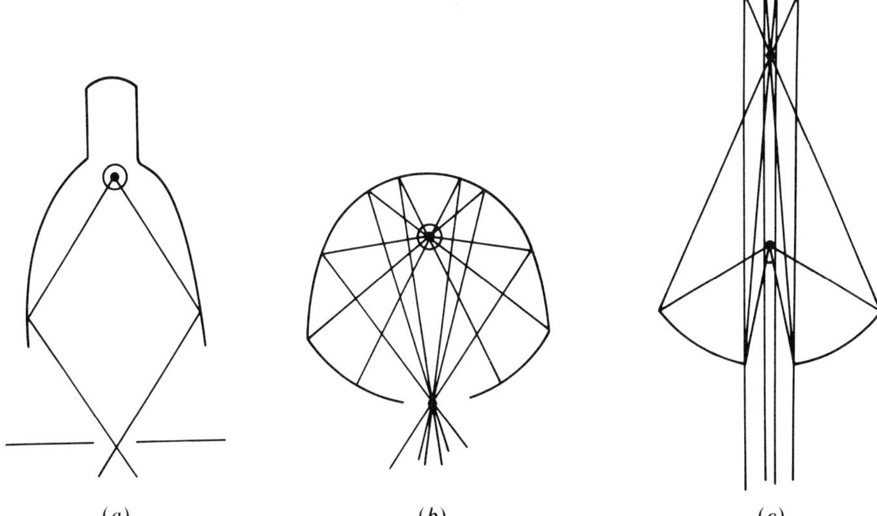

Figure 10.15 Applications of elliptical reflectors.

ellipse. Rays reflected from the ellipse pass through the second focal point, which is also the focal point of the parabola. An intense parallel-ray concentrated beam emerges from the parabola and passes through a hole in the ellipse and then to the operating table. A small shield blocks diverging direct rays from the source reaching the operating table.

The last basic reflector contour to be considered is the hyperbola, which can be used in place of the ellipse to produce a controlled diverging beam. A hyperbola is the locus of a point, the difference of whose distances from two fixed points is constant. In Fig. 10.16, the two fixed points (the two foci) are labeled F and F', with x coordinates $f + 2c$ and $-f$. For any point P on the right locus

$$FP - F'P = 2a \tag{10.40}$$

$$\sqrt{(x+f)^2 + y^2} - \sqrt{(x - 2c + f)^2 + y^2} = 2a \tag{10.41}$$

where

$$c = f + a \tag{10.42}$$

For the left locus, the quantity $2a$ becomes $-2a$ in Eq. (10.41).

Equation (10.41) can be algebraically simplified to yield

$$\frac{(x-a)^2}{a^2} - \frac{1}{c^2 - a^2}\, y^2 = 1 \tag{10.43}$$

A hyperbolic reflector can produce only a diverging beam. The rays from a point source of light at F' reflect from the hyperbola as though they had originated at F, as shown in Fig. 10.16. Thus, instead of passing through the second focal point, as in the ellipse, they appear to emanate from it.

It remains for us to show that the equations for the four basic reflector contours are related. Consider Eq. (10.24) for the ellipse. If the major and

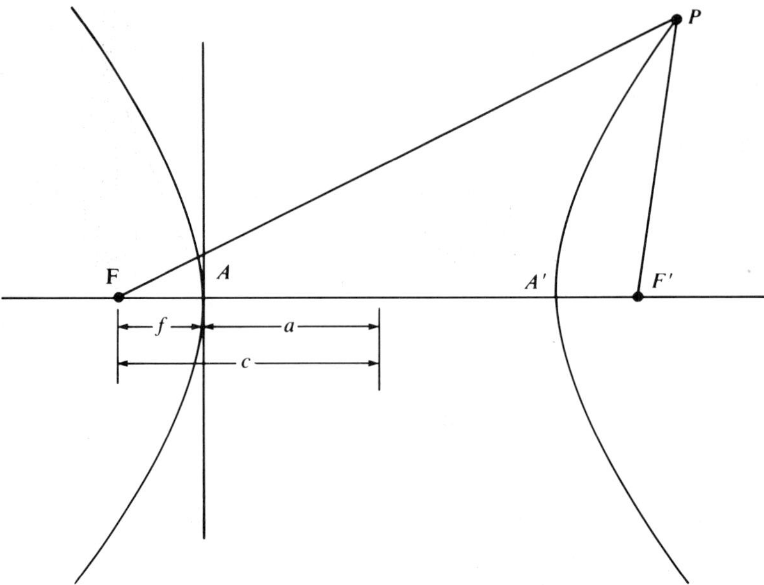

Figure 10.16 Hyperbolic contour.

minor axes are equal $a^2 = a^2 - c^2$ and thus $c = 0$ and $a = f$, from Eq. (10.21). The two focal points in Fig. 10.12 merge and Eq. (10.24) becomes Eq. (10.7), the equation of a circle.

The ellipse becomes a parabola if the focal point F' in Fig. 10.12 moves to infinity. This causes both a and c to become infinite. From Eqs. (10.21) and (10.24),

$$\frac{x^2}{a} - 2x + \frac{a}{a^2 - c^2}\, y^2 = 0 \qquad (10.44a)$$

$$\frac{x^2}{a} - 2x + \frac{a}{f(a + c)}\, y^2 = 0 \qquad (10.44b)$$

As a, and thus c, approach infinity, Eq. (10.44b) becomes

$$-2x + \frac{1}{2f}\, y^2 = 0 \qquad (10.45)$$

which is identical with Eq. (10.9) for the parabola.

Last, Eq. (10.43) for the hyperbola is identical in form to Eq. (10.24) for the ellipse, except that $c > a$ for the hyperbola and $c < a$ for the ellipse.

10.5 DESIGN OF NON-CONIC SECTION REFLECTORS

In situations where basic conic section reflectors are unsuitable, there are mathematical-graphical procedures for determining the appropriate reflector contour. We will describe one such procedure here, adapted from the method of Jolley, Waldram, and Wilson (1931).

In a general reflector design, the given quantities are the intensity distribution curve of the lamp and the desired intensity distribution curve of the lamp-reflector combination. The design involves determining what reflector contour will produce the latter from the former. It is accomplished by treating the reflector as an assemblage of small flat reflecting elements, with each element oriented so as to reflect the flux incident on it into the appropriate zone. The procedure will be developed in a step-by-step manner with the aid of a numerical example for a reflector with a 60° cutoff angle.

Step 1 Tabulate the intensity distribution data for the bare lamp and the desired intensity distribution data for the lamp-reflector combination by zones. For the example to be worked, this information is presented in the first three columns of Table 10.1. The choice of angular zone size is dependent on the available lamp data. The use of 5°, or $2\frac{1}{2}°$, zones can, of course, improve the accuracy of the design.

Step 2 For each zone, the bare lamp lumens and the required total lumens from the lamp-reflector combination are calculated, using the method of Sec. 2.6 and the zonal constants of Table 2.1. The results appear in columns 5 and 6 of Table 10.1.

For the procedure to work, it is necessary that the bare lamp lumens in each of the zones out to cutoff (60° in this case) be less than the required total lumens in those zones. This is the case in the example we have chosen. If it is not the case, there are two remedies possible: Choose a lamp of lesser intensity in each of the zones out to cutoff or scale the desired lamp-reflector intensity data by a factor $k > 1$, which will increase the lamp-reflector intensity values in each zone but not alter the shape of the distribution curve.

Step 3 For each zone out to cutoff, subtract the bare lamp lumens from the total lumens to obtain the reflected lumens required in each zone (column 7 in Table 10.1). Then sum these to obtain the cumulative reflected lumens (column 8). For the example, the total reflected lumens required is 1681.3 lm.

Step 4 For each of the zones from cutoff to zenith, multiply the bare lamp zonal lumens by the reflectance ρ of the reflector (column 9) and sum the results to obtain cumulative bare lamp lumens multiplied by reflectance (column 10).

It is necessary that the last number in column 10 equal the last number in column 8, so that the total reflected lumens provided equal the total number required. In equation form, the requirement is

$$\phi_{CO\text{-}180}\,(\rho) = \phi_{LR} - \phi_{0\text{-}CO} \tag{10.46}$$

where $\phi_{CO\text{-}180}$ is the total lamp lumens from cutoff to zenity, ϕ_{LR} is the

TABLE 10.1 GENERAL REFLECTOR DESIGN

(1) Zone (degrees)	(2) Bare lamp I_{mid} (cd)	(3) Lamp reflector I_{mid} (cd)	(4) Zonal constant (Table 2.1)	(5) Zonal bare lamp lumens	(6) Zonal total lumens	(7) Zonal reflected lumens	(8) Cumulative reflected lumens	(9) Zonal bare lamp lumens × ρ	(10) Cumulative bare lamp lumens × ρ
0–10	90	1536	0.095	8.6	145.9	137.3	137.3		
10–20	114	1350	0.283	32.3	382.1	349.8	487.1		
20–30	152	1094	0.463	70.4	506.5	436.1	923.2		
30–40	188	856	0.628	118.1	537.6	419.5	1342.7		
40–50	216	570	0.774	167.2	441.2	274.0	1616.7		
50–60	228	300	0.897	204.5	269.1	64.6	1681.3		
60–70	228		0.993	601.1 226.4	2282.4	1681.3		192.4	192.4
70–80	228		1.058	241.2				205.0	397.4
80–90	228		1.091	248.7				211.4	608.8
90–100	228		1.091	248.7				211.4	820.2
100–110	228		1.058	241.2				205.0	1025.2
110–120	228		0.993	226.4				192.4	1217.6
120–130	228		0.897	204.5				173.8	1391.4
130–140	165		0.774	127.7				108.5	1499.9
140–150	182		0.628	114.3				97.2	1597.1
150–160	143		0.463	66.2				56.3	1653.4
160–170	117		0.283	33.1				28.1	1681.5
170–180	44		0.095	4.2				3.6	1685.1
				1982.6				1685.1	

desired total lamp-reflector lumens, and $\phi_{0\text{-}CO}$ is the total lamp lumens from nadir to cutoff. Solving Eq. (10.46) for reflectance gives

$$\rho = \frac{\phi_{LR} - \phi_{0\text{-}CO}}{\phi_{CO\text{-}180}} \tag{10.47}$$

For the given example,

$$\rho = \frac{2282.4 - 601.1}{1982.6} = 0.848$$

A value of 0.85 was used in calculating the values in columns 9 and 10 in Table 10.1.

It is possible that Eq. (10.47) will give a required reflectance greater than unity. When this happens, it becomes necessary to reduce the desired intensity values in column 3. If each of these values is multiplied by a single constant $k < 1$, the final intensity distribution will be unaltered. Alternatively, if a slightly changed distribution and cutoff can be tolerated, the rim and neck of the reflector can be altered.

Step 5 A reflector pattern is chosen. There are four basic ray patterns for reflectors, as shown in Fig. 10.17. The contours in Fig. 10.17*a* and *b* reflect flux across the plane at the opening of the reflector; those in Fig. 10.17*c* and *d*, do not. For the contours in Fig. 10.17*a* and *c*, flux reflected within the 170 to 180° zone is directed to the 0 to 10° zone; for the contours in Fig. 10.17*b* and *d*, into the 80 to 90° zone.

The patterns in Fig. 10.17*c* and *d* require larger diameter reflectors, as shown. The pattern in Fig. 10.17*b* has the disadvantage that a considerable amount of flux passes back through the bulb. Thus the pattern in Fig. 10.17*a* is preferred, yielding a small reflector and minimum flux reflected through the lamp. This pattern will be used in the current example.

Step 6 Curves of required cumulative reflected lumens and of available cumulative bare lamp lumens times ρ are plotted from the data in columns 8 and 10 in Table 10.1. These curves are shown in Fig. 10.18. The lower curve is plotted first, from the data in column 8 and with lumens increasing to the

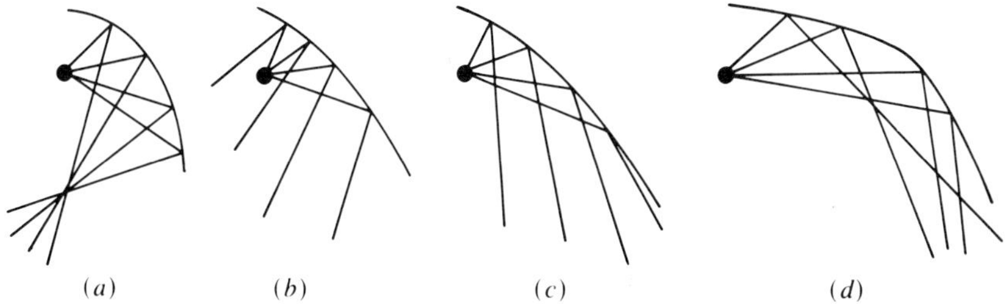

Figure 10.17 Four basic reflector patterns. [*Source:* J. E. Kaufman (ed.), *IES Lighting Handbook,* Reference volume, 1981 ed., IES, New York, 1981, fig. 6-29, with permission.]

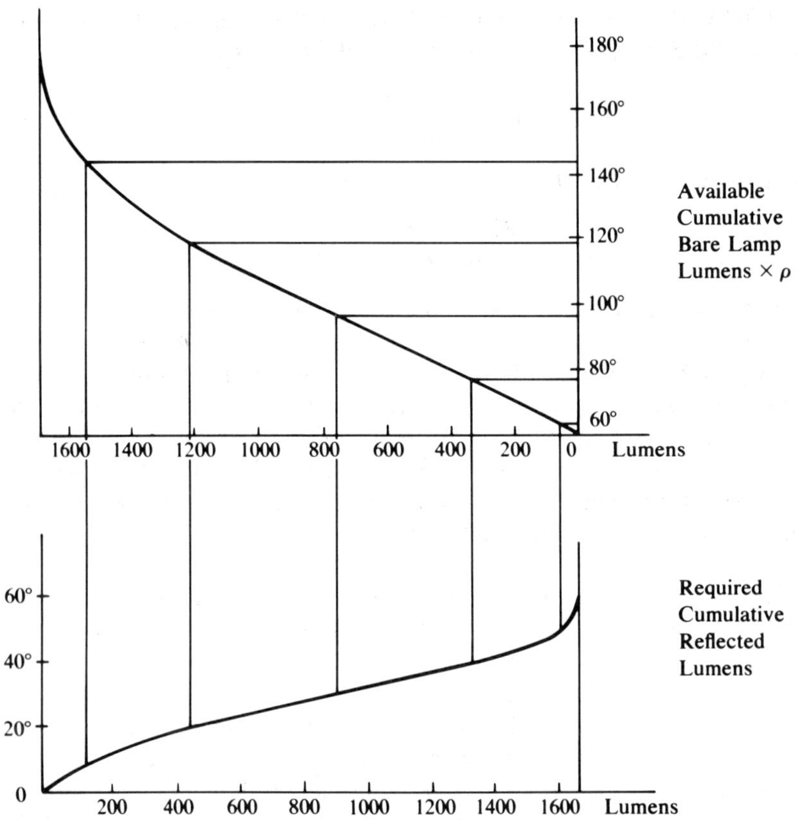

Figure 10.18 Graphical determination of reflector zones.

right along the x axis. The plot ends when 1681.3 lm have been accumulated. Next the upper curve is plotted from the data in column 10, with lumens increasing to the left. The reason for this is that we have chosen the pattern in Fig. 10.17*a*, which causes flux in the near-zenith zones to be redirected into the near-nadir zones.

Next vertical lines are drawn from the lower curve to the upper curve at the 10°, 20°, 30°, 40°, and 50° points. At the points of intersection with the upper curve, lines are drawn horizontally to the upper right vertical axis. To explain these lines, consider the leftmost one. It is drawn vertically at the 137.3-lm point. Its intersection with the upper curve is at a zonal angle of 143°. This tells us that a flat reflector element of 37° angular width should be inserted between 180 and 143° and oriented so as to direct the 137.3 lm reflected from it into the 0 to 10° zone. This construction is shown in Fig. 10.19, and proceeds until all zones are covered. Then a smooth curve can be drawn through the flat reflector segments.

The procedure just presented should be considered as providing a first estimate of the desired reflector contour. It does not consider finite source

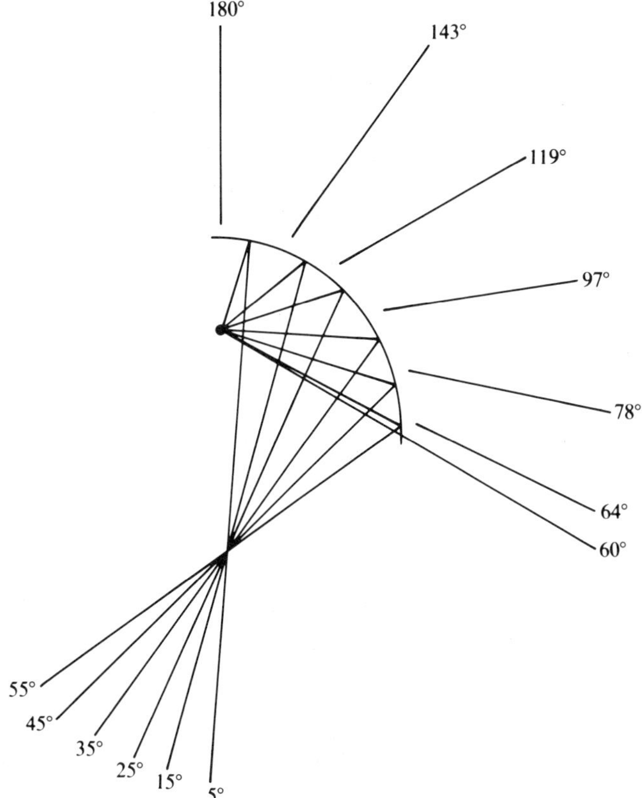

Figure 10.19 Development of reflector contour.

size and, as was shown in connection with conic sections, this can lead to errors in reflected ray paths. Also, as was discussed in Chap. 2, intensity is a questionable parameter in the near field and thus is not a reliable indicator of lamp performance close to the lamp. The technique works best for noncompact filaments but is likely to provide too high an overall intensity near the nadir.

10.6 REFRACTION

Refraction is the bending of light as it passes from one medium to another. As in the case of reflection, Huygens' principle may be applied to describe the situation. In Fig. 10.20*a*, a plane wavefront impinges at an angle ϕ on the boundary between two transparent media. Assuming the lower medium to be denser than the upper medium, the wavefront will be bent toward the normal and enter the lower medium at an angle $\phi' < \phi$.

The result in Fig. 10.20*a* may be quantified. First define the index of

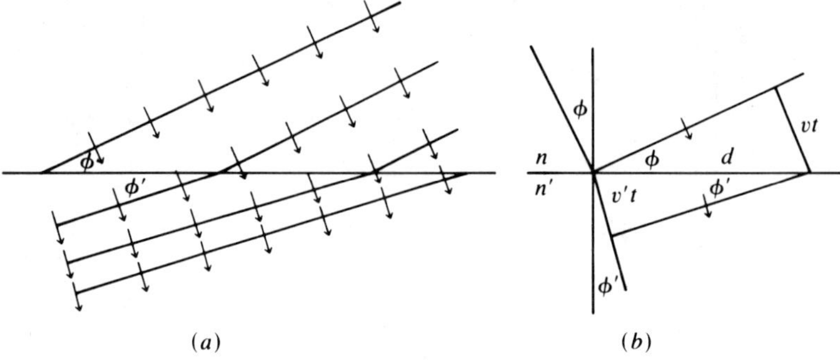

Figure 10.20 Refraction at a plane surface.

refraction of a medium as the ratio of the velocity of light in a vacuum to that in the medium. Thus

$$n = \frac{c}{v} \tag{10.48}$$

and is a number greater than one. Indexes of refraction are a function of the wavelength of the incident light. This phenomenon is called dispersion and will be discussed in connection with prisms later. Values of n at 589 nm (low-pressure sodium line) for common materials are shown in Table 10.2.

With n defined, consider Fig. 10.20*b*. A wavefront advancing at a velocity v in a medium of refractive index n impinges on a second medium of refractive index $n' > n$ at an angle ϕ. The velocity in the second medium is $v' < v$. In the time it takes for the extreme right of the wavefront in the upper medium to advance a distance $s = vt$, the extreme left of the

TABLE 10.2 INDEXES OF REFRACTION

Crown glass	1.48–1.61
Flint glass	1.53–1.96
Soda lime glass	1.51
Lead glass	1.53–1.67
Borosilicate glass	1.48
Aluminosilicate glass	1.54
Acrylic	1.49
Quartz	1.54
Diamond	2.42
Calcium carbonate	2.66
Calcium fluorite	1.43
Sodium chloride	1.54
Water	1.33
Carbon disulfide	1.63
Carbon dioxide	1.00045
Air	1.00029

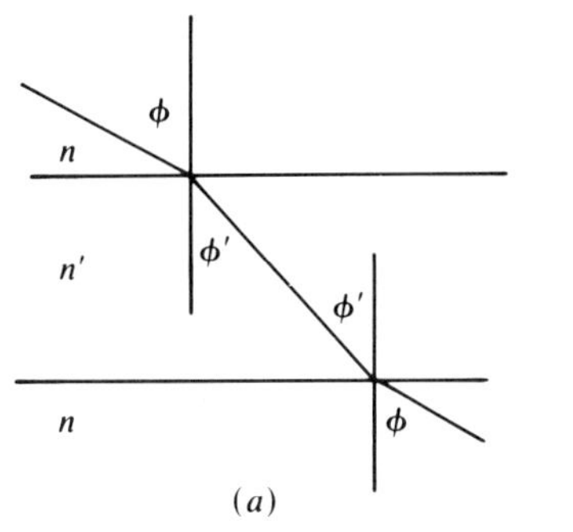

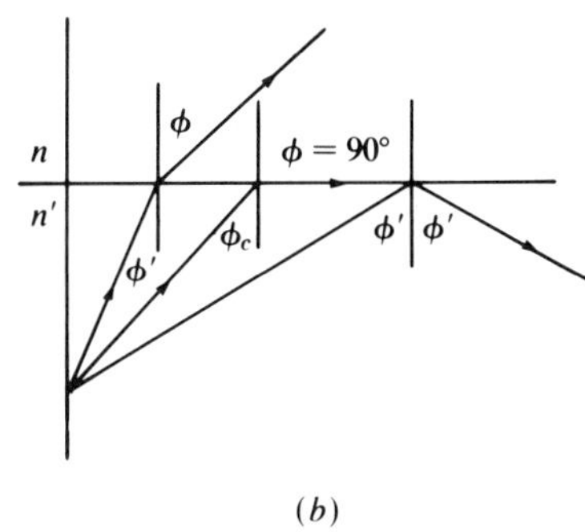

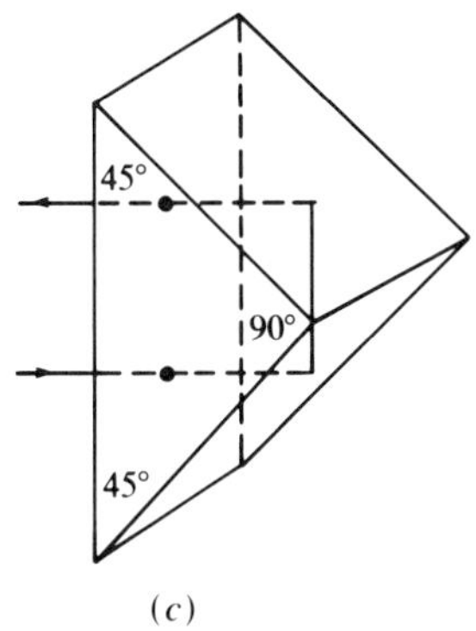

Figure 10.21 Applications of Snell's law.

wavefront in the lower medium will advance a distance $s' = v't$. From Fig. 10.20*b* we can write

$$\sin\,\phi = \frac{vt}{d} \qquad \sin\,\phi' = \frac{v't}{d} \tag{10.49}$$

Then, from Eqs. (10.48) and (10.49),

$$\frac{\sin\,\phi}{\sin\,\phi'} = \frac{v}{v'} = \frac{n'}{n} \tag{10.50}$$

The result in Eq. (10.50) is known as Snell's law.*

Velocity, frequency, and wavelength are related by Eq. (1.9). In passing from one medium to another, the frequency of the light wave does not change. Thus, as velocity changes, wavelength must change. In particular, in going from air to a denser medium, such as glass, velocity decreases, and thus so does wavelength, within the glass.

Refraction by a flat glass plate $(n' > n)$ with parallel faces is illustrated in Fig. 10.21*a*. Here the net effect is to displace the light ray rather than al-

*Named for Willebrord Snell in 1621. Snell studied refraction but died at the age of 35 with the results of his work unpublished. Rene Descartes published the law of refraction 11 years later.

ter its direction. In Fig. 10.21*b* (again $n' > n$), the possibility of total internal reflection is examined. At the critical angle ϕ_c, determined from Eq. (10.50) as

$$\sin \phi_c = \frac{n}{n'} \tag{10.51}$$

the refracted ray will graze the boundary. For all angles greater than ϕ_c, the ray will be totally reflected internally.

For an air-glass surface, with the latter having a typical index of refraction of 1.50, $\phi_c = 42°$. The fact that this angle is less than 45° makes it possible to construct totally reflecting prisms, as shown in Fig. 10.21*c*. These have the advantages over mirrored reflector surfaces of being 100% reflecting and of having no degradation over time.

There are two properties of prisms that should be discussed, namely deviation and dispersion. Consider the isosceles prism in Fig. 10.22*a*. A ray of light is incident on the left face at such an angle ϕ that it passes through the prism horizontally and exits symmetrically from the right face. The apex angle α of the prism is shown and we desire to calculate the deviation δ_m of the ray in passing through the prism. From the geometry and Eq. (10.50) and noting that half the deviation takes place at each surface, we can write

$$\phi' = \frac{\alpha}{2} \qquad \delta = \frac{\delta_m}{2} \tag{10.52a}$$

$$\phi = \phi' + \delta = \frac{\alpha + \delta_m}{2} \tag{10.52b}$$

$$\sin \phi = \sin \frac{\alpha + \delta_m}{2} = n' \sin \phi' = n' \sin \frac{\alpha}{2} \tag{10.52c}$$

$$n' = \frac{\sin [(\alpha + \delta_m)/2]}{\sin (\alpha/2)} \tag{10.52d}$$

Since it is possible to measure the angles α and δ_m very accurately, Eq. (10.52*d*) can be used to determine indexes of refraction with considerable precision.

The reason for the symbology δ_m requires explanation. As the angle of

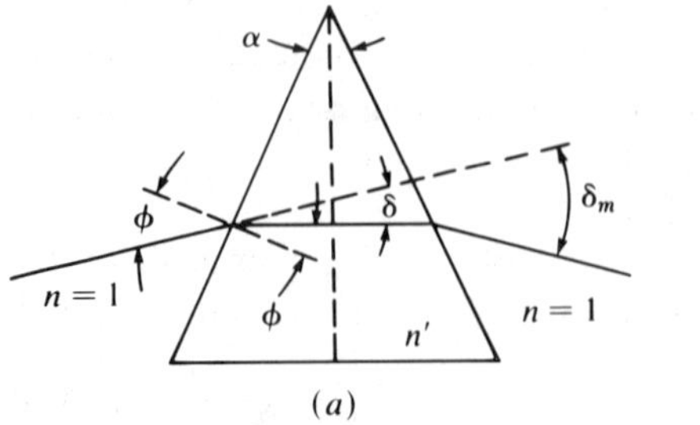

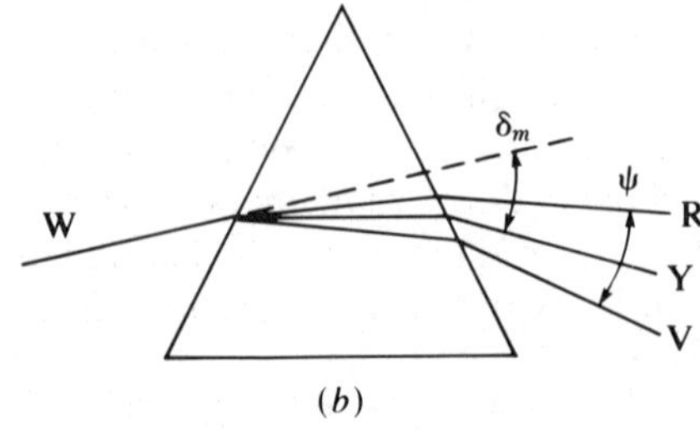

Figure 10.22 Deviation by a prism.

incidence ϕ in Fig. 10.22*a* is decreased from a very large value, the deviation first decreases and then increases. It is a minimum when the ray passes through the prism symmetrically, as is the case in Fig. 10.22*a*. It is this minimum deviation that is customarily measured for a prism.

A second parameter of interest with prisms is dispersion. As noted earlier, the index of refraction is a function of wavelength. Within the visible spectrum, $n_{\text{violet}} > n_{\text{red}}$ and thus the shorter visible wavelengths undergo greater deviation is passing through a prism than do the longer wavelengths. Because of dispersion, it is necessary to specify the wavelength at which prism deviation is measured. Customarily 589 nm is chosen, as in Table 10.2.

Prism dispersion is illustrated in Fig. 10.22*b*. White light (W) impinges on the left prism face, is dispersed, and emerges from the right prism face with the spectral components separated. The measure of dispersion is usually taken as the angular spread between the red (R) and violet (V) wavelengths.

For small prism apex angles, simple expressions for both dispersion and minimum deviation can be derived. From Eq. (10.52*d*), if α is small, we can replace the sine of the angle by the angle itself and obtain

$$\delta_m = (n_y - 1)\alpha \tag{10.53}$$

$$\psi = (n_v - n_r)\alpha \tag{10.54}$$

where y, v, and r denote yellow, violet, and red, respectively. δ_m and ψ are indicated in Fig. 10.22*b*. Typical values for δ_m and ψ for prisms with 10° apex angles are 6.2° and 0.50° for flint glass, 5.4° and 0.16° for quartz, and 5.1° and 0.18° for crown glass.

10.7 LENSES

Lenses involve refraction at two curved surfaces, generally air to glass and glass to air. As a preliminary to their discussion, it is necessary to consider refraction at a single spherical surface, as diagrammed in Fig. 10.23. An object AX is located a distance p from the vertext O of a spherical refractor of radius r. As in the spherical mirror in Fig. 10.3, three rays are shown emanating from point A. One is directed normally through the center of the spherical surface. The second is directed parallel to the refractor's axis and is bent upon entering the refractor. The third is directed toward the refractor's vertex and is also bent upon entering the refractor. All refractions obey Snell's law. For small ϕ, the three rays intersect at point B, producing the image of A. Thus the object AX is imaged as BY, an inverted image reduced in size.

To quantify the relations in Fig. 10.23, we note that

$$\frac{AX}{BY} = \frac{XC}{CY} = \frac{p + r}{q - r} \tag{10.55}$$

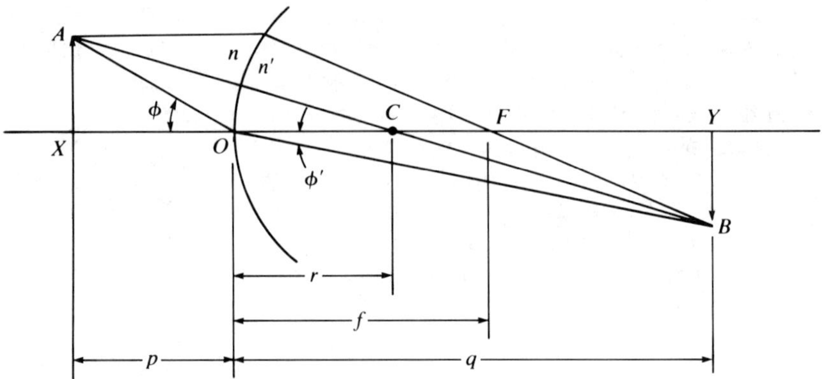

Figure 10.23 Refraction at a convex spherical surface.

$$\frac{AX}{OX} = \tan \phi \qquad \frac{BY}{OY} = \tan \phi' \tag{10.56}$$

For small angles, the sine and tangent are approximately equal and Eqs. (10.56) can be combined with the aid of Snell's law to yield

$$\frac{AX}{BY} = \frac{OX \sin \phi}{OY \sin \phi'} = \frac{pn'}{qn} \tag{10.57}$$

Then Eqs. (10.55) and (10.57) give

$$\frac{n}{p} + \frac{n'}{q} = \frac{n' - n}{r} \tag{10.58}$$

and

$$M = \frac{BY}{AX} = -\frac{qn}{pn'} \tag{10.59}$$

Sign conventions for Fig. 10.23 and Eqs. (10.58) and (10.59) are a bit different than those for a spherical mirror. In the present case, the following rules apply:

1. Object distances are positive if the object is to the left of the refractor vertex.
2. Image distances and radii are positive if the image and the center of spherical surface are to the right of the refractor vertex.
3. Images are positive in size if they lie above the horizontal axis.

As in the mirror case, the focal point of a refractor is the location of the image when the object is located at infinity. From Eq. (10.58)

$$\frac{n'}{f} = \frac{n' - n}{r} \qquad f = \frac{r}{1 - (n/n')} \tag{10.60}$$

The focal point and the focal length are labeled as F and f, respectively, in Fig. 10.23. Note that for $n' > n$, $f > r$.

EXAMPLE 10.3

For a spherical refractor with $n' = 2$ and $r = 3$ cm, find the image distance, focal length, and magnification for object distances of $p = 6$ cm, 1 cm, and -6 cm.

Solution. From Eqs. (10.58), (10.59), and (10.60), we can make the following tabulation; where all distances are in centimeters:

p	q	f	M
6	12	6	-1
1	-3	6	$+\frac{3}{2}$
-6	4	6	$+\frac{1}{3}$

In the first case, the image is inverted and to the right of the vertex. In the second case, the image is erect and to the left of the vertex. In the third case, the object is inside the refractor, placed there by a previous refraction. The image is erect and to the right of the vertex.

A lens may be considered as the superposition of two spherical refractors. Thus problems involving lenses can be solved by two applications of Eq. (10.58), as follows:

$$\frac{1}{p_1} + \frac{n'}{q_1} = \frac{n' - 1}{r_1}$$

$$\frac{n'}{p_2} + \frac{1}{q_2} = \frac{1 - n'}{r_2} \tag{10.61}$$

If we assume that the lens has negligible thickness compared with the other distances involved, $p_2 = -q_1$ and Eqs. (10.61) may be added to obtain

$$\frac{1}{p_1} + \frac{1}{q_2} = (n' - 1)\left(\frac{1}{r_1} - \frac{1}{r_2}\right) \tag{10.62}$$

There are two focal lengths associated with a lens. From Eq. (10.62), with either p_1 or q_2 infinite,

$$\frac{1}{f} = (n' - 1)\left(\frac{1}{r_1} - \frac{1}{r_2}\right) \tag{10.63}$$

Then substituting Eq. (10.63) into Eq. (10.62) gives

$$\frac{1}{p} + \frac{1}{q} = \frac{1}{f} \tag{10.64}$$

where we have now dropped the intermediate subscripts used in the development.

The ray-tracing construction for both convex and concave lenses is shown in Fig. 10.24. The location of the object for an image at infinity is called the first focal point ($F1$); the location of the image for an infinitely

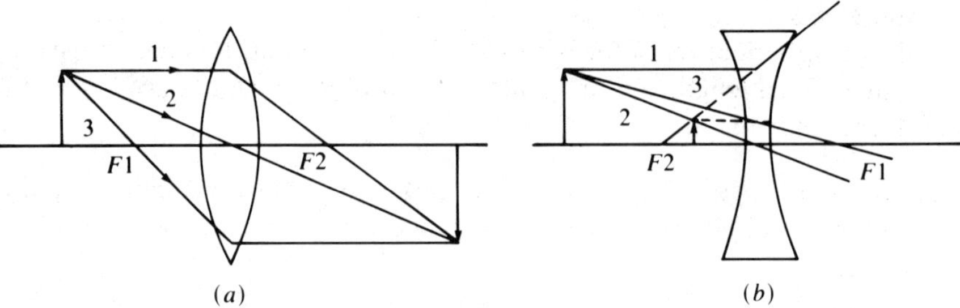

Figure 10.24 Ray tracing for lenses.

distant object is named the second focal point ($F2$). These points are labeled in Fig. 10.24.

The three rays used in graphically determining the image location are also shown in Fig. 10.24. In either diagram, ray 1 from an object at infinity, exits the lens through the second focal point. Ray 2 passes undeviated through the vertex of the lens. Ray 3, from an image at infinity, emerges from the lens through the first focal point. The intersection of the three rays marks the location of the image.

In Fig. 10.24*a*, the image is real and inverted. In Fig. 10.24*b*, it is virtual and not inverted. The virtual image is the apparent location of the object when viewed from the right side of the lens.

EXAMPLE 10.4

An object lies 6 cm left of a convex lens with $r_1 = 3$ cm, $r_2 = 4$ cm, and $n' = 2$. Find the location of the image produced.

Solution. The problem will be worked in two ways, first by use of Eq. (10.62) for thin lenses and second by two applications of Eq. (10.58) for a refracting surface. Each solution is illustrated in Fig. 10.25.

Application of Eq. (10.62) yields

$$\frac{1}{6} + \frac{1}{q_2} = (1)\left(\frac{1}{3} - \frac{1}{-4}\right) \qquad q_2 = \tfrac{12}{5} \text{ cm}$$

From Eq. (10.63),

$$\frac{1}{f} = (1)\left(\frac{1}{3} - \frac{1}{-4}\right) \qquad f = \tfrac{12}{7} \text{ cm}$$

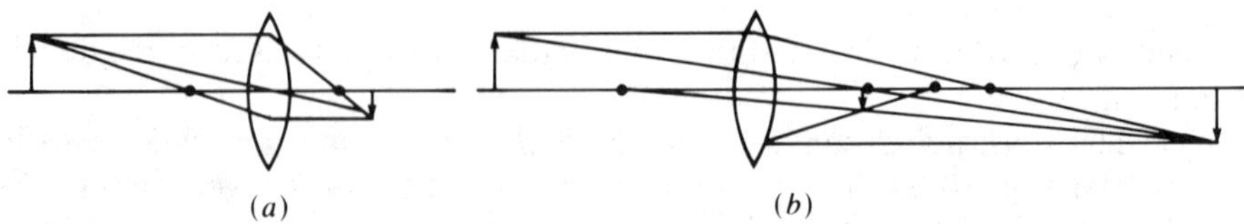

Figure 10.25 Graphical solution of Example 10.4.

f could also be obtained from Eq. (10.64):

$$\frac{1}{6} + \frac{5}{12} = \frac{1}{f} \qquad f = \tfrac{12}{7}\ \text{cm}$$

Using Eq. (10.58) gives

$$\frac{1}{6} + \frac{2}{q_1} = \frac{2-1}{3} \qquad q_1 = 12\ \text{cm}$$

Also Eq. (10.60) yields

$$f_1 = \frac{3}{1-\frac{1}{2}} = 6\ \text{cm}$$

A second application of Eq. (10.58), this time with an object distance of -12 cm, yields

$$\frac{2}{-12} + \frac{1}{q_2} = \frac{1-2}{-4} \qquad q_2 = \tfrac{12}{5}\ \text{cm}$$

From Eq. (10.60)

$$f_2 = \frac{-4}{1-\frac{2}{1}} = 4\ \text{cm}$$

The diagram in Fig. 10.25*a* is the thin-lens construction and duplicates that in Fig. 10.24*a*. The diagram in Fig. 10.25*b* is based on considering the problem as two refracting surfaces in which the image from the first refraction serves as the object for the second refraction.

We have yet to consider the magnification provided by a thin lens. It is simply the product of the magnifications at each of its surfaces. From Eq. (10.59)

$$M = \frac{q_1 n}{p_1 n'}\left(\frac{q_2 n'}{p_2 n}\right) = -\frac{q_2}{p_1} \equiv -\frac{q}{p} \tag{10.65}$$

where we have made use of the fact that $p_2 = -q_1$ for a thin lens.

10.8 LENS APPLICATIONS AND DESIGN

In this section we will consider a few applications of lenses in the field of light and vision and develop a graphical procedure for the design of a refractive surface.

Perhaps the most common use of lenses is in eyeglasses. The lens system of a normal human eye forms an image on the retina of an object at infinity. Two common deviations from this are hyperopia (farsightedness) and myopia (nearsightedness). In the hyperopic eye, the image of an object at infinity lies behind the retina; in a myopic eye, in front of it.

The far point of an eye is the farthest point from the eye at which an object will be imaged on the retina. The near point is the nearest point at which this occurs. In the myopic eye, the far point is nearer than infinity and the near point is closer to the eye than for a normal eye. In the hyperopic eye, the near point is farther from the eye than for a normal eye.

Both hyperopia and myopia can be corrected by eyeglasses, as can presbyopia, the increase in the near point of a normal eye with aging, which is described in Sec. 5.9. We will illustrate these corrections through an example.

EXAMPLE 10.5

(*a*) The near point of an eye is 50 cm in front of the cornea. What lens should be used to move the near point to 20 cm in front of the cornea?

(*b*) If the far point of an eye is 2 m from the cornea, what lens should be used to move the far point to infinity?

Solution. Equation (10.64) is the appropriate equation to use for both parts of the problem.

(*a*) We need a convex (converging) lens to make it appear to the eye that the object is at 50 cm. Thus $p = 20$ cm, $q = -50$ cm, and we have

$$\frac{1}{f} = \frac{1}{20} + \frac{1}{-50} \qquad f = 33 \text{ cm}$$

(*b*) A concave (diverging) lens is required which will make an object at infinity appear to the eye to be located at 2 m. Thus $p = \infty$, $q = -200$ cm, and the calculation yields

$$\frac{1}{f} = \frac{1}{\infty} + \frac{1}{-200} \qquad f = -200 \text{ cm}$$

Perhaps the second most common use of lenses is in photography. Actually, pictures can be taken without lenses using a pinhole camera, but one is restricted to stationary objects because the pinhole admits very little light and thus exposures of seconds, or even minutes, are required. If the opening is enlarged, more light is admitted but the picture becomes out of focus. With a lens, the opening can be enlarged to admit more than 50,000 times the amount of light passed by a pinhole. Thus the primary function of a camera lens is to focus the light rays so that the diaphragm opening can be large enough to avoid long exposures.

The fundamental parameters of a camera lens are its focal length and diameter. From Eqs. (10.64) and (10.65), for a given object size and distance, the longer the focal length of the lens, the larger the image. Thus, whether the lens is telephoto, normal, or wide-angle, the image size is determined by the focal length.

TABLE 10.3

p(in)	q(in)
∞	2.00
200	2.02
100	2.04
50	2.08
25	2.17
10	2.50
5	3.33
4	4.00
3	6.00

The diameter of a camera lens governs its light-gathering power and, along with focal length, establishes its maximum angular field of coverage. The actual field of coverage is set by the field stop and is generally stated in terms of the area of acceptable sharpness and definition, rather than the total area seen by the lens. A normal lens used in still photography covers a field of about 50°.

It is customary to provide a single number to describe a camera lens. This is the f-number (or f-stop) and is the ratio of focal length to diaphragm diameter.

$$f_n = \frac{f}{d} \tag{10.66}$$

The f-numbers commonly used form a series 1, 1.4, 2, 2.8, 4, 5.6, 8, 11, 16, and 22. Each succeeding f-number provides one-half of the light-gathering power of the one preceding it. For example, for a given lens, the ratio of diameters for $f_{5.6}$and f_8 is 8:5.6. The ratio of areas is $64{:}32 = 2$. Thus an $f_{5.6}$ lens admits twice the luminous flux as an f_8 lens of the same focal length in a given setting.

A 35-mm slide camera generally has a focal length of 2 in. From Eq. (10.66), for $f_{1.4}$, its diaphragm opening is 1.4 in and, at the other extreme, for f_{22}, the diaphragm opening is 0.09 in. Thus the range of its light-gathering power is about 250:1. The image location for different object distances for this lens is shown in Table 10.3. Note that until the object gets quite close to the camers (<10 in), the image distance is very nearly constant. At these closer distances, the lens can be moved farther from the film or a lens of slightly different focal length can be used. For example, at $p = 10$ in, a focal length of 1.67 would be required to place the image 2 in from the lens.

A third application of lenses, as well as of prisms, is in conjunction with lamps and luminaires in lighting systems. One example is the Fresnel* lens,

*Named for Augustin Fresnel (1788–1827), French engineer and mathematician, who made several contributions in geometric and physical optics.

illustrated in Fig. 10.26*a*. This is essentially a plano-convex lens in which portions of the convex surface have been eliminated to reduce weight and cost while still providing essentially the same optics. It can also be constructed in cylindrical form, as shown in Fig. 10.26*b*.

The use of a clear prismatic lens in a fluorescent luminaire is illustrated in Fig. 10.26*c*. The prisms redirect lamp lumens downward and upward and reduce luminances in and near the 45° glare zone. Clear prismatic lenses, in either glass or plastic, transmit 70 to 92% of the flux incident on them with very little diffusion. Several prism patterns are used, two common ones being hexagonal pyramids and circular cones, with densities of 16 to 64 prisms per square inch.

To close this section, we will now present a graphical procedure which has been developed (Committee on Light Control and Equipment Design, 1959, 1967, 1970) for the design of a refractive surface to redirect a ray from a light source. The procedure is illustrated in Fig. 10.27 and will be presented in steps.

Step 1 Establish the known quantities. In Fig. 10.27, the refractive plate has a flat bottom surface and its top surface is to be formed into a sawtooth pattern. One such "tooth," inclined at an angle θ (to be determined) is shown. The light source is located such that a ray from it to the tooth makes an angle α with the normal to the plate. It is desired that the refracted ray emerge from the bottom surface at an angle ϕ_2 with the normal. The index of refraction of the plate (n') is known and the medium above and below the plate is air $(n = 1)$. In Fig. 10.27, a value of 1.5 for n' is used.

Step 2 The graphical construction begins at point A. With A as a center, arcs are struck with radii proportional to n and n'. The line AB is drawn at the given angle ϕ_2. From point B a perpendicular is dropped to point C. Then the line CAE is drawn, defining the angle ϕ_2', which is the angle of the refracted ray within the plate. This construction is justified by noting that

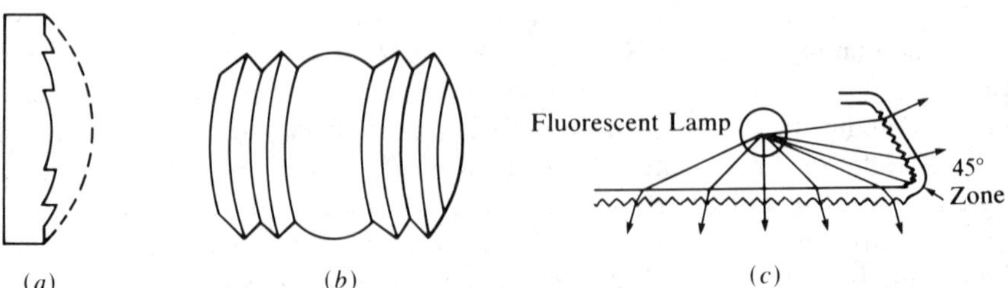

Figure 10.26 Lens and prism applications. (*Source* of (*c*): General Electric Co., "Light Measurement and Control," Bulletin TP-118, Nela Park, Cleveland, Ohio, 1971, with permission.)

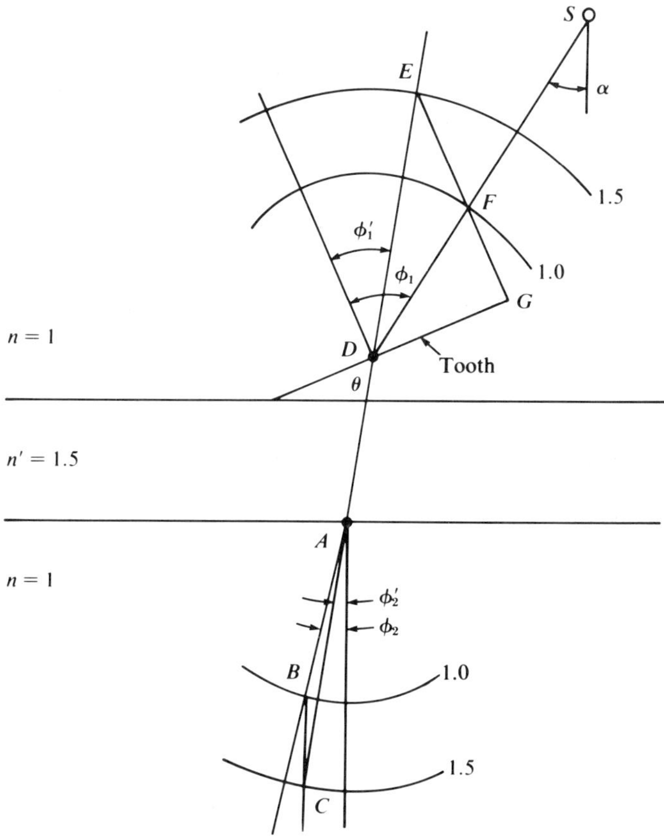

Figure 10.27 Design of a refractor surface.

$$\frac{\sin \phi_2'}{\sin \phi_2} = \frac{AB}{AC} = \frac{n}{n'} \tag{10.67}$$

in accordance with Snell's law.

Step 3 Estimate the angle θ required for the tooth and locate the point D on line CAE. With D as a center, strike two additional arcs whose radii are again proportional to n and n'. Locate point F on line SD and draw line EFG, which is the required normal to the slant face of the tooth. If angle FGD is not 90°, repeat the step with a new estimate of θ.

Step 4 To design additional teeth repeat steps 1 through 3.

The design can also be done analytically, again by repeated estimates. ϕ_2 is known and thus ϕ_2' can be found from Eq. (10.67). Then, using Snell's law and the known geometry, we can write

$$\phi_1' = \phi_2' + \theta$$
$$\phi_1 = \alpha + \theta$$
$$\sin\phi_1 = \frac{n'}{n}\sin\phi_1' \tag{10.68}$$

Combining these equations yields

$$\sin(\alpha+\theta) = \frac{n'}{n}\sin(\phi_2'+\theta) \tag{10.69}$$

All quantities except θ are known in Eq. (10.69). Since θ does not appear explicitly in that equation, it must be solved for by trial and error.

10.9 INTERFERENCE

The previous sections of this chapter dealt with geometric optics; the remainder of the chapter will deal with physical optics, treating the subjects of interference, diffraction, polarization, and fiber optics.

What is the distinction between these two branches of optics? In geometric optics we dealt with rays of light reflected or refracted at definite angles at a boundary between two media. The problems were geometrical and we considered only the amplitude of the light, not its phase.

In physical optics, we must abandon the notion of rays and return to the more fundamental idea that light is electromagnetic wave motion and that the superposition of more than one coherent wavefront depends on the time phase relationship between the waves, as well as their amplitudes. If, upon superposition, the amplitude of the resultant wave is greater than that of either of the original waves, the waves are said to constructively interfere. If it is less, destructive interference is said to occur. The extreme case of destructive interference is annulment, where the original waves are equal in amplitude and 180° out of time phase.

The concept of constructive interference of two light waves is illustrated in Fig. 10.28. Two coherent monochromatic point sources of light are placed a certain distance apart (3λ) at S_1 and S_2. The requirement for the wavefronts from these sources to constructively interfere at point P is

$$PS_1 - PS_2 = \frac{m\lambda}{2} \tag{10.70}$$

where m is an even integer (for destructive interference, m should be an odd integer). The loci in Fig. 10.28 are hyperbolic.

As pointed out in Sec. 1.3, one of the earliest demonstrations of interference was Thomas Young's famous experiment in 1827. The experiment is diagrammed in Fig. 10.29. Light from the source S falls on the slit at the left. By Huygens' principle, secondary wavefronts emanate from this slit and travel equal distances to the slits at A and B, arriving in phase at these

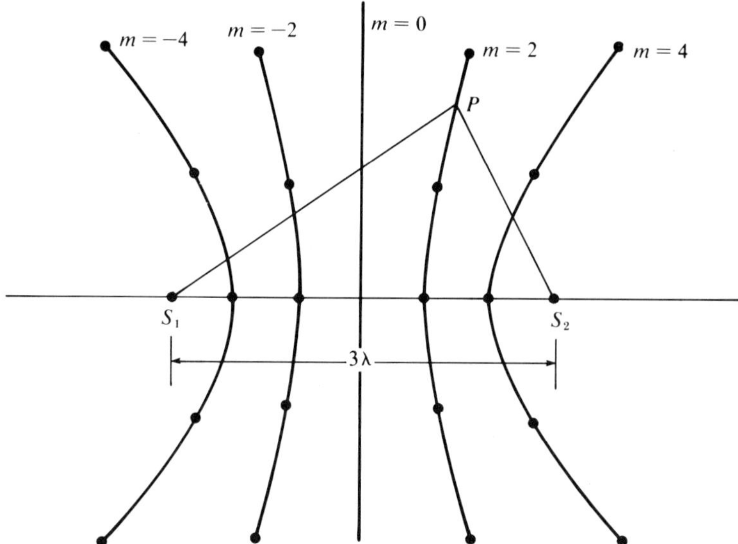

Figure 10.28 Interference loci.

points. Thus the three slits serve to provide coherent radiation at points A and B, even though the source S is not coherent.

Again by Huygens' principle, second sets of secondary wavefronts leave A and B and travel toward the screen. Alternating light and dark bands appear on the screen, as a function of the difference in the path lengths AP and BP. To derive an expression for the location of these bands, we observe that, if $R >> d$, the arc BC centered at P is essentially a straight line. The triangles ACB and DOP are similar and thus angle $ABC = \theta$. It follows that, for constructive interference

$$AC = d \sin \theta = m\lambda \qquad \sin \theta = \frac{m\lambda}{d} \tag{10.71}$$

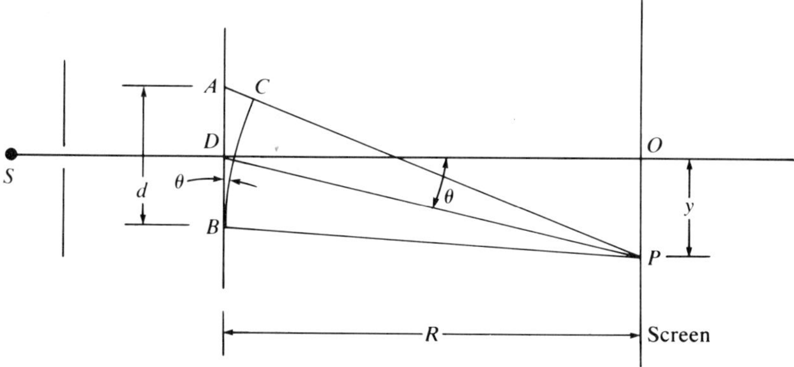

Figure 10.29 Young's interference experiment.

where m is an integer. A typical value for d is 10^{-2} cm. Generally 5 to 10 interference bands can be seen, so $m \leq 10$. For a λ at the middle of the visible spectrum

$$\sin \theta = \frac{10(550 \times 10^{-9})}{10^{-4}} = 0.055 \qquad \theta \simeq 3^\circ$$

Thus θ is very small and

$$y = R \tan \theta \simeq R \sin \theta \tag{10.72}$$

From Eqs. (10.71) and (10.72),

$$y = \frac{mR\lambda}{d} \qquad \lambda = \frac{yd}{mR} \tag{10.73}$$

The last Eq. (10.73) permits λ to be determined from simple length measurements.

EXAMPLE 10.6

Assume in Fig. 10.29 that the two slits are separated by 0.15 mm. The screen is 2 m from the slits and the angle θ of the fifth bright band is 1.2°. What is the wavelength of the (monochromatic) light source?

Solution. From Eq. (10.72),

$$y = 2 \sin 1.2 = 41.9 \text{ mm}$$

Then, from Eq. (10.73)

$$\lambda = \frac{(41.9 \times 10^{-3})\,(0.15 \times 10^{-3})}{5(2)} = 0.630 \times 10^{-6} \text{ m}$$

$$= 630 \text{ nm}$$

The phenomenon of interference permits an explanation of thin-film dielectrics, semiconductors, and metals that are used as optical coatings to control spectral and overall reflectance and transmittance in lamps and lighting control equipment. Consider the thin film of thickness t in Fig. 10.30a. A light beam (a) is incident on the film's upper surface. A portion is transmitted (b) and another portion is reflected (c). The transmitted portion impinges on the lower surface of the film where again a portion is transmitted (d) and another portion is reflected (e). The reflected portion, upon reaching the upper surface of the film is partly transmitted (f) and reflected (g), and so on.

For certain wavelengths, the thickness of the film may be such that the reflected waves (c and f) interfere (the transmitted waves can interfere also). From Eqs. (1.9) and (10.50), wavelengths and refractive indexes are related in general by

$$\lambda n = \lambda' n' \tag{10.73}$$

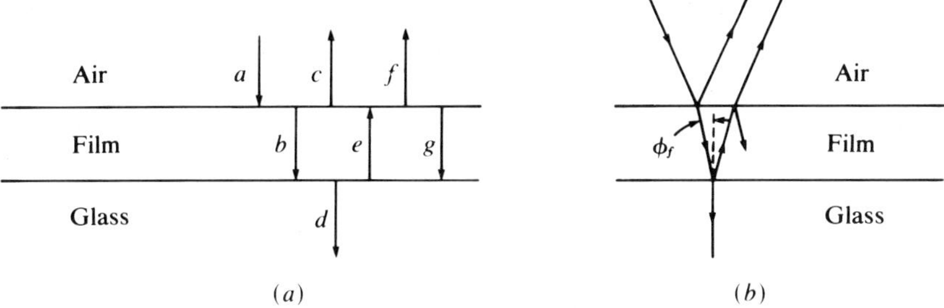

Figure 10.30 Thin-film interference.

For the air–thin film interface in Fig. 10.30*a*,

$$\lambda = \lambda_f n_f \qquad \lambda_f = \frac{\lambda}{n_f} \tag{10.74}$$

The number of wavelengths in two film thicknesses is (assuming normal incidence)

$$\text{No. of } \lambda_f\text{'s} = \frac{2tn_f}{\lambda} \tag{10.75}$$

If there were no other factors involved, the requirement for constructive interference would be that the quantity in Eq. (10.75) be an integer. However, it has been shown, analytically from Maxwell's equations or experimentally with the aid of a device known as Lloyd's mirror (Jenkins and White, 1937), that, when a wavefront impinges on the surface of a medium of higher refractive index than that in which it is traveling, there is a 180° shift in phase between the incident and reflected waves. This corresponds to one-half wavelength, and thus the requirement for constructive interference in Fig. 10.30*a* is that the quantity in Eq. (10.75) be an odd number of half-wavelengths. For destructive interference, an even number of half-wavelengths (an integral number of wavelengths) is required.

An application of the principles just presented is nonreflecting glass, illustrated in Fig. 10.30*b*. The film is generally a dielectric magnesium fluoride coating, either single or multilayered, and is chosen so that its refractive index lies between that of air and the glass. Thus there will be a 180° phase reversal at both the upper and lower film boundaries. If, from Eq. (10.75), the thickness of the film is

$$t = \frac{\lambda_f \cos \phi_f}{4} = \frac{\lambda \cos \phi_f}{4n_f} \tag{10.76}$$

the reflected waves will destructively interfere with each other. Since t can only satisfy the constraint in Eq. (10.76) at one wavelength, only partial annulment occurs. Generally t is designed for 550 nm. Also, Eq. (10.76) assumes that all waves strike the surface at the same angle, which is not usually the case. Finally there is the matter that the amplitudes of the two reflected waves are not necessarily equal. To explore the latter issue we must calculate the reflectances at the air-film and film-glass surfaces.

The minimum reflection from a boundary between two media occurs when the angle of incidence is 0° (at highly grazing angles, the reflectance may approach 100%). It can be shown (Jenkins and White, 1937; Waymouth and Levin, 1980) that the minimum reflectance is given by

$$\rho_{\min} = \left(\frac{n_1 - n_2}{n_1 + n_2}\right)^2 \tag{10.77}$$

For an air-glass surface, with an average refractive index of glass of 1.5, $\rho_{\min}$ is 4%.

For the present air-film-glass problem, we desire the reflectances at the two boundaries to be equal. From Eq. (10.77), the requirement is

$$\frac{n_a - n_f}{n_a + n_f} = \frac{n_f - n_g}{n_f + n_g} \tag{10.78}$$

which yields

$$n_f^2 = n_a n_g \qquad n_f = \sqrt{n_g} \tag{10.79}$$

since $n_a = 1$ for air. For $n_g = 1.5$, this requires that $n_f = 1.225$. In practical cases, both surfaces of a glass plate or lens are optically coated and it is possible to reduce the overall reflectance to a fraction of 1%.

Another application of thin films is the cool-beam PAR lamp. In this lamp, the vaporized aluminum normally present on the inside of the parabolic portion of the glass bulb is replaced by a multiple layer of metallic salts. The coating is spectrally selective, reflecting visible energy and transmitting infrared energy. The beam emerging from the front lens has only about one-third of the radiated power of a conventional PAR lamp of the same wattage.

The cool-beam lamp is one example of the use of so-called dichroic mirrors. Such mirrors are of two types, cold and hot. The cold type, as used in the cool-beam lamp, reflects visible radiation and transmits infrared. The hot type does the reverse. It is used in projection systems, in front of the lamp-reflector combination, to reflect the infrared radiation away from the film.

Dichroic mirrors are also used in colored PAR lamps. The interference film coating is placed on the inside of the cover lens and produces a given color by selective spectral transmission. Dichroic PAR lamps are available in blue, green, red, yellow, and amber colors.

10.10 DIFFRACTION

When electromagnetic waves pass through an aperture or pass the edge of an obstacle, they spread somewhat into the region which one would expect to

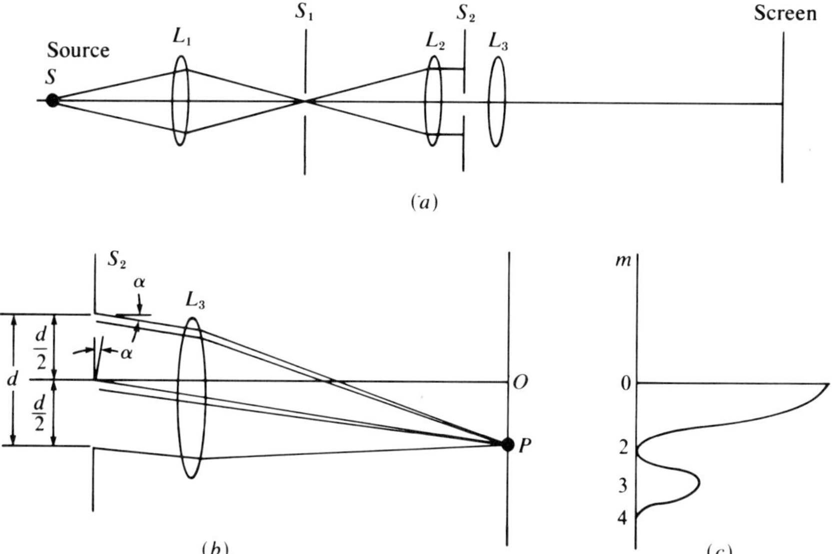

Figure 10.31 Fraunhofer diffraction.

be in shadow. This is the phenomenon of diffraction, the bending of light around edges of opaque objects.

An explanation of diffraction requires Huygens' principle and the concepts of interference. Historically, diffraction effects are divided into two classes: those in which the light source and the screen on which the diffraction pattern is observed are infinitely separated from the aperture producing the diffraction and those where they are not. The first class is known as Fraunhofer* diffraction; the second as Fresnel diffraction. The former is much simpler to treat theoretically and can be simulated quite easily in the laboratory through the use of lenses which produce parallel rays, thus making the source and screen appear to lie at infinity.

Fraunhofer diffraction by a single slit, as shown in Fig. 10.31, will be considered first. In Fig. 10.31*a*, light from a monochromatic source S is converged by lens L_1 through a slit S_1, located at the focal point of lens L_2. Thus the rays from L_2 passing through the slit are horizontal. They are converged by lens L_3 to the screen, which lies at the focal point of lens L_3.

The portion of Fig. 10.31*a* to the right of S_2 is shown expanded in Fig. 10.31*b*. Two very narrow wave trains of light are shown emerging from the top and center of S_2 and progressing at an angle α with the horizontal. They are converged by lens L_3 to point P on the screen, thus simulating a screen at infinity.

The two wave trains depart from S_2 in phase but the upper train travels

*Joseph Fraunhofer (1787–1826), a Bavarian glass grinder and specialist in experimental optics.

an additional distance $(d/2) \sin \alpha$, in reaching the screen. If this path difference is $\lambda/2$ for a given source, annulment occurs at P. The same is true for the next two narrow wave trains immediately below the first two, and for all other consecutive pairs of wave trains. Thus the light from each wave train in the upper half of slit S_2 is canceled by that in the lower half and point P is dark. The condition is that

$$\frac{d}{2} \sin \alpha = \frac{\lambda}{2} \qquad \sin \alpha = \frac{\lambda}{d} \tag{10.80}$$

Suppose now we consider the slit divided into four segments, instead of two. By the same reasoning as before, segments 1 and 2 will annul, as will segments 3 and 4 if

$$\frac{d}{4} \sin \alpha = \frac{\lambda}{2} \qquad \sin \alpha = \frac{2\lambda}{d} \tag{10.81}$$

In general, annulment will occur if

$$\sin \alpha = \frac{m\lambda}{2d} \qquad m \text{ an even integer} \tag{10.82}$$

The resulting light and dark band diffraction pattern is shown in Fig. 10.31*c*. The maximum constructive interference is for $m = 0$ (point O). The first annulment occurs for $m = 2$ (point P); the second for $m = 4$. For $m = 3$, constructive interference occurs, but the brightness is only one-third of that for $m = 0$. For this case, the slit is divided into three segments. The upper two will cancel for α such that Eq. (10.82) is satisfied for $m = 3$, leaving only the third segment to produce the bright band.

EXAMPLE 10.7

In Fig. 10.31*b*, assume the slit width is five wavelengths. What is the angular width of the central bright band of the diffraction pattern? What is the angle for the second null point?

Solution. From Eq. (10.82), with $m = 2$

$$\sin \alpha = \frac{2\lambda}{2(5\lambda)} = 0.2$$

$$\alpha = 11.5°$$

giving an angular width of 23°.

From Eq. (10.82), with $m = 4$

$$\sin \alpha = \frac{4\lambda}{2(5\lambda)} = 0.4$$

$$\alpha = 23.6°$$

Note that if the slit width were reduced to one wavelength, the central angle would increase to 180° and no diffraction pattern would be observed. If the slit width were increased to 100 wavelengths, the central angle would decrease to about 1° and the pattern would be very constrained. Generally a slit width of 5 to 10 wavelengths is considered optimal for observing.

Single-slit diffraction suffers in that the boundaries of the light and dark bands are not sharply defined. Much better resolution can be obtained with a diffraction grating, first introduced by Fraunhofer, which is an assemblage of parallel slits, equal in width and separation. The diffraction grating is widely used in the quantitative analysis of spectra.

A diffraction grating is diagrammed in Fig. 10.32. Only five slits are shown, whereas the usual grating contains several thousand. Their spacing is d, which is usually of the order of 2 μm. A wave train is incident normally on the grating. The slits are assumed to be narrow enough so that the diffracted beam from each slit has sufficient angular breadth to interfere with all other diffracted beams. As in single-slit diffraction, the lens is inserted to make the diffraction pattern on the screen appear to be at infinity.

Consider first the light in very narrow elements at the lower extremity of each slit and leaving the slits at an angle α. Let α be such that the distance $ab = \lambda$. Then $cd = 2\lambda$, $ef = 3\lambda$, and $gh = 4\lambda$. The waves in these small elements reach point P in phase and thus constructively interfere. The same applies to all other corresponding elements in each slit.

If α is increased very slightly, near annulment occurs, assuming a large number of slits. Thus the cutoff of the bright bands on the screen is very sharp. As α is increased further, eventually $ab = 2\lambda$ and a second bright band appears on the screen. The same is true for $ab = 3\lambda$, 4λ, etc. The angles for which maxima occur are obtained from

$$ab = d \sin \alpha = m\lambda, \; m \text{ an integer} \qquad \sin \alpha = \frac{m\lambda}{d} \tag{10.83}$$

If white light is incident on the grating, its spectral components are

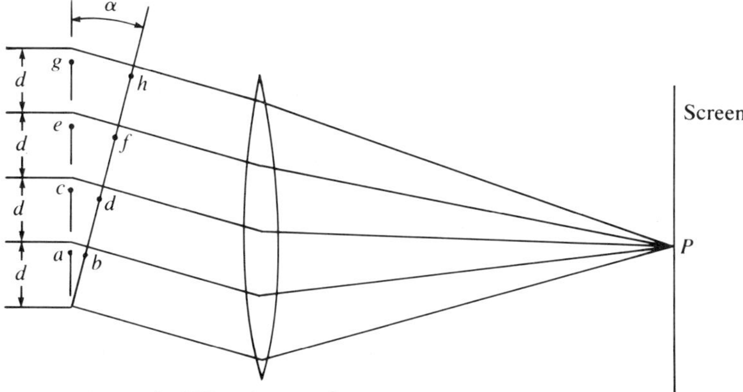

Figure 10.32 A diffraction grating.

displayed on the screen. However, unlike a prism which produces one spectrum, a diffraction grating produces several on each side of the normal. These are called *first order*, *second order*, . . . , spectra, corresponding to $m = 1, 2, \ldots$. For $m = 0$ in Eq. (10.83), no spectrum is produced since the deviation (α) is zero for all wavelengths. The result is a white image on the screen.

For appreciable deviation, the spacing of the grating should be the same order of magnitude as the wavelength of the light. Commercial gratings customarily have 600 or 1200 lines per millimeter. Fraunhofer's first gratings were made by wrapping fine wire around two parallel screws. Gratings are now made by ruling the equidistant grooves on a glass or metal surface with a diamond stylus. Replicas can then be made on aluminized glass blanks for research and scientific use. Inexpensive 2×2 in card-mounted diffraction gratings are also available for nonquantitative spectral analysis.

EXAMPLE 10.8

Find the angular width of the first-order diffraction pattern produced by a grating having 600 lines per millimeter when lighted with white light.

Solution. The grating spacing is 1.67 μm. We will assume that white light extends from 400 to 700 nm. Then Eq. (10.83) yields

$$\sin \alpha = \frac{1(4 \times 10^{-7})}{1.67 \times 10^{-6}} = 0.239$$

$$\alpha = 13.9^\circ \qquad \text{for violet}$$

$$\sin \alpha = \frac{1(7 \times 10^{-7})}{1.67 \times 10^{-6}} = 0.419$$

$$\alpha = 24.8^\circ \qquad \text{for red}$$

Thus the angular breadth is 10.9°.

10.11 POLARIZATION

In Sec. 1.3, we pointed out that electromagnetic waves consist of electric and magnetic fields vibrating at 90° to each other in space, at 90° to the direction in space of the wave propagation, in time phase, and of the same frequency. Such waves are called *transverse*, in contrast to, for example, sound waves, which are *longitudinal*, in that they vibrate along the direction of propagation. Interference and diffraction phenomena can be observed with both types of waves but polarization effects can be observed only in transverse waves.

A radio antenna is a source of polarized waves. As shown in Fig. 10.33*a*, it provides an $\mathscr{E}$ field parallel and an H field perpendicular to the an-

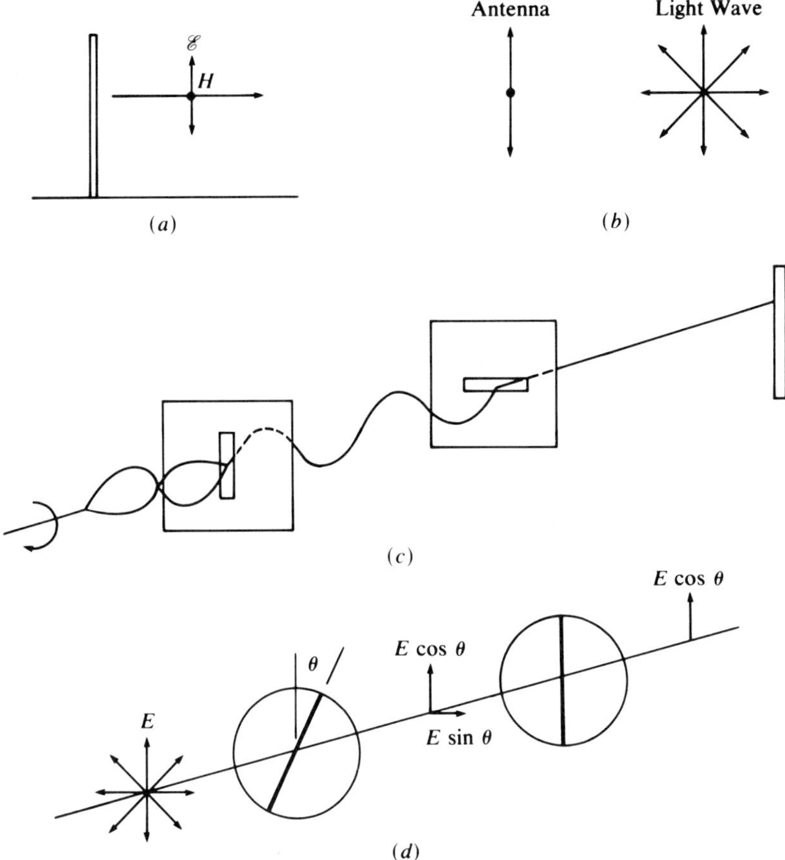

Figure 10.33 Pictorial representation of polarized light.

tenna, generated by oscillations, up and down, of electrons within the antenna. Such waves are called linearly polarized or plane polarized.

A light wave, unlike an antenna, normally has its $\mathscr{E}$ and H fields vibrating in all directions in the transverse plane. It can be thought of as being produced by an assemblage of antennas, oriented uniformly in all directions in the vertical plane in Fig. 10.33*a*. The resulting $\mathscr{E}$ field (which was shown by 0. Wiener in 1890 to be the field that produces the optical effects of polarized light) will also be oriented in all directions in this plane. The distinction between the $\mathscr{E}$ field of the antenna and those of the light wave are noted diagrammatically in Fig. 10.33*b*. The light wave is seen to be a superposition of linearly polarized waves in all possible transverse directions.

To produce a polarized light wave requires some device to pass only those $\mathscr{E}$ vectors vibrating in a certain direction. Such a device is called a polarizer, and its effect is indicated in Fig. 10.33*c*. In this diagram, a piece of clothesline is threaded through two slits, one vertical and one horiziontal, and tied to a post. A person at the left rotates the clothesline circularly as one would rotate a jump rope. The first polarizing slit (polarizer) removes all horizontal oscillations so that in between the slits the clothesline is moving only in a vertical direction. This is linearly polarized wave motion. Now if a second polarizing slit (analyzer) is inserted horizontally, the clothesline will be motionless from there to the post.

If the slits in Fig. 10.33*c* are replaced by crystals, such as tourmaline, or plastics with imbedded crystals, such as Polaroid film, and if the rope twirler is replaced by a source of unpolarized light and the post by a photosensor, we obtain the diagram in Fig. 10.33*d* in which light is being polarized by directional selective absorption. Assume the polarizer is oriented an angle θ from the vertical and produces a linearly polarized wave (single vector) of amplitude E, which may be broken into a vertical component $E \cos \theta$ and a horizontal component $E \sin \theta$. The analyzer is oriented vertically so only the vertical component passes through. The transmission is maximum when $\theta = 0°$ and zero when $\theta = 90°$. Since light intensity is proportional to the square of the electric field intensity, we can write

$$I = I_m \cos^2\theta \qquad (10.84)$$

where I_m is the intensity when polarizer and analyzer are aligned. This relation is called Malus's law.*

Suppose the source in Fig. 10.33*d* is partially polarized so that its $\mathscr{E}$ field amplitude is not the same in all directions. Place the sensor between the polarizer and analyzer and slowly rotate the polarizer, noting the reading of the sensor at each position. The degree of polarization of the source is given by

$$\text{Percent polarization} = \frac{I_{\text{max}} - I_{\text{min}}}{I_{\text{max}} + I_{\text{min}}} (100) \qquad (10.85)$$

where I_{max} and I_{min} are the maximum and minimum intensities recorded by the sensor as the polarizer rotates through 360°.

There are four common methods of producing polarized light. We have already considered selective absorption. The other techniques are by reflection, refraction, and double refraction. A fifth (natural) technique, scattering, is present in the earth's atmosphere. Incident unpolarized sunlight, striking the atmosphere is scattered, with blue light being scattered more than red and becoming linearly polarized.

Creating polarized light by reflection is perhaps the simplest method and was the first method historically, having been discovered by Malus in 1808. Malus observed that if a beam of unpolarized light was incident at a particular angle on a polished glass plate, the reflected beam was linearly polarized. The situation is diagrammed in Fig. 10.34*a*. Unpolarized light, which is indicated by arrows parallel and perpendicular to the plane of in-

*Named for E. L. Malus, who discovered it experimentally in 1809.

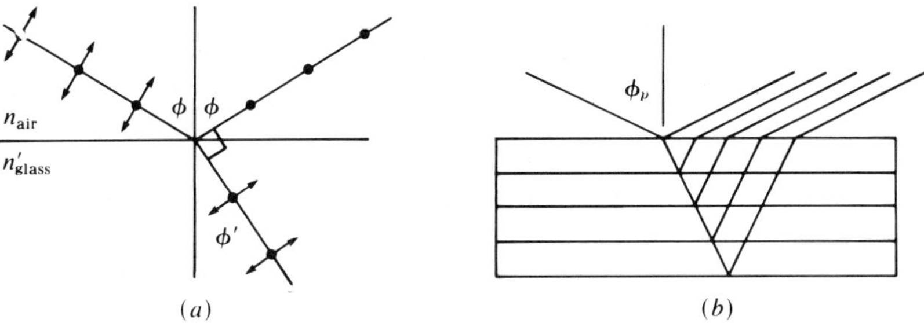

Figure 10.34 Polarization by reflection and refraction.

cidence between the ray and the normal to the surface, impinges on the glass plate at an angle ϕ. At one particular value of ϕ, given by

$$\tan \phi_p = \frac{n'}{n} \tag{10.86}$$

15% of the incident component perpendicular to the plane of incidence is reflected whereas none of the component parallel to this plane is reflected. Thus the refracted ray contains 85% of the perpendicular component and all the parallel component. The reflected ray is completely linearly polarized but weak; the refracted ray is strong but only partially polarized.

The relation in Eq. (10.86) is known as Brewster's law* and the angle ϕ_p is called the Brewster angle. It is about 57° for ordinary glass. Combining Brewster's law with Snell's law gives

$$\phi_p + \phi'_p = 90° \tag{10.87}$$

Thus, at the Brewster angle, the reflected and refracted rays are separated by 90°.

If the single glass surface is replaced by a pile of thin glass plates, as shown in Fig. 10.34*b*, the intensity of the reflected ray is increased and the refracted ray becomes nearly completely polarized. What happens is that the refracted ray loses 15% of its perpendicular component at each refraction and ultimately becomes linearly polarized. The effect of 10 such plates is to reduce the intensity of the perpendicular component to about 4% of its original valuc.

The final technique for producing polarized light is by double refraction, also called birefringence. This requires that the medium be anisotropic, that is, that the velocity of a light wave within the medium be a function of the direction of travel. Common refractive materials, such as glass and plastic, are not doubly refracting, but certain crystals, notably calcite ($CaCO_3$) and quartz (SiO_2), exhibit this property.

Two sets of Huygens' wavefronts, one spherical and the other ellipsoidal, emanate from each light wave surface in a doubly refracting crystal.

*Named for Sir David Brewster, who discovered it experimentally in 1912.

This results in two rays, an ordinary ray (O ray) from the spherical waves and an extraordinary ray (E ray) from the ellipsoidal waves, which propagate through the crystal. These two wavefronts are tangent to each other in only one direction in the crystal, called the *optic axis*.

Three orientations of E and O rays are shown in Fig. 10.35. In Fig. 10.35*a*, the incident ray is parallel to the optic axis, there is no double refraction and the two wavefronts travel at equal velocities through the crystal. In Fig. 10.35*b*, the incident ray is perpendicular to the optic axis. Again there is no double refraction but the E ray travels at a higher velocity through the medium than the O ray. In Fig. 10.35*c*, which is the usual situation, the incident ray is inclined to the optic axis and both double refraction and differing velocities are present. If the crystal is rotated about the axis of the incident ray, the E ray rotates about the fixed O ray.

Snell's law holds for the O ray but not for the E ray, since the velocity, and thus the index of refraction, for the latter is a function of direction. For calcite $n_O = 1.658$ and $n_E = 1.486$, where the latter is customarily given for travel perpendicular to the optic axis, which is the direction of maximum velocity of the E ray. For quartz, the values are $n_O = 1.544$ and $n_E = 1.553$.

The most significant attribute of doubly refracting crystals is that the E ray and the O ray are each linearly polarized, with the E ray oscillations at 90° in space to the O ray oscillations. Also the two rays emerge from the crystal out of time phase, with the amount of phase shift a function of path lengths within the crystal.

To explain this result, consider Fig. 10.36. In Fig. 10.36*a*, the two polarized rays are shown (they are not separated, but are shown that way to make the diagram clearer). The O ray is linearly polarized perpendicular to the page; the E ray in the plane of the page.

If the E and O rays are in time phase as they emerge from the crystal, the resultant ray is linearly polarized, as shown in Fig. 10.36*b*, and lies at an angle of 45°, if the E and O ray amplitudes are assumed equal. If the two rays emerge 180° apart in time phase. Fig. 10.36*c* results. Again the resultant is linearly polarized. Time shifts of 90° or 270° yield the result in Fig. 10.36*d*. This is circularly polarized light. The resultant vector does not oscillate along a given direction, as in Fig. 10.36*b* and *c*, but rather rotates in time,

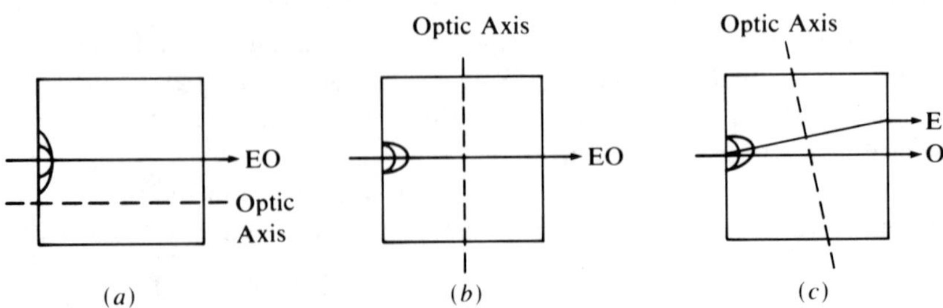

Figure 10.35 E and O rays in a crystal.

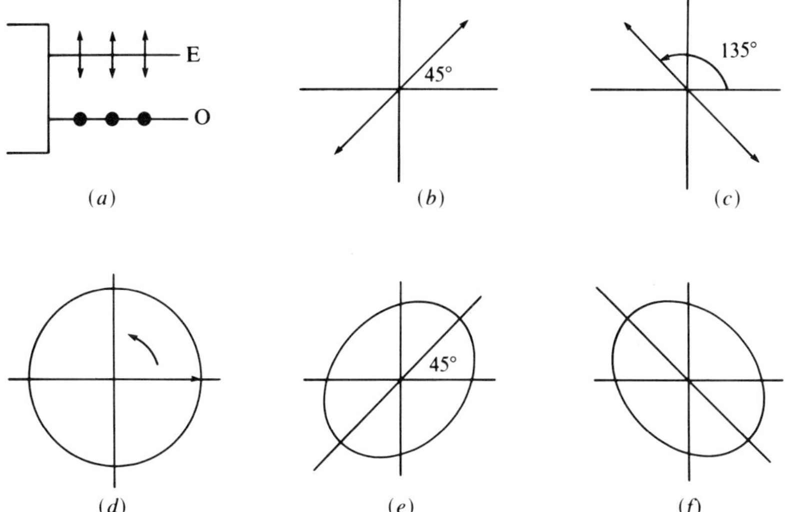

Figure 10.36 E and O ray polarization.

with constant amplitude, thus describing a circle locus. Figure 10.36*e* and *f* represent the situation where the time phase shift is 45° or 135°. Again the resultant vector rotates in space, but not with fixed amplitude. Rather it describes an elliptical locus, as shown.

For a given crystal, the pattern that is obtained depends on the frequency of the incident light, the indexes of refraction of the crystal and the crystal's thickness. If the crystal provides a phase shift of 90°, it is called a *quarter-wave plate;* if 180°, a *half-wave plate.* Because phase shift is a function of frequency, these descriptions are valid only for a single color (wavelength) of light.

Assume the incident radiation is white light and that a crystal which is a half-wave plate for green light is inserted between the polarizer and analyzer in Fig. 10.33*d*. Linearly polarized white light emerges from the polarizer, strikes the plate and is doubly refracted. The green component emerges from the plate rotated 90° in space and is linearly polarized. Other components emerge elliptically polarized. If the analyzer is oriented at 90° to the polarizer, the green component passes through unattenuated, whereas the other components, which are rotating elliptical vectors, are partially attenuated. If the analyzer is aligned with the polarizer, the green component is rejected and the other components again partially attenuated.

Ordinary cellophane tape is doubly refracting and striking color effects can be observed by inserting various thicknesses of such tape between a polarizer and an analyzer. Another way of achieving color effects is through dichroism. Some doubly refracting materials, such as tourmaline, exhibit this property in which one of the linearly polarized rays is absorbed appreciably in passing through the crystal.

EXAMPLE 10.9

A beam of linearly polarized light at 589 nm is incident on a calcite crystal at right angles to the optic axis. Find the wavelengths of the ordinary and extraordinary rays in the crystal.

Solution. The indexes of refraction for the O and E rays are given in the text as $n_O = 1.658$ and $n_E = 1.486$. From Snell's law,

$$nv = n'v'$$
$$n\lambda f = n'\lambda' f$$
$$n\lambda = n'\lambda'$$

Thus

$$1(589) = 1.658\lambda'_O \qquad \lambda'_O = 355 \text{ nm}$$
$$1(589) = 1.486\lambda'_E \qquad \lambda'_E = 396 \text{ nm}$$

EXAMPLE 10.10

Derive an expression for the phase difference θ between the O and E rays after traversing a crystal plate whose optic axis is perpendicular to the incident ray. Use this expression to find the thickness of a quarter-wave calcite plate for 400-nm radiation.

Solution. The phase difference in radians is given by

$$\theta = 2\pi f(t_O - t_E)$$

where t_O and t_E are the times for the O and E rays to pass through the plate. Inserting expressions for times gives

$$\theta = 2\pi f\left(\frac{s}{v_O} - \frac{s}{v_E}\right)$$

where s is the plate thickness, and v_O and v_E are the velocities of the O and E rays in the plate. Using Snell's law

$$\theta = 2\pi fs\left(\frac{n_O}{nv} - \frac{n_E}{nv}\right)$$

But $n = 1$ for air and $\lambda = v/f$. Thus

$$\theta = \frac{2\pi s}{\lambda}(n_O - n_E) \tag{10.88}$$

For a quarter-wave plate, $\theta = \pi/2$ and

$$s = \frac{\lambda}{4(n_O - n_E)} = \frac{400}{4(1.658 - 1.486)} = 581 \text{ nm}$$

We will conclude this section by discussing a few applications of polarized light. Perhaps the most well-known use is in sunglasses. We noted earlier that when unpolarized light strikes a horizontal surface, the $\mathscr{E}$ field component parallel to the plane of incidence is attenuated in the reflected ray. Thus the $\mathscr{E}$ field of the reflected ray is predominantly horizontal. In polarized sunglasses, the Polaroid film functions as an analyzer and is oriented vertically. Thus it eliminates the bulk of the reflected sunlight from ground surfaces. Sunglasses also absorb about 50% of the incident light, polarized or not. Thus they serve as a filter as well an an analyzer.

As a second example, many years ago there was considerable interest in polarized automobile headlights. The proposed scheme is shown in Fig. 10.37. The headlight lenses and windshield of each automobile would be of polarized material oriented at 45°. The driver would see the light from his own headlights, but not that from approaching headlights. What he would see is the reflection of the approaching car's headlights from the road, which would be polarized horizontally. The proposal was not adopted for several reasons: First, all cars would have had to convert to the new system at the same time and this would have been costly and difficult logistically. Second, the intensity provided by the filament of the headlight would have had to be increased by a factor of 3 to provide enough light on the road for the driver to see. This would have necessitated larger car generators. Last, a motorist approaching another at right angles at an intersection would have seen the full intensity of the other's headlights through his side window.

Another use of polarized light is in improving the CRF of a visual task. When unpolarized light from a ceiling-mounted luminaire reflects from a horizontal surface, particularly a visual task, the reflected light is predominantly horizontally polarized. Such light is quite effective in producing veiling reflections, particularly at large viewing angles from the vertical.

Multilayered polarized light panels which reduce this effect are available for use with fluorescent luminaires. The horizontal component of the unpolarized light emitted by the fluorescent lamps is reflected back to the source by the panel, whereas the vertical component passes through. Little absorption occurs because the reflected horizontal component is depolarized when it reflects from the inside of the luminaire. It then returns to the polarized panel, its vertical component passes through and its horizontal component returns for a second reflection and depolarization. The result at the visual task is that the incoming light is largely vertically polarized. Such light tends to be absorbed by the task and reradiated, rather than reflected.

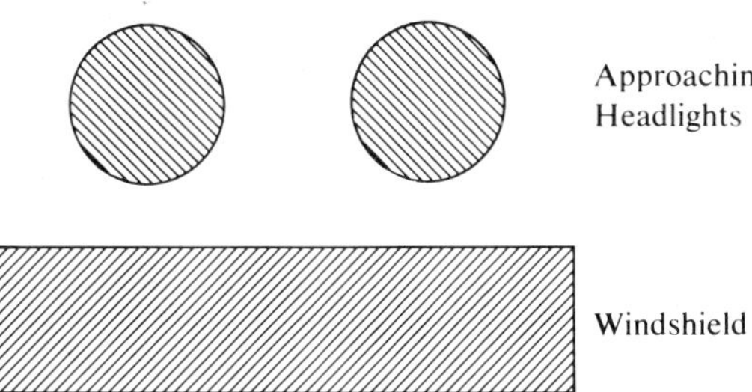

Figure 10.37 Proposal for polarized headlights.

Thus it does not result in the objectionable veiling reflections produced by horizontally polarized light.*

Polarized light is also used in industry, particularly in the detection of stress patterns in glass and plastic. These applications are based on photoelasticity, which is the phenomenon that certain materials, which are normally singly refracting, become doubly refracting when under stress. The material to be tested is placed between a crossed polarizer and analyzer. When the material is stressed, light and dark interference bands can be seen through the analyzer. The location of the bands, and their density, can be used to draw a stress diagram of the material.

10.12 FIBER OPTICS

In the late 1800s, John Tyndall, a British physicist, demonstrated that light could be transmitted along a curved path. Tyndall's apparatus consisted of an internally lighted water tank with a small hole in the side. The light emerging from the hole followed the curved stream of water downward.

It was not until the 1950s that light-channeling became feasible with practical optical devices. The light conductors are thin, flexible glass or plastic fibers, coated with glass or plastic sheaths (claddings) and assembled in bundles. Their operation depends on the phenomenon of total internal reflection, discussed earlier in this chapter.

Light is not actually bent as it travels through a fiber. Rather it follows a straight-line, zig-zag path, ricocheting off the sides of the rod. Because the index of refraction of the rod is greater than that of the surrounding medium and because most the light strikes the sides of the rod at angles of incidence greater than the critical angle, nearly total internal reflection occurs, limited only by the smoothness of the rod's surface. In a straight rod, the angle of incidence is the same from one bounce to the next and is determined by the angle at which light enters the end of the rod. In a curved rod, the angles of incidence are greater for a given entry angle and thus the degree of curvature is limited by the critical angle of the glass.

A single fiber usually ranges in diameter from 100 to 300 μm, although fibers as small as 2 μm and as large as several millimeters are available. It has a relatively high index of refraction and is surrounded by a cladding of lower index of refraction. This protects the polished conducting surface and separates the conducting cylinders when they are bundled. Fibers are available with critical angles of 50°, which permit a cone of light 80° wide to be gathered at the end of the fiber while still insuring that total reflection will occur within the rod.

Two recent technological advances have been instrumental in the

*In polarized luminaires, the polarization is referred to as *circular*. Actually this is the geometric distribution of light about the nadir rather than the phase-shifted circular polarization discussed earlier.

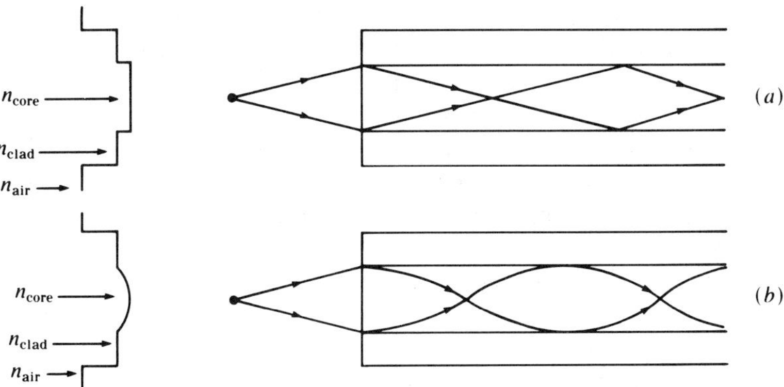

Figure 10.38 Principal types of fiber optics.

growth of fiber optics, particularly in communication systems. Prior to the late 1960s, fibers contained so many inpurities and surface defects that the beam was attenuated to 1% in only 10 m of fiber. In 1970, a high-purity glass was developed by Corning Glass Works which could transmit light over 1 km without amplification. More recently, fibers have become available with only 5% loss over 1 km. Most fibers are currently made in lengths of 10 km. To splice two lengths together, the ends are cut with a diamond stylus, are aligned, and are fused by an electric arc.

The second factor in fiber optic growth is the development of semiconductor lasers and light-emitting diodes which are smaller than a pinhead. Some of these can be pulsed at rates of 10^9 per second and thus can modulate a light beam, opening the door to optical and infrared communication.*

There are two principal types of fibers in use today. These are the step-index type and the graded-index type, both diagrammed in Fig. 10.38. In the step-index type shown in Fig. 10.38*a* the index of refraction of the core material is constant across its diameter (50 to 200 μm) and is 1 to 5% greater than the index of refraction of the cladding. The light rays propagate in straight zig-zag lines within the core. This type is available in glass cladding/glass core, plastic cladding/glass core, and plastic cladding/plastic core varieties, with the first variety the least lossy and the last the most lossy.

The other type of fiber, shown in Fig. 10.38*b*, is the graded-index In this type, the refractive index of the core is related parabolically to the core's radius. The rays follow a curved path within the core. As they move outward from the core axis, the index of refraction decreases and they are bent back inward. The parabolic profile provides a greater bandwidth than the step profile and thus is preferable when optical pulses are to be transmitted.

As we have noted, optical fibers may be used singly or in bundles. Single-fiber cables are more efficient but are difficult to align because of their

*Direct optical communication by a laser beam through the atmosphere is impractical because of extreme scattering during inclement weather.

small diameter. Bundles are easier to work with but have higher losses. Thus it appears that single-fiber cables will be used for long distance communication and bundles for shorter distances.

The applications of fiber optics are myriad and increasing daily. Their first use was in the late 1950s when doctors developed a fiber optic probe, called an endoscope, to observe inside the body. Many improvements have been made in the device since. Basically it consists of two fiber optic bundles, one to deliver light into the body and the other to return an image to the physician. The device is used for examination, to detect ulcers, cancerous tissues, polyps, etc., and to enable the doctor to see to remove foreign objects and growths or collect tissue samples.

Perhaps the most important application of fiber optics is in optical communication. It is possible to transmit 10,000 conversations simultaneously using a single pair of glass fibers. Each voice signal is sampled 8000 times a second and the sampled signals are encoded in 8-bit numbers (256 signal levels), time multiplexed with the bits from other conversations, and sent over the fibers at the rate of 44.7×10^6 bits per second. At the receiving end the signals are unscrambled and reconstructed.

Fiber optic systems are free of the cross talk that plagues conventional cable transmission systems and are unaffected by random electrical disturbances such as lightning. Because they can be slipped into ducts alongside existing copper cables, they do not require space beyond what is now used for conventional cables. Due to these advantages, more than 100 telephone systems around the world now use fiber optics.

In the field of lighting, applications of fiber optics are less abundant. A scheme is being developed to replace an array of light sensors with a fiber optic array feeding a single light sensor. This will eliminate the difficult problem of maintaining and aligning the calibration of many light sensors over time. There is also interest in using large fiber optic arrays to pipe daylight deep into multistoried buildings, possibly through centrally located atria.

REFERENCES

Baker, G. A., R. V. Heinisch, and I. Lewin: "New Techniques for Reflector Design and Photometry," *Journal of IES,* July 1977.

Committee on Light Control and Equipment Design of the IES: "IES Guide to Design of Light Control," Parts I–IV, *Illuminating Engineering,* November–December 1959, August 1967, August 1970.

Elmer, W. B.: *The Optical Design of Reflectors,* 2nd ed., Wiley, New York, 1979.

General Electric Company: "Light Measurement and Control," Bulletin TP-118, Nela Park, Cleveland, Ohio, 1971.

Jenkins, F. A., and H. E. White: *Fundamentals of Physical Optics,* McGraw-Hill, New York, 1937.

Jolley, L. B. W., J. M. Waldram, and G. H. Wilson: *Theory and Design of Illuminating Engineering Equipment,* Chapman-Hall, London, 1931.

Kao, C. S.: *Optical Fiber Systems,* McGraw-Hill, New York, 1978.

Kaufman, J. E.: *IES Lighting Handbook,* Reference volume, 1981 ed., Illuminating Engineering Society, New York, 1981.

Myodo, O., and M. Karino: "A New Method for Computer-Aided Design of Luminaire Reflectors," *Journal of IES,* January 1982.

Sears, F. W., and M. W. Zemansky: *University Physics*, Addison-Wesley, Cambridge, Mass., 1955.

Waymouth, J. F., and R. E. Levin: *Designers Handbook: Light Source Applications,* GTE Products Corp., Danvers, Mass., 1980.

Wey, R. A.: "Fiberoptic Communications," 1980 Laser Focus Buyer's Guide, Newton, Mass., 1980.

Chapter 11

Exterior Lighting

11.1 INTRODUCTION

Most of the luminaire descriptions and lighting calculation procedures presented thus far have been applicable to interior lighting situations, where the goal has been to provide a prescribed average horizontal illuminance level or a certain horizontal illuminance at a particular location. Although the same lighting fundamentals apply to exterior lighting, their application is quite different for several reasons.

1. Except for the possibility of ground reflection or reflection from nearby structures, there is no reflected component to include in exterior lighting design.
2. Exterior lighting is often done for a variety of persons and/or tasks. For example, in sports lighting, light must be provided for the players, the referees, the spectators, and the TV cameras. Each of these viewers has different visual requirements.
3. The viewing of vertical, or perhaps oblique, surfaces is often of primary importance in exterior lighting. For example, it is more important for a tennis player to be able to see the side of the ball as it approaches than to be able to see its top.
4. There is no standard 25° viewing angle in exterior lighting. Viewing is often in all directions and thus the problem of providing glare-free seeing is increased.
5. Quite often, especially in sports lighting, the object being viewed is moving. This also occurs sometimes indoors, particularly in industrial situations, but is nearly absent from school and office lighting.

6. Illuminance levels are generally lower for outdoor activities, although some of the seeing tasks can be as difficult as those indoors. The reason for this is that most outdoor lighting covers large areas and is accomplished with luminaires mounted on poles whose height is much greater than the average 8-ft ceiling height in a room. For such installations to provide 50 fc is often impractical, from both cost and energy points of view.
7. Safety is often of great concern in exterior lighting. If an office worker makes an error because of inadequate lighting, it is bad enough, but if a motorist is unable to see a pedestrian, it is much worse.

In the remainder of this chapter we will cover the general concepts of exterior lighting, the lamps and luminaires used, and the particular calculation procedures employed. We will illustrate these concepts and procedures by considering two areas of exterior lighting in some detail, namely roadway and sports lighting.

11.2 FLOODLIGHT LUMINAIRES

Outdoor luminaires fall into two categories, those that may be aimed and those that have fixed aiming. Floodlights are in the first category; roadway and area luminaires in the second. We will examine the photometric characteristics of each type.

The National Electrical Manufacturers Association (NEMA) has defined four classes of floodlight luminaires (Kaufman, 1981). A heavy-duty (HD) floodlight is a weatherproof unit consisting of a metal housing and a separate and removable reflector. It is provided with a hinged door and a cover glass whose light opening is equal (or greater) in diameter to that of the reflector. A general-purpose (GP) floodlight is also a weatherproof unit with a cover glass, but it is constructed so that the housing forms the reflecting surface. A ground-area open (O) floodlight is a unit providing a weatherproof enclosure for the lamp socket and housing, but not for the lamp itself. It does not include a cover glass. Lastly, a ground-area open floodlight with a reflector insert (OI) is constructed so that the housing forms a portion of the reflecting surface. The unit is weatherproof, has no cover glass, and utilizes an auxiliary reflector.

There are several pieces of data that the floodlight manufacturer should provide to aid the user in choosing the proper luminaire. One of these is the beam spread of the unit. NEMA has established seven floodlight types based on their beam spreads. These are shown in Table 11.1. The NEMA definition of beam spread ψ is the total angle between the 10% of maximum intensity points on opposite sides of the floodlight axis. The maximum intensity may or may not lie on this axis. Beam spread of some floodlighting units changes from one vertical plane to another. For such asymmetric units, two

TABLE 11.1 NEMA FLOODLIGHT TYPES

Type	Beam spread (degrees)
1	$10 = \psi < 18$
2	$18 = \psi < 29$
3	$29 = \psi < 46$
4	$46 = \psi < 70$
5	$70 = \psi < 100$
6	$100 = \psi < 130$
7	$\psi \geq 130$

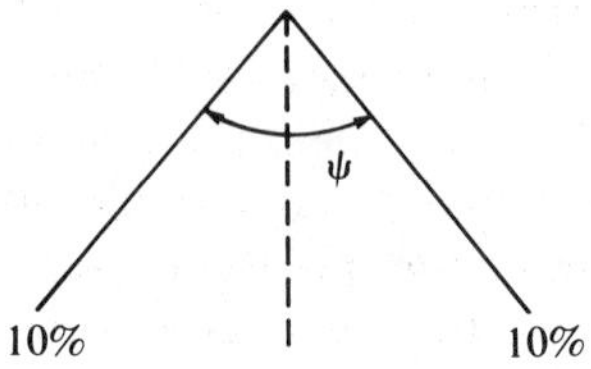

beam spreads are given. Thus a floodlight with a horizontal beam spread of 125° and a vertical beam spread of 90° is designated as a Type 6 × 5 floodlight.

A manufacturer's photometric data sheet for a typical floodlight is shown in Fig. 11.1. General information about the unit is given in the lower right box, on a per 1000 lumen basis. It is common practice to present floodlight data in this relative manner. Then, for whatever lamp size is used, the data can simply be scaled by the actual number of lamp lumens divided by 1000. In Fig. 11.1, a 400-W high-pressure sodium lamp is specified. From Table 6.11, its initial lumen rating is 50,000 lm.* Thus a factor of 50 should be applied to the lumen and candela values in Fig. 11.1.

In the lower left section of Fig. 11.1, the lumen distribution of the floodlight is shown by angular zones. Traditionally, floodlights have been photometered on a so-called Type B goniometer, illustrated in Fig. 11.2*a*. The photocell is located in a fixed position facing the floodlight, as shown. The floodlight is rotated about one axis with the other held constant and vice-versa, which produces intensity traces on a sphere about the floodlight, as shown in Fig. 11.2*b*. In this procedure, horizontal angles are measured about the vertical axis and vertical angles about the horizontal axis.

If the floodlight is mounted directly over the 0,0 point with its axis perpendicular to the *xy* plane, as shown in Fig. 11.2*c*, the intensity traces project onto the plane as shown by the dashed lines in that figure. Note that each box subtends equal horizontal and vertical angular increments, in this case 10°, with the angles for the 20 to 30° box indicated in Fig. 11.2*c*. The actual horizontal area included within the two horizontal and two vertical angles is not a square. Rather it has two straight sides, defined by the constant vertical angles, and two curved sides, defined by the constant horizontal angles. Thus the grid in Fig. 11.1 is not the actual horizontal plane, but a transformation of it from the horizontal plane of Fig. 11.2*c*.

The various angles involved in this discussion are shown in Fig. 11.2*d*. θ_V is the vertical angle, along the *y* axis, but the angle ϕ along the *x* axis is not called the horizontal angle. Rather, the horizontal angle is θ_H, which indeed

*Initial values will be used throughout the discussion. Light loss factors for floodlights are roughly 0.75 for enclosed units and 0.65 for open units, based on cleaning once a year.

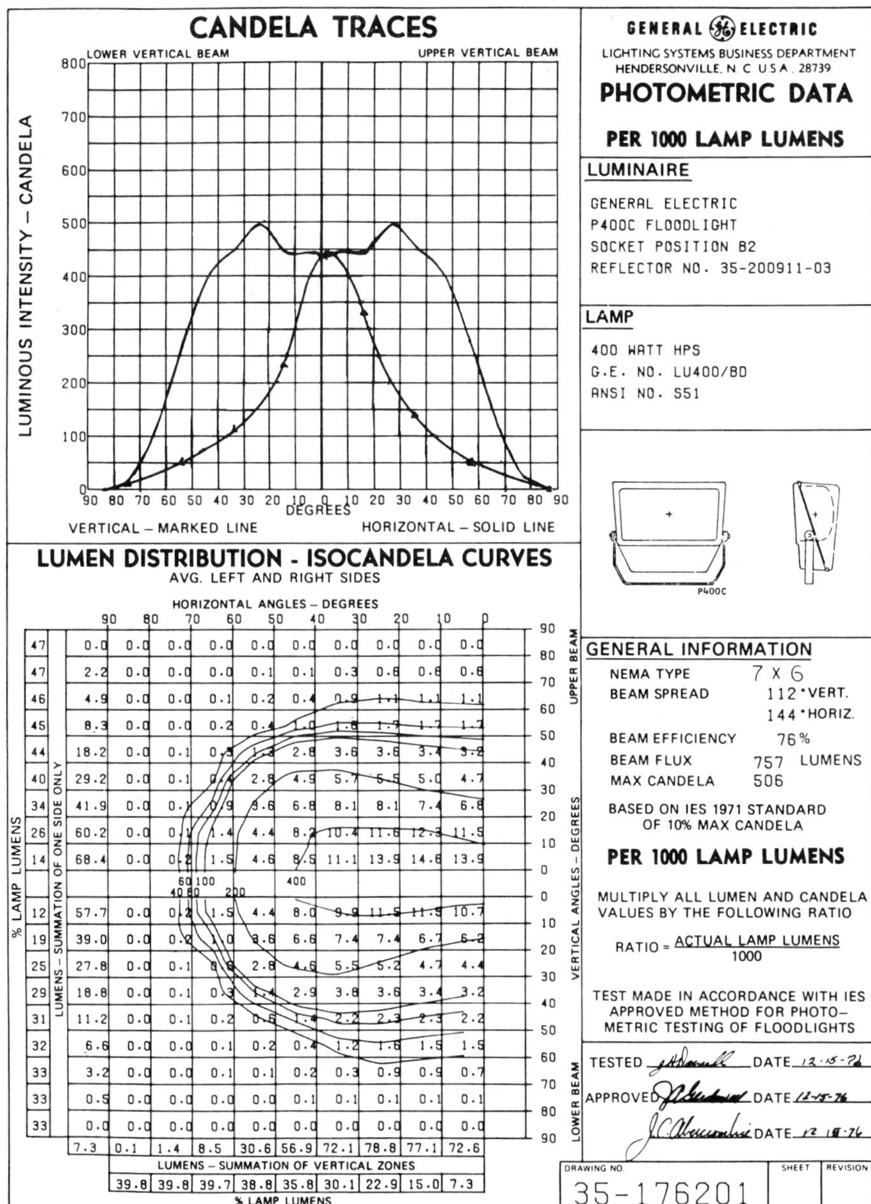

GENERAL ELECTRIC
LIGHTING SYSTEMS BUSINESS DEPARTMENT
HENDERSONVILLE, N. C. U.S.A. 28739

PHOTOMETRIC DATA

PER 1000 LAMP LUMENS

LUMINAIRE

GENERAL ELECTRIC
P400C FLOODLIGHT
SOCKET POSITION B2
REFLECTOR NO. 35-200911-03

LAMP

400 WATT HPS
G.E. NO. LU400/BD
ANSI NO. S51

LUMEN DISTRIBUTION - ISOCANDELA CURVES
AVG. LEFT AND RIGHT SIDES

HORIZONTAL ANGLES – DEGREES

% LAMP LUMENS	LUMENS – SUMMATION OF ONE SIDE ONLY	90–80	80–70	70–60	60–50	50–40	40–30	30–20	20–10	10–0	VERTICAL ANGLES – DEGREES
47	0.0	0.0	0.0	0.0	0.0	0.0	0.0	0.0	0.0	0.0	UPPER BEAM 90–80
47	2.2	0.0	0.0	0.0	0.1	0.1	0.3	0.6	0.6	0.6	80–70
46	4.9	0.0	0.0	0.1	0.2	0.4	0.9	1.1	1.1	1.1	70–60
45	8.3	0.0	0.0	0.2	0.4	1.0	1.8	1.7	1.7	1.7	60–50
44	18.2	0.0	0.1	0.5	1.3	2.8	3.6	3.6	3.4	3.2	50–40
40	29.2	0.0	0.1	0.4	2.8	4.9	5.7	5.5	5.0	4.7	40–30
34	41.9	0.0	0.1	0.9	3.6	6.8	8.1	8.1	7.4	6.8	30–20
26	60.2	0.0	0.1	1.4	4.4	8.2	10.4	11.6	12.3	11.5	20–10
14	68.4	0.0	0.2	1.5	4.6	8.5	11.1	13.9	14.6	13.9	10–0
12	57.7	0.0	0.2	1.5	4.4	8.0	9.9	11.5	11.5	10.7	0–10
19	39.0	0.0	0.2	1.0	3.6	6.6	7.4	7.4	6.7	6.2	10–20
25	27.8	0.0	0.1	0.6	2.8	4.6	5.5	5.2	4.7	4.4	20–30
29	18.8	0.0	0.1	0.3	1.4	2.9	3.8	3.6	3.4	3.2	30–40
31	11.2	0.0	0.1	0.2	0.6	1.4	2.2	2.3	2.3	2.2	40–50
32	6.6	0.0	0.0	0.1	0.2	0.4	1.2	1.6	1.5	1.5	50–60
33	3.2	0.0	0.0	0.1	0.1	0.2	0.3	0.9	0.9	0.7	60–70
33	0.5	0.0	0.0	0.0	0.0	0.1	0.1	0.1	0.1	0.1	70–80
33	0.0	0.0	0.0	0.0	0.0	0.0	0.0	0.0	0.0	0.0	LOWER BEAM 80–90
LUMENS – SUMMATION OF VERTICAL ZONES	7.3	0.1	1.4	8.5	30.6	56.9	72.1	78.8	77.1	72.6	
% LAMP LUMENS		39.8	39.8	39.7	38.8	35.8	30.1	22.9	15.0	7.3	

GENERAL INFORMATION

NEMA TYPE 7 X 6
BEAM SPREAD 112°VERT.
144°HORIZ.
BEAM EFFICIENCY 76%
BEAM FLUX 757 LUMENS
MAX CANDELA 506

BASED ON IES 1971 STANDARD OF 10% MAX CANDELA

PER 1000 LAMP LUMENS

MULTIPLY ALL LUMEN AND CANDELA VALUES BY THE FOLLOWING RATIO

$$\text{RATIO} = \frac{\text{ACTUAL LAMP LUMENS}}{1000}$$

TEST MADE IN ACCORDANCE WITH IES APPROVED METHOD FOR PHOTOMETRIC TESTING OF FLOODLIGHTS

TESTED [illegible] DATE 12-15-76
APPROVED [illegible] DATE 12-15-76
[illegible] DATE 12-18-76

DRAWING NO. 35-176201 SHEET REVISION

Figure 11.1 Typical floodlight photometric data sheet. (*Source:* General Electric Company, "Lighting System Product Application Guide," Hendersonville, N.C., 1980, reprinted with permission.)

is a horizontal angle from the point of view of the photometry but is a lateral angle from the point of view of floodlighting geometry. The angle ψ is the angle of incidence at point B on the horizontal plane. The several angles may be related as follows:

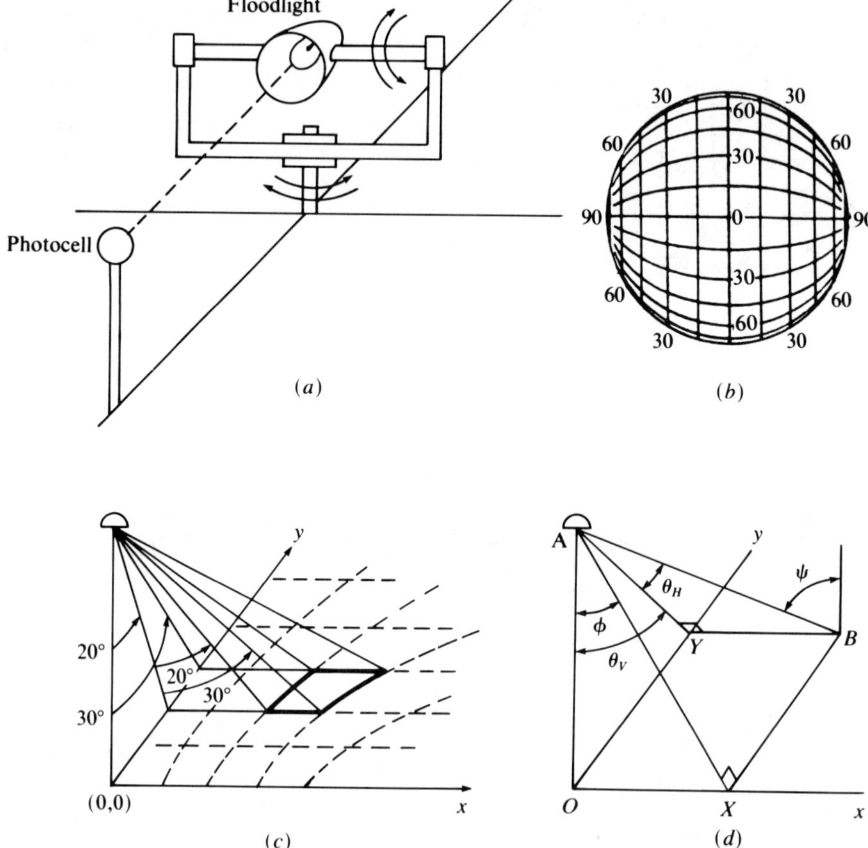

Figure 11.2 Typical floodlight angles and orientation. [*Source;* Part *b* is from J. E. Kaufman (ed.), *IES Lighting Handbook,* Reference volume, 1981 ed., IES, New York, 1981, fig. 4-9b, with permission.]

$$\cos\,\theta_V = \frac{AO}{AY}$$

$$\cos\,\psi = \frac{AO}{AB}$$

$$\cos\,\theta_H = \frac{AY}{AB} = \frac{\cos\,\psi}{\cos\,\theta_V} \tag{11.1}$$

The lumen data in each box in Fig. 11.1 are for one side of the floodlight only, it being assumed that the floodlight has left-right symmetry. In testing, the average between the two halves is taken. The summations of lumens for all horizontal zones on one side are given in the second column from the left; those for all vertical zones on one side in the second row from the bottom. Cumulative percentages of lamp lumens for both sides appear in the left-most column and for one side in the bottom most column.

The percent total efficiency of a floodlight is 100 times the ratio of total floodlight lumens to total lamp lumens. For the unit in Fig. 11.1, 47% of the lamp lumens are in the upper zones; 33% in the lower zones. Thus the total efficiency is 80%. Alternatively, 39.8% of the lamp lumens are in the left zone, and doubling this gives about 80% efficiency.

Percent beam efficiency is less than percent total efficiency because only the lumens in the beam are included. In Fig. 11.1, the beam spread is 112° × 144°. Thus the lumens out to ±56° vertically and ±72° horizontally are included in the beam (10% of maximum intensity is 50.6 cd), giving a beam efficiency of 76%.

Shown superimposed on the lumen data in Fig. 11.1 are isocandela curves. These, together with the candela traces in the upper left section of Fig. 11.1, describe the intensity distribution pattern of the floodlight. The isocandela curves are generally drawn so that each successive curve has twice the intensity of the one before it, except near the periphery of the beam where more curves are given to improve the accuracy of illuminance calculations. Candela traces are usually drawn in the θ_H = constant and θ_V = constant planes in Fig. 11.2. For rotationally symmetric units, only one candela trace need by drawn.

Because a floodlight is usually not aimed perpendicularly to a surface, it is desirable for the manufacturer to provide information in addition to that given in Fig. 11.1. Particularly useful are isoilluminance diagrams and charts such as shown in Fig. 11.3 for the floodlight in Fig. 11.1. In the diagrams, horizontal distances and aiming points are expressed in terms of mounting heights. This makes the shapes of the isoilluminance curves independent of mounting height and permits them to be labeled with identifying letters. Thus, for a given mounting height and lamp lumen output, the illuminance represented by each curve can be tabulated. This has been done for the 400-W sodium lamp for three mounting heights in Fig. 11.3.

EXAMPLE 11.1

Calculate the illuminance at the labeled points P and Q in Fig. 11.3 for a mounting height of 30 feet and a 400-W high pressure sodium lamp.

Solution. The situation is shown in Fig. 11.4. The floodlight is aimed 1 MH directly in front of the pole. For the point P, 2 MH's in front of the pole, the vertical angle difference is 18.4°. From the vertical candela trace in Fig. 11.1, $I = 300$ cd per 1000 lm. Thus $I = 300 \times 50 = 15{,}000$ cd.

The form of the ISL in Eq. (8.22) is useful.

$$E = \frac{I \cos^3 \theta}{(\mathrm{MH})^2} \tag{11.2}$$

Inserting values gives

$$E = \frac{15{,}000 \cos^3 63.4}{30^2} = 1.5 \text{ fc}$$

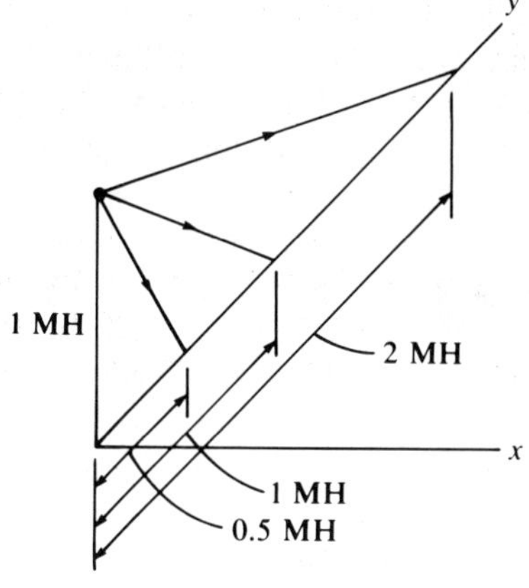

400 W Sodium
NEMA 7 × 6

Initial fc
Table

	Mounting Height		
	30	40	50
A	.18	.1	.06
B	.36	.2	.13
C	.89	.5	.32
D	1.8	1.0	.64
E	3.6	2.0	1.3
F	8.9	5.0	3.2

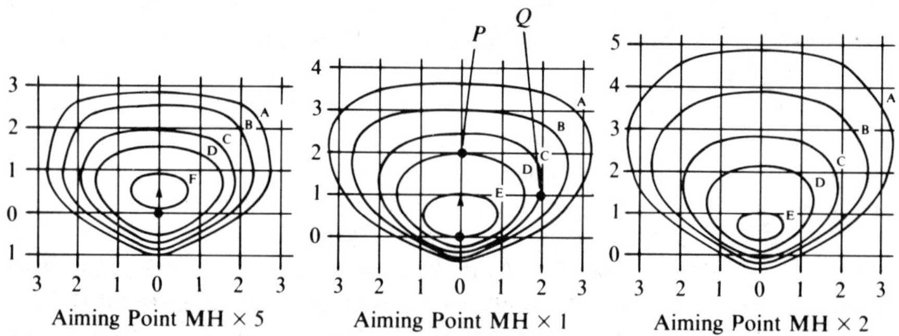

Figure 11.3 Isoilluminance diagrams. (*Source:* General Electric Company, "Lighting System Product Application Guide," Hendersonville, N.C., 1980, reprinted with permission.)

For the point Q, 1 MH in front of the pole and 2 MH to the side, the angles θ_H and ψ in Fig. 11.2*d* are needed.

$$\cos \psi = \frac{1}{\sqrt{6}} \qquad \psi = 65.9°$$

$$\cos \theta_H = \frac{\sqrt{2}}{\sqrt{6}} \qquad \theta_H = 54.7°$$

From the horizontal candela trace in Fig. 11.1 for a horizontal angle of 54.7°, $I = 50 \times 290 = 14{,}500$ cd. Then

$$E = \frac{14{,}500 \ \cos^3 65.9}{30^2} = 1.1 \text{ fc}$$

The illuminance values differ slightly from those indicated in Fig. 11.3. The isofootcandle diagrams are intended only for quick reference. For

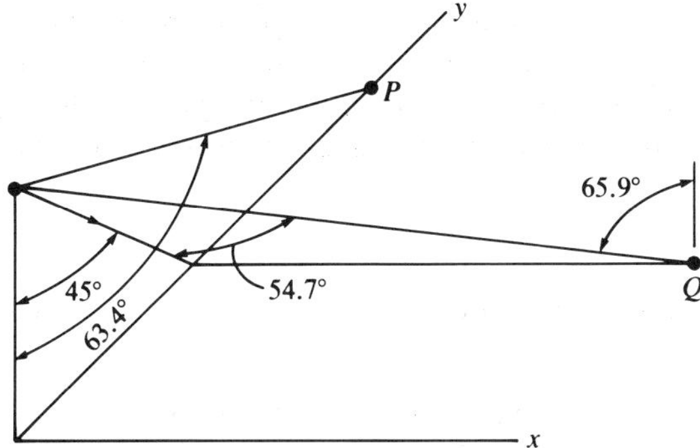

Figure 11.4

greater precision, the photometric data sheets, such as Fig. 11.1, should be used.

Isoilluminance diagrams are useful in estimating the coverage of a given floodlight and the effects of overlapping two or more floodlights. The curves are drawn so that a given curve provides approximately twice the illuminance of the curve immediately beyond it, for a given mounting height and lamp lumens. Thus curve *C* in Fig. 11.3 provides about twice the illuminance of curve *B*, etc. The centermost curve should be approximately the maximum illuminance for that aiming point. The outermost curve should be 1 to 2% of this value.

Another useful diagram is a set of utilization curves, shown in Fig. 11.5 for the floodlight in Fig. 11.1. These present the same summation of lumens

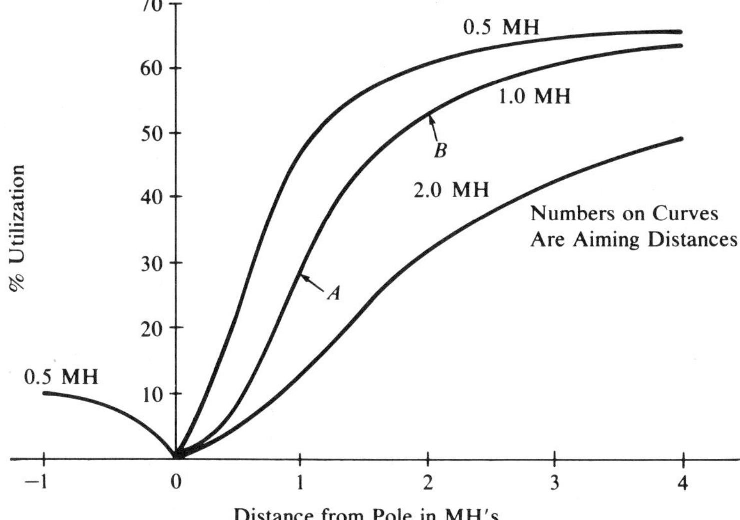

Figure 11.5 Utilization data for Fig. 11.1.

data as was given in Fig. 11.1 but as a function of aiming angle. In Fig. 11.5, the ordinate is the cumulative percent of lamp lumens, for a given aiming angle, which fall in horizontal strips in front of and behind the pole, with the width of each strip expressed in mounting heights. An example will illustrate the calculation.

EXAMPLE 11.2

Verify the percent lamp lumen utilization at the labeled points in Fig. 11.5.

Solution. The aiming point is at 1 MH, so Fig. 11.4 may be used. We desire the total lumens falling in a horizontal strip 45° wide immediately behind the aiming point and in one 18.4° wide immediately in front of the aiming point. From Fig. 11.1, we obtain

$$\phi_{0 \text{ to } -45^\circ} = 2[57.5 + 39.0 + 27.8 + 18.8 + 0.5(11.2)]$$
$$= 297.8 \text{ lm}$$
$$\phi_{0 \text{ to } 18.4^\circ} = 2\,[68.4 + 0.84(60.2)]$$
$$= 237.9 \text{ lm}$$

Since there are 1000 lamp lumens, the percent utilizations are

At A: 29.8%

At B: 53.6%

The selection of the aiming point is based on both utilization and uniformity (Frier and Frier, 1980). Ideally, the utilization curve should appear as in Fig. 11.6*a* to optimize both, where the quantity k is the width of the area to be lighted in numbers of mounting heights.

If the curve in Fig. 11.6*a* could be obtained, all of the floodlight flux would fall in the lighted area, the percent utilization would equal the total floodlight efficiency, and the flux distribution throughout the area would be uniform. Unfortunately, if the utilization curve is nearly linear within the lighted area, it is likely to continue rising outside that area, as in Fig. 11.6*b*. This wastes lumens. Or, if most of the lumens are contained within the area, it is likely that the utilization curve will have considerable curvature, providing nonuniform illumination, as in Fig. 11.6*c*.

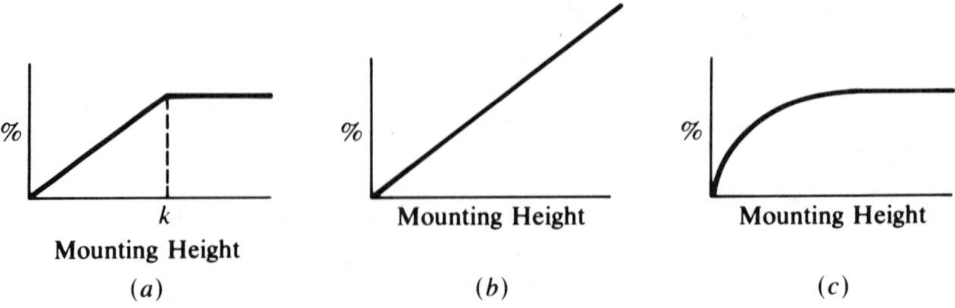

Figure 11.6 Utilization curves.

Thus a compromise must be reached and the aiming point chosen accordingly. Utilization increases as the aiming point moves toward the pole, but uniformity decreases. The highest utilization (for a symmetrical vertical candela trace) occurs when the aiming line bisects the angle subtended at the floodlight by the near and far sides of the area to be lighted. For example, if the area to be lighted is 3 MH's wide with the pole on one of its boundaries, the aiming point should be at 0.72 MH for optimum utilization. However uniformity, as can be seen from Fig. 11.5, would be less than optimum.

The issue of uniformity can be examined further by solving Eq. (11.2) for I and expressing horizontal distance in numbers of mounting heights, k. This gives

$$I = E(\mathrm{MH})^2\ (1 + k^2)^{3/2} \equiv I_0(1 + k^2)^{3/2} \qquad (11.3)$$

where I_0 is the intensity directly under the unit for a given E and MH. Keeping these latter quantities constant, the I required at 1 MH, 2 MH, and 3 MH from the pole would be $2.8I_0$, $11.2I_0$, and $31.6I_0$, respectively. Now if the floodlight is aimed, for example, at the 2-MH point, the illuminance at the base of the pole would be the same as that at the 2-MH point if the I toward the base of the pole were $(1/11.2)I$, or roughly 10% of I_0. As we have seen, this is about the limit of coverage of a floodlight beam if we are to achieve reasonably uniform illumination. Thus aiming distances should not be greater than 2 MH's horizontally.

From the inverse-square law, the maximum illuminance produced by a floodlight occurs nearer the pole than the aiming point. It is desirable that this maximum illuminance not be greater than three times the average illuminance of the floodlight, if the area is to appear uniformly lit. It is also important that the beams from adjacent floodlights overlap sufficiently so that the minimum illuminance is not far different from the average illuminance. As a rough guideline, the beams should overlap at their 25% of maximum intensity points. This will provide 50% of maximum intensity at these intersections, and produce sufficient illuminance to maintain reasonable uniformity. For more careful assessment of maximum and minimum illuminances, point-by-point calculations should be made by computer.

11.3 CONSIDERATIONS IN FLOODLIGHTING DESIGN

For horizontal areas, such as parking lots, floodlights are generally located on poles around the perimeter of the area to be lighted. For vertical areas, such as building fronts or signs, they are usually located on the ground some distance from the vertical surface.

For a horizontal area, three configurations are most common, as shown in Fig. 11.7. In Fig. 11.7*a* floodlights are shown at the four corners of the area. This is probably the most effective layout for three reasons. First, most of the floodlight beam can be contained in the area being lighted. Second, the floodlight beams criss-cross, providing more uniform illumination and mask-

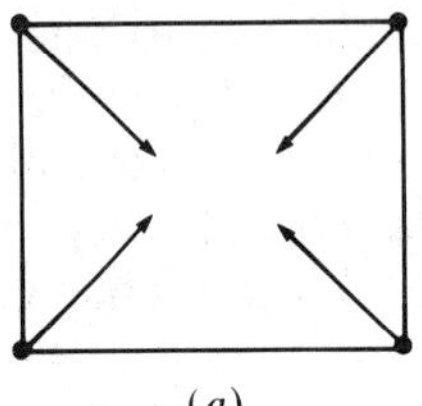
(a)

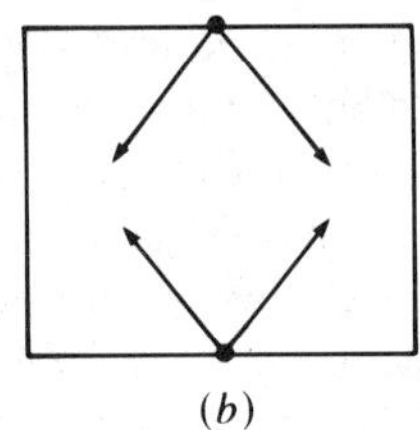
(b)

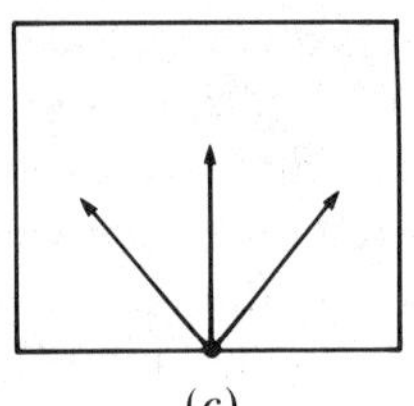
(c)

Figure 11.7 Floodlight configurations.

ing shadows produced by vertical objects. Last, probably only one floodlight per pole will be needed to achieve reasonable uniformity. The primary disadvantage, of course, is more poles.

The layout in Fig. 11.7*b* is a compromise. Two floodlights will be needed on each pole. There is more spill light into surrounding areas and uniformity is more difficult to achieve. The arrangement in Fig. 11.7*c* is the least satisfactory. It is particularly difficult to reach the far corners of the area to be lighted with this layout.

In addition to pole layout, pole height must be selected. Here the issues of uniformity and utilization discussed in the previous section must be addressed. Each floodlight can effectively light an area 2 MH wide. This determines pole height in Fig. 11.7*b* and *c*. Thus the area in Fig. 11.7*b* can be 4 MH wide, that in Fig. 11.7*c*, 2 MH wide. In Fig. 11.7*a* the long dimension of the area should be no more than 4.5 MH or intermediate poles will be needed.

We have yet to consider the types of lamps to use. Generally the choice is between tungsten halogen, metal halide, and high-pressure sodium. A comparison of these three source types is given in Table 11.2 (IES Sports and Recreational Areas Committee, 1969), where L, M, and H refer to low, medium, and high, respectively. For further details of specific lamps, the reader is referred to Chap. 6.

11.4 FLOODLIGHTING CALCULATIONS

There are two common methods for calculting the illuminance produced by a floodlight. One is the point-by-point method, using the inverse-square law in the form of Eq. (11.2). Though tedious to perform manually, the

TABLE 11.2 LAMPS FOR EXTERIOR LIGHTING

Lamps	Lumens	lm/W	Life	Color	Control
TH	M	L	L	H	H
MH	H	M	M	M	M
HPS	H	H	H	L	M

method is easy to program on a computer or programmable calculator and is as accurate and complete as the number of calculations the user wishes to make. Generally the area to be lighted is divided into a square grid, like a checkerboard. The illuminance is calculated at the center of each square and at a few other points that may be deemed critical, for example, along the boundary of the area. If the resulting values are felt to be inadequate, the computer may be asked quite easily to recalculate illuminances for new mounting heights, lamp sizes, aiming points, and/or pole locations.

The second floodlight calculation procedure is the beam lumen method, which has certain similarities to the zonal cavity method of interior lighting design. In this method the average maintained illuminance is obtained from

$$E_m = \frac{n(\phi_B)(\text{CBU})(\text{LLF})}{A} \tag{11.4}$$

where n = number of floodlights
ϕ_B = beam lumens
CBU = coefficient of beam utilization
LLF = light loss factor
A = area to be lighted

The coefficient of beam utilization is the ratio of the lumens incident on the floodlit area to the total beam lumens. The determination of this parameter will be illustrated by an example (adapted from IES Sports and Recreational Areas Committee, 1969).

A playing field 360 × 160 ft is shown in Fig. 11.8. It is to be lighted by floodlights on poles, one of which is shown. The pole is located at point O and the floodlight is mounted 50 ft up and aimed at point H. The floodlight is an enclosed symmetric unit, NEMA Type 5. The lamp emits 33,300 lm; the beam contains 18,000 lm.* Thus the beam efficiency is 54%.

The isocandela curves and the lumen distribution by angular zones are also shown in Fig. 11.8 for the unit. The problem is to determine how many of these lumens strike the playing surface, in other words, to find the projection of the boundary of the playing field onto the lumen distribution grid for the given mounting height and aiming point. The reason a projection of the playing field is needed is because the data at the bottom of Fig. 11.8 were taken with the floodlight aimed straight down, whereas the floodlight is aimed at an angle to the playing field in the top of Fig. 11.8.

To perform the needed projection requires that we obtain the vertical and lateral angles subtended by the playing field, as illustrated in Fig. 11.9. A floodlight at P is aimed at H. A plane $QRST$ is passed through point H and is perpendicular to line PH. This is the plane of the lumen grid in Fig. 11.8.

*Prior to 1971, the beam was taken out to 10% of the average maximum candlepower, which was the average of the 10 highest candlepower readings during photometric testing. For this unit, the maximum candlepower is 46,500 cd and the maximum average candlepower is 26,370 cd. The latter value was used in determining the beam limits in this example.

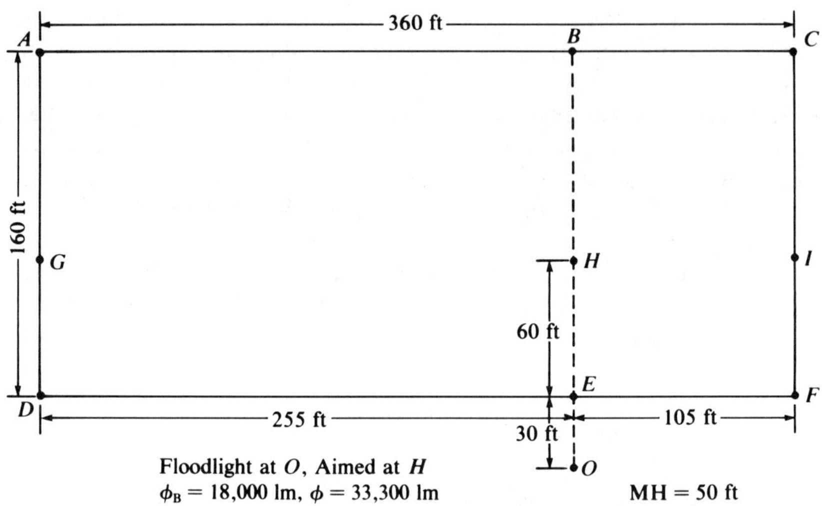

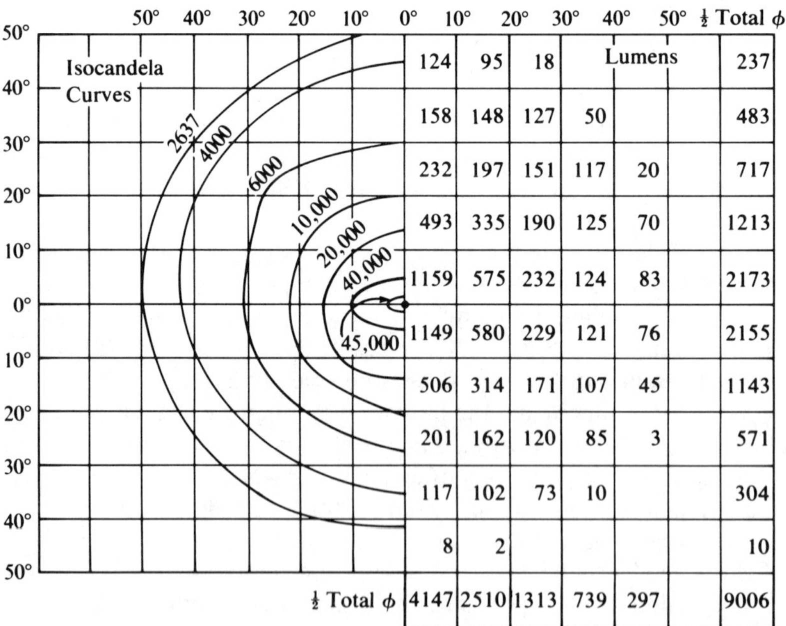

	0°–10°	10°–20°	20°–30°	30°–40°	40°–50°		½ Total ϕ
50°–40°	124	95	18				237
40°–30°	158	148	127	50			483
30°–20°	232	197	151	117	20		717
20°–10°	493	335	190	125	70		1213
10°–0°	1159	575	232	124	83		2173
0°–10°	1149	580	229	121	76		2155
10°–20°	506	314	171	107	45		1143
20°–30°	201	162	120	85	3		571
30°–40°	117	102	73	10			304
40°–50°	8	2					10
½ Total ϕ	4147	2510	1313	739	297		9006

Figure 11.8 Floodlighting design. (*Source:* IES Committee on Sports and Recreational Areas, *Recommended Practice for Sports Lighting*, IES, New York, 1969, fig. C-2, with permission.)

The right-hand extremity of the playing field CIF is also shown in Fig. 11.9. Its projection onto the lumen grid is shown dashed through the points K, I, and L. Consider point C. It is determined by the vertical angle HPB and the lateral angle CPB. Each of these angles is preserved in going from the play-

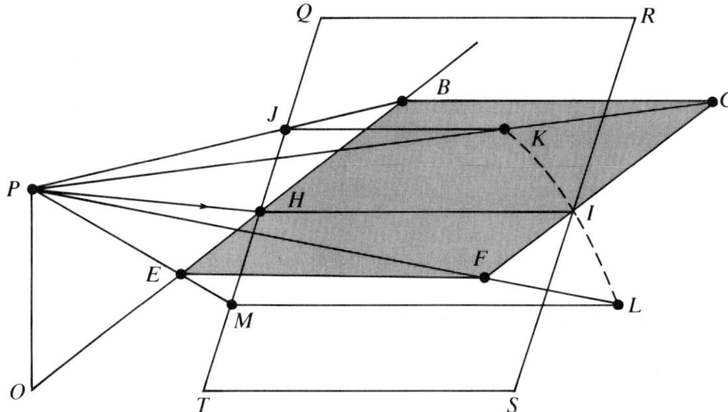

Figure 11.9 Projection of surface to be lighted.

ing field to the lumen grid. Thus, if we can find the angles HPB and CPB for the point C, we can locate this point on the lumen grid.

The necessary triangles for points A, B, and C are shown in Fig. 11.10. First θ_B and PB are found from Fig. 11.10a. The values are $\theta_B = 14.2°$ and $PB = 196.5$ ft. Then the angles ϕ_A and ϕ_C are found from Fig. 11.10b, giving $\phi_A = 52.4°$ and $\phi_C = 28.1°$. Similar calculations can be made for points G, H, and I, yielding $\theta_H = 0°$, $\phi_G = 68.0°$, and $\phi_I = 45.5°$. Finally, for points D, E, and F, $\theta_E = -30°$, $\phi_D = 77.1°$, and $\phi_F = 61°$.

The angles just calculated can be located directly on the lumen grid and the outline of the playing field sketched. Alternatively, a template can be

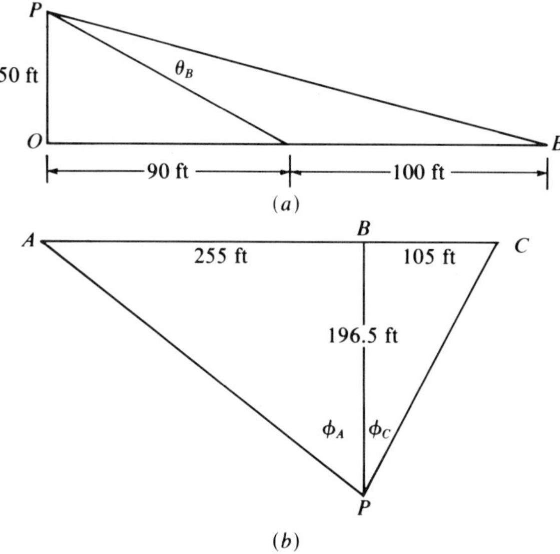

Figure 11.10 Locating points A, B, and C on the lumen grid.

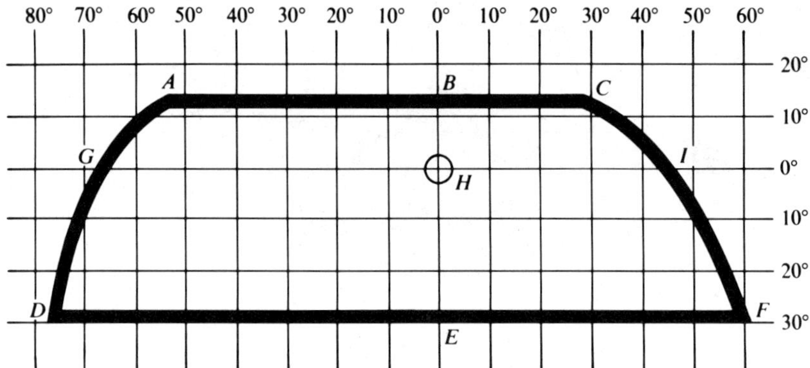

Figure 11.11 Template for Fig. 11.8.

prepared to the same scale as the lumen grid, as shown in Fig. 11.11. This shows clearly the projection of the field on the grid.

With the aid of the template, those beam lumens which strike the field can be determined. The template is placed over the grid and the following summations are made:

$$\text{Right side:}\quad 0.42(493 + 335 + 190) + 15 + 2173 - 65 - 10 + 2155 - 8 + 1143 + 571 = 6402$$

$$\text{Left side: } 0.42(1213) + 2173 + 2155 + 1143 + 571 = 6551$$

The numbers 15, 65, 10, and 8 are estimates of lumens hitting the fields in those zones where the projected field boundary slants across the zone. The total lumens striking the field are 12,953 and the coefficient of beam utilization is

$$\text{CBU} = \frac{12{,}953}{18{,}000} = 0.72$$

Assume four poles like the one in Fig. 11.8 are placed symmetrically with respect to the field, as shown in Fig. 11.12*a*. Assume further that five floodlights are placed atop each pole and aimed as in Fig. 11.8. Then the template in Fig. 11.11 may be used for all floodlights and, assuming a light

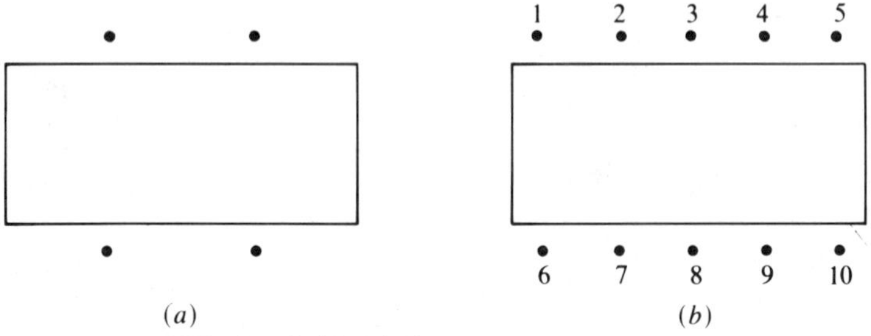

Figure 11.12 Playing field pole layouts.

loss factor of 0.75, the average maintained illuminance on the field is, from Eq. (11.4),

$$E_M = \frac{4(5)(18{,}000)(0.72)(0.75)}{360(160)} = 3.4 \text{ fc}$$

A better arrangement of poles from a uniformity point of view is shown in Fig. 11.12*b*. The previous template applies to poles 2, 4, 7, and 9, assuming the floodlights are all aimed the same. New templates would have to be made for poles 1, 5, 6, and 10 and for poles 3 and 8, that is, a given template is valid only for one mounting height, aiming point, and pole location.

11.5 ROADWAY AND AREA LUMINAIRES

Unlike floodlights, roadway and area luminaires are mounted horizontally and thus have fixed vertical aiming. The only flexibility permitted is a rotation of the luminaire about its vertical axis, which is possible in some luminaires designed for installation on tall poles (high-mast luminaires).

Roadway luminaires have particular intensity distributions which are designed to light long narrow horizontal strips on one side of the luminaire, while minimizing the intensities on the other side of the luminaire. The intensity distributions up and down the narrow strip are generally the same. Any fixed-aimed luminaire which does not have this type of intensity distribution is called an area luminaire.

There is a bit of jargon in connection with roadway and area luminaires which should be learned at the outset. Consider Fig. 11.13*a*. A roadway lighting unit is shown in a typical orientation with its pole set back from the curb and its luminaire overhanging the roadway. The point of 0 vertical

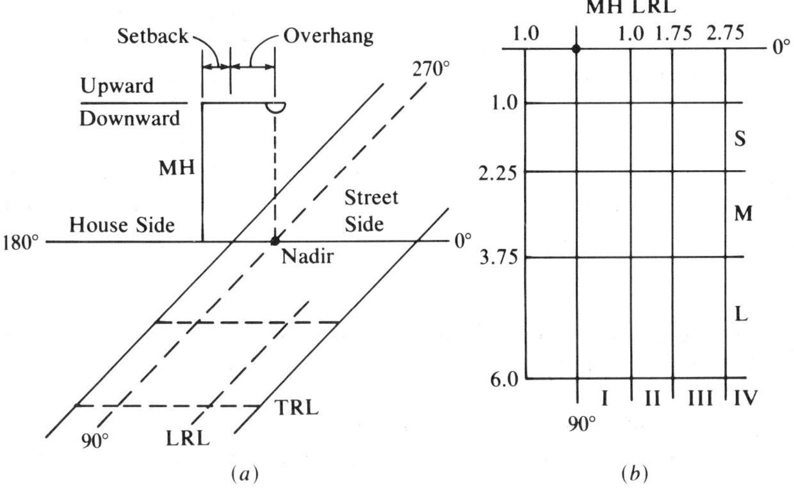

Figure 11.13 Roadway luminaire photometric nomenclature.

degrees is directly under the unit and labeled *nadir*. The line of 0 lateral degrees extends directly in front of the luminaire, toward the street side. The line of 180 lateral degrees faces the house side, behind the luminaire.

Transverse (TRL) and longitudinal (LRL) roadway lines are shown dashed in Fig. 11.13*a*, and are repeated on a more detailed grid pattern in Fig. 11.13*b*. These are given in terms of MH and have to do with three classifications of roadway luminaires established by the IESNA (IES Roadway Lighting Committee, 1983), based on the luminaire's vertical intensity distribution, lateral intensity distribution, and intensity distribution near 90° vertical angle.

The vertical classification of a roadway luminaire is determined by the location of its maximum intensity. If it falls between the 1.0 and 2.25 MH TRL lines, the vertical distribution is said to be *short* (S); between the 2.5 and 3.75 MH TRL lines, *medium* (M); and between the 3.75 and 6.0 MH TRL lines, *long* (L). The 1.0, 2.25, 3.75, and 6.0 MH TRL lines correspond to angles of incidence on the roadway of 45°, 66°, 75°, and 80°, respectively. Thus, if a luminaire has a medium distribution, its maximum intensity lies at a vertical angle between 66 and 75°. An example of such a luminaire is shown in Fig. 11.14.

The lateral classification of a luminaire is a measure of the width of the beam and is determined by the location on the roadway of an isocandela curve of one-half the maximum luminaire intensity. This is called the *one-half maximum candlepower isocandela trace* and four luminaire types are designated based on this trace, as shown in Fig. 11.13*b*.

Type I This type is for luminaires mounted near the center of the roadway, rather than near its edge. A luminaire is Type I if its one-half candela trace does not cross either of the 1.0 MH CRL lines while it is within the zone of maximum intensity (S, M, or L).

Type II This type and the two to follow are for luminaires mounted near the edge of a roadway. A luminaire is Type II if its one-half candela trace does not cross the 1.75 MH LRL line in the zone where the maximum intensity is located.

Type III A luminaire is Type III if its one-half candela trace crosses the 1.75 MH LRL line but does not cross the 2.75 MH LRL line in the zone where the maximum intensity is located. An example of a Type III luminaire is shown in Fig. 11.14.

Type IV A luminaire is Type IV if its one-half candela trace crosses the 2.75 MH LRL line in the zone where the maximum intensity is located.

There is a fifth intensity distribution (Type V) not shown in Fig. 11.13*b*. This is for luminaires mounted at the center of a four-way intersec-

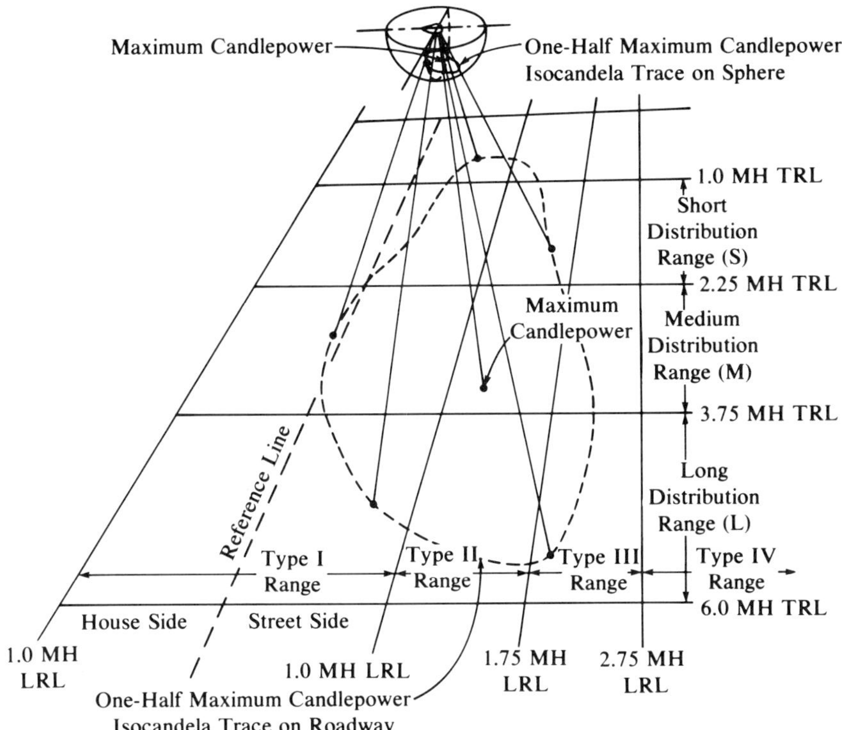

Figure 11.14 Roadway luminaire, medium, Type III. [*Source:* J. E. Kaufman (ed.), *IES Lighting Handbook*, Application volume, 1981 ed., IES, New York, 1981, fig. 14-5, with permission.]

tion and describes an intensity distribution which is the same at all lateral angles.

The final categorization of a roadway luminaire is based on its intensity distribution at large vertical angles and gives a measure of the potential for glare. A luminaire's distribution is described as cutoff if its intensity is not greater than 100 cd per 1000 lm above a vertical angle of 80° and not greater than 25 cd per 1000 lm above 90°, at all lateral angles. A luminaire is classed as semicutoff if these numbers are 200 cd per 1000 lm above 80° and 50 cd per 1000 lm above 90°. It is called noncutoff if there are no intensity limitations.

A typical photometric data sheet for a roadway luminaire is shown in Fig. 11.15. The luminaire is medium, Type III, semicutoff, and designed to house a high-pressure sodium lamp. In the general information box, the maximum intensity is given as 466 cd per 1000 lamp lumens and occurs at a vertical angle of 72.5°, placing it in the medium vertical distribution range. The intensities at vertical angles of 80° and 90° are 174 and 39 cd per 1000 lamp lumens, respectively, which is within the semicutoff range. The nadir

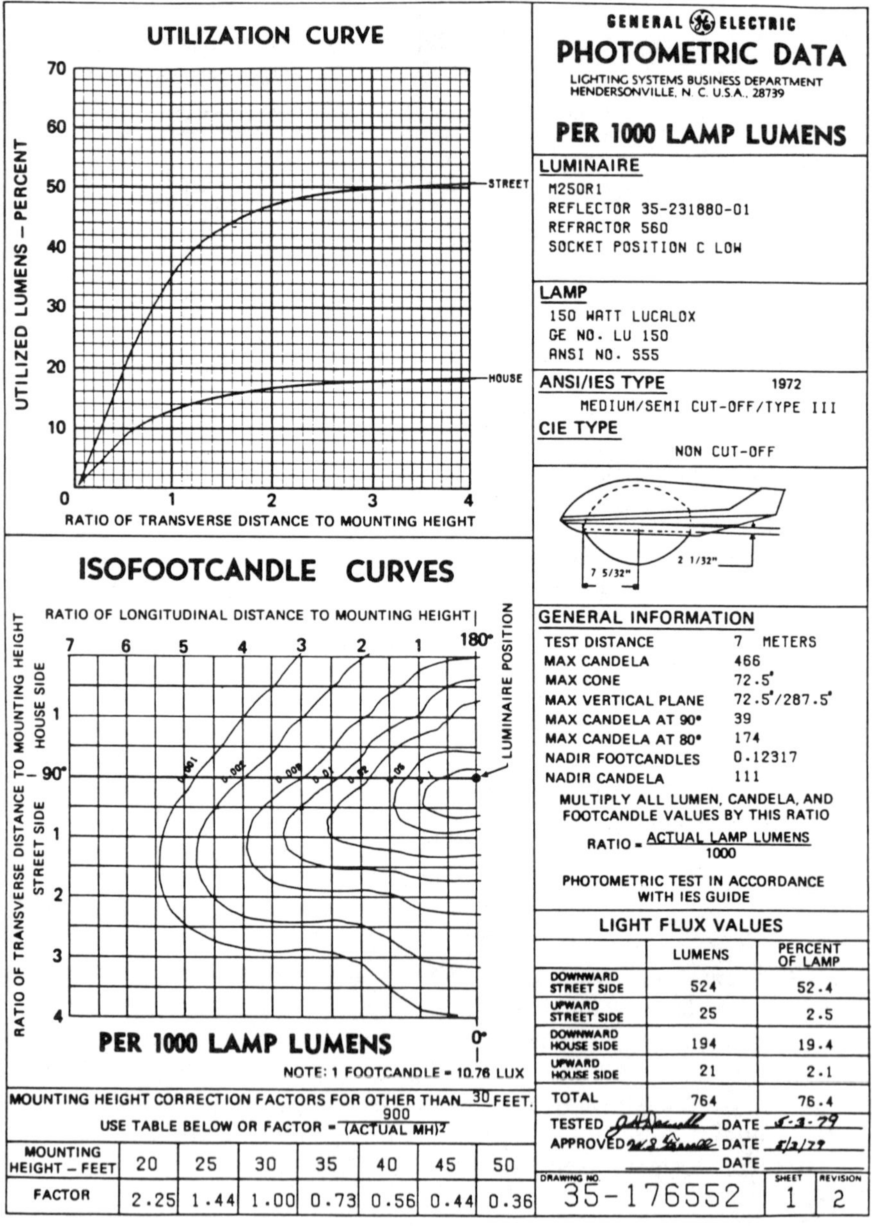

GENERAL ELECTRIC

PHOTOMETRIC DATA

LIGHTING SYSTEMS BUSINESS DEPARTMENT
HENDERSONVILLE, N. C. U.S.A., 28739

PER 1000 LAMP LUMENS

UTILIZATION CURVE

UTILIZED LUMENS – PERCENT

RATIO OF TRANSVERSE DISTANCE TO MOUNTING HEIGHT

STREET

HOUSE

LUMINAIRE
M250R1
REFLECTOR 35-231880-01
REFRACTOR 560
SOCKET POSITION C LOW

LAMP
150 WATT LUCALOX
GE NO. LU 150
ANSI NO. S55

ANSI/IES TYPE 1972
MEDIUM/SEMI CUT-OFF/TYPE III

CIE TYPE
NON CUT-OFF

7 5/32"
2 1/32"

ISOFOOTCANDLE CURVES

RATIO OF LONGITUDINAL DISTANCE TO MOUNTING HEIGHT

RATIO OF TRANSVERSE DISTANCE TO MOUNTING HEIGHT

HOUSE SIDE

STREET SIDE

LUMINAIRE POSITION

180°

90°

0°

PER 1000 LAMP LUMENS

NOTE: 1 FOOTCANDLE = 10.76 LUX

GENERAL INFORMATION

TEST DISTANCE	7 METERS
MAX CANDELA	466
MAX CONE	72.5°
MAX VERTICAL PLANE	72.5°/287.5°
MAX CANDELA AT 90°	39
MAX CANDELA AT 80°	174
NADIR FOOTCANDLES	0.12317
NADIR CANDELA	111

MULTIPLY ALL LUMEN, CANDELA, AND FOOTCANDLE VALUES BY THIS RATIO

$$\text{RATIO} = \frac{\text{ACTUAL LAMP LUMENS}}{1000}$$

PHOTOMETRIC TEST IN ACCORDANCE WITH IES GUIDE

LIGHT FLUX VALUES

	LUMENS	PERCENT OF LAMP
DOWNWARD STREET SIDE	524	52.4
UPWARD STREET SIDE	25	2.5
DOWNWARD HOUSE SIDE	194	19.4
UPWARD HOUSE SIDE	21	2.1
TOTAL	764	76.4

MOUNTING HEIGHT CORRECTION FACTORS FOR OTHER THAN 30 FEET.

USE TABLE BELOW OR FACTOR $= \frac{900}{(\text{ACTUAL MH})^2}$

MOUNTING HEIGHT – FEET	20	25	30	35	40	45	50
FACTOR	2.25	1.44	1.00	0.73	0.56	0.44	0.36

TESTED DATE 5-3-79
APPROVED DATE 5/3/79
DATE

DRAWING NO. 35-176552 SHEET 1 REVISION 2

Figure 11.15 Roadway luminaire, photometric data. (*Source:* General Electric Company, "Lighting System Product Application Guide," Hendersonville, N.C., 1980, reprinted with permission.)

candelas and footcandles, taken directly beneath the luminaire, are 111 and 0.123 per 1000 lm, respectively. The nadir footcandles are based on a 30-ft MH. Thus the 0.123 is obtained by dividing 111 by 900.

In the light flux box, the lumen distribution of the luminaire is given, on a per 1000 lamp lumen basis. 71.8% of the lamp lumens are directed downward, 52.4% to the street side and 19.4% to the house side. 4.6% of the lamp lumens are directed upward. The efficiency of the luminaire is 76.4%.

The utilization curve for the luminaire appears in the left upper box. This is identical in form to the curves for a floodlight except that, instead of a curve for each aiming point, there is one curve for the street side and one for the house side. Again the issues of uniformity and utilization must be addressed. The discussion in Sec. 11.2 still applies except that, with a roadway luminaire, we cannot alter the aiming angle to improve either of these parameters. Rather we need to set guidelines on the width of roadway that can be reasonably covered with a given luminaire.

From Fig. 11.15, on the street side, 35% of the lamp lumens impinge on the first MH of road width, 12% more on the second MH, 3% more on the third MH, and 1% more on the fourth MH. Saturation has certainly begun to occur by the 2-MH mark. Thus, as a general guideline, the mounting height of a roadway luminaire should be one-half the width of the street (minus the overhang if one exists). The only exception would be if a greater mounting height is needed to project lumens further up and down the street. This would improve uniformity while reducing utilization.

Isofootcandle curves on a per 1000 lamp lumen basis are shown in the lower left of Fig. 11.15. As in the case of floodlights, the values of illuminance are chosen such that a change of about 2:1 in illuminance is achieved between adjacent curves. A bulge in the illuminance curves occurs at a horizontal angle of 72.5° (also at the symmetrical angle of 287.5°). This is noted in the general information box under the listing "maximum vertical plane."

The isofootcandle values are based not only on 1000 lamp lumens but also on 30-ft MH. Multiplying factors for other mounting heights are given at the bottom of Fig. 11.15.

The average illuminance from a roadway lighting installation is given by a formula very similar to that used in floodlighting design:

$$E_M = \frac{\phi_L(\text{CU})(\text{LLF})}{S(W)} \tag{11.5}$$

where ϕ_L = initial lamp lumens
CU = coefficient of utilization
LLF = light loss factor
W = width of the road (or area to be lighted)
S = spacing between luminaires

The coefficient of utilization is the ratio of the lumens falling on the area to be lighted to the toal lamp lumens, both on an initial basis. CU values can be obtained directly from the utilization curve. For example, in Fig. 11.15 if the road width is 2 MH and their is no luminaire overhang, 47% of the lamp lumens reach the street and thus CU = 0.47.

Recommended illuminances for roadways are given in Table 11.3. In

TABLE 11.3 RECOMMENDED AVERAGE MAINTAINED ROADWAY ILLUMINANCES
In Lux (Footcandles)

Roadway classification	Area classification	Pavement classification R1	R2, R3	R4
Freeway	C	6(0.6)	9(0.8)	8(0.7)
	I	4(0.4)	6(0.6)	5(0.5)
	R	—	—	—
Expressway	C	10(0.9)	14(1.3)	13(1.2)
	I	8(0.7)	12(1.1)	10(0.9)
	R	6(0.6)	9(0.8)	8(0.7)
Major	C	12(1.1)	17(1.6)	15(1.4)
	I	9(0.8)	13(1.2)	11(1.0)
	R	6(0.6)	9(0.8)	8(0.7)
Collector	C	8(0.7)	12(1.1)	10(0.9)
	I	6(0.6)	9(0.8)	8(0.7)
	R	6(0.6)	6(0.6)	5(0.5)
Local	C	6(0.6)	9(0.8)	8(0.7)
	I	5(0.5)	7(0.7)	6(0.6)
	R	3(0.3)	4(0.4)	4(0.4)

Source: IES Roadway Lighting Committee, "Proposed American Standard Practice for Roadway Lighting." Reprinted from the April 1983 issue of the *Journal* of *IES* with permission of the Illuminating Engineering Society of North America.

this table C, I, and R stand for commercial, intermediate, and residential, thus describing the area through which the roadway passes. R1 through R4 are roadway reflectance classifications. R1 is a mostly diffuse surface, such as concrete; R2 is a diffuse-specular asphalt surface with at least a 60% gravel base; R3 is a slightly specular rough-textured asphalt surface; R4 is a mostly specular smooth asphalt surface.

The usual situation is that the average maintained illuminance is prescribed by Table 11.3 and thus spacing and lamp size are the unknowns. To illustrate this, Eq. (11.5) can be rewritten as

$$S = \frac{\text{CU(LLF)}}{W(E_M)} \, (\phi_L) \tag{11.6}$$

We note that if a larger lamp is used, spacing may be increased, thus saving poles and luminaires. There is a limit to this, however, for uniformity may deteriorate as spacing increases. It is desirable that the ratio of average to minimum illuminance not exceed 3:1, thus placing an upper bound on spacing. The IESNA (Kaufman, 1981) suggests that spacing be less than 4.5 MH for a short distribution luminaire, less than 7.5 MH for a medium distribution luminaire, and less than 12.0 MH for a long distribution luminaire. This will insure that the maximum intensity lines of two adjacent luminaires will at least meet on the roadway surface.

The spacing criteria apply for each of the three typical types of pole placement along a roadway, namely, on one side, aligned on opposite sides, or staggered on opposite sides. Better uniformity can be achieved with units on both sides of the road, either aligned or staggered, but wiring costs are increased. If the road is wide, as in the case of major highways, poles on both sides may become necessary to avoid excessive mounting heights.

For area lighting, S:MH ratios of 4.5:1 are recommended. To explain why this is so, refer to Fig. 11.16, in which a luminaire is shown mounted at the corner of a square area whose sides are each 4.5 MH long. The most difficult point to light in this situation is at the area's center (A) and thus the maximum intensity of the luminaire should be directed there. The vertical angle is 72.6°, which requires a luminaire with a Class M vertical distribution (66 to 75°). This is the class of luminaire generally used for area lighting. For the illuminance at A to equal that at C, directly under the pole, requires that the intensity toward A be 37 times as great as toward C. This is about the upper limit to be expected of a luminaire. At B, the vertical angle is 66.0° and the intensity required is 15 times that to produce the same illuminance at C.

The light loss factor for a roadway or area luminaire consists of two parts, the lamp lumen depreciation factor (LLD) and the luminaire dirt depreciation factor (LDD). The former can be found from the lamp tables in Chap. 6. The latter, as with interior luminaires, depends on the environment and the period between cleanings. Representative values are given in Table 11.4.

11.6 ROADWAY LIGHTING DESIGN

The approach to roadway lighting design will be illustrated through an example. Suppose it is desired to light a major commercial artery 50 ft wide with an R4 pavement using the luminaire in Fig. 11.15 and a 30-ft MH. The

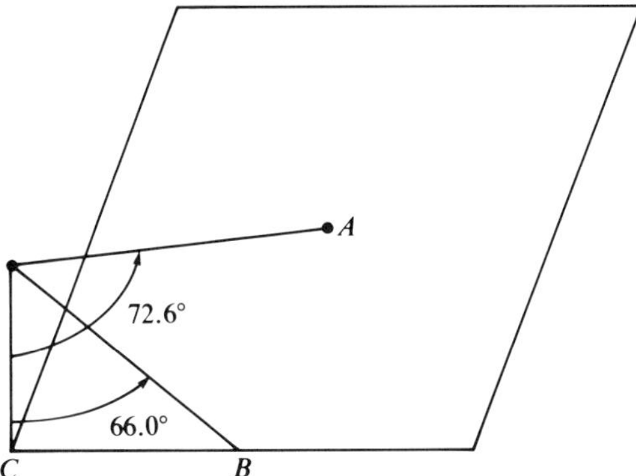

Figure 11.16 Pertaining to S:MH for area lighting. (*Source:* Adapted from J. P. Frier and M. E. G. Frier, *Industrial Lighting Systems*, McGraw-Hill, New York, 1980.)

TABLE 11.4 LDD FACTORS FOR ROADWAY LUMINAIRES

	Time between cleanings (years)		
Environment	1	2	3
Very clean	0.98	0.94	0.93
Clean	0.95	0.92	0.90
Moderate	0.92	0.87	0.84
Dirty	0.87	0.81	0.75
Very dirty	0.72	0.63	0.57

Source: J. E. Kaufman (ed.), *IES Lighting Handbook*, Application volume, 1981 ed., IES, New York, 1981, fig. 14-20, with permission.

luminaires are to be mounted along one side of the street and are to overhang the curb by 5 ft.

From Table 11.3, the desired illuminance is 1.4 fc. It is 45 ft transversely from the luminaire to the far side of the street (1.5 MH). From the utilization curve in Fig. 11.15, the CU on the street side is 0.43. On the house side, the distance is $\frac{5}{30} = 0.17$ MH. The CU for this side is 0.03. Thus the total CU is 0.46.

The LLF is composed of two parts. We will assume a clean environment with units cleaned annually. From Table 11.4, LDD = 0.95. The luminaire in Fig. 11.15 uses a high-pressure sodium lamp. Assume for the moment that the lamp will be 250 W. Then from Table 6.11, LLD is 24,750/27,500 = 0.90. This is based on the lumens at 70% life, which, for the sodium lamp, is at about 17,000 h.

We now have sufficient data to use Eq. (11.6)

$$S = \frac{0.46(0.95)(0.90)}{50(1.4)}(\phi_L) = 0.00562\phi_L$$

If the 250-W lamp is chosen,

$$S = 0.00562(27{,}500) = 155 \text{ ft}$$

This is $\frac{155}{30} = 5.2$ MH, well within the spacing guideline for a medium vertical distribution of less than 7.5 MH.

To check uniformity, the isofootcandle diagram in Fig. 11.15 is used in conjunction with the layout of the roadway in Fig. 11.17. It is customary to find the illuminance levels at three points, labeled *A*, *B*, and *C* in Fig. 11.17. This has been done in chart form in Table 11.5.

The values for the *A* column are found as follows: For luminaire 1, proceed horizontally to the left 5.2 MH on the isofootcandle plot in Fig. 11.15. Then proceed vertically downward 1.5 MH. This locates a point between the 0.001 and 0.002 isofootcandle curves. The value is estimated at 0.0013. The same value applies for luminaire 3. For luminaire 2, move ver-

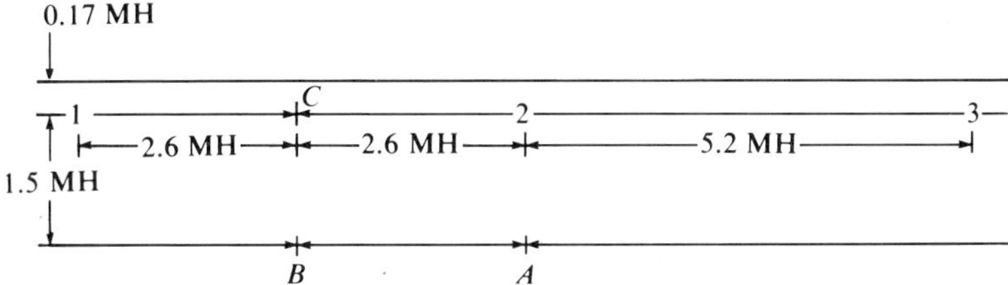

Figure 11.17 Checking roadway lighting uniformity.

tically downward 1.5 MH which locates the illuminance between the 0.01 and 0.02 isofootcandle curves. A value of 0.015 is estimated.

The isofootcandle values are for 1000 initial lamp lumens. They must be multiplied by the actual initial kilolumens times the light loss factor. The result is 0.42 fc for the *A* location. Points *B* and *C* yield 0.47 and 0.42 fc, respectively. The ratio of average to minimum illuminance is then found to be 3.3. This is slightly poorer than the guideline of 3:1. As a remedy, the spacing could be reduced slightly or the poles could be staggered. However pursuing the latter course does not help in this situation because the illuminance at point *A* will be less with staggered poles. This is because, with poles 1 and 3 on the opposite side of the street in Fig. 11.17, their contribution to the illuminance at point *A* is reduced, since advantage is no longer taken of the bulge in the isofootcandle curves at 72.5° and 287.5°.

11.7 AREA LIGHTING DESIGN

As in the previous section, an example will be used to illustrate the design procedure. An unusually shaped hospital parking lot is shown in Fig. 11.18. Two $2\frac{1}{2}$-ft-high square pedestals spaced 178 ft apart are already present in the parking area, and the luminaire poles are to be placed on these. Since patient windows are on the front of the hospital building, it is desired to minimize the luminaire intensity in that direction.

TABLE 11.5

Luminaire	*A*	*B*	*C*
1	0.0013	0.01	0.009
2	0.015	0.01	0.009
3	0.0013	—	—
	0.018	0.02	0.018
× 27.5 × 0.855	0.42 fc	0.47 fc	0.42 fc

$$\frac{\text{Average}}{\text{Minimum}} = \frac{1.4}{0.42} = 3.3$$

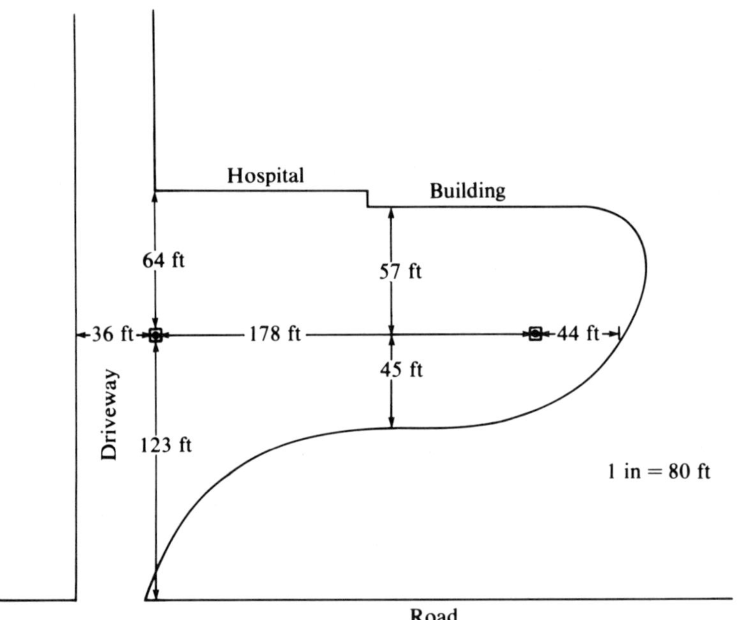

Figure 11.18 Hospital parking lot.

The roadway luminaire shown in Fig. 11.19 had been chosen previously by the hospital administrator; thus the design was constrained to choosing a lamp, a mounting height, and an orientation of the luminaires on the poles. This is perhaps more often the case than not; the designer is not given complete freedom of choice but rather is restricted by choices made by others prior to his arrival on the scene.

The first step is to select a tentative mounting height. Generally, as was discussed previously, spacing between luminaires should be less than 4.5 MH. Thus, with the 178-ft spacing predetermined, MH should be at least 40 ft. It was decided to use 40-ft poles which, with the $2\frac{1}{2}$-ft pedestal height, makes the mounting height 42.5 ft.

The next step is to select a lamp size. The IESNA recommended illuminance levels for open parking areas are given in Table 11.6. It was judged that the area being considered had a low level of activity and that the major issue was pedestrian safety, simply providing adequate horizontal illuminance for hospital employees and visitors to go safely to and from their cars. However it was felt that some security should be provided. A compromise was made to provide 0.5 fc and a 400-W high-pressure sodium lamp was chosen tentatively.

From Table 6.11, a 400-W high-pressure sodium lamp emits 50,000 lm initially and 45,000 mean lumens (at 70% life). From Fig. 11.19, the initial illuminance directly under a pole is

$$E = \frac{149(50)}{42.5} = 4.1 \text{ fc}$$

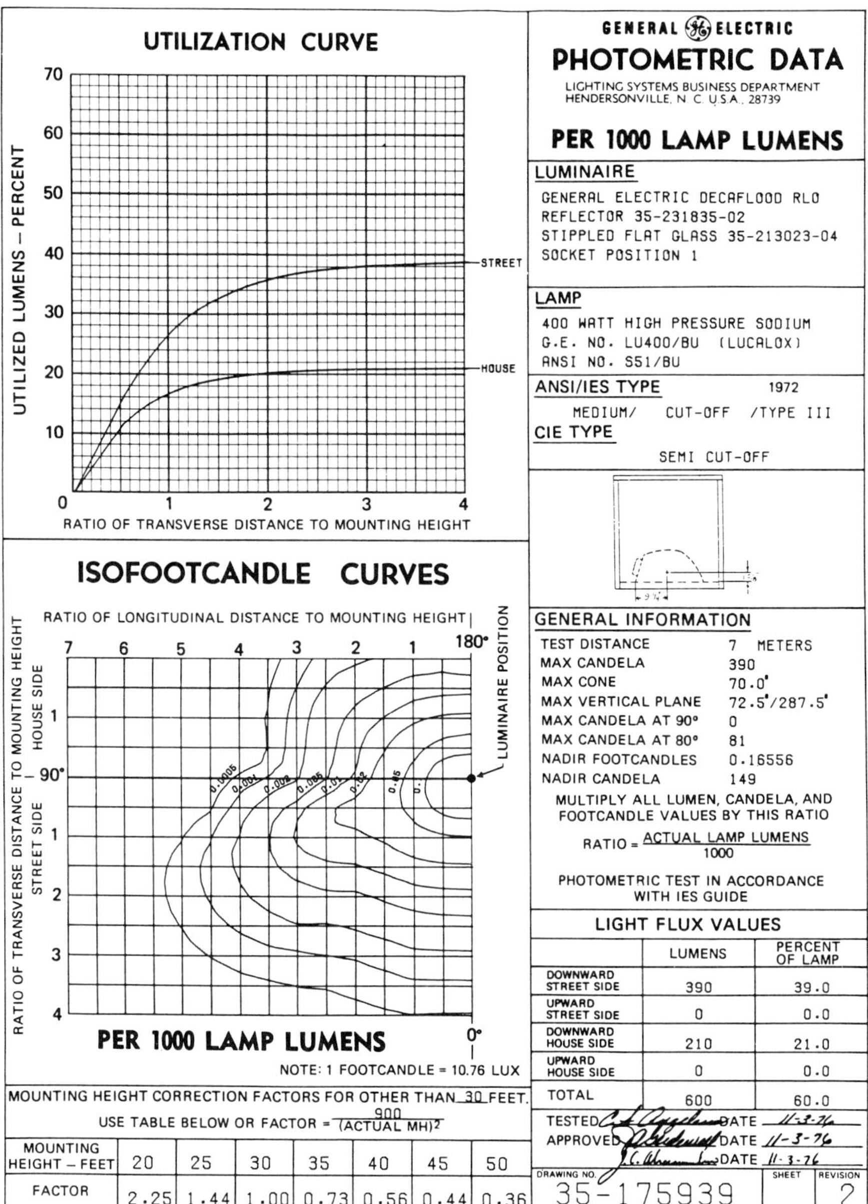

MOUNTING HEIGHT CORRECTION FACTORS FOR OTHER THAN 30 FEET.
USE TABLE BELOW OR FACTOR = $\frac{900}{(\text{ACTUAL MH})^2}$

MOUNTING HEIGHT – FEET	20	25	30	35	40	45	50
FACTOR	2.25	1.44	1.00	0.73	0.56	0.44	0.36

GENERAL ELECTRIC
PHOTOMETRIC DATA
LIGHTING SYSTEMS BUSINESS DEPARTMENT
HENDERSONVILLE, N. C. U.S.A., 28739

PER 1000 LAMP LUMENS

LUMINAIRE
GENERAL ELECTRIC DECAFLOOD RLO
REFLECTOR 35-231835-02
STIPPLED FLAT GLASS 35-213023-04
SOCKET POSITION 1

LAMP
400 WATT HIGH PRESSURE SODIUM
G.E. NO. LU400/BU (LUCALOX)
ANSI NO. S51/BU

ANSI/IES TYPE 1972
MEDIUM/ CUT-OFF /TYPE III
CIE TYPE
SEMI CUT-OFF

GENERAL INFORMATION

TEST DISTANCE	7 METERS
MAX CANDELA	390
MAX CONE	70.0°
MAX VERTICAL PLANE	72.5°/287.5°
MAX CANDELA AT 90°	0
MAX CANDELA AT 80°	81
NADIR FOOTCANDLES	0.16556
NADIR CANDELA	149

MULTIPLY ALL LUMEN, CANDELA, AND FOOTCANDLE VALUES BY THIS RATIO
RATIO = $\frac{\text{ACTUAL LAMP LUMENS}}{1000}$
PHOTOMETRIC TEST IN ACCORDANCE WITH IES GUIDE

LIGHT FLUX VALUES

	LUMENS	PERCENT OF LAMP
DOWNWARD STREET SIDE	390	39.0
UPWARD STREET SIDE	0	0.0
DOWNWARD HOUSE SIDE	210	21.0
UPWARD HOUSE SIDE	0	0.0
TOTAL	600	60.0

TESTED DATE 11-3-76
APPROVED DATE 11-3-76
DATE 11-3-76
DRAWING NO. 35-175939 SHEET REVISION 2

Figure 11.19 Luminaire for parking lot lighting. (*Source:* General Electric Company, "Lighting System Product Application Guide," Hendersonville, N.C., 1980, reprinted with permission.)

In using the isofootcandle curves in Fig. 11.19, a multiplier of $50(30^2/42.5^2) = 25$ is required. Midway between the poles (2.25 MH) the initial illuminance from each luminaire is

$$E = 25(0.01) = 0.25 \text{ fc}$$

TABLE 11.6 RECOMMENDED MAINTAINED ILLUMINANCES FOR OPEN PARKING FACILITIES
Lux (Footcandles)

Level of activity	For vehicular traffic	For pedestrian safety	For pedestrian security
Low	5 (0.5)	2 (0.2)	9 (0.8)
Medium	11 (1)	6 (0.6)	22 (2)
High	22 (2)	10 (0.9)	43 (4)

Source: J. E. Kaufman (ed.), *IES Lighting Handbook*, Application volume, 1981 ed., IES, New York, 1981, fig. 14-18, with permission.

assuming the poles are mounted facing the street. Thus it appears that the 400-W lamp will be sufficient to provide the desired illuminance.

In a similar fashion, the illuminances at various other locations produced by a single luminaire were calculated. These appear in Table 11.7. The parameter *d* is distance horizontally from the pole in feet. The data from Table 11.7 were used to prepare the template shown in Fig. 11.20. The template is drawn to the same scale (1 in = 80 ft) as the layout of the parking area in Fig. 11.18.

Two of the templates shown in Fig. 11.20 should be made on transparent paper. These can then be overlaid on the parking area at the pole locations to determine the overall footcandle distribution. Because the curves have a bulge at 72.5° and 287.5°, it was decided to rotate the luminaire on the

TABLE 11.7 INITIAL HORIZONTAL ILLUMINANCES PRODUCED BY THE LUMINAIRE IN FIG. 11.19
400-W HPS Lamp, 42.5 ft MH

Location		Isofootcandles				
		2.5	1.25	0.5	0.25	0.125
Laterally (90°)	MH	0.80	1.2	1.75	2.10	2.5
	d	34	51	74	89	106
House side (180°)	MH	0.40	0.75	1.1	1.4	1.7
	d	17	32	47	60	72
Street side (0°)	MH	0.65	1	1.45	1.85	2.3
	d	28	43	62	79	98
45° Toward street	MH	0.78	1.13	1.77	2.26	2.83
	d	33	48	75	96	120
45° Toward house	MH	0.61	0.92	1.34	1.69	2.12
	d	21	39	57	72	90
17° Toward street	MH	—	—	2.4	3.05	3.5
	d	—	—	102	130	149
17° Toward house	MH	—	—	1.5	1.9	2.3
	d	—	—	64	81	98

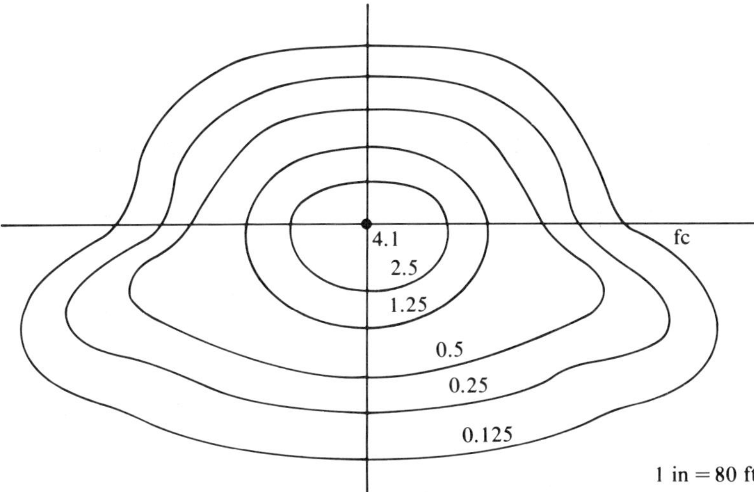

Figure 11.20 Isofootcandle curves (initial) for luminaire in Fig. 11.19. (400-W high-pressure sodium lamp, 42.5 ft MH)

left pole 20° to the right and that on the right pole 20° to the left. This increases the illumination level between the poles and also on the driveway entering the lot. It has the disadvantages of spilling a bit more light onto the area adjacent to the right front of the lot. This turned out not to be a problem, other than the loss of utilization. The illuminances between the poles for both cases of luminaire orientation are given in Table 11.8. It is seen that the uniformity ratio of E_{max}/E_{min} is considerably improved when the luminaires are rotated by 20°.

The average illuminance throughout the area was not calculated but it was judged that it would be adequate. In fact, it is possible that lamp size could have been reduced to 250 W. However this would have reduced all

TABLE 11.8 ILLUMINANCES BETWEEN POLES

Luminaires Facing Street			Luminaires Rotated 20°		
MH, ft	*d*, ft	fc	MH, ft	*d*, ft	fc
0	0	4.1	0	0	4.1
0.8	34	2.5	0.85	36	2.8
1.2	51	1.3	1.30	55	1.6
1.75	74	0.6	2.30	98	1.3
2.1	89	0.5	3.25	138	2.8
2.5	106	0.6	3.70	157	3.3
3.0	127	1.3			
3.4	144	2.5			
$\frac{E_{max}}{E_{min}} = \frac{4.1}{0.5} = 8.2$			$\frac{E_{max}}{E_{min}} = \frac{4.1}{1.3} = 3.2$		

illuminance levels by 45 percent and it was decided that this would produce a marginal installation.

11.8 LUMINANCE METHOD OF ROADWAY LIGHTING DESIGN

Until very recently, roadway lighting in North America was based on horizontal illuminance, as outlined in Sec. 11.6. However, it has been recognized for some time that pavement luminance, combined with the veiling luminance produced at the eye by stray light from roadway luminaires, would provide better criteria than horizontal illuminance for estimating visibility of objects or features on or near a roadway.

In 1982, a new American Standard Practice for Roadway Lighting (IES Roadway Lighting Committee, 1983) was approved which is based on the luminance of roadway surfaces. However, because determination of pavement luminances and their subsequent use in roadway lighting design is a more difficult procedure than determining illuminances and because use of illuminance criteria can result in adequate visibility if good design judgment is used, the Standard Practice retains the previously presented illuminance method as an acceptable alternative procedure. It is expected that, as measurement techniques and computer procedures for determining luminance become more common and accessible, the luminance method will become the dominant procedure.

The measurement of roadway luminance in the new procedure is based on Figs. 11.21 and 11.22. Figure 11.21 depicts the geometry of the driver,

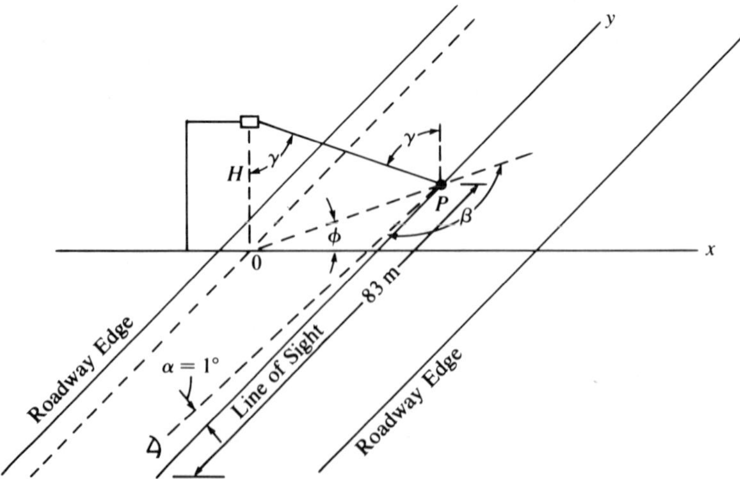

Figure 11.21 Roadway reflectance angles. (*Source:* IES Roadway Lighting Committee, "Proposed American Standard Practice for Roadway Lighting." Reprinted from the April 1983 issue of the *Journal of IES* with permission of the Illuminating Engineering Society of North America.)

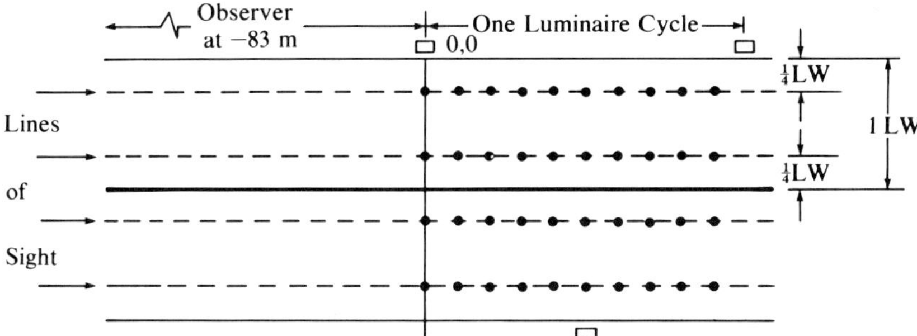

Figure 11.22 Roadway luminance geometry. (*Source:* IES Roadway Lighting Committee, "Proposed American Standard Practice for Roadway Lighting." Reprinted from the April 1983 issue of the *Journal of IES* with permission of the Illuminating Engineering Society of North America.)

the luminaire, and the roadway. The driver's eyes are located 1.45 m (about 5 ft) above the roadway with the line of sight 1° down from the horizontal. This places the driver 83 m (270 ft) from the point at which a luminance measurement is to be made.

As shown in Fig. 11.22, measurements are made along lines parallel to the edge of the roadway at the $\frac{1}{4}$- and $\frac{3}{4}$-lane-width (LW) points of each roadway lane. For each luminaire cycle, defined as the distance between luminaires on one side of the roadway, luminances are to be obtained at at least 10 equally spaced points in each lane, with no more than 5 m between points. For each point, the observer is located 83 m to the left, as shown in Figs. 11.21 and 11.22. Figure 11.21 also shows the geometry for a typical luminaire, making a vertical angle γ and horizontal angle ϕ with the measurement point P. In calculating the luminance at P, at least one luminaire behind P and at least three luminaires ahead of P should be included.

The basis for the new roadway luminance method is much the same as that described in Sec. 8.6. Referring to Fig. 11.21, the horizontal illuminance at P due to the single luminaire is

$$E_h = \frac{I(\gamma,\phi)\cos^3\gamma}{H^2} \tag{11.7}$$

The luminance at point P in the direction of the observer is

$$L = \frac{1}{\pi} E_h\, q(\gamma,\beta)^* \tag{11.8}$$

where $q(\gamma,\beta)$ is defined as a directional reflection coefficient and is identical in form to the luminance factor in Sec. 8.6, except that the latter was for viewing at 25° from the vertical. It should be noted that $\beta = 90 + \phi$.

*The π factor is omitted if E_h is in footcandles and L is in footlamberts.

Combining Eqs. (11.7) and (11.8) yields

$$L = \frac{1}{\pi}\left(\frac{q(\gamma,\beta)\ I(\gamma,\phi)\cos^3\gamma}{H^2}\right)$$

$$\equiv \frac{1}{\pi}\left(\frac{r(\gamma,\beta)\ I(\gamma,\phi)}{10{,}000H^2}\right) \tag{11.9}$$

The quantity

$$r(\gamma,\beta) = 10{,}000\ q(\gamma,\beta)\cos^3\gamma \tag{11.10}$$

is called a reduced luminance coefficient. Values of r have been measured as a function of γ and β for the four roadway surfaces previously identified (IES Roadway Lighting Committee, 1983). They are given for surface R3, which is the surface that is assumed if the pavement classification is unknown, in Table 11.9. The 10,000 in Eqs. (11.9) and (11.10) is simply to avoid small decimals in the presentation of the r-value data in Table 11.9.

The luminance in Eq. (11.9) is for a single luminaire. To obtain the overall luminance, a summation is performed.

$$L_p = \sum_{1}^{n} \frac{r(\gamma,\beta)\ I(\gamma,\phi)}{10{,}000\pi H^2} \tag{11.11}$$

Equation (11.11) gives the initial pavement luminance at point P. It should be depreciated by an appropriate light loss factor for the lamp and luminaire in question.

Assume now that the designer had determined the roadway and pavement classifications and has assessed the environment through which the roadway will pass. The next step is to consult Table 11.10. In that table, recommended luminance values and ratios are given for the various road and area classifications, where, as in Table 11.3, C, I, and R mean commerical intermediate, and residential, respectively. In the first data column, the desired average maintained luminance of the roadway surface is shown. This is the average of the luminances calculated or measured at the points shown in Fig. 11.22. The second and third data columns are design guidelines so that reasonable uniformity will be achieved. The fourth column is a measure of disability glare. L_v is the veiling luminance at the observer's eyes in candelas per square meter and may be obtained from the IES Roadway Lighting Committee (1983).

$$L_v = \frac{10E_v}{\theta^2 + 1.5\theta} \tag{11.12}$$

where E_v is the vertical illuminance in the plane of the driver's pupil in lux and θ is the angle in degrees between the line of sight and the line from the observer to the luminaire. The line of sight is a horizontal line 1.45 m above the roadway, parallel to the curb, and along one of the dashed lines in Fig. 11.22.

The luminance in Eq. (11.12) is for one luminaire. A summation over

TABLE 11.9 r-VALUES FOR SURFACE R3 (All Values Multiplied by 10^4)

tan γ \ $\beta°$	0	2	5	10	15	20	25	30	35	40	45	60	75	90	120	150	180
0	294	294	294	294	294	294	294	294	294	294	294	294	294	294	294	294	294
0.25	326	326	321	321	317	312	308	308	303	298	294	280	271	262	253	244	240
0.5	344	344	339	339	326	317	308	298	289	276	262	235	217	204	199	199	194
0.75	357	353	353	339	321	303	285	267	244	222	204	176	158	149	149	136	140
1	362	362	352	326	276	249	226	204	181	158	140	118	104	100	100	100	100
1.25	357	357	348	298	244	208	176	154	136	118	104	83	73	70	74	77	78
1.5	353	348	326	267	217	176	145	117	100	86	78	72	60	57	60	60	62
1.75	359	335	303	231	172	127	104	89	79	70	62	51	45	44	46	45	47
2	326	321	280	190	136	100	82	71	62	54	48	39	34	34	35	36	38
2.5	289	280	222	127	86	65	54	44	38	34	25	23	22	23	24	24	25
3	253	235	163	85	53	38	31	25	23	20	18	15	15	14	15	16	17
3.5	217	194	122	60	35	25	22	19	16	15	13	9.9	9.0	9.0	11	12	13
4	190	163	90	43	26	20	16	14	12	9.9	9.0	7.4	7.0	7.1	8.3	9.0	9.9
5	145	109	60	24	16	12	9.0	8.2	7.7	6.8	6.1	4.3	3.2	3.3	4.3	6.5	7.1
6	113	77	36	15	11	9.0	8.0	6.5	5.1								
8	83	47	17	6.1	4.4	3.6	3.1										
10	65	32	9.0	3.3	2.4	2.0											
12	53	22	5.6	2.1	1.8												

Source: IES Roadway Lighting Committee, "Proposed American Standard Practice for Roadway Lighting." Reprinted from the April 1983 issue of the *Journal of IES* with permission of the Illuminating Engineering Society of North America.

TABLE 11.10 MAINTAINED LUMINANCE CRITERIA FOR ROADWAYS

Roadway classification	Area classification	cd/m²	$\frac{L_{avg}}{L_{min}}$	$\frac{L_{max}}{L_{min}}$	$\frac{L_v}{L_{avg}}$
Freeway	C	0.6	3.5	6	0.3
	I	0.4	3.5	6	0.3
	R	—	—	—	—
Expressway	C	1.0	3	5	0.3
	I	0.8	3	5	0.3
	R	0.6	3.5	6	0.3
Major	C	1.2	3	5	0.3
	I	0.9	3	5	0.3
	R	0.6	3.5	6	0.3
Collector	C	0.8	3	5	0.4
	I	0.6	3.5	6	0.4
	R	0.4	4	8	0.4
Local	C	0.6	6	10	0.4
	I	0.5	6	10	0.4
	R	0.3	6	10	0.4

Source: IES Roadway Lighting Committee, "Proposed American Standard Practice for Roadway Lighting." Reprinted from the April 1983 issue of the *Journal of IES* with permission of the Illuminating Engineering Society of North America.

the n luminaires must be performed to obtain the overall veiling luminance. Then the ratio of L_v to L_{avg} can be calculated and compared with the guidelines in Table 11.10.

The next step in the design process is to obtain the recommended average maintained illuminance from Table 11.3. Then tentative selections of a lamp, a luminaire, and a mounting height are made. From the luminaire utilization curve, the coefficient of utilization is determined, and a light loss factor is calculated from the mean lumen data of the lamp and from Table 11.4.

With these quantities in hand, it is possible to calculate the pole spacing from Eq. (11.6). Then a tentative layout is drawn, tying down such factors as overhang, setback, stagger, etc. For each layout selected, horizontal illuminances should be calculated and compared with Table 11.3 and luminances should be calculated and compared with Table 11.10. The design should be iterated if the values or their ratios do not fulfill the criteria in Tables 11.3 and 11.10.

11.9 OUTDOOR SPORTS LIGHTING

Outdoor sports lighting is one of the most fascinating areas of lighting design, largely because, as was noted earlier, of the diverse constituencies that must be served. Essentially, there are three:

1 The Participants The lighting must be sufficient for the players and officials to perform their visual tasks at an appropriate level of skill and accuracy. Four levels of competition are defined, in order of decreasing illuminance requirements.

Professional

College, tournament

High school, club

Grade school, recreational

In sports, the visual task often involves seeing a three-dimensional moving object. There is often no "work plane," especially for aerial games. Thus not only horizontal, but also vertical, illuminance is of importance.

2 The Spectators Lighting requirements must also take into account the needs of spectators, particularly those spectators sitting in the farthest row of seats from the playing area. The lighting levels needed may be higher than those required for the players, and, as with the players, vertical, as well as horizontal, illuminance is necessary.

A classification of stadia for lighting purposes is given in Table 11.11.

3 Television Requirements Television coverage of nighttime sports activities has increased dramatically in the last 20 years, particularly since the development of the metal halide lamp, which has made widespread color television coverage feasible. The needs of television are much like those of spectators—higher levels of both horizontal and vertical illuminance. The IESNA minimum illuminance recommendations for color television range from 150 to 300 fc (1650 to 3300 lux), depending on the type of play and the intensity of action on given areas of the field. The lower sensitivity of the color TV camera, and especially of the zoom lens for close-ups, requires 1.5 to 3 times the illuminance needed by the participants' eyes.

Recommended illuminance levels for several outdoor sports are shown in Table 11.12. These are minimum recommended maintained values on a horizontal plane 3 ft (0.9 m) above the ground. Although it is recognized that

TABLE 11.11 CLASSES OF STADIA

Class	Distance from nearest sideline to farthest spectator row, ft	Spectator seating capacity
I	Over 100	Over 30,000
II	50–100	10,000–30,000
III	30–50	5000–10,000
IV	Less than 30	Under 5000
V	No fixed seating facilities	

TABLE 11.12 RECOMMENDED ILLUMINANCE LEVELS IN fc (lux)

Sport	Level	Area	Illuminance
Baseball	Professional	Infield	150 (1650)
		Outfield	100 (1100)
	High school	Infield	50 (550)
		Outfield	30 (330)
	Recreational	Infield	15 (165)
		Outfield	10 (110)
Croquet	Tournament		10 (110)
	Recreational		5 (55)
Golf		Tee	5 (55)
		Fairway	3 (33)
		Green	5 (55)
Racing	Auto		20 (220)
	Bicycle		30 (330)
	Dog		30 (330)
	Horse		20 (220)
Football and soccer	Class I		100 (1100)
	Class II		50 (550)
	Class III		30 (330)
	Class IV		20 (220)
	Class V		10 (110)
Swimming	Exhibitions		20 (220)
	Recreational		10 (110)
Tennis	Tournament		30 (330)
	Club		20 (220)
	Recreational		10 (110)
Volley ball	Tournament		20 (220)
	Recreational		10 (110)

Source: Sylvania Outdoor Lighting, "Sports Lighting Made Easy," Bulletin SP-2, GTE Sylvania, Fall River, Mass., with permission.

vertical illuminance in sports lighting is often as significant, if not more so, than horizontal illuminance, it has been found that the vertical illuminances will be adequate if the horizontal illuminances in Table 11.12 are met and if the lighting equipment is located and aimed appropriately.

In outdoor sports lighting, it is important to have reasonable uniformity of illuminance on the playing area and throughout the entire space above the playing area. Sharp changes in illuminance from one part of the playing area to another through which a ball is traveling can make it seem as though the ball is accelerating or decelerating. This occurs when there is inadequate overlapping of floodlight beams. Acceptable uniformity requires a ratio of maximum to minimum horizontal illuminance of no more than 3:1.

Also, light must be provided from enough directions to a given point to avoid harsh shadows. Often semidirectional vertical lighting is used to provide shading and modeling. It is becoming common, in stadium lighting, to have more luminaires on one side of the field (the TV camera side) than on the other side, by as much as 2:1. For unidirectional sports, such as driving golf balls and archery, it is desirable to provide much higher illuminance levels in one direction. This makes it possible to locate luminaires so that they are almost completely shielded from the field of view.

Glare control is very important in outdoor sports. Not only will participants be adversely affected if glare is present, especially in aerial games, but the spectators will also suffer. In addition, glare can be annoying to inhabitants of buildings nearby.

Several steps can be taken to reduce the effects of glare produced by floodlights:

1. Lamps of small source size should be used to permit accurate optical control. For example, color-corrected mercury lamps should not be used, because, for those lamps, the source is the outer, phosphor-coated bulb.
2. The NEMA beam spreads given in Table 11.1 should be consulted. As shown in Fig. 11.23*a*, if beam spread is too great, there is wasted light and increased glare in spectators' eyes. As the distance of the luminaire from the surface to be lighted increases, beam spread must be decreased if the same playing surface area is to be lighted.
3. For floodlighting systems used in outdoor sports, the angle between the horizontal playing surface and a line drawn through the lowest mounted floodlight and a point one-third the distance across the playing field should not be less than 30°. In equation form,

$$\mathrm{MH} = 0.577\left(\mathrm{SB} + \frac{W}{3}\right) \tag{11.13}$$

where SB is the setback and W is the field width.

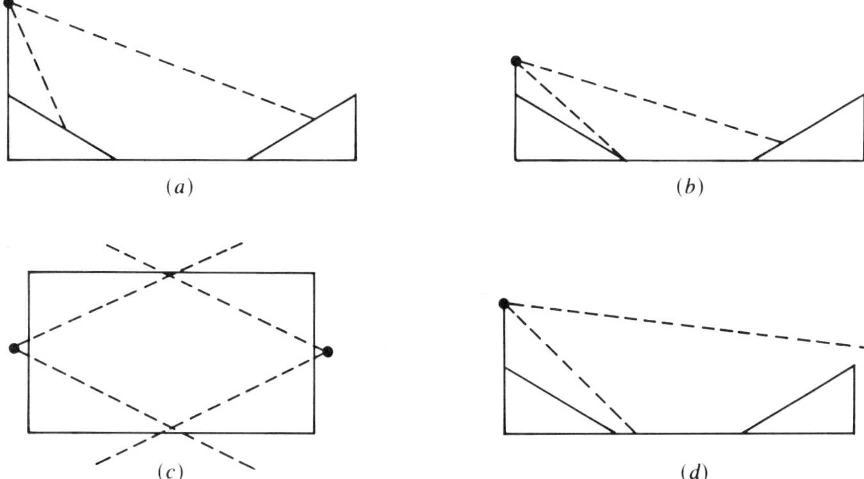

Figure 11.23 Factors affecting floodlight glare. (*a*) Beam spread too great. (*b*) Mounting height too low. (*c*) Improper pole location. (*d*) Incorrect aiming.

In addition to Eq. (11.13), the minimum mounting height should not be less than 20 ft (6.1 m) for ground sports and 30 ft (9.1 m) for aerial sports. If the floodlight mounting height is too high, there is loss in beam utilization and nonuniformity of lighting. If it is too low, there is wasted light, glare in spectators' eyes and to areas adjacent to the field, and air balls become invisible. This is shown in Fig. 11.23*b*.

4. The effects of glare are reduced if luminaires are located away from the normal lines of sight of players, spectators, and TV cameras. An example of improper location is shown in Fig. 11.23*c*.
5. As shown in Fig. 11.23*d*, incorrect aiming of floodlights can produce glare and wasted light. Also, as was pointed out in Sec. 11.2, the aiming distance should not exceed 2 MHs horizontally if reasonable uniformity is to be achieved.
6. Increasing the luminance of surrounding areas reduces the glare effect. For outdoor sports, adequate light in the stands, on fences, and on the ground around the playing field will improve the situation.

Sports played at night before thousands of spectators in large stadia have become a way of American life in recent years. The basic requirements for a light source for stadium lighting are that it permit good optical control, provide adequate color rendition, and operate economically. Prior to 1965, the only light source satisfying these three requirements well was incandescent, and in particular, the 1000-W PS52 incandescent lamp was very widely used in outdoor sports lighting.

An experiment in combined high-intensity discharge and incandescent lighting was tried in only one major U.S. stadium prior to 1965. This was Shea Stadium in New York, which was lighted with 953 clear 1000-W mercury lamps and 612 1500-W tungsten-halogen lamps in the early 1960s. Color rendering was satisfactory, but not good.

The advent and growth of color television sports broadcasting from 1965 onward could not have occurred without two major developments. One was the introduction of the metal-halide lamp, particularly the 1000-W size, with its high efficacy, small source size, and good color-rendering properties. It made available 90,000 lm per lamp, 50% more than its 1000-W mercury counterpart.

The second major development in the mid-1960s was that computer technology reached a point where it could be used to simplify the design of complex lighting systems. Computer technology provides the only economical (in money and time) method of precisely calculating aiming angles and light distributions.

The first major U.S. stadium to be relighted with 1000-W metal-halide lamps was the Municipal Stadium in Cleveland, Ohio. 1318 incandescent units were replaced by 920 1000-W metal-halide units. The kilowatt load was decreased from 2306 to 1790 and the lumen output increased from 58 to 100 million.

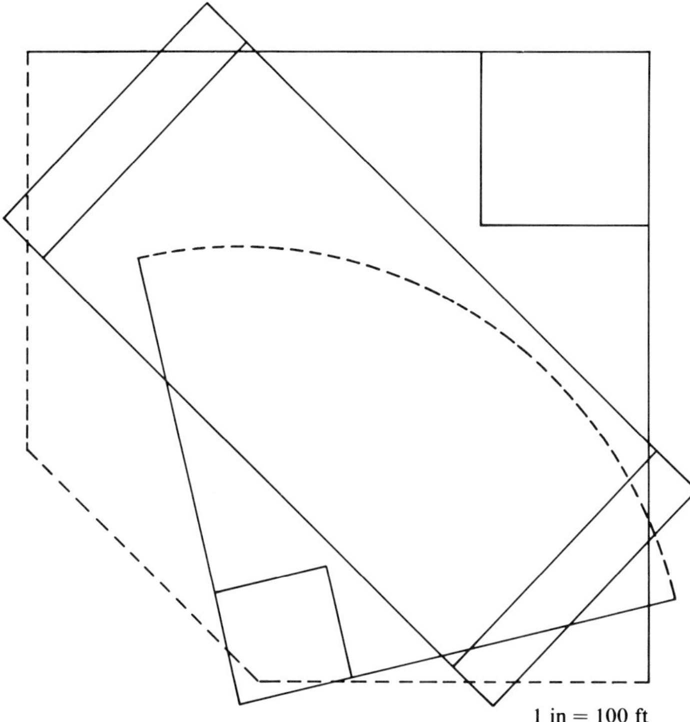

Figure 11.24 Neighborhood playfield area.

It had long been customary, when using incandescent lamps for sports lighting, to operate them at 10% overvoltage. This gave 30% more lumens for 16% more watts but reduced life to 20% of rated. This life amounted to about one playing season, after which the entire installation was relamped. The economics of this procedure were very favorable.

The same idea was used with mercury lamps. One thousand watt lamps were operated on 1500-W ballasts, resulting in an increase of 50% in light output accompanied by a severe reduction in lamp life. The same idea was tried in the early 1000-W metal-halide installations. However lamp life was too radically reduced. The lamp was modified, but also a short-life 1500-W metal-halide lamp was introduced, and it has been the standard for outdoor stadium lighting since 1970. It gives 150,000 initial lumens and has a life of 3000 h. The first U.S. stadium to be relighted with the new lamp was Baltimore's Memorial Stadium, used for baseball. The 512 1500-W metal-halide units used provided 500 fc (5500 lux) on the infield and 300 fc (3300 lux) on the outfield.

From the huge stadium, we turn to the neighborhood playfield shown in Fig. 11.24.* This is a multiuse area, typical of what is found in many small

*This example is adapted from a lighting design done by Donald L. Umbricht, Outdoor Lighting Specialist, GTE Lighting Products, Fall River, Mass.

communities. It contains a soccer-football field 160 × 360 ft, a little league softball field with 60-ft basepaths, and a Babe Ruth baseball field with 90-ft basepaths. It is desired to light the area so that all of these sports may be played at night (not simultaneously!).

The first step in the design is to select appropriate illuminance levels. From Table 11.12, the caliber of activity is somewhere between recreational and high school for baseball and is Class V for soccer-football. The levels chosen are 20 fc for the softball and baseball infields, 15 fc for the softball and baseball outfields, and 15 fc for the entire soccer field.

The next step is to examine standard layouts which have been developed for softball, baseball, and soccer fields (IES Sports and Recreational Areas Committee, 1969; Sylvania Outdoor Lighting, 1977). Two of these are reproduced in Fig. 11.25.

In the baseball layouts, eight poles are used, as indicated in Fig. 11.25*a*. For the 60-ft softball field, $V = 185$ ft, $W = 20$ to 30 ft, $X = 30$ to 50 ft, $Y = 5$ to 15 ft, and $Z = 90$ to 110 ft. For the 90-ft baseball field, these numbers are $V = 400$ ft, $W = 30$ to 60 ft, $X = 40$ to 80 ft, $Y = 20$ to 30 ft, and $Z = 130$ to 180 ft.

For the soccer layout, the number of poles and their spacing and setback depend on the class of play. A 10-pole layout is suggested for Classes IV and V, an 8-pole layout for Class III, either a 6- or 8-pole layout for Class II, and a 6-pole layout for Class I, with spacings and setbacks as shown in Fig. 11.25*b*.

Because the area being considered is multiuse, some compromises will have to be made. Even though a 10-pole layout is indicated for soccer, poles cannot be mounted close to the soccer field because they would interfere with baseball and softball. Thus an 8-pole layout is chosen, representing a Class II or III layout for soccer. The poles are located as shown in Fig. 11.26.

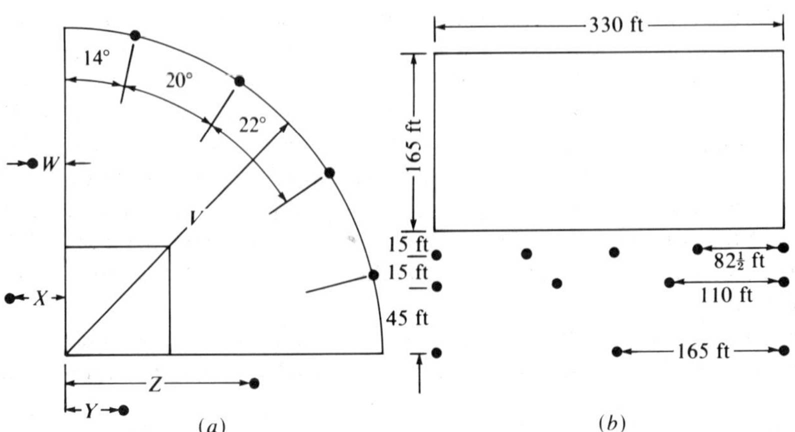

Figure 11.25 Standard baseball and soccer lighting layouts.

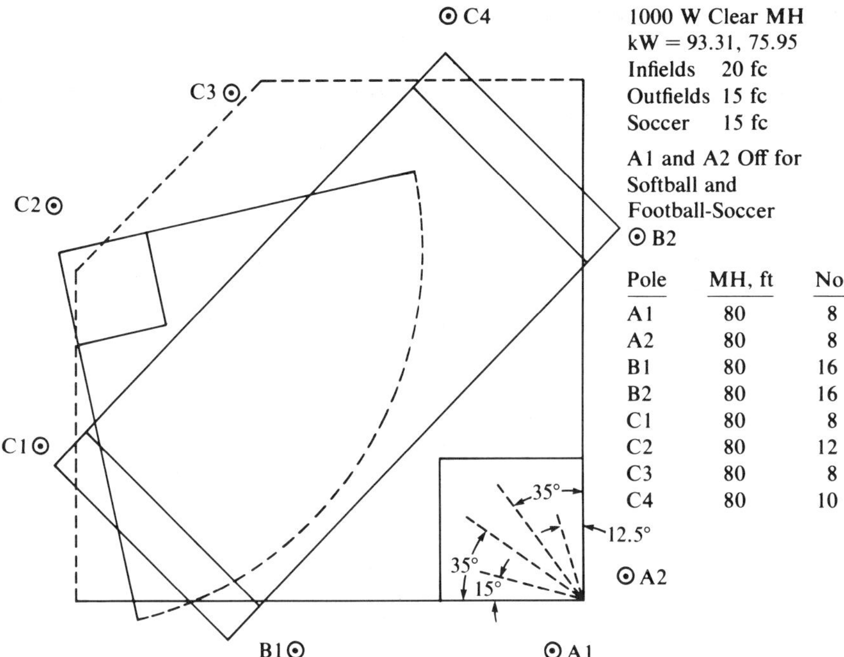

Figure 11.26 Lighting layout for playfield.

The choice of mounting height is next. For rectangular areas such as the soccer field, the guideline given in Eq. (11.12) should be followed, giving

$$\text{MH} = 0.577\ (90 + \tfrac{160}{3}) = 82.7\ \text{ft}$$

The setback used in this calculation is an average setback for poles C2 and C3 in Fig. 11.26.

For the softball and baseball fields, minimum mounting heights are given with the standard layouts (Sylvania Outdoor Lighting, 1977). The recommended minimum values are 40 to 50 ft for softball and 70 ft for baseball. Because the poles for softball are set back further than normally and because only one pole is behind the plate, it was decided to make all poles 80 ft high.

The next task is to choose a lamp and luminaire and to determine the number of luminaires needed per pole. A 1000-W metal-halide lamp was chosen, which, from Table 6.11, provides 115,000 initial lumens, 92,000 mean lumens, and 10,000 h life. Together with its ballast, it consumes 1.085 kW.

A general-purpose floodlight was chosen. The number of floodlights, their NEMA types, and the resulting kilowatt load are shown in Table 11.13. The total number of luminaires is 86 and they consume 93.31 kW, which is reduced to 75.95 kW when soccer or softball are played by turning off the units on poles A1 and A2.

TABLE 11.13 FLOODLIGHT SUMMARY FOR FIG. 11.26

Pole	Number of units	NEMA type	Kilowatts
A1	5	4	5.425
	3	6	3.255
A2	5	4	5.425
	3	6	3.255
B1	12	4	13.02
	4	6	4.34
B2	12	4	13.02
	4	6	4.34
C1	5	4	5.425
	3	6	3.255
C2	6	4	6.51
	6	6	6.51
C3	5	4	5.425
	3	6	3.255
C4	7	4	7.595
	3	6	3.255
	86		93.31

REFERENCES

Frier, J. P., and M. E. G. Frier: *Industrial Lighting Systems*, McGraw-Hill, New York, 1980.

General Electric Company: "Lighting System Product Application Guide," Hendersonville, N.C., 1980.

———: "Building Floodlighting," Bulletin TP-115, Nelà Park, Cleveland, Ohio, 1970.

Helms, R. N.: *Illumination Engineering for Energy Efficient Luminous Environments*, Prentice-Hall, Englewood Cliffs, N.J., 1980.

IES Roadway Lighting Committee: "Proposed American National Standard Practice for Roadway Lighting," *Journal of IES*, April 1983.

IES Sports and Recreational Areas Committee: *Sports Lighting*, Illuminating Engineering Society, New York, 1969.

Kaufman, J. E. (ed.): *IES Lighting Handbook*, Reference and Application vols., 1981 ed., Illuminating Engineering Society, New York, 1981.

Loch, C. H.: "Driver's View of the Roadway," *Journal of IES*, January 1982.

Lowell, J. M., and L. J. Maloney: "Illuminance Selection Procedure for Sports Lighting," *Lighting Design and Application*, May 1983.

Sylvania Lighting Center: "Outdoor Lighting," GTE Sylvania, Danvers, Mass. 1977.

Sylvania Outdoor Lighting: "Sports Lighting Made Easy," GTE Sylvania, Fall River, Mass., 1977.

Problems

CHAPTER 1

1.1 An electric field intensity is given by

$$\mathscr{E} = 100 \sin 2\pi\left(5 \times 10^5 t - \frac{s}{600}\right) \qquad \text{V/m} \qquad (s \text{ in meters, } t \text{ in seconds})$$

(a) What are the period and wavelength?
(b) Sketch $\mathscr{E}$ versus s for a fixed t and $\mathscr{E}$ versus t for a fixed s.

1.2 An electron is at rest at point a. It accelerates toward point c. Find the following:
(a) Work done, KE gained, and PE lost as the electron moves from a to b; from a to c
(b) Force on the electron and acceleration of the electron
(c) Velocity of the electron at points b and c and the time to reach each point

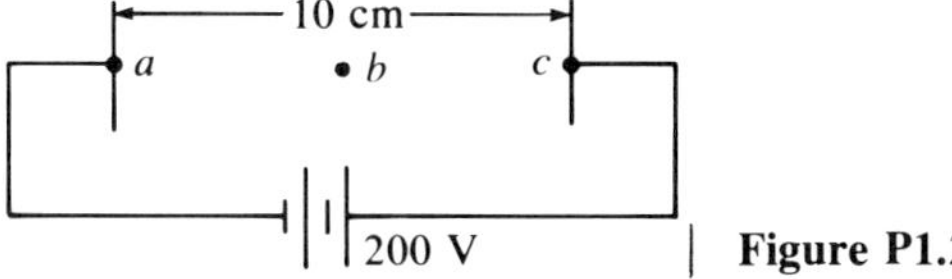

Figure P1.2

1.3 **(a)** Given that 1 J = 9.48×10^{-4} Btu, how many kilowatts per square meter does 1 Btu per hour per square foot represent?
(b) A sodium lamp emits a single spectral line at 0.59×10^{-6} m. Express this wavelength in microns, nanometers, and Angstroms.

1.4 An electron with a KE of 2.0×10^{-18} J collides with a hydrogen atom. What is the highest energy level to which the electron in the hydrogen atom can be raised? What wavelength radiation will be emitted when the electron reverts to its normal state from this level?

1.5 A perfectly diffusing spherical glass globe is 14 in in diameter and has a luminance of 3 cd/in^2 and a transmittance of 60%. What is the intensity (assumed uniform) of the incandescent lamp at its center?

1.6 An incandescent lamp bulb, assumed spherical, has a uniform intensity of 130 cd and a diameter of $2\frac{5}{8}$ in. Find the luminance and luminous exitance of the bulb.

1.7 An incandescent lamp whose intensity is 1000 cd is placed inside a perfectly diffusing globe 0.3 m in diamter. The luminance of the globe is measured as 10^4 cd/m^2. Find the globe's transmittance and the illuminance in lux on the inside of the globe.

CHAPTER 2

2.1 A long straight street is lighted by incandescent lamps on poles 20 ft high and spaced 50 ft apart. Each lamp has an intensity of 2000 cd, uniform in every direction. Find the illuminance on the street midway between any two poles. Find the illuminance on a small vertical sign 5 ft off the ground at the same location.

2.2 A 150-W floodlamp has the following data:

Degrees from lamp axis	Intensity in candelas
0	3500
5	3210
10	2820
15	2040
20	1360
30	290
40	240
50	190
60	100

One of these lamps is mounted on a pole 20 ft high and aimed at a point on the ground 6 ft from the pole. Find the illuminance at a second point on the ground 8 ft from the first point and 10 ft from the pole.

2.3

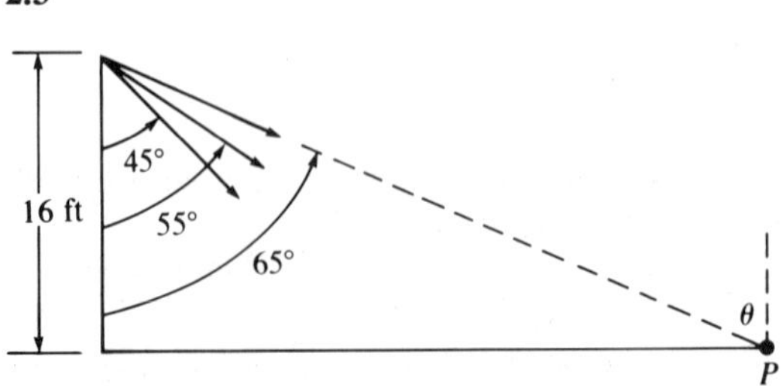

Figure P2.3

Angle, degrees	Intensity, cd
0	16,000
10	14,000
20	10,000
30	2,000
40	500
50	0

A shuffleboard runway is to be lighted by three PAR lamps on a pole aimed as shown in Fig. P2.3. Derive a general expression for the illuminance at point P in terms of θ and the intensities. Find the illuminance at point P when $\theta = 65°$.

2.4 In a basement recreation room with an 8-ft ceiling, a remodeling contractor has specified 150-W recessed incandescent units to light the ping-pong table. The table is 9 × 5 ft and is 3 ft high. Three 120-V, 150-W lighting units are to be used—one centered over each end of the table and one over the middle of the table. Two types of lenses are available for the lighting units—a prismatic clear glass lens and an opal diffusing lens. Find the illuminances on the table center line at the points shown. Which lens would you choose?

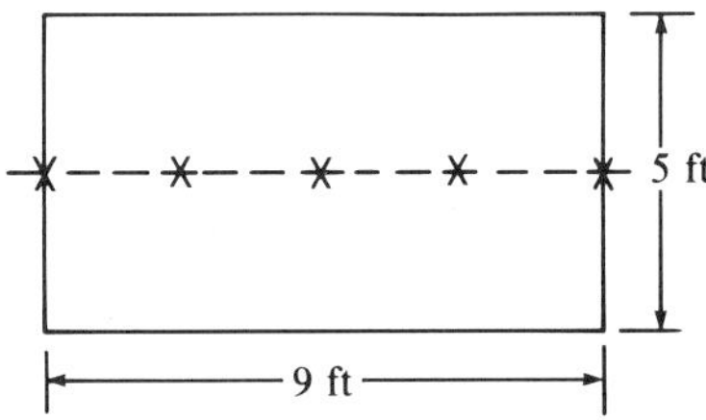

Figure P2.4

Prismatic lens		Opal lens	
Polar angle, degrees	Candelas	Polar Angle, degrees	Candelas
85	7.0	85	44.7
75	41.2	75	86.5
65	67.5	65	135
55	112	55	179
45	248	45	222
35	457	35	259
25	635	25	289
15	755	15	309
5	805	5	320
0	810	0	321

2.5 The luminance of a perfectly diffusing translucent square luminous wall panel 10 ft² in area is 0.5 cd/in². The panel has a transmission factor of 45%. Find

(a) The apparent intensity of the panel when viewed by an observer at point A (assumed far away from the panel)

(b) The illuminance on the panel at B

(c) The illuminance on the plane at point C

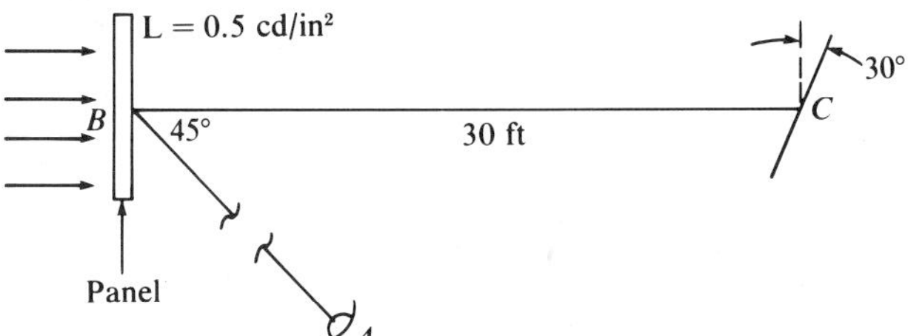

Figure P2.5

2.6 The candlepower distribution curve of a hypothetical batwing lamp-reflector unit is shown. How many lumens does the unit emit?

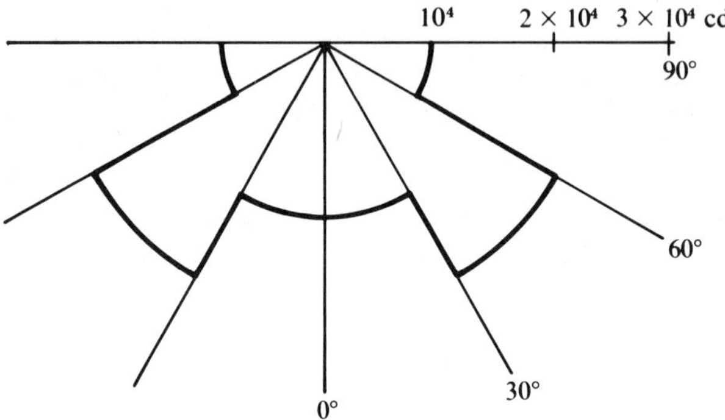

Figure P2.6

2.7 A company has installed a large array of 400-W high-pressure sodium lamp and reflector units in its greenhouse to provide electric light for rose growing. The installation consists of 225 lamps and reflectors mounted in a grid arrangement, as shown, an average of 5 ft above the buds of the plants. Each lamp has an initial average output of 47,000 lm and a percent mean lumens with continuous burning of 90%. The candlepower distribution data for the lamp-reflector combination is given below.

Figure P2.7

Polar angle, degrees	Initial I, cd	Polar angle, degrees	Initial I, cd
0	4500	50	9000
5	4650	55	8980
10	4800	60	8400
15	5200	65	6900
20	5700	70	4900
25	6300	75	2700
30	6950	80	1400
35	7530	85	750
40	8100	90	400
45	8650		

(a) What is the horizontal illuminance on a rose bud directly beneath a lamp in the center of the array?

(b) What is the efficiency of the reflector in delivering lamp lumens to the plants?

(c) At what distance, horizontally, from directly under a lamp-reflector unit is the horizontal illuminance on a bud due to that single unit down to 10% of its maximum value?

2.8 An incandescent lamp of 1000 cd uniform intensity is mounted in a small reflector whose reflectance is 80%. All of the lumens in the 90 to 180° zone are intercepted by the reflector. Of those lumens reflected, 60% enter the 0 to 20° zone. What percent of the total lamp lumens end up in the 0 to 20° zone? What is the apparent candlepower of the lamp-reflector combination in the 0 to 20° zone? Assume the viewer is at least 10 times the reflector diameter away from the source.

2.9 A surface with an area of 5 in^2 has an intensity distribution curve between the limits of $-\pi/3$ and $\pi/3$ rad of $I = 1000 \cos\theta + 500 \cos 3\theta$ cd, where θ is the angle with respect to the normal to the surface. Sketch, in polar form between the given limits, the curve of luminance versus θ for the surface.

2.10 A circular fluorescent lamp lights a point P in a horizontal plane 4 ft directly below the lamp's center. The lamp is a 1-in-diameter tube and has a luminance of 4.4 cd/in^2. Find the approximate illuminance at point P and the apparent intensity of the lamp in the direction of P. In working the problem, assume the inverse-square law holds and thus that the lamp is located on a circular locus $11\frac{1}{2}$ in from the lamp's center.

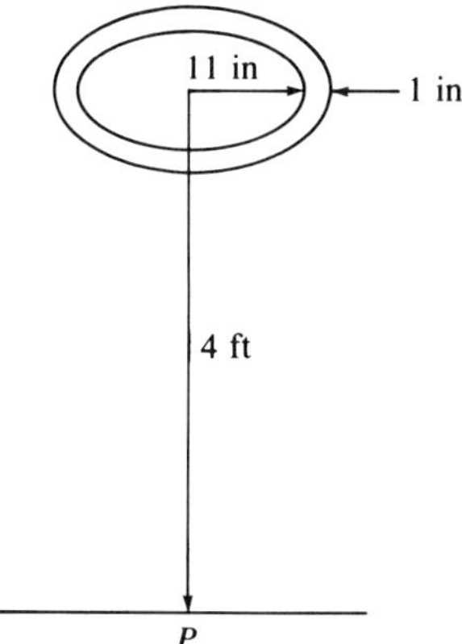

Figure P2.10

2.11 A standard cool white fluorescent lamp has the following data: 40 W, 48 in long, $1\frac{1}{2}$ in in diameter, $L = 4.2$ cd/in^2. The lamp is mounted 5 ft above a desk. What is the approximate illuminance at a point on the desk directly under the lamp's center? In working the problem, break the lamp into four 1-ft-long pieces, assume each is a point source, find the approximate illuminance produced by each using the inverse-square law, and sum the results.

Repeat the problem with two 2-ft pieces and with one 4-ft piece. Compare results.

2.12 The diagram shows a proposed bowling alley lighting installation (bowler at left, pins at right). The circles represent 4-ft, $1\frac{1}{2}$-in-diameter, 40-W fluorescent lamps, centered over the alley. Each lamp has a luminance of 4.2 cd/in^2. Neglecting ceiling and floor reflectances, find the horizontal illuminance at point P.

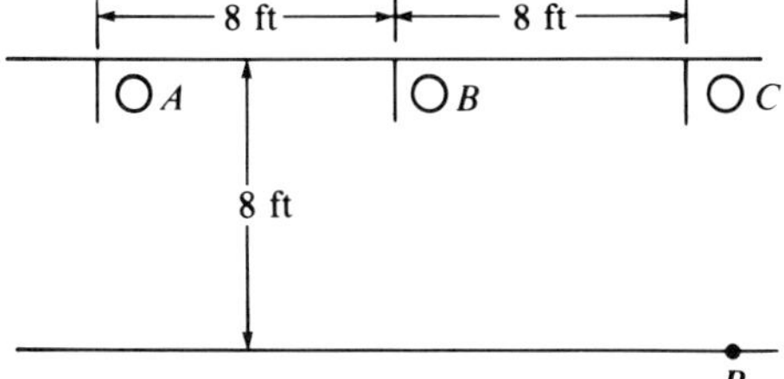

Figure P2.12

2.13 A room has three rows of fluorescent lamps on its ceilings. The rows are each 32 ft long and are separated by 8 ft. Find the illuminance on a desk top 6 ft below the ceiling and located
(a) Beneath the middle of the middle row
(b) Beneath one end of the middle row
(c) Beneath the middle of one outside row
Assume that the lamps are $1\frac{1}{2}$ in in diameter with a luminance of 4 cd/in^2.

2.14 A top view of two rectangular luminaires is shown. They are perfectly diffusing with $L = 630/\pi$ cd/ft^2. Find the horizontal illuminance at P, 10 ft below the luminaires.

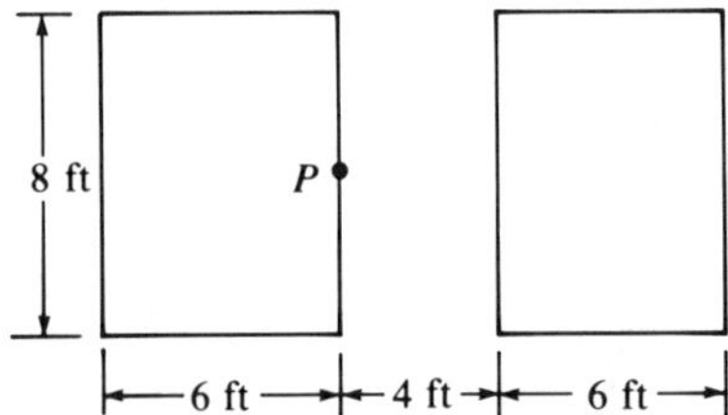

Figure P2.14

2.15 A room 40 × 20 ft with a 10-ft ceiling height has a wall exitance of 30 lm/ft^2 and a ceiling exitance of 40 lm/ft^2. All surfaces are perfectly diffusing. Find the horizontal illuminance at the center of the floor.

CHAPTER 3

3.1 The spectral power density curve of an incandescent radiator in milliwatts per square centimeter per 10 nanometers is shown. Find the following:
(a) Total milliwatts per square centimeter in the 500 to 510 nm range
(b) Total milliwatts per square centimeter in the 600 to 640 nm range
(c) Total milliwatts per square centimeter in the 400 to 700 nm range

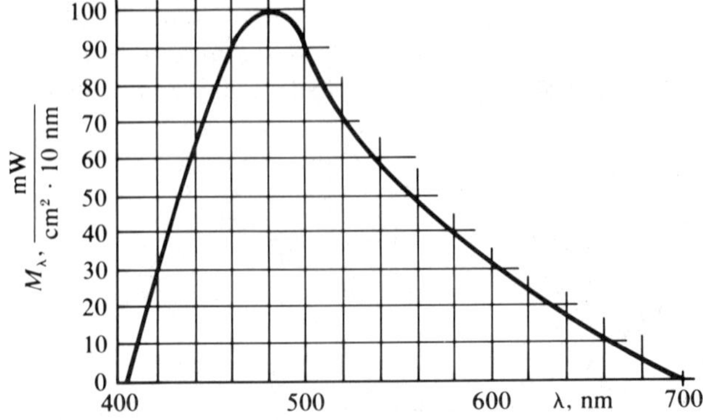

Figure P3.1

3.2 The filament of an incandescent lamp is 0.006 cm in diameter and 60 cm long. It consumes 100 W. Assuming the filament may be considered as a blackbody radiator, what is its operating temperature in kelvins? At what wavelength does the maximum value of M_λ occur? Is this filament an efficient light source? Explain.

3.3 Assume the straight-line approximation shown represents the spectral power density of a blackbody radiator. Determine the absolute value in W/(cm² · μm) of $M_{\lambda,max}$. At what temperature is the blackbody operating?

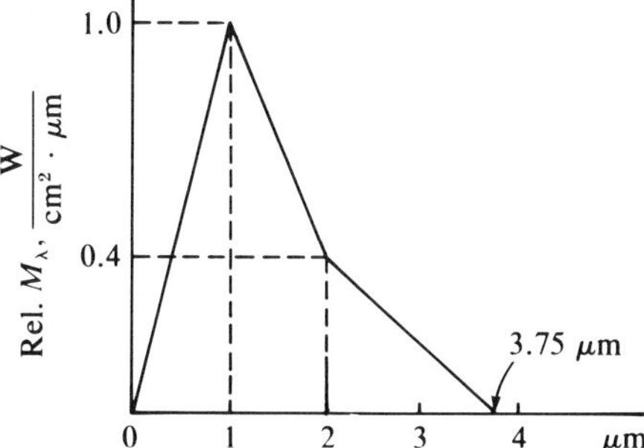

Figure P3.3

3.4 Relative values of M_λ in W/(cm² · μm) in the visible spectrum are given below for a blackbody radiator at 2800 K. What percent of the total power density is in the visible spectrum at this temperature?

λ, μm	Relative M_λ	λ, μm	Relative M_λ	λ, μm	Relative M_λ
0.38	0.075	0.51	0.531	0.64	1.311
0.39	0.093	0.52	0.585	0.65	1.373
0.40	0.113	0.53	0.640	0.66	1.433
0.41	0.137	0.54	0.697	0.67	1.492
0.42	0.163	0.55	0.756	0.68	1.550
0.43	0.193	0.56	0.816	0.69	1.608
0.44	0.225	0.57	0.876	0.70	1.663
0.45	0.261	0.58	0.938	0.71	1.718
0.46	0.299	0.59	1.000	0.72	1.770
0.47	0.340	0.60	1.062	0.73	1.822
0.48	0.384	0.61	1.125	0.74	1.871
0.49	0.431	0.62	1.188	0.75	1.919
0.50	0.480	0.63	1.250	0.76	1.965

The absolute value of M_λ at 0.59 μm is 88.39 W/(cm² · μm).

3.5 Planck's equation for a blackbody radiator may be plotted on log-log paper to facilitate calculations. Taking the log of Eq. (3.3) yields

$$\log M_\lambda = \log[C_1 \, (e^{C_2/\lambda T} - 1)^{-1}] - 5 \log \lambda$$

For values of λ and T such that λT is constant, the equation has the form $y = mx + b$. Thus each line of slope −5 on the log-log plot is a line of constant λT.

Assume two temperatures, T_1 and T_2, and two values of λT, k_1 and k_2, as shown. Show that

$$\log M_3 - \log M_1 = \log M_4 - \log M_2$$

and that

$$\log \lambda_1 - \log \lambda_3 = \log \lambda_2 - \log \lambda_4$$

This proves that we can use a single log-log curve in blackbody radiator calculations and shift the coordinate axes to accommodate various temperatures.

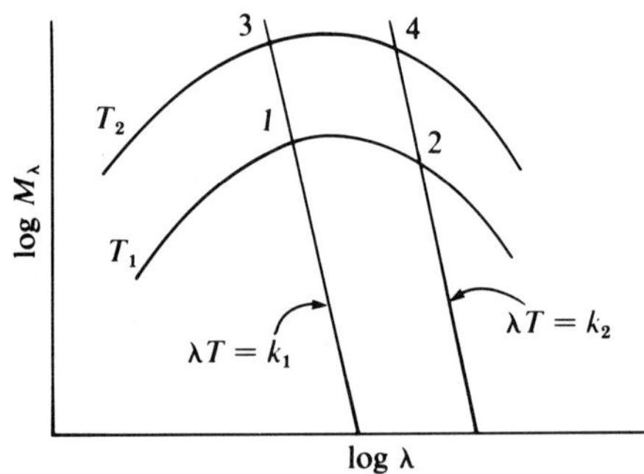

Figure P3.5

3.6 A tungsten filament has the following data: 75 W, 120 V, 1150 initial lumens, $l = 16$ in, $d = 0.0018$ in, 18.5% losses. Its effective emissivity versus temperature is shown below. Find its operating temperature (by trial and error).

T, K	$\bar{\epsilon}$
2000	0.254
2100	0.263
2200	0.286
2300	0.296
2400	0.308
2500	0.316
2600	0.328
2700	0.332
2800	0.336
2900	0.342
3000	0.348

3.7 The graphs in Fig. P3.7 are straight-line approximations of the spectral luminous efficiency curve of the eye and an incandescent lamp spectral power density distribution curve. Find the following:

(a) Watts per square centimeter emitted by the lamp

(b) Light watts per square centimeter emitted by the lamp

(c) Lumens per square centimeter emitted by the lamp

(d) Luminous efficacy of the lamp

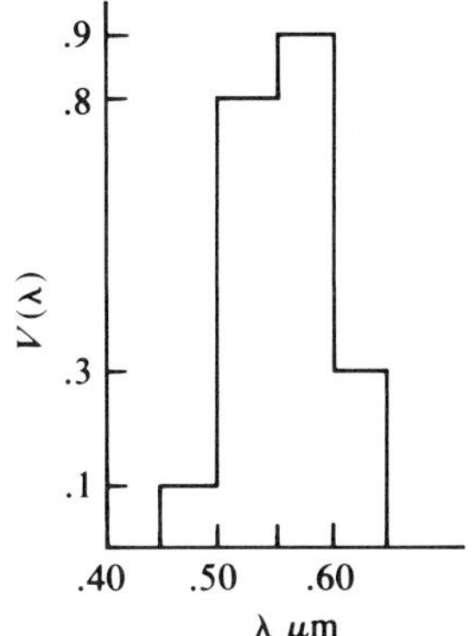

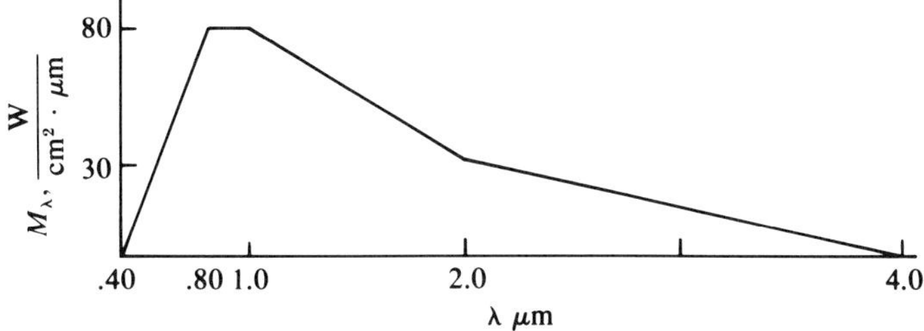

Figure P3.7

3.8 Find the luminance of the perfectly diffusing surface for which ρ_λ is the spectral reflectance, E_λ is the incident spectral radiant power density, and $V(\lambda)$ is an appoximation to the spectral luminous efficiency curve of the eye.

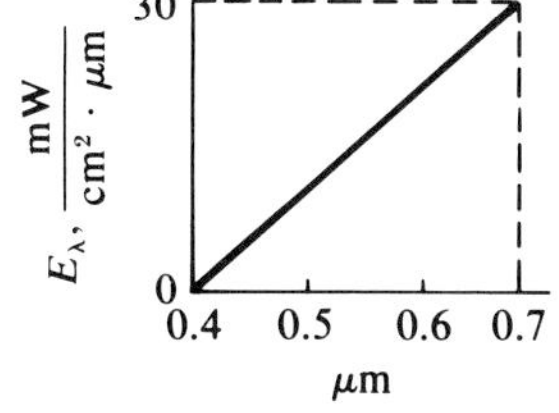

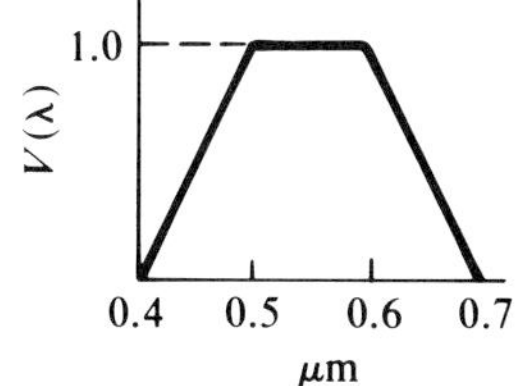

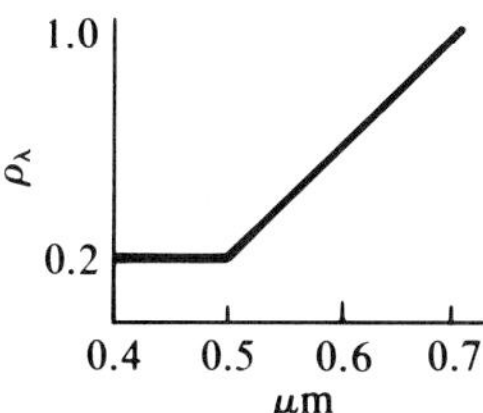

Figure P3.8

3.9 A gaseous discharge lamp emits two visible lines, one at 500 nm, the other at 600 nm. If 300 W of power goes into these lines and they give a combined output of 8900 lm, find the lumens produced by each line.

3.10

λ, nm	P_i, W
404.7	80
435.8	40
546.1	10
578.0	20

Figure P3.10

A color-corrected mercury lamp is a mercury lamp to which a phosphor has been added to the inside of the outer bulb to augment the orange and red portions of the spectrum. Assume a 400-W color-corrected mercury lamp has its four visible lines present with the given wattage values and further assume that the spectral power distribution of the phosphor may be represented as shown in Fig. P3.10. Calculate the efficacy of the lamp.

3.11 Values of solar irradiation in W/(cm^2 · μm) on a horizontal surface for a clear day with the sun at zenith are given in the table below. Find the luminous efficacy of the sun under these conditions.

λ	E_λ	λ	E_λ	λ	E_λ
0.30	0	0.65	0.132	1.00	0.065
0.35	0.042	0.70	0.125	1.20	0.045
0.40	0.080	0.75	0.108	1.40	0.032
0.45	0.132	0.80	0.096	1.60	0.024
0.50	0.147	0.85	0.085	2.00	0.012
0.55	0.139	0.90	0.075	2.20	0.007
0.60	0.139	0.95	0.068	2.40	0.006

3.12 Given are two single-line-spectrum uniform point sources at 500 nm and 650 nm with equal radiant intensities of 50 W/sr. The sources are mounted on poles 8 ft above the ground and are separated by 12 ft. At what point on the ground on a line connecting the poles will the horizontal illuminances provided by the sources be equal? What is the total horizontal illuminance at this point?

CHAPTER 4

4.1 Find the maximum speed at which electrons will be ejected from a cesium surface whose work function is 1.8 eV when that surface is irradiated by monochromatic sodium light at 589 nm. Would the maximum speed increase or decrease if the radiation were changed to a mercury line at 253.7 nm?

4.2 An electron leaves the filament of a fluorescent lamp and is accelerated to a velocity of 1.35×10^6 m/s, at which time it collides with a mercury atom, exciting the latter.

(a) What is the wavelength of the emitted photon?
(b) What is the kinetic energy of the departing electron?
(c) What is the wavelength of the electron just prior to collision?
(d) What is the mass of the emitted photon?

4.3 Argon is used in fluorescent lamps to facilitate starting. Assume an electron leaves the cathode of a 48-in fluorescent lamp with zero initial velocity. Assume a lamp voltage of 120 V dc and a positive arc column of 46 in with a gradient of 2 V/in.

(a) To what velocity must the electron be accelerated if it is to be just able to ionize an argon atom and help initiate the arc?
(b) What is the least distance from the filament at which the electron has sufficient energy to ionize an argon atom?

(c) How many 253.7-nm exciting collisions could the electron make with the mercury atom if it had the energy of part (a)?

Assume negligible voltage drop at the anode.

4.4 A photon of wavelength 140 nm is absorbed by a mercury atom in its normal state. Two photons are emitted, one at 185 nm. What is the wavelength of the other photon? What percent of the input energy produces visible radiation in this process?

4.5 A quantum of energy at 253.7 nm impinges on a phosphor and results in the emission of a quantum of energy at 579 nm. What percent of the input energy is lost in this conversion? Could a quantum of energy at 700 nm produce the 579-nm photon? Explain.

4.6 What is the angular momentum in joule-seconds of the $n = 4$, $l = 3$ energy level in an atom?

4.7 A neon atom in its first excited state collides with a mercury atom in its rest state. A 253.7-nm photon is observed. Immediately thereafter, the neon atom collides with another mercury atom in its rest state, ionizing the latter. What is the maximum velocity of the electron that is freed by the second collision?

4.8 The voltage distribution versus distance from one electrode of a 4-ft fluorescent lamp is given by

$$v = 15x \qquad 0 < x < 1$$
$$v = 1.8x + 13.2 \qquad 1 < x < 48$$

where x is the distance from the electrode in inches.

(a) Find the distance from the electrode at which an electron, starting from rest at the electrode, will have the highest probability of ionizing a mercury atom. What is the electron's velocity at that point?

(b) Assume that, due to prior collisions, an electron arrives at $x = 1$ in with zero velocity. How much further must it travel to have the highest probability of producing a 185-nm mercury photon?

(c) What is the maximum number of excitation collisions immediately yielding radiation that the electron in (b) can make from $x = 1$ in to $x = 48$ in?

4.9 A gaseous discharge lamp emits 33.5 mW and 10^{17} photons per second at a single wavelength. What is that wavelength?

4.10 Assume an electron with a mass M_1 and traveling with a velocity v_{1i} collides with an atom of mercury of mass M_2 which is standing still. If the collision is elastic, find expressions for v_{1f}, the final velocity of the electron, and v_{2f}, the final velocity of the atom, in terms of v_{1i}, assuming that $M_2 >> M_1$ and that the collision is "head on," rather than glancing.

4.11 A gas in a discharge tube has $\alpha = 100$ electrons per meter and $\beta = 2 \times 10^{-4}$ secondary electrons per positive ion. What electrode spacing is required for a self-maintained discharge?

4.12 Assume a gas atom has a diameter of 2.5×10^{-10} m. In a container of these atoms at $p = 10^{-2}$ mm of mercury and $T = 273$ K, find

(a) The density of the gas atoms

(b) The mean free path of an electron

4.13 A gas discharge tube is shown. Electrons leave the left electrode with a velocity of 0.6×10^6 m/s. Mercury atoms are located at $x = 4, 5, 12, 21$, and 25 cm. The voltage along the tube is given by

$$V = 3x \qquad 0 < x < 5 \text{ (Crooke's dark space)}$$
$$V = 0.4x + 13 \qquad 5 < x < 10 \text{ (negative glow)}$$
$$V = -0.2x + 19 \qquad 10 < x < 15 \text{ (Faraday dark space)}$$

The remainder of the tube is positive column. Collision rules are

1. If an electron has enough energy to ionize a mercury atom, ionization will occur.
2. If an electron has enough energy to excite a mercury atom, it will excite it to the highest possible level.

Describe what happens at each collision. If a photon is emitted, find its wavelength.

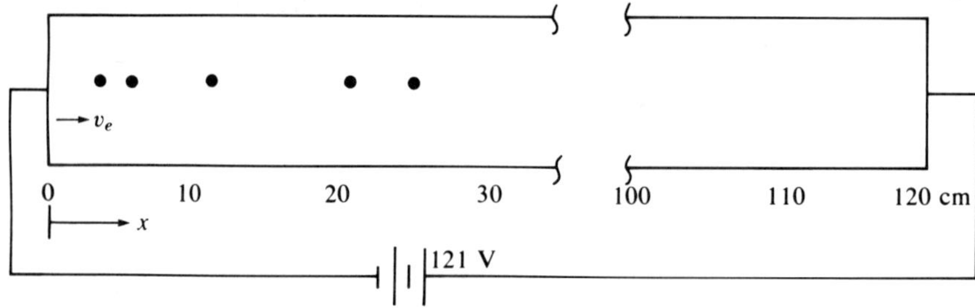

Figure P4.13

CHAPTER 5

5.1 What is the threshold size of the black object in Fig. 5.3 on a white background of luminance $L_b = 2$ fL? How large is the object if it is viewed from 14 in away?

5.2 What is the minimum perceptible contrast of a 4-min luminous disc at a background luminance of 10 fL? Compare the values obtained for MPC from Figs. 5.5 and 5.7. Explain the difference.

5.3 A white dot 0.02 in in diameter is just seen from 69 in away when its luminance is 1 fL. Find its contrast and the luminance of its background if the time for viewing is 1 s, using Fig. 5.7.

5.4 An L_b of 20 fL produces VL1 for a certain visual task in a given lighting environment. What change in contrast is required to produce VL8 in the same lighting environment if L_b is reduced to 10 fL?

5.5 A task has $\tilde{C}_0 = 0.8$ and $\rho_b = 0.75$. Find the background luminance in footlamberts and illuminance in footcandles to give a visibility of VL8 in a sphere.

The same task is now placed in a real lighting environment and its CRF is measured as 0.85. Find the background luminance and illuminance to give a visibility of VL8 in the real lighting environment.

5.6 A pencil task is viewed at an angle of 25° in a real lighting environment in which the illuminance is 100 fc. The task and background luminances at that viewing angle are $L_t = 48.6$ fL and $L_b = 57.0$ fL. Under sphere lighting, $\tilde{C}_0 = 0.1675$. Find the following:
(a) $\tilde{C}$ **(b)** CRF **(c)** New $\tilde{C}$ for VL8 **(d)** Actual VL **(e)** Footcandles in sphere for same VL

5.7 A lighting system provides 112 fc on a visual task. L_b at the 25° viewing angle is 95.2 fL. CRF has been measured as 0.804. Find the illuminance in the sphere which will provide the same visibility.

5.8 A car approaches a beige dog on a highway at dusk. Immediately past the dog, the road slopes at 5°. The vertical illuminance on the dog from the sky is 0.1 fc. The horizontal illuminance on the road from the sky is 0.2 fc. The dog has a reflectance of 50%, the road 20%, both in the direction of the car. The car's headlamps have a combined intensity of 90,000 cd in the direction of the dog. Using Fig. 5.7 and assuming the dog is a rectangle 2 ft high and 3 ft long, can it be seen by the driver?

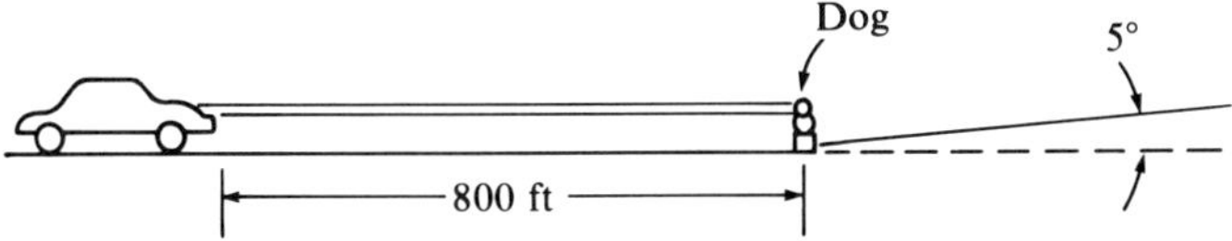

Figure P5.8

5.9 Rosie and Cyndi go to sea in a small boat whose engine dies. The rest of the class sets out in a large boat to rescue them. The small boat is 14 ft long and 3 ft above the water line. Its reflectance is 30%. On the rescue boat is a searchlight with end-on intensity of 1.4×10^6 cd. It is dusk and the sky luminance near the horizon is 0.02 fL. If the searchlight is aimed directly at the small boat when the two boats are 2500 ft apart, should the small boat be seen with 99% certainty? We must rescue Rosie and Cyndi!

5.10 A conference room is being remodeled to become an engineering office where the engineers and their clients will critically review drawings and specifications for design work. The room is finished in dark wood paneling and dark carpeting. Determine the desired illuminance on the conference room table.

5.11 A stairway in an elementary school is to be lighted. The wall surfaces are concrete block, painted white. Prescribe the illuminance level.

5.12 **(a)** The CIE spectral primaries are to be combined so as to match 6 W of pure spectral radiation at 600 nm. Find the number of watts and lumens of each primary required.
(b) Find R, G, and B for 6 W of pure spectral radiation at 600 nm. Find the luminous flux and chromaticity coordinates of this radiation.
(c) The same primaries are combined to match 6 W/nm of equal energy white. How many lumens of each primary are required? How many units?

5.13 Find the tristimulus values and the chromaticity coordinates of a line spectrum source having the following spectral data:

λ, nm	M_i, W/cm^2
490	1.70
550	3.20
680	7.60

Describe the color of the source.

5.14 It is desired to color match equal energy white using two pure spectral colors, one of which is 1 W at 480 nm. What must the other spectral color be and how many watts of it are required?

5.15 Relative spectral radiances from a gold fluorescent lamp are given in the table below. Find the tristimulus values and the chromaticity coordinates of the overall radiance.

λ, μm	Rel. L_λ	λ, μm	Rel. L_λ
0.51	1	0.61	21
0.53	4	0.63	11
0.55	12	0.65	6
0.57	19	0.67	2
0.59	24	0.69	1

In what color region of the CIE diagram does the lamp lie? What spectral color does the lamp most nearly match?

5.16 A color sample is illuminated by the source in Prob. 5.13. The spectral reflectances of the sample are

λ, nm	ρ_i
490	0.10
550	0.15
680	0.40

Find the chromaticity coordinates of the lighted sample and its dominant wavelength and purity.

5.17 Compute the chromaticity coordinates of the color that results from the combination of the following colors:

	x	y	L, cd/ft^2
1.	0.272	0.494	12.0
2.	0.329	0.218	9.5

What is the luminance of the resultant color?

5.18 Three colored fluorescent lamps have the following specifications:

Lamp	x	y	lm/W
A	0.600	0.300	10
B	0.500	0.400	20
C	0.200	0.300	30

All lamps have the same wattage. Find the chromaticity coordinates of the resultant color produced by four of lamp A, two of lamp B, and three of lamp C.

5.19 The colors from two sources are to be mixed. What is necessary in order that the chromaticity coordinates of the resultant color lie at the midpoint of the line connecting the chromaticity coordinates of the individual sources?

5.20 A surface has the following spectral reflectances:

λ, nm	ρ_λ	λ, nm	ρ_λ	λ, nm	ρ_λ
380	0.01	500	0.10	620	0.29
390	0.02	510	0.15	630	0.25
400	0.03	520	0.22	640	0.20
410	0.05	530	0.32	650	0.16
420	0.055	540	0.45	660	0.12
430	0.06	550	0.57	670	0.09
440	0.065	560	0.67	680	0.07
450	0.07	570	0.70	690	0.05
460	0.075	580	0.61	700	0.04
470	0.08	590	0.48	710	0.03
480	0.085	600	0.39	720	0.02
490	0.09	610	0.32	730	0.01

It is lighted by illuminant C. Find the chromaticity coordinates of the reflected radiation.

5.21 An area is lighted with luminaires, each of which contains one 400-W metal-halide lamp and one 400-W high-pressure sodium lamp. What is the chromaticity of the resulting illumination?

	HPS	MH
CCT(K)	2100	4500
x	0.512	0.362
y	0.420	0.371
ϕ, lm	50,000	34,000

CHAPTER 6

6.1 A vacuum incandescent lamp with an uncoiled filament operates at a temperature of 2570 K. Use the curves in Fig. 6.2 to find

(a) Life of the filament
(b) Watts consumed by the filament
(c) Lumens emitted by the filament
(d) Lumens per watt efficacy of the filament
(e) Operating voltage of the filament

The filament is 22.2 in long and 0.0012 in in diameter. Assume losses represent 6% of the input power. Tungsten has a density of 19.35 g/cm^3.

6.2 Repeat Prob. 6.1 for a gas-filled incandescent lamp with a coiled filament operating at 2870 K. The filament has an uncoiled length of 22.8 in, a coiled length of $\frac{15}{16}$ in, an uncoiled diameter of 0.0025 in, and a coiled diameter of $\frac{1}{30}$ in. Losses represent 18% of input power. *Note:* With gas filling and coiling, γ is reduced by a factor of 50 at a given temperature.

6.3 An incandescent lamp is rated at 100 W, 120 V, and 3000 K filament T. If the applied voltage is raised to 142 V, what is the new filament T? For temperatures greater than 1800 K, the resistance of an incandescent lamp filament, expressed as a percent of the cold resistance at 300 K, is $\%R = 0.65(T - 500)$, T in kelvins.

6.4 A log-log plot of lumens versus volts for a 75-W, 1150-initial lumen, 120-V incandescent lamp is shown in Fig. P6.4. Find the number of initial lumens emitted if the lamp is operated at 130 V.

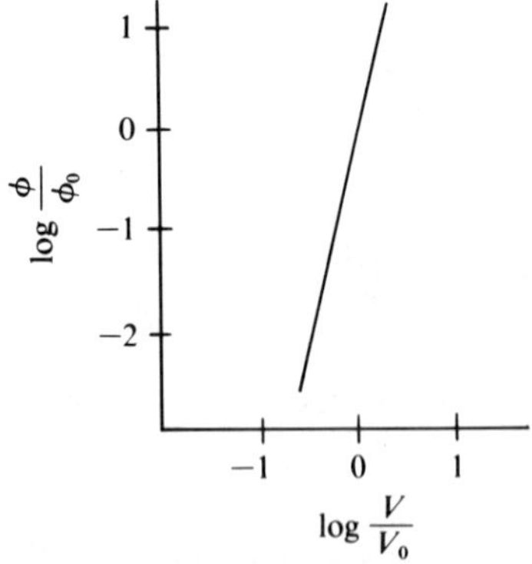

Figure P6.4

6.5 At what voltages should a 120-V incandescent lamp rated at 1000 h life and 1500 initial lumens be operated to give

(a) 1500 h life
(b) 1950 initial lumens

6.6 Find the cost of energy of producing one million lumen hours of light by an incandescent lamp under the following conditions:

Line voltage = 121 V
Rated lamp voltage = 115 V
Energy cost = 8¢/kWh

Rated initial lamp power = 100 Watts

Rated lamp life = 750 h

Rated initial luminous flux = 1630 lm

Lumen depreciation factor = 0.82

6.7 The mortality and lumen depreciation curves for a large installation of experimental incandescent lamps are shown in Fig. P6.7. What life rating should the manufacturer place on these lamps? What is the percentage of survivors at 100% life? What percentage of the total lumen hours of the installation occurs before 100% life is reached?

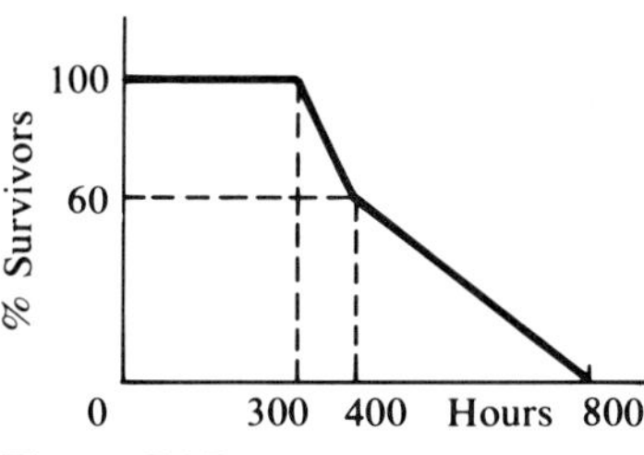

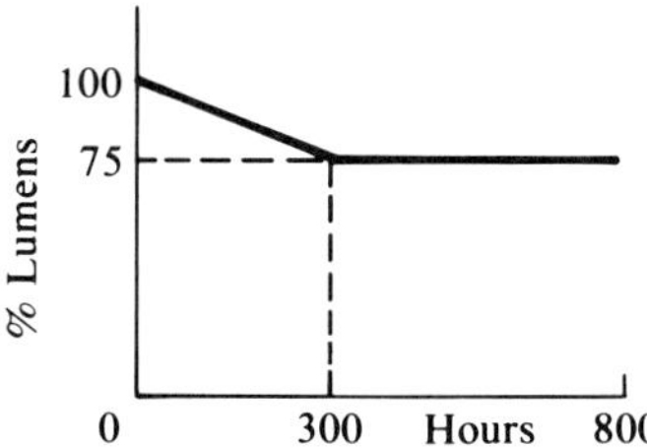

Figure P6.7

6.8 A fluorescent lamp and inductive ballast operate in series from a 200-V 60-Hz secondary of an autotransformer. The lamp and ballast data are

Lamp watts = 39.0

Lamp volts = 102.0

Lamp amperes = 0.42

Ballast watts = 8.5

(a) What are the L and R of the ballast?
(b) If the primary voltage is 118 V, what value of C placed across the line (primary) will produce a power factor of 90% lag?

6.9 A fluorescent lamp has the following specifications:

Lamp watts = 26.0

Lamp amperes = 0.49

Lamp volts = 60

Two of these lamps are operated in a lead-lag circuit directly from a 115-V, 60-Hz supply. Each ballast consumes 6 W and has the same inductance. The overall circuit operates at unity power factor. Find L and C.

6.10 The relation between the instantaneous values of voltage across and current through a T12 fluorescent lamp is given by

$$e = \frac{2.37l}{0.56 + i} + 12.5$$

where l is tube length in inches, e is instantaneous lamp voltage in volts, and i is instantaneous lamp current in amperes. If the starting voltage is twice the operating voltage for 0.1 A maximum instantaneous lamp current, what is the maximum length lamp that can be started from a 120-V RMS source? If l is 0.75 of this maximum length, what value of series resistance must be used to limit the maximum instantaneous lamp current to 1 A?

6.11 A fluorescent lamp is to be designed to meet the following specifications:

Loading = 0.25 W/in^2
Cathode fall = 16 V
Positive column slope = 2 V/in
Cathode fall distance = 1 in
$V_{lamp} = 206$ V
$I_{lamp} = 800$ mA
$P_{lamp} = 90$ W

Find the length and diameter of the lamp.

6.12 An 8-ft T12 fluorescent lamp has the following data:

Lamp current = 1.5 A
Loading = 0.422 W/in^2
Positive column gradient = 1.656 V/in
$V_{start} = 2(V_{run})$

Two of these lamps are to operate at 60 Hz in a lead-lag circuit using an autotransformer. RMS readings on the primary side of the autotransformer are

$$V = 240 \text{ V} \qquad I = 2.104 \text{ A} \qquad P = 505 \text{ W}$$

Find the inductance and resistance of each inductor, the capacitance of the lead-circuit capacitor, and the autotransformer turns ratio. Assume the inductors are identical.

CHAPTER 7

Note: Assume work-plane height is 30 in unless otherwise stated.

7.1 A school room has the following data:

Dimensions: 28 × 32 × 10 ft
Reflectances: 80%, 50%, 30%
Illuminance level: 50 fc average maintained
Luminaire: No. 6, two lamps
Lamp: F40 CW
LLF: 0.85

Find the number of luminaires needed and the watts per square foot required.

7.2 A room 40 × 30 × 10 ft has the following reflectances:

Ceiling	70%
Walls	50%
Floor	30%
Windows	5%
Chalkboards	15%

Windows occupy 10% of the wall area, chalkboards 5%. The room is lighted with F40 CW lamps in luminaire 6. Assuming a light loss factor of 0.85, how many lamps are needed to provide 50 average maintained footcandles?

7.3 A lobby area is shown in Fig. P7.3. The entrance is on the far left. The 40-ft wall is entirely glass. The lobby is to be lighted to an average maintained level of 30 fc using F40 CW fluorescent lamps and luminaire 7. A light loss factor of 0.75 is to be assumed. Calculate the number of luminaires required and sketch a possible lighting layout. Assume the work plane is the floor.

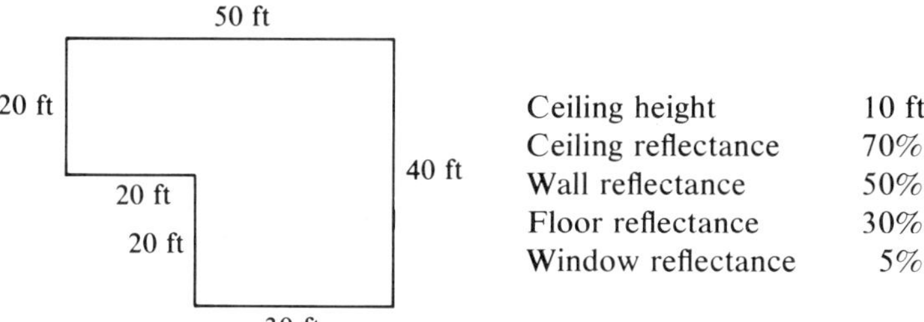

Figure P7.3

7.4 An elementary classroom is shown in Fig. P7.4. It is 20 × 30 ft and has a 10-ft ceiling. The room reflectances are ceiling 80%, walls 50%, and floor 30%. The tack board and chalkboard are each 60 ft^2. Tackboard reflectance is 30%; chalkboard reflectance is 15%.

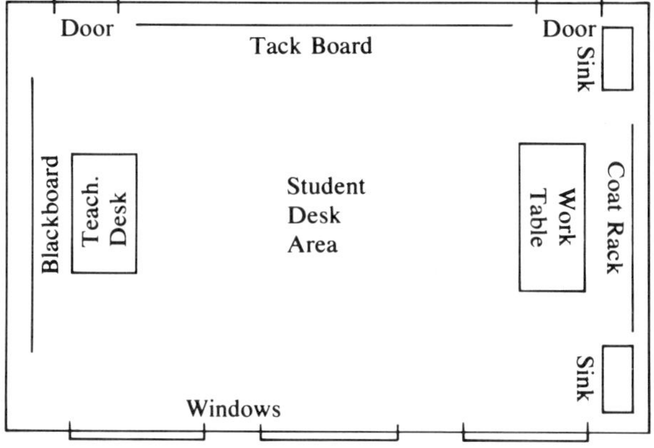

Figure P7.4

(a) Design a fluorescent lighting installation using luminaire 8 to give 50 average maintained footcandles, using F40 WW fluorescent lamps.
(b) Design a 175-W metal-halide lighting installation using luminaire 3 to give 50 average maintained footcandles.
In each case, calculate an appropriate light loss factor, show fixture layout, and calculate the watts per square foot and watts per footcandle required.

7.5 Describe a room which has RCR = 0. Assume luminaire 5 is placed in such a room. Show that the CU for $\rho_c = 80\%$, $\rho_w = 50\%$, and $\rho_f = 20\%$ is 0.75. *Hint:* $1 + k + k^2 + \cdots = 1/(1-k)$, $k < 1$.

7.6 A large industrial area, 200 × 200 ft with a 20-ft ceiling height, is to be lighted by 8-ft, 1500-mA fluorescent lamps in luminaire 4. Estimate the number of lamps to provide 100 average maintained footcandles.

7.7 An old mill building has been converted into a discount house. It is to be lighted to an initial average illuminance of 50 fc using unit 4 and 60-W high-output fluorescent lamps. The luminaires are mounted on the undersides of the roof trusses. How many are needed?

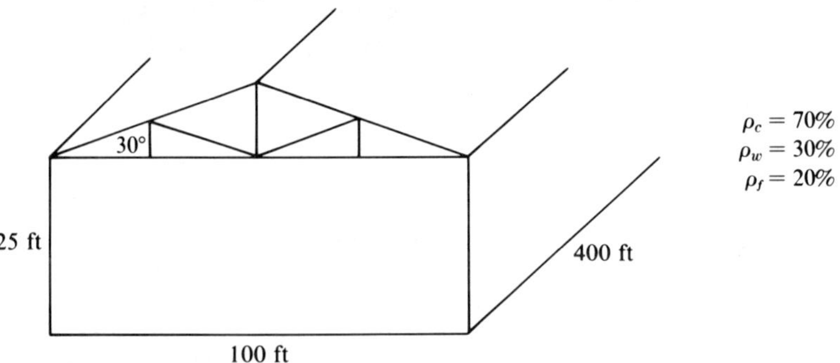

Figure P7.7

7.8 The room shown has the following data:

Ceiling reflectance	70%
Reflectance of two long walls and beams	55%
Reflectance of two short walls	35%
Floor reflectance	30%

A row of eight F40CW two-lamp luminaires (unit 6) is surface mounted on the bottom of each beam. Find the average initial illuminance.

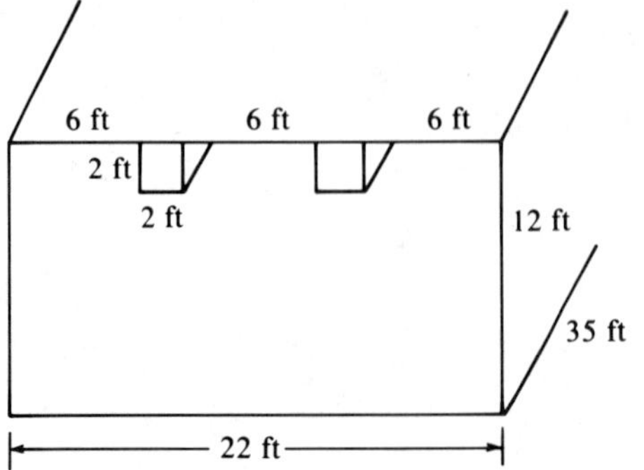

Figure P7.8

7.9 The room shown is 12 ft deep and has reflectances of

Ceiling	80%
Walls	40%
Floor	30%

It is lighted by six fluorescent luminaires (unit 7) each containing two F40CW fluorescent lamps (three rows, two luminaires per row). The luminaires are suspended from the ceiling as shown. Find the average initial illuminance on the work plane.

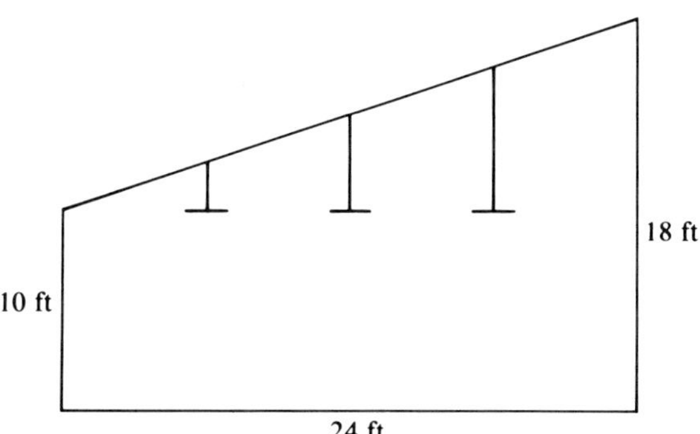

Figure P7.9

7.10 The room is 30 × 60 × 11½ ft. It is lighted with six rows of 14 luminaires each, whose coefficients are given in the tables. Each luminaire has two F40CW fluorescent lamps. The ceiling is composed of 4 × 4 ft coffers, each 18 in deep, surrounded by 12-in-wide beams. Reflectances are: beams and coffers 80%, walls 50%, and floor 20%. Find the following:

(a) Average initial footcandles on the work plane
(b) Average initial footcandles on the walls
(c) Floor cavity luminous exitance
(d) Ceiling cavity luminous exitance

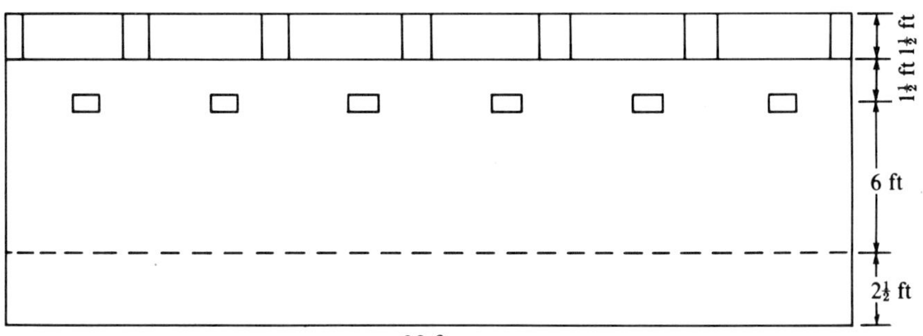

Figure P7.10

ρ_{cc}	70			50		
ρ_w	50	30	10	50	30	10
RCR	% CU for 20% ρ_{fc}					
1	61	59	57	46	44	43
2	54	50	47	41	38	36
3	47	43	39	36	33	30
4	42	37	34	32	29	26
5	37	32	29	28	25	23
	% Wall exitance coefficient					
1	18	10	3	13	8	2
2	17	9	3	12	7	2
3	15	8	2	11	6	2
4	14	7	2	10	6	2
5	13	7	2	10	5	1
	% Ceiling exitance coefficient					
1	56	55	54	39	38	37
2	56	54	52	38	37	37
3	55	53	51	38	37	36
4	55	52	50	38	37	35
5	55	52	50	38	36	35

7.11 A sketch of an ice arena is shown in Fig. P7.11*a*. It is a segment of a cylinder (the dashed line extensions show that it is not half a cylinder). The arena is 219 ft long and 175 ft wide. Its maximum height is 62 ft and the lighting units are mounted 32 ft above the ice. The ice dimensions are 200 × 85 ft (assume the ice area is a rectangle). The lighting consists of nine rows of two-lamp fluorescent luminaires (unit 4) with 13 luminaires per row. An additional four luminaires are mounted in the corners of the arena (one in each corner). A section of the ceiling is shown in Fig. P7.11*b*. The 11 main beams are 11 in wide, 12 in deep, and extend vertically from the ground around the entire roof. Reflectances are

Beams	20%
Dome in between beams	60%
End walls	30%
Floor and seats around rink	50%
Ice area	70%

The lamp used is F96T12CWHO. The LLD factor is 0.81.

Assuming that the area is cleaned once every 2 years and that a maximum of 10 lamps are left in a burned-out condition before replacement, compute the average maintained footcandle level in the arena.

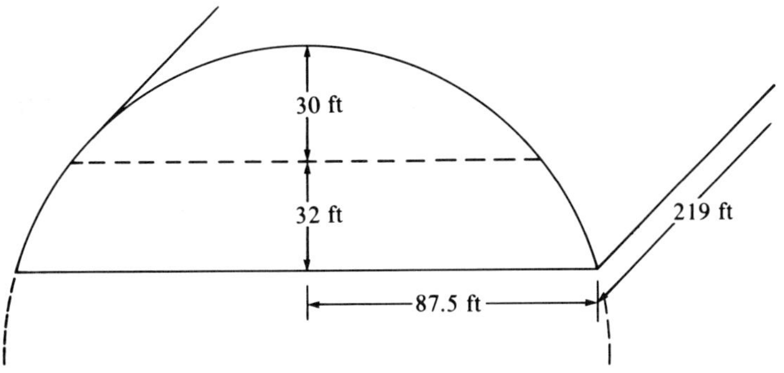

Figure P7.11*a*

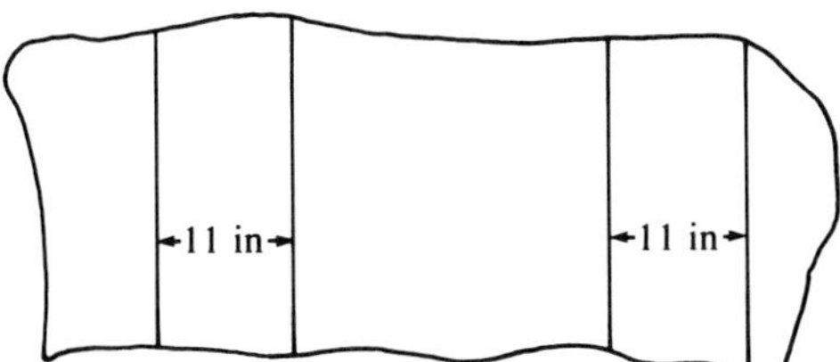

Figure P7.11*b*

7.12 For the parallel rectangles shown in Fig. P7.12, show that

$$f_{14} = \frac{f_{(12)(34)} - f_{13} - f_{24}}{1 + (A_3/A_2)}$$

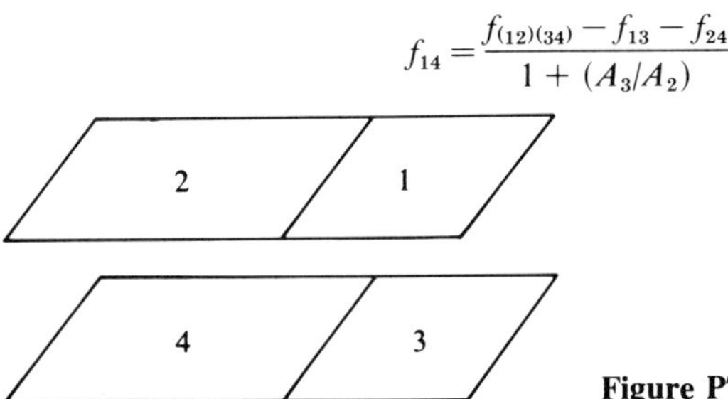

Figure P7.12

7.13 The form factor for two parallel circular discs is given in Eq. (7.40), that for two parallel rectangles in Eq. (7.41). If $l = w = a\sqrt{\pi}$, making the disc areas equal, show that the form factors for the two geometries are essentially the same for $\alpha \geq 1$.

7.14 A cylindrical ceiling recess is shown. Its top and sides are perfectly diffusing. It is lighted by a wide-beam floodlamp of negligible dimensions. The lamp delivers 2000 initial lumens to the sides of the recess and 4000 initial lumens to its top. Find the final luminous exitances of the top and sides of the recess.

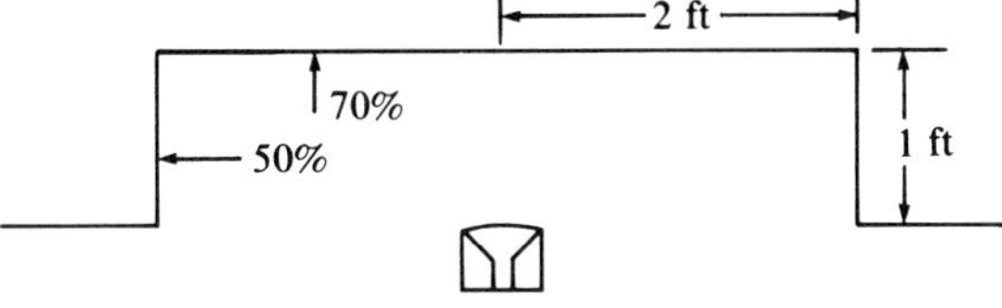

Figure P7.14

CHAPTER 8

8.1 Using Eq. (8.5), show that the configuration factor at point P on a horizontal plane caused by a perfectly diffusing surface A which is generated by an infinitely long line moving parallel to itself between the limits θ and ϕ is given by $c = \frac{1}{2}(\cos\theta - \cos\phi)$.

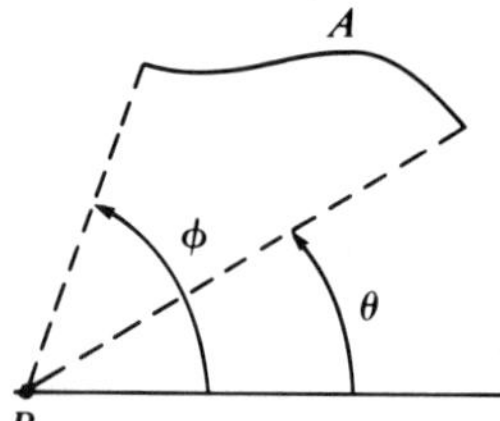

Figure P8.1

8.2 Show that the configuration factor at point P on a horizontal plane caused by a perfectly diffusing circular disc source of radius R is given by [Eq. (8.8)]

$$c = \frac{1-\cos\gamma}{2} = \frac{1}{2}\left[1 - \frac{D^2+S^2-R^2}{\sqrt{(D^2+S^2+R^2)^2-4R^2S^2}}\right]$$

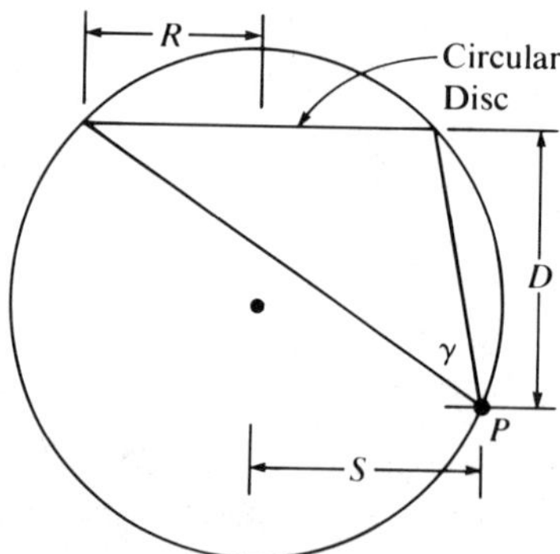

Figure P8.2

8.3 Use Eq. (8.5) to find an exact value for the configuration factor for the situation in Fig. 8.9. What percent error results from using the configuration factor method for strip sources in finding this configuration factor?

8.4 Let the point P in Fig. 8.11 be located such that the 6-ft and 11-ft distances become 8 ft and 16 ft, respectively. What must the maximum intensity I_0 become so that the horizontal illuminance at P is unchanged?

8.5 A room 32×20 ft has an 8-ft ceiling height. It is lighted by two perfectly diffusing 12×8 ft ceiling elements with a luminous exitance of 300 lm/ft². Use the configuration factor method for rectangular sources to find the horizontal illuminance at point P on the floor.

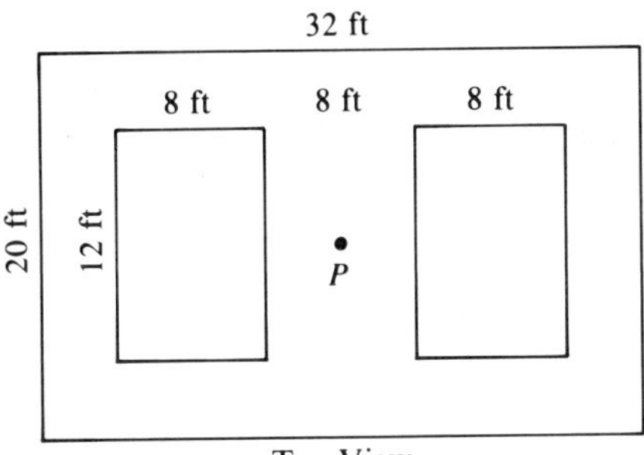

Figure P8.5

8.6 A short, vertical, 70-W, 5800-lm, high-pressure sodium lamp is located at the center of a right circular cylinder as shown.

(a) Assuming the lamp is a Lambertian cylinder, find its intensity at $\theta = 90°$.

(b) Find the luminous fluxes to the top and walls of the cylinder. Use these to find the initial luminous exitances of these surfaces.

(c) Find the various form factors.

(d) Calculate the final luminous exitances from the top and sides of the cylinder.

(e) Find the configuration factors to point P and the horizontal illuminance at point P.

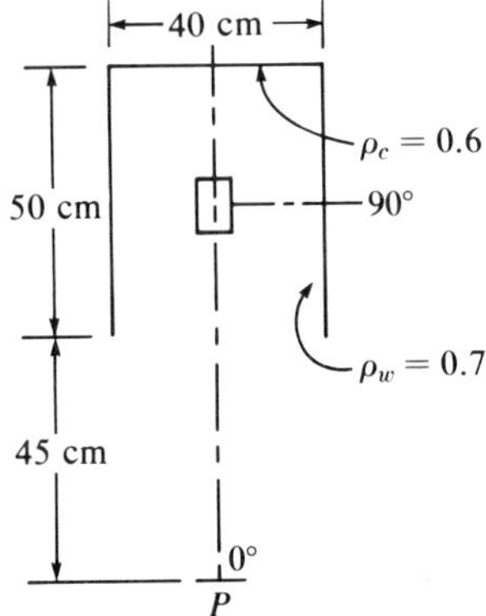

Figure P8.6

8.7 A room 36 × 24 ft with a 10-ft ceiling height is lighted by three rows of fluorescent luminaires, with eight luminaires per row. The luminaire is no. 7 in Table 7.4 and houses two F40 CW lamps. Room reflectances are $\rho_c = 80\%$, $\rho_w = 50\%$, and $\rho_f = 20\%$. Calculate the reflected component of initial horizontal illuminance on the work plane at a point 10.8 ft from a short wall and 9.6 ft from a long wall.

8.8 Using the data in Table 8.3, find the intensity at $\psi = 30°$ and $\theta = 40°$.

(a) Using the parabolic interpolation procedure.

(b) Using the fourth-power equation procedure.

Compare the results and comment.

8.9 Calculate the zone multiplier for $\alpha = 30$ to $40°$ and $\beta = 25°$ and verify its value in Table 8.7. Use the result to calculate the horizontal component of illuminance produced by an element of the luminaire in this zone whose intensity data is given in Table 8.6. Verify the result in Table 8.5.

8.10 Three rows of the prismatic wraparound luminaires of Table 8.5 are shown. Find the horizontal illuminance at point P.

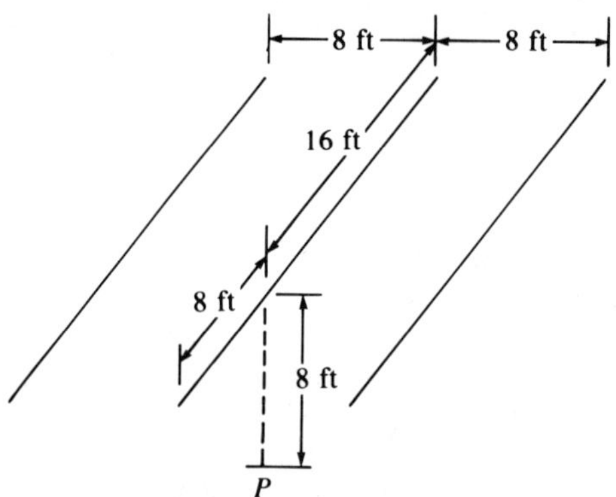

Figure P8.10

8.11 Intensity distribution data in candelas is given for a batwing fluorescent unit, where $I_{90°}$ represents intensity values in a vertical plane perpendicular to the fixture axis, $I_{0°}$ parallel to the fixture axis, and $I_{45°}$ in between the two.

Four of these 4-ft units are mounted as shown in Fig. P8.11 on a ceiling which is 7 ft above the work plane. The direction of viewing is shown, the viewing angle is 25°, and the task is the standard #2 pencil task.

Subdivide the lighting units into appropriate lengths and calculate the horizontal illuminance on the task, and the CRF and VL provided. Use linear interpolation to obtain intensity values.

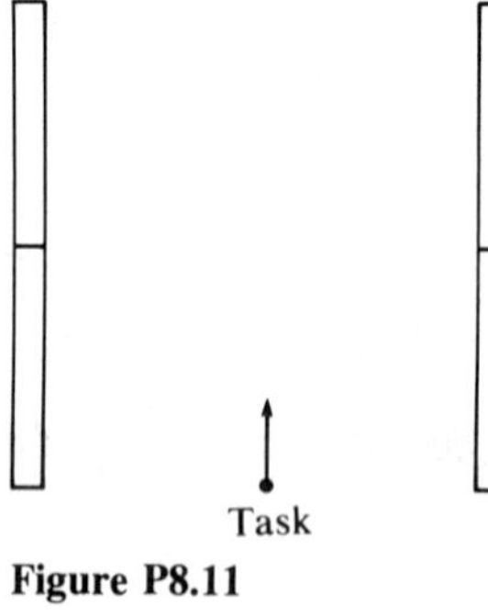

Figure P8.11

$\theta°$	$I_{90°}$	$I_{45°}$	$I_{0°}$
0	349	348	348
5	390	368	349
10	533	437	351
15	730	551	351
20	941	699	354
25	1047	808	326
30	1088	872	307
35	1054	859	285
40	935	778	260
45	685	648	217
50	402	493	179
55	305	344	159
60	276	254	146
65	250	193	127
70	223	158	106
75	185	128	78
80	141	92	48
85	32	32	13

8.12 Two glare sources are 31° up from and 45° to the left and right of the line of sight of an observer. Each source has a luminance of 3085 cd/m² and subtends a solid angle of 0.015 steradian at the observer. The field luminance is 137 cd/m². Find DGR and VCP.

8.13 Six 4-ft-long fluorescent luminaires are mounted in a row on the ceiling of a room as shown in Fig. P8.13. Each luminaire has a lighted area of 4 ft² and a luminance of 1720 cd/m². The observer is at point *A*, directly under the end of the first luminaire. The luminance of wall *B* is 240 cd/m². Calculate the *M* of each luminaire, the DGR, and the VCP.

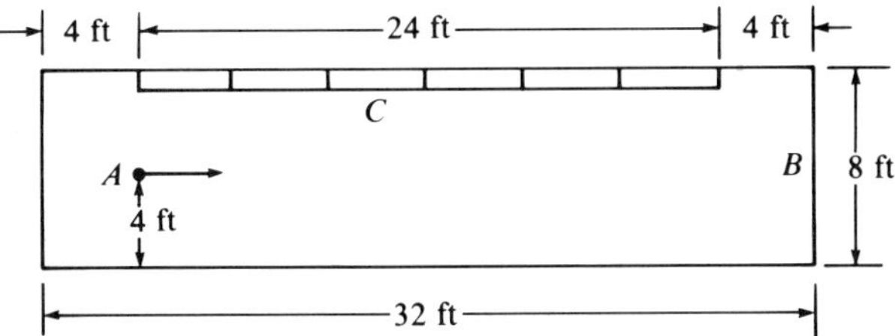

Figure P8.13

8.14 Note the *M* of source *C* from Prob. 8.13. Now split source *C* into two sources of 2 ft² each and calculate the DGR. Repeat for source *C* split into four sources of 1 ft² each. Compare the results. Is the DGR procedure consistent?

CHAPTER 9

9.1 A classroom is shown in Fig. P9.1. It has three windows with dimensions and spacings given. The windows have a transmittance of 85%. Assume an overcast sky with an average luminance of $1000/\pi$ cd/ft². Find the direct components of horizontal and vertical illuminance at point *P*, 10 ft perpendicular from the center window at sill height.

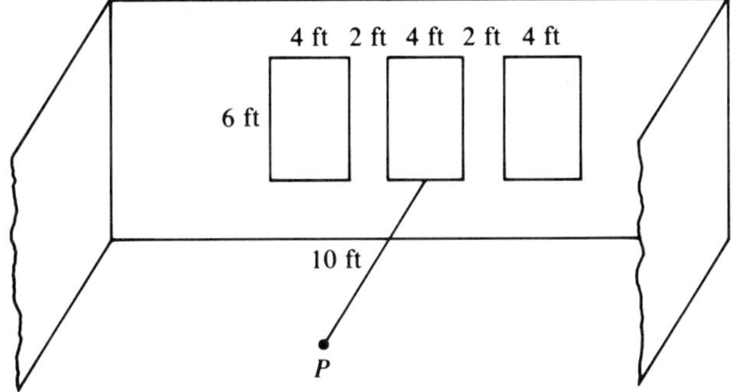

Figure P9.1

9.2 Figure P9.2 shows a proposed bowling alley lighting installation with the bowler at the left and the pins at the right. The small arcs represent 4-ft-long, 1-ft-wide cylindrical skylights, centered over the alley and at right angles to it. Each skylight has a luminance of 5 cd/in². Find the direct components of horizontal and vertical illuminance at point P on the alley.

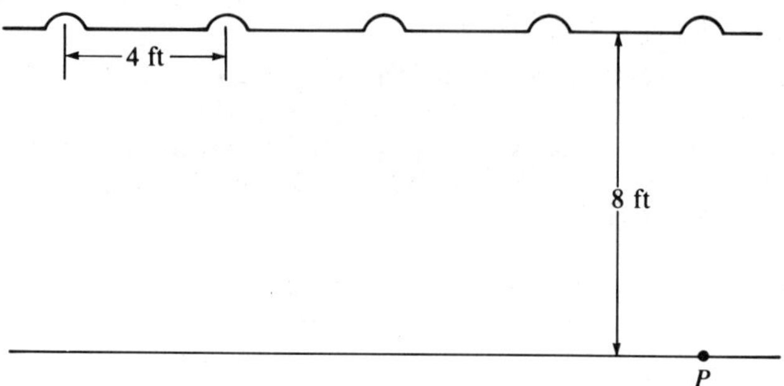

Figure P9.2

9.3 An astrodome with a perfectly diffusing plastic skylight having a transmittance of 90% is shown in Fig. P9.3. Also shown is the distribution of illuminance on the skylight on an overcast day. This illuminance varies with polar angle, but not with azimuth angle.

Assume second base is at the center of the astrodome. Find the direct component of horizontal illuminance on second base and the direct component of vertical illuminance on a runner standing on second base.

What would these illuminances become if the illuminance on the dome were constant at 400 fc?

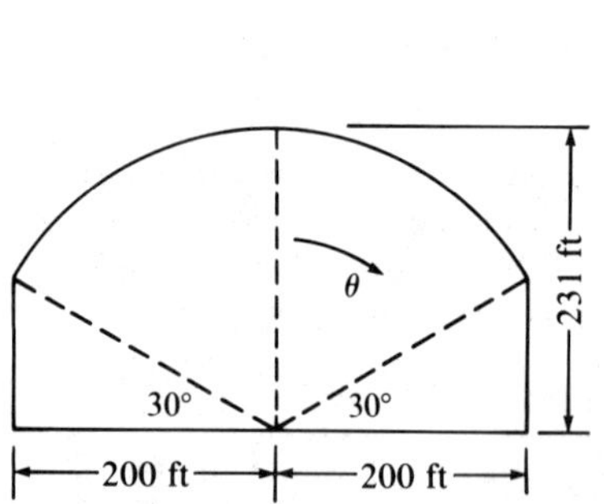

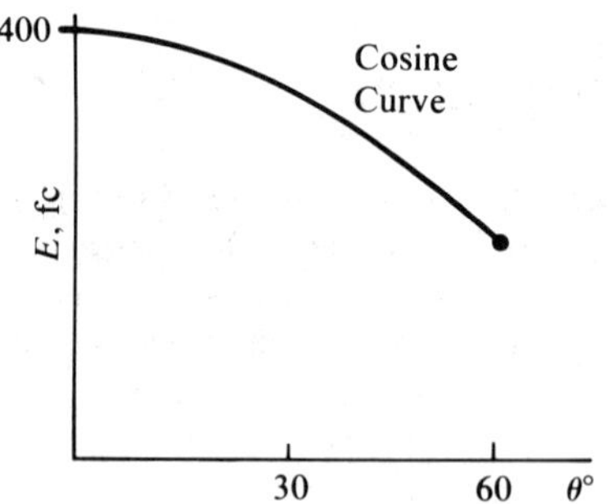

Figure P9.3

9.4 The luminance distribution for an overcast sky at a solar altitude of 40° is shown below, as a function of the polar angle θ, where a relative $L = 1.0$ corresponds to $1860/\pi$ cd/ft². Calculate the horizontal illuminance on the ground.

$\Delta\theta$, degrees	Relative L
0–10	1.00
10–40	0.95
40–50	0.85
50–60	0.75
60–70	0.65
70–80	0.55
80–85	0.45
85–90	0.40

9.5 It is desired to calculate the horizontal illuminance at point P on the floor of a room due to light coming in through the window.

The illuminance on the outside of the window is made up of three components:

1. Direct sunlight
2. Clear sunlight
3. Reflected ground light

Assume it is noon at 46° north latitude on June 21. The solar altitude is 67°. The direct illuminance from the sun on a plane perpendicular to the sun is 8800 fc. The clear sky illuminance on the window is 1400 fc. The ground area around the window is grass with a reflectance of 6%. The illuminance on the window from reflected ground light may be calculated as one-half of the ground luminous exitance. The window has a transmittance of 85% and a light-loss factor of 90%.

Find the following:

(a) Luminance of the inside of the window in candelas per square inch
(b) Horizontal illuminance at point P on the floor in footcandles

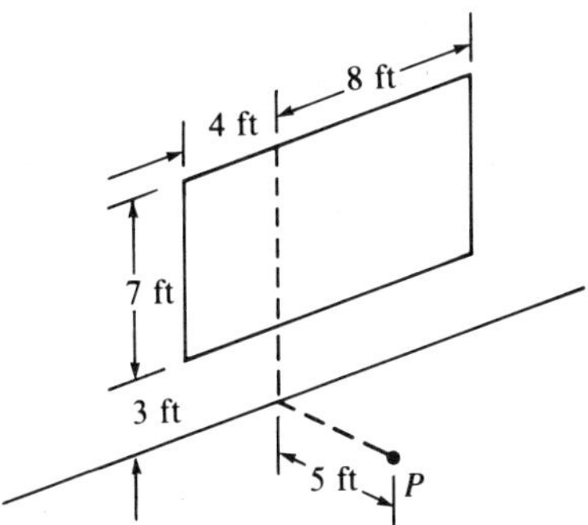

Figure P9.5

9.6 Find the solar altitude and azimuth angles at 9:30 A.M. EST on April 21 at 80°W longitude and 40°N latitude.

9.7 For the situation in Prob. 9.6, find the horizontal illuminance and the illuminances on vertical surfaces facing east and south produced by the sun on a clear sky day.

9.8 A circular horizontal field is completely surrounded by trees, whose tops are at an angle of 30° with the horizontal at point P. The sky is overcast and the sun is at an altitude of 22.7°. If the sky is a CIE sky, find the horizontal illuminance at P. Repeat if the sky is based on the Kimball and Hand data. Neglect reflections from the trees.

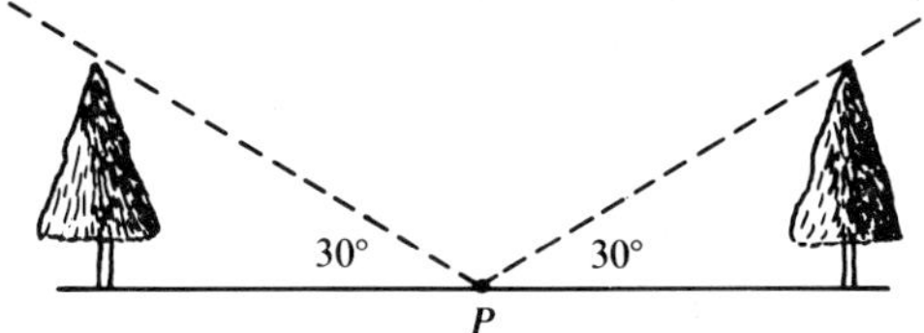

Figure P9.8

9.9 Find the average luminance, the zenith luminance, and the luminance at the north horizon for the clear sky in Example 9.4.

9.10 A large office is 100 × 40 ft with a 10-ft ceiling. Its reflectances are: ceiling 75%, walls 50%, and floor 20%. It is to be lighted by 4 × 4 ft double-domed (transparent over translucent) skylights whose transmittances are:

	T_d	T_D
Transparent	0.75	0.90
Translucent	0.40	0.50

Each skylight has a 15-in-high curb, a 3-in-wide retaining frame, and a 60% well reflectance. Depreciation factors are RSDD = 0.95, SDD = 0.80. The sky is overcast.

Find the number of skylights to provide an average maintained horizontal illuminance of 30 fc at 11 A.M. at 42° north latitude on March and September 21. How many average maintained horizontal footcandles will the installation provide if the sky is clear?

9.11 A warehouse lighted by skylights is shown in Fig. P9.11. There are 10 translucent flat skylights on each side of the roof. Their transmittance is 40% and their underside reflectance is 45%. The exterior illuminance on the skylights on the right side of the roof is 5000 fc, that on the left side 2500 fc. SDD = 0.75 and RSDD = 0.95. Room surface reflectances are ceiling 70%, walls 30%, and floor 20%. Find the average maintained horizontal illuminance on the work plane.

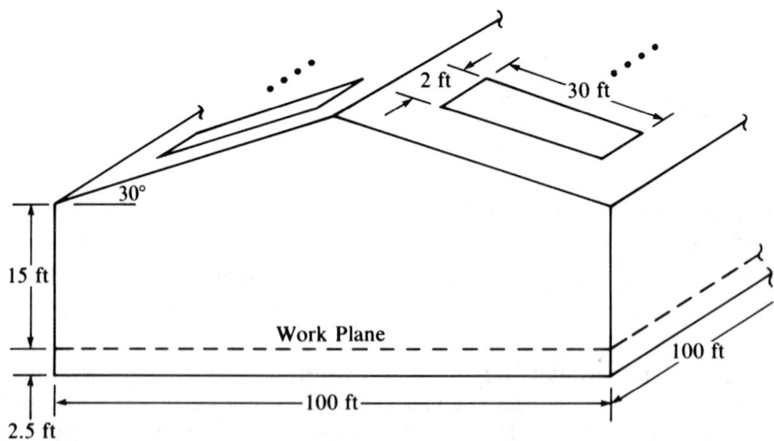

Figure P9.11

9.12 Using the lumen method of sidelighting, find the horizontal illuminance at the three points in Fig. P9.12 under the following constraints:

Clear summer day

Solar altitude = 60°

Solar azimuth = 45° east of south

Room: 40 × 30 × 12 ft
$\rho_c = 80\%$, $\rho_w = 50\%$, $\rho_f = 30\%$
Office area, WDD = 0.83, RSDD = 0.95

Ground surface: Grass, $\rho = 15\%$

Windows: $\frac{1}{4}$-in clear glass
Sill height: 3 ft
Direction: Facing east
Controls: None
Frame: 20% of opening

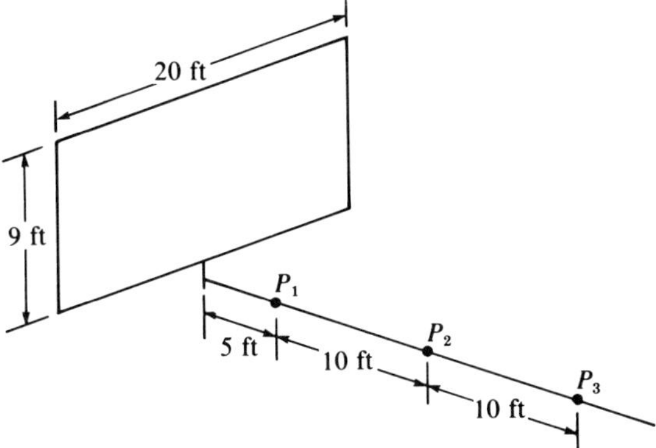

Figure P9.12

9.13 In Fig. 9.15, the following values apply: $H = 20$ ft, $D = 50$ ft, $q = 8$ ft, $h = 6$ ft, $x = 20$ ft, $y = 40$ ft, $z = 10$ ft. The window is of clear glass with a width of 16 ft. Room reflectances are ceiling 70%, walls 50%, and floor 30%. Find the overall daylight factor at P produced by the CIE overcast sky.

CHAPTER 10

10.1 The prism shown in Fig. P10.1 has a refractive index of $\sqrt{2}$. What is the angle between the two light rays as they emerge from the prism? What would this angle be if the prism were immersed in water?

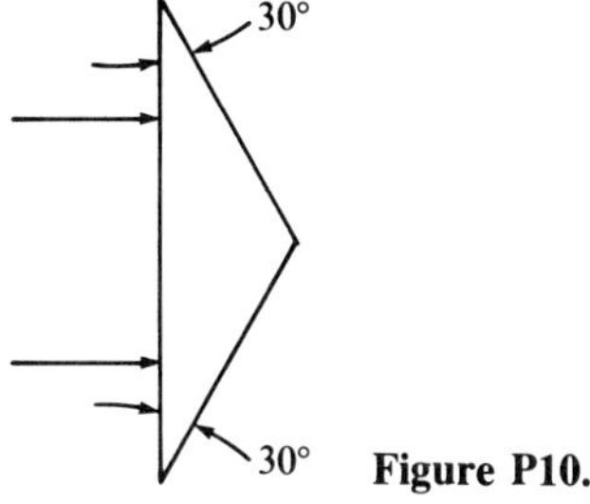

Figure P10.1

10.2 A concave mirror is to form the image of the filament of a PAR lamp on a screen 4 m from the mirror. The filament is 5 mm high and the desired image height is 40 cm. Find

(a) Radius of curvature of the mirror

(b) Location of the filament

10.3 A solid glass hemisphere is placed with its flat face down on a tabletop. The hemisphere has a radius of 10 cm and a refractive index of 1.5. It is lighted from above by a cylindrical beam 1 cm in diameter centered on the hemisphere axis. What is the diameter of the circle of light on the table?

10.4 The focal length of a convex lens is to be measured. An object is imaged by the lens on a screen 12 cm from the lens. When the lens is moved 2 cm further from the object, the screen must be moved 2 cm closer to the object to obtain an image. What is the focal length of the lens?

10.5 **(a)** The near point of an eye is 100 cm in front of the cornea. What type and focal length of lens should be used so that the person can see clearly an object 25 cm in front of the cornea? Draw a ray diagram.

(b) The far point of an eye is 1 m in front of the cornea. What type and focal length of lens should be used so that the person can see clearly an object at infinity? Draw a ray diagram.

10.6 A spotlight system consists of a circular reflector, two identical plano-convex lenses, a diaphragm, and a double-convex lens. The source is a 400-W tungsten-halogen lamp emitting 7500 lm. Assume it is a uniform point source. The reflector has a reflectance of 90%. The lenses have a transmittance of 92% and an index of refraction of 1.85. The curved lens surfaces are circular and the lenses may be assumed thin. Find the following:

(a) Lumens leaving the double-convex lens

(b) Radius of each plano-convex lens

(c) Diameter of the diaphragm if it is to just allow all of the lumens from the second plano-convex lens to pass through

(d) Radius of the double-convex lens

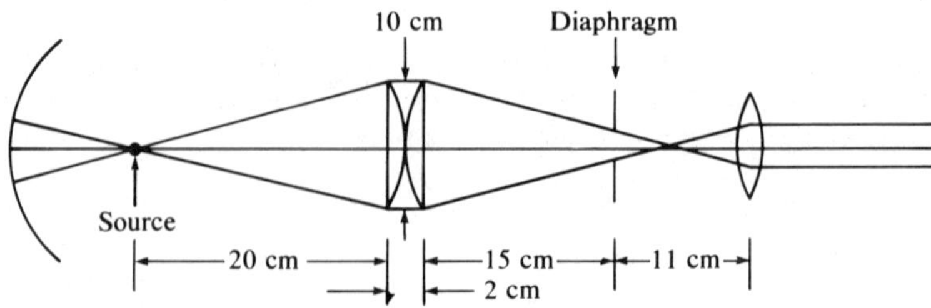

Figure P10.6

10.7 A reflector consists of a parabola and an ellipse. Its reflectance is 80%. The lamp, which may be assumed to be a uniform point source, has an intensity of 100 cd and is located at the focal point of the parabola and at one of the focal points of the ellipse. Find the average illuminance on the circular sign.

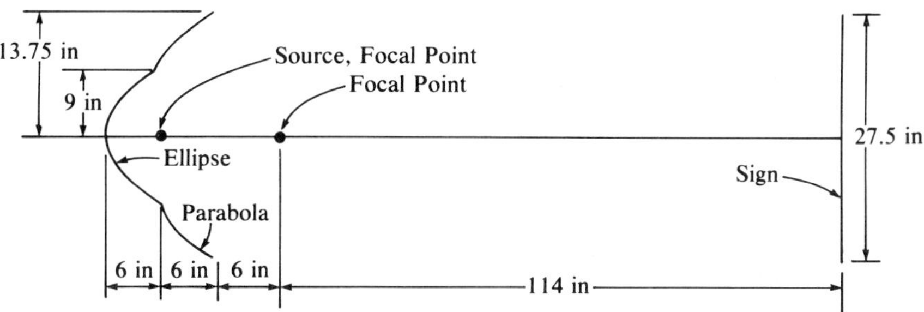

Figure P10.7

10.8 A surgical spotlight is shown. It consists of an elliptical reflector, with a 10° hole in its center, and a small parabolic reflector. Lumens from the source (assumed point and uniform) reflect from both reflectors before becoming a concentrated parallel beam. Find the following:

(a) Lumens in the parallel beam

(b) Illuminance on the circular plate

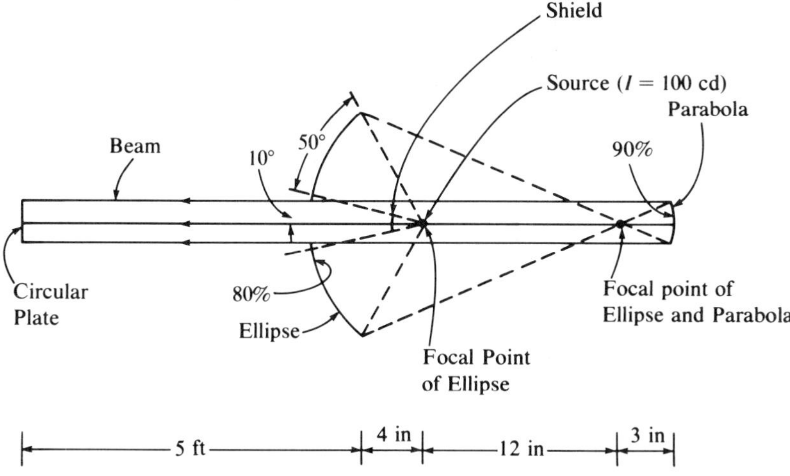

Figure P10.8

10.9 **(a)** A parabolic reflector is to be designed to fit into a cylindrical enclosure 5 in deep and 4 in in radius. Find the focal length of the parabola and accurately draw its shape.

(b) An elliptical reflector is to be designed to fit into the same enclosure, such that exactly half of the complete ellipse is inside the enclosure. Find the two focal points of the ellipse and accurately draw its shape.

10.10 The lamp whose intensity values are given in Table 10.1 is to be used to design a reflector with a 70° cutoff angle such that the lamp-reflector combination has the following intensity distribution:

Zone, degrees	Lamp reflector I_{mid}, cd
0–10	720
10–20	780
20–30	840
30–40	900
40–50	580
50–60	380
60–70	240

Prepare a table similar to Table 10.1 for this data. Determine the required ρ of the reflector. Assume the pattern in Fig. 10.17*a* is to be used. Prepare graphs similar to those in Fig. 10.18. Then develop the reflector contour, as in Fig. 10.19.

10.11 Two slits are spaced 0.3 mm apart and 50 cm from a screen. Find the distance on the screen between the second and third dark lines of the interference pattern when the slits are illuminated by 600-nm radiation.

10.12 What is the thinnest film of refractive index 1.20 in which constructive interference of violet light (400 nm) can occur by reflection? What is the thinnest film for destructive interference? What index of refraction should the glass have if the destructive interference is to produce near-annulment?

10.13 What wavelength of light will produce a maximum at an angle of 20° when passed through a diffraction grating having 6000 lines per centimeter?

10.14 Radiation from a mercury lamp is to be passed through a diffraction grating. How many lines per centimeter should the grating have so that the first- and second-order visible spectra of mercury are separated by 5°? The visible mercury lines are 405, 454, 546, and 579 nm.

CHAPTER 11

11.1 What aiming angle of a floodlight will produce the maximum horizontal illuminance at a given distance from the pole base? Assume the maximum candela point of the floodlight is along the aiming line.

11.2 For an aiming distance of 0.5 MH, verify the values of % utilization given in Fig. 11.5 for distances from the pole of 0.5 MH, 1 MH, and 2 MH.

11.3 The floodlight in Fig. 11.1 is aimed at a point 1.5 MH horizontally from the base of the pole.

- **(a)** What percent of the total lamp lumens lie in the horizontal strip from the pole out to 1 MH?
- **(b)** If the mounting height is 40 ft, find the initial horizontal illuminance at a point 40 ft out from the pole in the direction of the aiming line and 23 ft to the right of the plane of the aiming line and the pole.

11.4 A portion of a lumen distribution diagram for a floodlight is shown in Fig. P11.4. For a 50-ft mounting height, sketch and dimension the zone on the ground containing the 500 lm, assuming the floodlight is aimed straight down.

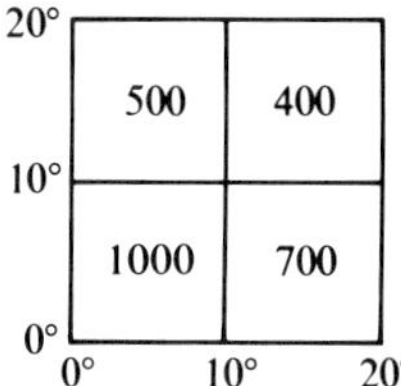

Figure P11.4

11.5 In the floodlighting design in Sec. 11.4, assume the mounting height is increased to 70 ft but that the aiming angle is unchanged.

(a) Find the new pole location.

(b) Calculate the new locations of points A through I on the lumen distribution grid.

(c) Determine the coefficient of beam utilization and the average maintained illuminance on the field.

11.6 Two of the floodlights whose data appear in Fig. 11.8 are used to light a large billboard, as shown in Fig. P11.6. The floodlights are located at points P and Q and aimed at points O and R on the billboard. The lines $FP$and QG are horizontal and perpendicular to the billboard. Find the average illuminance on the billboard.

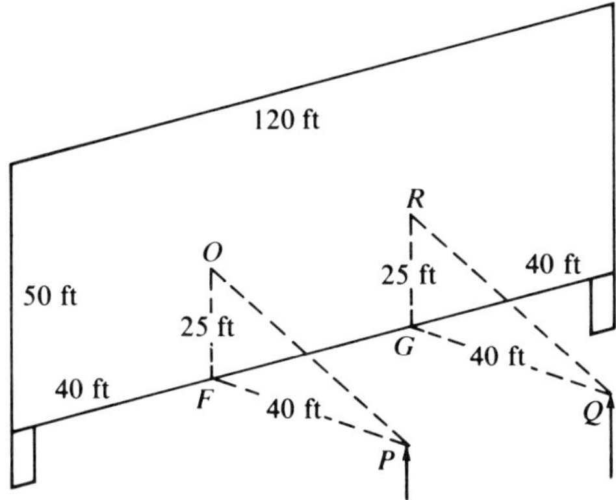

Figure P11.6

11.7 It is desired to light a 30-ft-wide road to an average maintained illuminance of 1.4 fc using the luminaire in Fig. 11.15. The mounting height is 30 ft and the luminaires are mounted directly over the curb. The lamp is a 250-W high-pressure sodium. Assume the light loss factor is 0.75. Find the minimum spacing between luminaires.

11.8 The luminaire in Fig. 11.19, housing a 250-W high-pressure sodium lamp, is mounted atop a 40-ft pole at A. What is the initial horizontal illuminance at B?

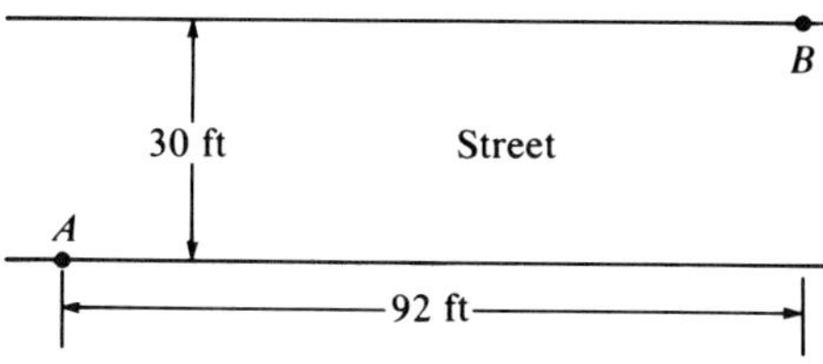

Figure P11.8

11.9 A commercial artery 50 ft wide is to be lighted to an average maintained illuminance of 2 fc using the luminaire in Fig. 11.19 with a 400-W high-pressure sodium lamp. Luminaires are to be mounted directly over the curb. Design the installation for $E_{ave}/E_{min} \leq 3$ and justify that this criterion is achieved.

11.10 A parking lot 240 × 480 ft is to be lighted to an average maintained horizontal illuminance of 1 fc using the luminaire in Fig. 11.19. Two luminaires are to be mounted back-to-back on each pole. Select mounting height, the number and location of poles, and the wattage size of high-pressure sodium lamps. Calculate CU and LLF and verify that 1 fc is achieved.

Answers to Selected Problems

1.2 **(a)** $W_{ab} = 1.6 \times 10^{-17}$ J, $W_{ac} = 3.2 \times 10^{-17}$ J
(b) $F = 3.2 \times 10^{-6}$ N, $a = 0.35 \times 10^{15}$ m/s^2
(c) $v_b = 0.59 \times 10^7$ m/s, $t = 0.017$ μs; $v_c = 0.84 \times 10^7$ m/s, $t = 0.024$ μs

1.4 $n = 3$, $\lambda = 103$ nm

1.6 $L = 24.0$ cd/in^2, $M = 75.5$ lm/in^2

1.7 $\tau = 71\%$, $E = 44{,}400$ lux

2.2 $E = 2.3$ fc

2.3 $E = \dfrac{\cos^3\theta}{256} \Sigma I$, 12 fc

2.5 **(a)** 509 cd **(b)** 502 fc **(c)** 0.7 fc

2.6 $\phi = 117{,}000$ lm

2.7 **(a)** $E = 560$ fc **(b)** 74% **(c)** 11 ft

2.8 27%, 9000 cd

2.11 11.02 fc, 11.18 fc, 12.10 fc

2.13 **(a)** 31 fc **(b)** 16 fc **(c)** 27 fc assuming tubular sources

2.14 114 fc

3.3 $M_{\lambda,\max} = 252$ $W/(\text{cm}^2 \cdot \mu\text{m})$, $T = 2878$ K

3.6 $T = 2725$ K

3.7 **(a)** 117 W/cm^2 **(b)** 3.325 light watts/cm^2 **(c)** 2271 lm/cm^2 **(d)** 19.4 lm/W

3.10 58 lm/W

3.12 9.3 ft from 500-nm source, 96 fc

4.2 **(a)** 253.7 nm **(b)** 4.95×10^{-20} J **(c)** 0.538 nm **(d)** 8.7×10^{-36} kg

4.4 574 nm, 24%

4.7 $v_{max} = 6.8 \times 10^5$ m/s

4.8 **(a)** 4.21 in, 2.70×10^6 m/s **(b)** 3.72 in **(c)** 17

4.9 593 nm

4.11 8.5 cm

4.12 **(a)** $n = 3.53 \times 10^{20}/m^3$ **(b)** $\lambda_3 = 5.77$ cm

5.1 0.654 min, diameter = 0.0027 in

5.3 $C = 1.75$, $L_b = 0.36$ fL

5.5 18 fL, 24 fc, 52.5 fL, 70 fc

5.6 **(a)** 0.147 **(b)** 0.88 **(c)** 0.673 **(d)** 1.75 **(e)** 41 fc

5.8 Yes

5.9 Just barely

5.12 **(a)** red: 492W, 1370 lm; green: 1.50 W, 1000 lm; blue: 0 W, 0 lm
(b) R = 0.106, G = 0.019, B = 0
(c) red: 6 units, 77,200 lm; green: 6 units, 354,000 lm; blue: 6 units, 4340 lm

5.13 $X = 1.7970$, $Y = 3.6668$, $Z = 0.8186$, $x = 0.286$, $y = 0.584$

5.14 580 nm, 0.77 W

5.15 $X = 75.003$, $Y = 66.339$, $Z = 0.504$, $x = 0.529$, $y = 0.468$

5.16 $x = 0.354$, $y = 0.563$, 571 nm, 41%

5.18 $x = 0.356$, $y = 0.319$

5.21 $x = 0.447$, $y = 0.399$

6.1 **(a)** 201 h **(b)** 45 W **(c)** 584 lm **(d)** 12.9 lm/W **(e)** 166 V

6.4 1580 lm

6.6 $5.27

6.8 **(a)** $L = 0.775$ *h*, $R = 48.5\ \Omega$ **(b)** $C = 8.75\ \mu F$

6.11 $l = 8$ ft, $d = 1$ in

6.12 $L = 0.366$ h, $R = 16.7\ \Omega$, $C = 6.48\ \mu F$ $N_1/N_2 = 1.46$

7.1 14 luminaires, 1.4 W/ft²

7.2 36 lamps

7.3 30 luminaires

7.4 **(a)** LLF = 0.75, 8 luminaires, 2.5 W/ft²
(b) LLF = 0.67, 5 luminaires, 1.8 W/ft²

7.7 314 luminaires

7.8 70 fc

7.9 29 fc

7.11 32 fc

7.14 $M_1 = 147$ lm/ft², $M_2 = 263$ lm/ft²

8.3 $c = 0.0297$

8.4 $I_0 = 5510$ cd

8.6 **(a)** 587.7 cd **(b)** 0.1649 lm/cm², 0.5692 lm/cm² **(c)** $f_{23} = 0.1230$
(d) 0.810 lm/cm², 1.225 lm/cm² **(e)** 172 fc

8.7 23 fc

8.10 120 fc

8.11 $E_H = 30.7$ fc, CRF = 0.96, VL = 1.7

8.13 DGR = 66.8, VCP = 78

9.1 $E_H = 30.6$ fc

9.3 $E_H = 210$ fc, $F_v = 49.6$ fc; $E_H = 270$ fc, $E_v = 70.4$ fc

9.5 **(a)** 8.6 cd/in^2 **(b)** 544 fc

9.6 43.2°, 64.8°

9.7 5910 fc, 5680 fc, 2680 fc

9.8 614 fc, 596 fc

9.10 26 skylights, 180 fc

9.12 $E_{P1} = 630$ fc, $E_{P2} = 282$ fc, $E_{P3} = 183$ fc

9.13 DF = 6.3%

10.1 30°, 4°

10.3 $\frac{2}{3}$ cm

10.5 **(a)** f = 33 cm **(b)** f = −100 cm

10.6 **(a)** 167 lm **(b)** 17 cm **(c)** 2.5 cm **(d)** 10.2 cm

10.8 **(a)** 219 lm **(b)** 5940 fc

10.10 $\rho = 81\%$

10.12 167 nm, 1.44

10.14 3650 lines/cm

11.1 54.8°

11.3 **(a)** 18.5% **(b)** 2.6 fc

11.6 3.5 fc

11.7 $S = 172$ ft

11.8 0.3 fc

11.9 MH = 35 ft, $S = 120$ ft

Index

Index